ALLE ZEIT WACH
1842

V. I. Babushkin G. M. Matveyev
O. P. Mchedlov-Petrossyan

Thermodynamics of Silicates

Editor in Chief O. P. Mchedlov-Petrossyan

Translated by
B. N. Frenkel and V. A. Terentyev

With 93 Figures and 135 Tables

Springer-Verlag
Berlin Heidelberg New York Tokyo 1985

Professor Dr. sc. techn. V. I. BABUSHKIN
Deputy Rector of the Kharkov Civil Engineering Institute
3 Sammerovski Lane, App. 14, 310003 Kharkov, USSR

Cand. Sci. G. M. MATVEYEV
Deputy Director of the Institute of Economics and Information
of the Building Material Ministry
4(2) Second Pugachyov Street, App. 105, 107061 Moscow, USSR

Professor Dr. sc. techn. O. P. MCHEDLOV-PETROSSYAN
Kharkov Civil Engineering Institute
Chief of the Department for Physico-chemical
Mechanics and Concrete Technology
14 Vorobyov Street, App. 5, 310057 Kharkov, USSR

Translated by:

Cand. Sci. B. N. FRENKEL
113 Vernadsky Avenue, App. 46, 117571 Moscow, USSR

Sci. Worker V. A. TERENTYEV
N35-2 Lublinskaya, App. 6, 109390 Moscow, USSR

Translation of: Thermodynamika Silikatow, 4th edn

ISBN-13:978-3-642-69322-9 e-ISBN-13:978-3-642-69320-5
DOI: 10.1007/978-3-642-69320-5

Library of Congress Cataloging in Publication Data. Mchedlov-Petrosian, O. P. (Otar Petrovich) Thermodynamics of silicates. Translation of: Termodinamika silikatov. Includes index. Bibliography: p. 1. Silicates. 2. Thermodynamics. I. Babushkin, V. I. (Vladimir Ivanovich) II. Matveev, G. M. (German Mikaĭlovich) III. Title. QD181.S6M34513 1985 546'.68324 84-5442 ISBN-13:978-3-642-69322-9 (U.S.)

Typesetting: K + V Fotosatz, Beerfelden

2132/3130-543210

Dedicated to the memory of

Professor Dr. Nikolai Landiya

Preface

The intensification of the production of silicate materials and products makes a detailed theoretical study of the processes underlying their manufacture and service more and more urgent.

The thermodynamic method is of great importance for studying chemical reactions of silicate technology. Together with a study of the rate and mechanism of substance transfer, it permits obtaining necessary data for the efficient operation of technological processes.

The progress of science in recent years has solved numerous problems in the field of the physical chemistry of silicates. The great progress in deciphering silicate structures, and working out methods of the synthesis of minerals and studying their properties must be mentioned.

New methods of determining thermic constants have appeared. In future these methods should be more widely used for determining the heats of the silicate formation and related compounds in crystalline and vitreous state. This concerns in particular the system – $CaO - Al_2O_3 - Fe_2O_3 - SiO_2 - H_2O$ – which is of great importance for the technology of cement and concrete, ceramics, refractories and glass.

In the first edition of this monograph (1962), the authors collected and generalized the material on the thermodynamic study of the reactions in silicate systems available at that time in Soviet and world literature.[1] In the second revised edition (1965) the range of the systems considered was broadened substantially. The book aroused interest outside the U.S.S.R.: a translation was twice published by the Verlag für Bauwesen in the GDR (1965–1966), and references and reviews in both Soviet and other literature emphasized the usefulness of the work. In the third edition, using the new data of periodicals, the reaction calculations in some systems were improved, the contents of some chapters revised, and remarks and comments from references and reviews were taken into consideration.

In the present fourth edition all the sections of the book dealing with the practical application of thermodynamics to silicate and concrete technology have been substantially revised. All the calculations have been carried out with refined interrelated thermodynamic data.

The material on pyrosilicate reactions has been prepared for the most part by G. M. Matveyev and that on hydration reactions and corrosion by V. I. Babushkin.

---

[1] A number of works should be mentioned dealing with the application of thermodynamics to technological and, in particular, geological problems [34, 56, 82, 83, 145, 162, 263, 291, 347, 550, 641].

The authors are profoundly grateful to all those who rendered assistance in writing the monograph and took part in discussing the main principles of it. All comments on the present book will be gratefully accepted by the authors.

O. P. MCHEDLOV-PETROSSYAN

Introduction

Silicates in the form of minerals and rocks dominate in the composition of the earth crust and the products of silicate technology – cement, ceramics, refractories and glass – have a very wide range of use in building, ferrous and non-ferrous metallurgy, science, technique and home.

The chemistry of silicates represents one of the most important theoretical and engineering disciplines. It is the most complicated section of inorganic and physical chemistry and its theoretical foundations are based on most varied modern investigation methods including thermodynamics, one of the most reliable.

Thermodynamics is a mathematical science dealing with the laws of mutual transformation of heat and other kinds of energy. Chemical thermodynamics is an application of thermodynamic methods of investigation to chemical processes.

Together with kinetic methods, thermodynamics enables us to view complicated phenomena taking place in chemical interactions and phase transformations as a whole. Unlike kinetics, which deals with the rate of chemical reactions, i.e., with the transfer of the mass of a substance with time, thermodynamics studies equilibria and leaves the problem of how solutions are obtained, their rate and mechanism of the process, to kinetics.

The thermodynamic method being statistical is applicable only to the study of macrosystems with large masses. It permits determination of:

1. energy possibility and direction of the run of reactions;
2. heat alterations entrained by reactions, allowing calculation of the heat balances of processes;
3. preferability of reactions[2] and stability of compounds formed;
4. maximum equilibrium concentrations of the products of reactions and their maximum output;
5. ways of suppressing undesirable reactions and eliminating by-products;
6. the choice of an optimum regime of the path of the reactions (temperature, pressure and concentration of the reacting substances).

The thermodynamic investigation method should be noted as possessing one important peculiarity: it allows the theoretical solution of most problems by the use of a relatively small number of thermic constants of reacting compounds and a simple

---

[2] As in previous works of the authors a theoretical permissibility of comparison has been shown, by the value of $\Delta G°$, of the probability of various reactions proceeding (in one and the same system on condition of the identity of the initial structural and kinetic factors), the term "preferability" is used in this case.

mathematical apparatus, without resorting to labour-consuming and sometimes technically unrealizable experiments for the equilibrium study.

For a thermodynamic analysis of reactions only the initial state of the system and conditions affecting it need be known. This substantially facilitates handling thermodynamic ideas, but at the same time it restricts the scope of the thermodynamic method, making no allowance for the time factor.

Most phenomena of nature may be considered as equilibrium ones only at first sight; in reality, however, they are connected with non-equilibrium states and irreversible alterations. For this reason, classical thermodynamics created chiefly for equilibrium reversible processes has been supplemented of late by the thermodynamics of irreversible processes, and we found it necessary to sketch, if only in general, its essence and application to the phenomena that interest us.

As initial data for thermodynamic calculations, heats of formation, heat capacities, entropies and some other data characterizing equilibria are used.

Apart from studying equilibria, chemical thermodynamics enables us to analyze silicate systems under consideration also from the point of view of the phase rule that gives qualitative and, in certain cases, quantitative characteristics of an equilibrium [151, 468].

M. M. SHUL'TS has put forward 3 directions of the applications of thermodynamics [574] in the silicate chemistry and technology: (1) amplification of the investigations of thermodynamic properties, (2) quantitative description of phase diagrams and (3) plotting and analysis of energy balances of production silicate materials.

According to chemical processes of manufacturing and life of materials and products all reactions in silicate systems may be divided in principle into two groups: pyrosilicate ones and those of hydration and corrosion.

The first one comprises a vast range of dehydration reactions and modification transformations, reactions in solid phases and melts; the second one, the entire diversity of reactions of silicates with water and aggressive media at normal and elevated temperatures and pressures. Apart from the well-developed domain of slag melts, the thermodynamic method has been rarely applied in the study of silicates. It concerns mostly solid-phase reactions of silicate compounds first studied in systematic works by one of the authors of this book. Reactions of hydration and corrosion of silicates under normal and hydrothermal conditions are of great practical importance, but have been little studied thermodynamically until recently.

The book comprises a theoretical part, in which general thermodynamic ideas are stated, experimental and approximate methods of obtaining the necessary initial thermodynamic data are discussed and the expediency and limits of the application of thermodynamics to the study of silicates are grounded, and an applied one, in which as far as the initial data are available thermodynamic calculation of reactions in silicate systems of theoretical and practical importance are carried out and their theoretical interpretation given.

O. P. MCHEDLOV-PETROSSYAN

Contents

List of Symbols

A	work, atomic weight, chemical affinity, frequency factor
A_k	activation coefficient
a	activity, radius of particle, coefficient of temperature conductivity
a^*	efficient coefficient of the substance transfer
a, b, c	coefficients in the equation of heat capacity as a function of temperature
Bi	the Biot criteria
C	true heat capacity, concentration
$\bar{C}$	the mean molar heat capacity
C_p	heat capacities at constant pressure
C_v	heat capacities at constant volume
C_{vap}	heat capacities of vapour
C_{liq}	heat capacities of liquid
C_{sol}	heat capacities of solid
1 cal	$\equiv$ 4,1868 Joule
D	diffusion coefficient, covalent bond value in compounds, permittivity
E	electrical potential, bond energy, activation energy
E_a	activation barrier
e	electric charge
F	Helmholtz free energy, the Faraday constant ($9.648456 \cdot 10^4$ C mol^{-1})
f	number of degrees of freedom, intensive value
G	the Gibbs free energy
H	enthalpy
H_{ev}	heat of evaporation
H_m	heat of melting
I	mechanical equivalent of heat, extensive value, nucleation rate
i	van't Hoff's coefficient
j	flux density
K	equilibrium constant
k	number of components, Boltzmann's constant ($1.380662 \cdot 10^{-23}$ J K^{-1})
L	latent heat of the water evaporation, phenomenological coefficient (the kinetic coefficient of proportionality), solubility
L_i	the ion hydration heat
M	molar mass
m	mass, number of water moles in crystalline hydrates
N	the Avogadro constant ($6.022045 \cdot 10^{23}$ mol^{-1}), molar fraction, serial number of cation

n	number of phases, concentration of ions of the solution, number of moles
n_w	concentration of the molecules of water
P	osmotic pressure
$\bar{P}$	the mean polimerization degree
p	pressure
p_w	dipole moment of the molecules of water
Q	heat effect of the process
q	heat of polymerization
R	gas constant (8.31441 J K^{-1} mol^{-1})
r	number of exterior parameters, thermochemical radius
S	entropy
Sp	solubility product
T	absolute temperature
T_g	vitrification temperature
T_m	temperature of melting
T_{tr}	temperature of transformation
U	internal energy
V	volume of the system
v	molar volume, rate of the substance transfer
W	amount of electric effect (of electricity)
W/C	water-cement radio
W/S	water-solid ratio
w	rate of the chemical reaction, component of the thermal flux
x	space variable
y	integration constant
Z	thermodynamic property, valence of the ions of the solution
z	ion charge
α	degree of the dissociation, external coefficient of heat exchange
α, β, γ	etc., particular phases
γ	coefficient of activity, components of the system
δ	variation derivative, the deformation values
Δ	finite change of a property
ε	mean bond energy, dielectric permittivity
λ	coefficient of heat conductivity
μ	chemical potential, ionic strength
ν	frequency, number of atoms, entropy criterion, number of ions
ϱ	density of the substance
σ	source in a unit of time per unit of volume
τ	time, time variable
Φ^*	reduced potential
Φ'	different reduced potential
φ	intensive variable
Ω	limiting surface, potential energy
ω	potential energy of a unit mass, amount of the chemically bound water

Chapter I. The Basic Concepts and Laws of Thermodynamics[3]

1 Basic Concepts of Thermodynamics

In order to formulate the main principles of the thermodynamic method more precisely, the chief concepts, terms and values used in thermodynamics must be defined.

For the thermodynamic analysis of a process individual groups or individual substance, or even separate parts of it, whose interaction is to be studied, are singled out of the multitude of the substances under consideration, the scope of the investigation thus being narrowed. Only these objects, called systems are submitted to thorough analysis. The substance not included in the system is the environment and may be taken into consideration only to the extent to which it interacts with the system in question.

Systems may occur as homogeneous and heterogeneous ones the latter having boundaries between the parts of the system differing in properties. In case homogeneous systems have the same composition and properties in all the points, they are moreover uniform, whereas heterogeneous systems comprise at least two phases. Therefore the term *phase* denotes a combination of homogeneous parts of a system having identical chemical and thermodynamic properties separated by a boundary from the other phases. Phases can be subdivided into separate parts but that does not increase their number in the system, e.g., in a saturated solution of dihydrate gypsum in water, when there is any amount of crystalline particles of gypsum on the bottom of the vessel and vapour above the solution, always only one crystalline, one liquid and one vaporous phase will exist.

Components are those kinds of particles which can exist independently beyond the system. Thus in a water solution of $CaSO_4$ at 293 K one can find the following kinds of particles: water, gypsum, anhydrite, hydrated ions: Ca^{2+}, SO_4^{2-}, H^+, OH^-. There are however only three components there: water, anhydrite and gypsum, as the ions cannot be extracted from the given system separately.

Components are independent constituents of a system. Their number corresponds to that of the constituents sufficient for forming all the phases of an equi-

[3] For a deeper study of the chemical thermodynamics one should consult special handbooks and monographs [1, 12, 66, 118, 149, 150, 151, 194, 234, 242, 274, 277, 302, 304, 311, 488, 527, 529, 550, 562, 603, 622, 658, 667, 672, 719]

librium system minus the number of the equations of reactions linking the concentrations of the substances in one of the phases.

The interaction of the system with the environment may manifest itself in interchange of substance and energy – open systems –. It is sometimes quite reasonable to consider a system in an idealized state, when its interaction with the environment is reduced to a minimum and its volume remains constant. In this case we deal with isolated systems in which the energy despite the processes and chemical transformations taking place remains invariable. A change of the energy of a system can be observed only in the case of its isolation being broken subsequently allowing energy interaction with the environment. If a system is not entirely isolated and can alter its volume and exchange energy with the environment, it is called (unlike the previous isolated one) closed. The values that characterize all the physical and chemical properties of a system, such as temperature, pressure, volume, internal energy, entropy, concentration etc., are denominated thermodynamic parameters of state.

For a thermodynamic analysis of the alteration of the state of a system the change of its thermodynamic parameters must be known. The number of the parameters of state depends however on the number of the degrees of freedom of a system[4]. An alteration of one of them results in an alteration of the state of the whole system, which causes a thermodynamic process in that system.

The process wherein a system having departed from the initial state and having undergone a few alterations returns to the initial state, is called a circular process or a cycle.

A process is called reversible when it allows the return of a system to the initial state with no alterations in the outer medium or in the system itself.

If the course of the process in a certain direction and in the reverse one results in some alterations in a system or in the environment, the process will be irreversible.

It should be borne in mind that all the processes actually taking place in Nature are irreversible. The alterations causing them go on in time with finite differences between the forces effecting the system, and they are accompanied by unavoidable losses.

Using the concept of a reversible process and other limiting notions (isolated system, ideal solution etc.) is very important, as it permits studying the processes in the most simplified idealized state which in turn simplifies the analysis. The results obtained in the limits of admissible experimental errors may be well applied to actual processes.

---

[4] The number of the parameters of state capable of freely altering without breaking the phase equilibrium is called the number of the degrees of freedom of a system or its variance. The number of the degrees of freedom is determined for each system by the difference between the number of the parameters of the system in question and that of the equations linking them

One of these idealized notions is the idea of an equilibrium state, which is characterized by a stable state of a system in time, in which no spontaneous processes take place.

The chemical equilibrium is mobile, for the rates of direct and reverse reactions at the moment of the equilibrium are equal.

Taking into consideration that an alteration of external conditions will effect the equilibrium state of a system and may result in breaking it, for the sake of study it is necessary to choose such a system in which all extraneous influences are reduced to a minimum, i.e., to isolate the system.

The thermodynamic process in which a system passes through a continuous series of equilibrium states, is called an equilibrium process. An equilibrium process is characterized by reversibility, absence of any losses, infinitesimal difference of forces effecting a system, equality of the temperatures of a system and the external medium, an infinitely great duration of any final alteration of the state of a system.

A process going on infinitely slowly is often called quasistatic.

In chemical processes reversibility and equilibrium generally concur, i.e., any equilibrium process is always reversible.

Unlike an equilibrium process, non-equilibrium exists when finite alterations have taken place in a system. A non-equilibrium process is unidirectional. In this case owing to the finite rate of the process and because of losses, the work done by a system will always be smaller than in an equilibrium process.

A system having performed an equilibrium process may return to the initial state, passing in the reverse process the same equilibrium states that it passed in the direct process. This property, as mentioned above, is called reversibility of an equilibrium process.

As chemical thermodynamics studies the processes bringing about a system's transition from one equilibrium state to another, differing from the initial one, the need arises for more detailed knowledge of the parameters, studying the alteration laws of which produces the goal of the thermodynamic method.

The total measure of the material motion with its transformation from one kind to another is called energy. A thermodynamic system consists of an enormous number of continuously moving and interacting particles, their energy constituting the total energy of the system. The latter is divided into external and internal ones. The external energy covers the energy of the motion of a system as a whole and the potential energy of a system in the field of the forces acting. The rest of the energy of a system is called its internal energy. In chemical thermodynamics stationary systems in the absence of an external field are considered, therefore the total energy of a system will equal the internal one.

The internal energy of a system consists of that of all types of motion and interaction of the particles entering the system: the energy of the forward and rotary motion of molecules and that of the oscillatory motion of atoms, the energy of the molecular interaction, the intra-atomic energy of the filled electron levels, the intranuclear energy etc.

At present it is not possible to calculate the absolute value of the internal energy U of any system, but for a thermodynamic analysis it is quite sufficient to measure the change of the energy U, taking place in one or another process. The internal energy depends both on the amount of the substance under consideration and on the conditions of its existence. When a thermodynamic system interacts with the environment two different ways of energy transfer from the system to the outer bodies are possible: with a change of the external parameters of the system and without such. The first method of energy transfer is called work, the second, heat, and the process itself, heat exchange.

A system that does not exchange heat with other bodies is called adiabatically isolated. The processes taking place at constant temperature are called isothermic, at constant pressure, isobaric, at constant volume, isochoric. The processes accompanied by heat release are called exothermic, those accompanied by absorption of heat, endothermic.

Enthalpy is an important thermodynamic function. Like the internal energy it is a parameter of state, i.e., its change is determined by the initial and final states of a system and does not depend on the process.

$$H = U + pV, \tag{1}$$

where p, pressure; V, volume of a system; U, internal energy.

Enthalpy is used in isobaric processes, change of the internal energy in isochoric ones.

The sum of the heat given off and the entire work over the environment excluding the work of expansion is called the heat effect Q of the process.

The heat of formation is the heat given off or absorbed during formation of a compound, and indicates the decrease or increase of internal energy of the elements of the compound.

The heat of polymorphic transformations, of dissolution, melting, sublimation etc. is that of the corresponding isothermic and reversible processes.

All energy values – internal energy, enthalpy, heat effects, heats of formation etc. – may be expressed in any energy units. They are mostly expressed in calories or joules and are referred to as moles, gram atoms or gram ions.

Heat capacity is the ratio of the amount of heat imparted: increase of the temperature.

As the heat capacity of substances depends on temperature, mean and true heat capacities are distinguished. The mean molar heat capacity of a system $\bar{C}$ is the heat Q, expended on heating one mole of the substance divided by the difference of temperatures ΔT of heating.

$$\bar{C} = Q/\Delta T. \tag{2}$$

Mean heat capacity is used chiefly in technical estimations. For a more precise calculation the concept of actual heat capacity is used, a ratio of infinitesimal increment of heat: infinitesimal increment of temperature:

$$C = dQ/dT\,. \tag{3}$$

Isobaric C_p and isochoric C_v heat capacities must be noted as well as molar, atomic, volume and weight heat capacities depending on the process and the mass units of the substance.

The pressure p of a system characterizes the rate of interaction with environment and is specified as a force normally affecting a surface unit of the system. If the system is in equilibrium, the pressure should be the same overall and equal to the exterior pressure.

A change of the volume ΔV of a system implies interaction between system and environment. If the volume is constant, there is no exchange of work between system and environment; if it alters, however, the exchange of work is certain.

The temperature of the system characterizes the rate of heat motion of the molecules.

The difference of temperature means the degree of deviation of heat equilibrium from one body to another under equal pressure. For quantitative determination of temperature the absolute thermodynamic scale of temperatures based on the second law of thermodynamics is used. The initial point of this universal scale is the absolute zero K that equals $-273.16\,°C$.

To define entropy and to explain its physical sense we shall first define the coordinates of state, potentials and the motive forces of a process [672].

The coordinates of state of a system are the parameters of the state, whose change serves as an indication of a certain interaction.

For any process a coordinate of state (a parameter of its own) may be found, a change of which is intrinsically connected with a rise in interaction. Thus, chemical interaction is always accompanied by a change of the mass of individual substances making up the system, and electrical interaction by a change of the electric charge (the amount of electricity); mechanical interaction gives a change of the volume. The absence of change in these parameters means total absence of interaction. Thus the mass of a substance is its chemical coordinate of state, the electric charge, is its electric coordinate of state, and the volume, is its mechanical coordinate of state.

Entropy is particular coordinate of state corresponding to the thermic interaction of a system with the environment. Entropy plays the same role in heat phenomena as an electric charge in electric phenomena, and may in a sense be regarded as a thermic charge. Thus thermic interaction only occurs if entropy (thermic charge) exchange exists between a system and the environment.

In system-environment interaction such parameters as pressure, electric potential and temperature play a completely different part from that of volume, electric charge and entropy.

It is clear that if the pressure difference between a system and the environment is zero, the system is in equilibrium with it; if, however, the pressure difference is not zero, then mechanical interaction is taking place, the greater the pressure difference, the higher the interaction rate. The direction of the process is predetermined by the sign of the difference of pressures. It is reasonable to call such parameters potentials.

The thermic interaction of a system with the environment has its own potential, temperature, the potential of chemical interaction is the chemical potential, and the potential of electric interaction is the electric potential. If these potentials in the systems and in the environments are equal, no interaction takes place and a system will be in a state of thermodynamic equilibrium with the environment. Interaction starts if there is a difference between the potentials of a system and the environment, becoming more intense the greater the difference. The direction of the interaction process depends on where the potential is greater: in the system or in the environment.

Thus the potentials play the part of motive forces. Each degree of freedom of a system has its own two parameters of state, a coordinate of state and a corresponding potential; the amount of the effect (it is to be noted that the term 'effect' in some contexts (e.g. heat effect etc.) may mean a different entity) is always determined as the product of the potential and the elementary change of the coordinate of state.

To state a general rule of signs for varying amounts of effect, the following rule should be used: the amount of effect is positive if in the interaction process of a system with the environment the corresponding coordinate of state increases. If the coordinate of state of a system decreases the amount of effect is held to be negative. The chemical effect – the work of chemical forces – is held to be positive if it is directed towards the system, i.e., the mass of the chemically interacting substance of the system augments. The amount of heat (the amount of heat effect) is held to be positive if heat is fed to the system (entropy increases), and negative if heat is removed from the system (entropy decreases).

2 The First Law of Thermodynamics: Thermochemistry

An alteration of the internal energy of a thermodynamic system is always compared with physical values: amount of work, heat effect, chemical or electric effect. Depending on the properties of a system or actual conditions it may either transmit part of its internal energy or receive environmental influence.

Heat and work are two forms of energy transfer from the environment to the system and vice versa. We have already demonstrated that any amount of effect is obtained by multiplying the potential by a coordinate of state. In chemical reactions the amount of chemical effect will equal the product of the chemical potential μ by an elementary alteration dm of the mass of the reactant:

$$dG = \mu\, dm\,. \tag{4}$$

In electric processes the amount of the electric effect is:

$$dW = E\, de\,, \tag{5}$$

where E, potential; de, alteration of the electric charge.

In heat effect the amount of heat is determined as follows:

$$dQ = TdS\,, \tag{6}$$

where T, absolute temperature; dS, elementary alteration of the entropy (thermic charge).

In mechanical processes the amount of mechanical action – work – will be:

$$dA = pdV. \tag{7}$$

In thermodynamics all the processes considered in the system are covered by the law of conservation and transformation of energy.

In the most general form, when a system has f degrees of freedom, the law of conservation and transformation of energy may be presented as an increment of the internal energy of the system equalled to the sum of all f amounts of effect:

$$dU = dQ - dA + dW + dG + \ldots \tag{8}$$

or by means of parameters of state

$$dU = TdS - pdV + Ede + \mu dm + \ldots\ . \qquad (9)^{5}$$

For thermomechanical and thermochemical systems having two degrees of freedom from general Eq. (8) we obtain:

$$dU = dQ - dA\,, \tag{10}$$

$$dU = TdS - pdV\,. \tag{11}$$

[5] It should be borne in mind that all the values enter Eq. (9) in corresponding units. The sign of minus before the work means that the rule of signs for it is the opposite one as compared to other effects

Equations (8) and (9) are the most general terms of the first law of thermodynamics which can be reduced to the expression that an alteration of the internal energy always equals the sum of effects of various kinds.

In isolated systems [($Q = 0$); ($A = 0$)]. In spite of probable transformations of energy within the system, the total amount of the internal energy will remain constant:

$$U = \text{const} \quad \text{and} \quad dU = 0 . \tag{12}$$

If, however, a certain amount of heat dQ is imparted to this isolated system, then, as the volume of isolated systems remains constant, all the heat will be spent on the increase of internal energy, i.e.,

$$dQ = dU . \tag{13}$$

In closed systems the heat imparted will be spent both on the increase of the internal energy dU and on performing mechanical work dA:

$$\delta Q = dU + dA = dU + pdV . \tag{14}$$

If the volume of a system is constant, when $V = \text{const}$ and $dV = 0$

$$dQ_v = dU \quad \text{or} \quad Q_v = U_2 - U_1 , \tag{15}$$

where Q_v is the heat under constant volume, equal to the increment of the internal energy of the system.

In case of constant pressure $p = \text{const}$ by way of integrating from Eq. (14) we obtain:

$$Q_p = U_2 - U_1 + p(V_2 - V_1) = (U_2 + pV_2) - (U_1 + pV_1) . \tag{16}$$

The expression $U + pV$ is the heat function of state of a system and, as said above, is called enthalpy H or

$$dH = dU + pdV . \tag{17}$$

Thus

$$Q_p = H_2 - H_1 \tag{18}$$

or

$$H_2 - H_1 = \int_{T_1}^{T_2} C_p dT , \tag{19}$$

where C_p is the heat capacity under constant pressure.

In a cyclic process the work of a system A is proportionate to the heat Q:

$$\oint \delta A = I \oint \delta Q\,, \tag{20}$$

where δA and δQ, infinitesimal amounts of work and heat and I, mechanical equivalent of heat (later we shall omit it, admitting that heat and work are expressed below in the same units).

Then Eq. (20) may be expressed in the following way:

$$\oint \delta A - \oint \delta Q = 0\,, \tag{21}$$

i.e., as a result of a cyclic process the energy reserve in a system does not alter and

$$\oint dU = \oint \delta A - \oint \delta Q\,. \tag{22}$$

Thus, the internal energy of a system is a single-valued function and its alteration does not depend on the path. The work in a cyclic process, however, is not equal to zero and because it occurs by absorption of heat the longer the path of the cycle, the greater it will be; therefore work and heat are not functions of state in this case.

The first law of thermodynamics applied to chemical processes is considered in thermochemistry, which treats heat measurement in chemical reactions.

The term "heat effect of a chemical reaction" denotes the heat absorbed or released in a transition from pure initial substances to pure final substances in quantities shown in the reaction equation, on the condition that no work is being performed but the work of expansion. This is called calorimetric or non-equilibrium heat [149].

The heat released in the reaction and the work performed thereby are proportional to the mass of the bodies reacting. Below we shall relate the heat effect to a gram-molecule or a mole. As chemical reactions which, if an alteration of the volume of the reacting substances is accompanied by the work of expansion, (other kinds of work are absent in view of the elctric, magnetic, surface and other forces) the equation of the first principle will be written in form (14).

In reactions with $V =$ const [see Eq. (15)] the heat effect under constant volume equals the loss of the internal energy of the system. With $p =$ const [see Eqs. (16, 19)] the heat effect of the reaction equals the loss of enthalpy.

Thus, the heat effect presents a change of the total energy of a system and, since U and H are functions of state of a system whose change depends not on the path of the process, but only on the initial and final state, the heat effect of the reactions both for $p =$ const and $V =$ const does not depend on whether these reactions take place in one or several stages, i.e., does not depend on the path of the process, and is determined only by the initial and final state of the system.

This statement was first formulated by the Russian chemist Hess on the strength of experimental data. Hess' law states that the amount of heat released (or absorbed) in chemical processes depends solely on the initial and final state of a system that has taken part in these processes. Using Hess' law one can calculate the heats of this or another reaction with the help of the heats of other reactions. This is particularly important for the reactions to find whose heat is experimentally hardly feasible. The heat of any reaction can be calculated on the basis of Hess' law by algebraic summation of heats of formation of all reactants, i.e., it is determined as a difference between the sum of the heats of formation of the reaction products and the sum of the heats of formation of the initial substances.

Taking into account that according to Eq. (19) the heat effect of a reaction depends on its heat capacity; the equations of heat capacity must be known. For most solids and gases the temperature dependence of the heat capacity is expressed by empirical equations in the form of power series, the coefficients of which are selected so that the equation should correspond to experimental data:

$$C_p = a + bT + cT^2 . \tag{23}$$

Equations such as Eq. (23) may be obtained both for mean values of $\bar{C}_p$ or $\bar{C}_v$ and for true C_p and C_v. While an equation of the type, Eq. (23) expresses the temperature dependence of the mean heat capacity, the dependence of the heat capacity C_p or C_v may be found by differentiating it:

$$C = \frac{dQ}{dT} = \frac{d(\bar{C}T)}{dT} = a + 2bT + 3cT^2 + \dots . \tag{24}$$

On the contrary, the temperature dependence of the mean heat capacity may be found from the true one by integrating it.

$$\bar{C} = \frac{Q}{T} = \frac{\int_0^T C\,dT}{T} = a' + \frac{b'}{2}T + \frac{c'}{3}T^2 . \tag{25}$$

In both cases the mean heat capacity is taken within the range from 0 to T. Within the range from T_1 to T_2 the equations will be somewhat different. It should be borne in mind that trinominal equations of $C_p = f(T)$ are valid with a mean error of 1% – 3% at temperatures from 0° to 2000 °C.

For gases the equations $C_p = f(T)$ are based mainly on spectroscopic data. For solids these equations, besides experimental methods, may be based precisely enough on the entropy method of Landiya [320]. Recently empirical formulae somewhat different from those in Eqs. (23) – (25) have often been used, namely:

$$C_p = a + bT + cT^{-2}, \tag{26}$$

which we also accepted.

With the help of temperature dependence equations of heat capacity of reactants, one can also calculate the amount of heat necessary to bring the initial substances to a certain temperature, as well as solve other urgent problems. The temperature dependence of the heat of a chemical reaction was first studied by Kirchhoff.

The ∂H of a reaction is known to equal the difference between the sum of the heat content of the reaction products and that of the substances entering the reaction, and the molar heat content is known to alter with the temperature according to the equation:

$$\left(\frac{\partial H}{\partial T}\right)_p = C_p .$$

Combining these equations we obtain Kirchhoff's:

$$\left(\frac{\partial H}{\partial T}\right)_p = \left(\frac{\partial \Sigma H_1}{\partial T}\right) - \left(\frac{\partial \Sigma H_2}{\partial T}\right) = \Delta C_p = \Sigma C_{\text{final product}} - \Sigma C_{\text{init. subst.}}, \tag{27}$$

where ΔC_p is the difference of the sum of heat capacities of the reaction products and of the initial substances, i.e., the total alteration of the heat capacity of a system as a result of reaction.

Considering that ΔH_1 and ΔH_2 are heat effects of the isothermic reaction in question at two different temperatures T_1 and T_2, on integrating Eq. (27) we derive:

$$\Delta H_1 - \Delta H_2 = \int_{T_1}^{T_2} \Delta C_p dT . \tag{28}$$

Equations (27) and (28) strictly apply only when the pressure over each component at the required temperature is the same as at the given temperature.

Kirchhoff's equation enables us to calculate the heat effect of a reaction at any temperature proceeding from a known heat effect at any given temperature and an alteration of the heat capacity of the process. For the process of evaporation we have:

$$\frac{dH_{\text{ev}}}{dT} = C_{\text{vap}} - C_{\text{liq}}, \tag{29}$$

where H_{ev} is heat of evaporation, and C_{vap} and C_{liq}, heat capacities of vapour and liquid respectively.

For the process of melting:

$$\frac{dH_{\mathrm{m}}}{dT} = C_{\mathrm{liq}} - C_{\mathrm{sol}}\,, \tag{30}$$

where H_{m}, heat of melting; C_{liq} and C_{sol}, heat capacities of a liquid and a solid respectively.

3 The Second Law of Thermodynamics

It has been noted above that the first law of thermodynamics enables one only to make up the energic balance of processes but does not indicate the possibility of their realization and direction. These problems can be solved by using the second law of thermodynamics resulting from many years' experience and the use of extensive experimental material.

All processes in Nature are irreversible, i.e., they proceed at a relatively great difference of potentials of the system and the environment. Under real conditions it is impossible to reverse a process with an infinitesimal alteration of the potential. In this case a complicated distribution of the corresponding potentials inevitably arises in the system, and studying these processes would seem to be impossible.

However, in order to investigate actual processes at least approximately we must simplify those occurring in the system and reduce them to a form suitable for analysis, by introducing the concepts of an equilibrium and quasistatic state.

An equilibrium within a system develops either whenever the potential on its surface retains the value equal to the potential of the environment for a long time, or when the system is isolated from it. It has been proven by tests that under such conditions an equilibrium state is established throughout the system after a certain period of time due to the levelling of the potentials. In this case to each degree of freedom – each coordinate of state – corresponds a single value of the second parameter of this kind, the potential. Thus, under the conditions of an equilibrium state it is very easy to express the amount of a certain effect. If, however, there is no equilibrium in the system, i.e., the potentials are distributed unevenly, we are uncertain as to which one of the enormous number of the values of the potentials of the system we are to employ for determining the amount of action. In this case it is quite expedient to establish such conditions for the interaction of the system and the environment in which the difference between the potentials should be so small that it could be neglected as compared with the absolute values of the potentials. Thus, an equilibrium alteration of state will be achieved. Such processes are called quasistatic ones.

In the course of equilibrium and quasistatic processes an infinitesimal alteration of the potential is enough to change the direction of the process, explaining why both equilibrium and quasistatic processes are always reversible. Nonstatic – irreversible – processes are studied by methods of thermodynamics of irreversible processes; this will be considered in a special chapter.

The second law of thermodynamics reads as follows: in a quite reversible process (to say nothing about irreversible ones) only a part of its heat is transformed into work, the other part however, depreciates passing from more heated parts of the system to colder ones without performing work.

For chemical reactions occurring at constant pressure and temperature the second law will be expressed as:

$$\Delta H = \Delta G + T\Delta S\,, \tag{31}$$

where ΔH, total heat of the process (enthalpy); ΔG, the Gibbs free energy of the process presenting the maximum part of the energy of the system which can be transformed into work; ΔS, alteration of the entropy of the system; T, temperature of the process.

Equation (31) may be derived by using the first law according to which the alteration of the energy of a system equals the sum of the amounts of action. For a chemical reaction according to the first law we can express:

$$\Delta H = T\Delta S + \Delta G\,.$$

In case of a reversible process

$$dS = \frac{\delta Q}{T}\,,$$

$$dU = dQ - dA + dG$$

as

$$dQ = TdS;\quad dA = pdV;\quad dG = \mu\, dm$$

then

$$dU = TdS - pdV + \mu\, dm$$

at $p = \text{const}$ $dU + pdV = dH$. Thus

$$dH = TdS + \mu\, dm \tag{32}$$

or substituting dG for μdm for the final conditions we shall have:

which gives on integrating:

$$\Delta S = \int_{T_1}^{T_2} \frac{\delta dQ}{T}. \tag{33}$$

If a system is heated from T_1 to T_2 at a constant volume and constant pressure, for the alteration of entropy we obtain

$$\Delta S_v = \int_{T_1}^{T_2} C_v \frac{dT}{T} \tag{34}$$

and

$$\Delta S_p = \int_{T_1}^{T_2} C_p \frac{dT}{T} \tag{35}$$

respectively.

If a system absorbs heat at a constant temperature, then

$$\Delta S = \frac{Q}{T}. \tag{36}$$

For example, for isothermic evaporation of a liquid

$$\Delta S = \frac{\Delta H_{\text{ev}}}{T}. \tag{37}$$

Calculating the alteration of entropy for an irreversible process one should theoretically substitute one or several subsequent reversible processes for it retaining the initial and final states of the system unaltered.

Entropy is an additive property of a system, i.e., the entropy of a system equals the sum of the entropies of the constituents.

Entropy characterizes the amount of the dispersed, depreciated energy per degree. With crystalline bodies it is expedient to present it as a logarithmic measure of the degree of deviation of the crystal structure from an ideal order.

If a system is isolated, entropy does not vary when reversible processes take place in it, as it has reached its maximum value, whereas in irreversible processes entropy grows. When an irreversible process brings an isolated system to a state of equilibrium, its entropy reaches its maximum. Thus, entropy is a criterion for the direction of a process. As criterion for the unrealizability of the process, the inequality $\Delta S < 0$ may serve, i.e., there can be no process which would take place in an isolated system with a decrease of entropy. In an unisolated system however both reversible and irreversible processes with a decrease of entropy can take place. Solving practical tasks by using entropy may seem impossible, because in the term $\Delta S < 0$ for actual irreversible processes the sign of inequality figures. However, as we have already shown, any irreversible process can always be presented as a quasistatic one, and in this case the calculation of ΔS is by no means difficult for the alteration of entropy does not depend on the character of transformations. That is the only reason for the thermodynamic method to be employed in the study of equilibrium and quasistatic processes.

Thus, as the first law introduces a function of state, internal energy, the second introduces a new function of state, entropy. An alteration of the entropy of a system is an indication of a thermodynamic interaction in it.

4 Methods of Thermodynamics

The thermodynamic method of investigating various processes is based on using the first and the second laws of thermodynamics: for solving actual problems two methods are used in thermodynamics: the method of circular processes – cycles – and that of thermodynamic potentials – characteristic functions –. When using the method of circular processes for establishing a certain regularity in one or the other phenomenon an appropriate reversible cycle is considered and to that cycle the equations of the first or the second laws are applied:

$$\oint \delta Q = U \quad \text{and} \quad \oint \frac{\delta Q}{T} = 0 \,. \tag{38}$$

If we theoretically make the system studied perform Carnot's cycle which is known to consist of two isothermic and two adiabatic processes, the equation $\oint \delta Q/T = 0$ can be used to express the efficiency of that cycle. We have already stated that one of the most important conditions of equilibrium is $dS = 0$. It should be noted that in many cases it is a too general criterion, it being not easy to study the alteration of entropy of reactants in all cases. Therefore for the calculation of equilibria numerous thermodynamic functions have been suggested, i.e., functions by means of which or derivatives whereof the thermodynamic properties of any system can be expressed. The chief thermodynamic functions are: the internal energy U, enthalpy H, the Helmholtz free energy F and the Gibbs free energy G.

The method of thermodynamic functions has been more widely accepted for the study of the conditions of the equilibrium of chemical systems than the method of cycles.

The internal energy is a characteristic function at independent variables V and S:

$$\left(\frac{\partial U}{\partial S}\right)_V = T \quad \text{and} \quad \left(\frac{\partial U}{\partial V}\right)_S = -p \,. \tag{39}$$

From Eq. (39) one can see that temperature is a measure of the increase of the internal energy with the increase of entropy at a constant volume, whereas pressure is that of the decrease of the internal energy with the increase of the volume at a constant value of entropy.

Enthalpy is a characteristic function at independent variables p and S:

$$\left(\frac{\partial H}{\partial S}\right)_p = T \quad \text{and} \quad \left(\frac{\partial H}{\partial p}\right)_S = V. \tag{40}$$

From Eq. (40) one can see that temperature is a measure of the increase of enthalpy of a system with the growth of entropy at $p =$ const, whereas volume is that of the increase of enthalpy with the growth of pressure at $S =$ const.

The Helmholtz free energy is a characteristic function at independent variables V and T:

$$\left(\frac{\partial F}{\partial V}\right)_T = -p \quad \text{and} \quad \left(\frac{\partial F}{\partial T}\right)_V = -S. \tag{41}$$

That is, pressure is a measure of the loss of the free energy of a system with the growth of the volume at a constant temperature, whereas entropy is that of the loss of the free energy with the growth of the temperature at a constant volume of a system.

The Gibbs free energy is a characteristic function at independent variables p and T:

$$\left(\frac{\partial G}{\partial p}\right)_T = V \quad \text{and} \quad \left(\frac{\partial G}{\partial T}\right)_p = -S. \tag{42}$$

That is, the volume of a system is a measure of the increase of the Gibbs free energy with the growth of pressure at a constant temperature, whereas entropy is that of the loss of the Gibbs free energy with the growth of temperature at a constant pressure. It is worth noting that besides $dS \leqslant 0$ the equilibrium may be formulated otherwise which will make other parts of the complicated interaction of a system with the environment more prominent. The general condition for the processes to take place in the direction of stable equilibria can be achieved from equations:

$$dF = -pdV - SdT \quad \text{at} \quad T \quad \text{and} \quad V = \text{const}; \tag{43}$$

$$dG = Vdp - SdT \quad \text{at} \quad T \quad \text{and} \quad p = \text{const}; \tag{44}$$

$$dF \leqslant 0 \quad \text{at} \quad T \quad \text{and} \quad V = \text{const}; \tag{45}$$

$$dG \leqslant 0 \quad \text{at} \quad T \quad \text{and} \quad p = \text{const}. \tag{46}$$

Thus, at a constant temperature and volume only such processes can spontaneously occur that are accompanied by a decrease of F, the equilibrium being characterized by its minimum value.

At a constant temperature and pressure only such processes can spontaneously occur that are accompanied by a decrease of G, the equilibrium being characterized by the minimum value thereof.

Similar formulations of equilibrium conditions can be made in respect to the thermodynamic functions U and H as well, but for them one of the variables, the entropy, is $(dU)_{VS}=0$; $(d^2U)_{VS}>0$, the constancy of which is difficult to be provided for in practice as compared to V, T and p for dF and dG.

The calculation of the alterations of the Gibbs and Helmholtz free energies in order to predict the directions of these processes is one of the main tasks of chemical thermodynamics.

The link between the heat effect of an irreversible chemical process and the work of the irreversible process in question is found by Gibbs-Helmholtz' equation.

For the Helmholtz free energy

$$\Delta F = \Delta U_V + T\left(\frac{\partial \Delta F}{\partial T}\right)_V; \tag{47}$$

for the Gibbs free energy

$$\Delta G = \Delta H_p + T\left(\frac{\partial \Delta G}{\partial T}\right)_p. \tag{48}$$

Chemical processes are characterized not only by energy transfer, but by substance transfer as well. Therefore considering the Gibbs free energy, one should bear in mind that an alteration of the quantity of moles of some component with retention of a constant amount of the rest components at constant T and p will to some extent alter the value of the Gibbs free energy. The increment of the amount of chemical energy during the reaction will be expressed as the product $\mu\, dm$ where μ is the chemical potential and dm is an alteration of the mass.

In mathematical presentation chemical potentials are partial derivatives of the characteristic functions according to the number of moles of the given component of a system at constant values of independent parameters and amount of the rest of components:

$$\mu_1 = \left(\frac{\partial U}{\partial m_1}\right)_{V,S,m_2,m_3,\ldots};$$

$$\mu_1 = \left(\frac{\partial H}{\partial m_1}\right)_{p,S,m_2,m_3,\ldots}; \tag{49}$$

$$\mu_1 = \left(\frac{\partial F}{\partial m_1}\right)_{V,T,m_2,m_3,\ldots}; \tag{49}$$

$$\mu_1 = \left(\frac{\partial G}{\partial m_1}\right)_{p,T,m_2,m_3,\ldots}.$$

A chemical potential may be considered to be the work of chemical forces, i.e., performed with an alteration of the amount of some component in the system, the parameters and mass of the rest of components being constant.

Being a potential, the chemical potential must characterize the state of a system. If several systems form one and have equal potentials equilibrium results in the system. Thus, the equality of chemical potentials of all components in all parts of the common system characterizes its chemical equilibrium:

$$\sum \mu_i dm_i = 0\,. \tag{50}$$

An alteration of the chemical potential of one of the components of a system will result in its leaving a phase and in a shift of the equilibrium.

However, as the absolute values of the thermodynamic potentials are unknown (according to the first law one can only determine the corresponding differences connected with alterations of the state of a system), one should choose an initial state to which later one could relate all the subsequent values of potentials.

Generally the state in which the enthalpy of all the elements at any temperature and at the pressure of 1 atm equals zero is considered to be standard. Then the enthalpies of compounds consisting of elements in this standard state will be equal to the alteration of the enthalpy that accompanies the formation of a mole of these compounds.

In a similar way a standard state for the rest of thermodynamic functions – potentials – is chosen. The values of the thus obtained thermic constants are tabulated for the temperature of 298.16 K. The tabulated values are called standard values of the corresponding thermodynamic functions – enthalpy, the Gibbs free energy, entropy etc. –; to differentiate them in the right upper corner of the corresponding symbols the index ^{0} is written (e.g., ΔH^0, ΔG^0, μ^0, S^0 etc.).

It is worth mentioning once more that a standard state may be chosen in any way convenient for calculations, it is only necessary that later all the calculations performed should relate to that arbitrarily chosen state.

For ideal solutions and gases

$$\mu_i = \mu_i^0 + RT \ln p_i\,. \tag{51}$$

For actual solids and liquids simple relations between the chemical potential and any property are unknown; therefore the chemical potential is employed mainly for drawing theoretical conclusions.

In actual solutions deviations from Eq. (51) are caused by the electrostatic interaction of ions, i.e., in actual solutions there is an active concentration which is determined as

$$a = \gamma C_k , \tag{52}$$

where γ: activity coefficient; C_k: concentration.

The ion activity is an effective concentration which manifests itself in chemical reactions. Ordinary concentrations show only the amount of substance contained by the solution. As far as gases are concerned, the term "fugacity" is substituted for "activity".

5 The Third Law of Thermodynamics

The ability of various substances to interact with one another forming new products is called chemical affinity. As a measure of the chemical affinity van't Hoff suggested using the maximum work of chemical forces, i.e., an alteration of free energy F at $V = \text{const}$ or an alteration of G at $p = \text{const}$:

$$A_p = -\Delta G .$$

The chemical affinity does not depend on the path and stages of reactions' course, but only on the initial and final state. Its sign determines the direction of a reaction and its value, that of the reaction constant. If we take the initial concentrations or the initial pressures of all the reactants as equal to one, then

$$A_p^0 = -\Delta G^0 = RT \ln K_p , \tag{53}$$

$$A_v^0 = -\Delta F^0 = RT \ln K_v . \tag{54}$$

The two equations characterize the distance of the processes from the equilibrium state. The temperature dependence of the equilibrium constant is expressed by the following equations:

$$\frac{d(\ln K_p)}{dT} = -\frac{Q_p}{RT^2} \quad \text{at} \quad p = \text{const} . \tag{55}$$

$$\frac{d(\ln K_v)}{dT} = -\frac{Q_v}{RT^2} \quad \text{at} \quad V = \text{const} . \tag{56}$$

The qualitative direction of a process can often be predicted with the help of the Le Chatelier principle which states that any physico-chemical system being in the equilibrium state tends to retain it and responds to any effect from without by an onset of the processes within the system that tend to overcome that effect.

On integrating the isobar Eq. (55) and the isochore Eq. (56) of a chemical reaction one can obtain equations that enable calculating the heat effect of the reaction from the measurements of the two equilibrium constants at two temperatures:

$$d(\ln K) = -\frac{Q}{RT^2}\,dT\,. \tag{57}$$

Bearing in mind that the heat effect of a reaction depends on temperature

$$Q_T = Q_0 - \Delta aT - \tfrac{1}{2}\Delta bT^2 + \Delta cT^{-1} + \ldots$$

on integrating Eq. (57) we obtain the equation of the isochore-isobar:

$$\ln K = \frac{Q_0}{RT} + \frac{\Delta a}{R}\ln T + \frac{\Delta b}{2R}T + \frac{\Delta c}{2R}T^{-2} + \ldots y \tag{58}$$

or by substituting A/RT for $\ln K$ in the Eq. (58) we derive the Haber equation:

$$A = Q_0 + \Delta aT\ln T + \frac{\Delta b}{2}T^2 + \frac{\Delta c}{2}T^{-1} + RyT\,. \tag{59}$$

Using Eqs. (58) and (59) one can calculate any equilibrium as well as the normal affinity of a reaction. For this purpose, however, besides the heat effect Q_0 of a reaction and the equations of the temperature dependence of heat capacity one should also know the integration constant y. In order to determine it one measures the equilibrium at least at some temperature. In order to solve this problem without resorting to a study of chemical equilibrium using only thermic data (heat effects of the reactions and heat capacities) first Nernst and then Planck formulated theorems and postulates which became the foundation of the third law of thermodynamics. The Nernst theorem may be expressed in a formula valid for reactions in condensed systems:

$$\lim\left(\frac{dQ}{dT}\right) = \lim_{T\to 0}\left(\frac{dA}{dT}\right), \tag{60}$$

i.e., the curves $Q = f(T)$ and $A = f(T)$ osculate at $T^0 = 0$ K (Fig. 1)

This statement enables one to simplify the Haber equation and to obtain a formula that does not contain integration constant and precisely determine the shape of the curve of the temperature dependence of the chemical affinity:

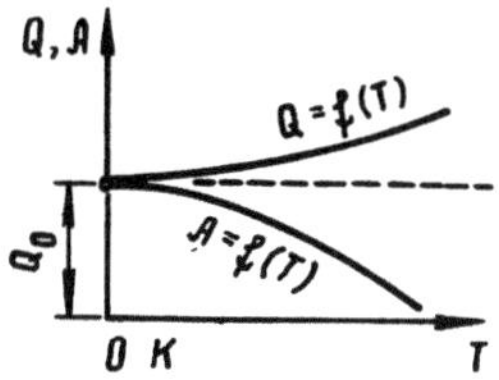

Fig. 1. The position of $Q=f(T)$ and $A=f(T)$ according to the Nernst heat theorem

$$A = Q_0 + \Delta a T \ln T + \frac{\Delta b}{2} T^2 + \frac{\Delta c}{2} T^{-1} \tag{61}$$

as for condensed systems the difference between A_v and A_p may be neglected,

$$A = RT \ln K = Q_0 + \Delta a T \ln T + \frac{\Delta b}{2} T^2 + \frac{\Delta c}{2} T^{-1} . \tag{62}$$

It should be borne in mind that Eq. (62) is exactly applicable only to the temperatures close to the absolute zero. However, one can employ this equation in the range of higher temperatures as well, given that curves $f(T)$ and $A=f(T)$ have no discontinuities and inflexions up to the temperature of polymorphic transformations and melting.

Planck expressed his postulate in the following form:

$$\lim_{T \to 0} S = 0 \quad \text{or} \quad S_{0\,\mathrm{K}} = 0 , \tag{63}$$

i.e., close to the absolute zero the entropy of a correctly formed pure crystal equals zero whereas at any other state of a substance its entropy is greater than zero. It follows from Planck's postulate that:

1. the properties U, H, F, G, C_p, C_v and others of condensed systems near the abolute zero depend no more on temperature;

2. before reaching $T = 0$ K a system changes to a state in which reaching the absolute zero is out of question in principle.

6 Chemical Equilibrium and the Principles of Calculating It

Of great importance in the application of the thermodynamic and thermochemical calculations to various chemical reactions is the knowledge of the chief principles of chemical equilibria that determine the limits and scope of the calculation.

For the chemical reaction

$$lB + cC = lL + mM \tag{64}$$

taking place at a constant pressure and temperature the total alteration of the Gibbs free energy is determined by the equation

$$\Delta G_{p,T} = \Delta G^0_{p,T} + RT \ln \left(\frac{a^l_L a^m_M}{a^b_B a^c_C} \right), \tag{65}$$

where $\Delta G^0_{p,T}$, standard value of an alteration of the Gibbs free energy of a process found according to the standard values of $\Delta G^0_{p,T}$ of the reactants; a, respective values of the activities of reactants related to any possible moment.

In case of equilibrium $\Delta G_{p,T} = 0$ Eq. (65) takes the following form:

$$\Delta G^0_{p,T} = -RT \ln \left[\frac{(a^l_L)_{\text{equ}} (a^m_M)_{\text{equ}}}{(a^b_B)_{\text{equ}} (a^c_C)_{\text{equ}}} \right], \tag{66}$$

where the corresponding activities with the index "equ" relate to the equilibrium state.

The ultimate alteration of the Gibbs free energy for any possible moment of a reaction can be determined by a generalized equation:

$$\Delta G_{p,T} = -RT \ln \left[\frac{(a^l_L)_{\text{equ}} (a^m_M)_{\text{equ}}}{(a^b_B)_{\text{equ}} (a^c_C)_{\text{equ}}} \right] + RT \ln \left(\frac{a^l_L a^m_M}{a^b_B a^c_C} \right) \tag{67}$$

or in abridged form

$$\Delta G_{p,T} = -RT \ln K_a + RT \ln A_k,$$

where

$$K_a = \frac{(a^l_L)_{\text{equ}} (a^m_M)_{\text{equ}}}{(a^b_B)_{\text{equ}} (a^c_C)_{\text{equ}}}$$

and

$$A_k = \frac{a^l_L a^m_M}{a^b_B a^c_C}.$$

This equation, called the equation of the reaction isotherm, enables one to determine the probable direction of a process under given conditions. For the process to take place in the direction desired it is necessary that ΔG should be less than 0, i.e., K_a should be greater than A_k. This can be achieved by either an increase of K_a with the retention of the constant activation coefficient A_k (e.g., in case of endothermic processes, such as dissociation, elevation of temperature) or a decrease of A_k at a constant value of K_a (e.g., by an alteration of the pressure or the composition of the system).

It is also necessary to discriminate between a standard alteration ΔG^0 of a substance in its standard state – pure solid, pure liquid or a gas under the pressure of 1 atm – and the value of the alteration ΔG of a substance in any state chosen. It is needed because the Gibbs free energy alters with pressure, temperature and concentration. A resulting interdependence exists between any arbitrary and standard values of the Gibbs free energy:

$$\Delta G = \Delta G^0 + RT \ln a \tag{68}$$

where a, activity of a substance.

The activities of a pure liquid and a pure solid at mean pressures and temperatures are taken for 1. The activities of gases are taken for equal to a partial pressure of a gas at low pressures (not more than several atmospheres).

The activities of components in solid and liquid solutions are equal to the equilibrium pressures of their vapours over the solution related to the vapour tension of a pure substance. For higher temperatures such data are not practically available. Therefore one must find approximate methods of determining activities. This approximation is made possible by Raoult's law according to which the pressure of the vapour of any constituent equals its mole fraction multiplied by the vapour tension of a pure constituent at the same temperature. It is obvious therefore that the activities in solid and liquid solutions may be taken as equal to their mole fraction in the solutions.

In accordance with the equation of the normal affinity Eq. (53) at $p = \text{const.}$

$$A_p^0 = -\Delta G_p^0 = RT \ln K_p , \tag{68a}$$

hence

$$\ln K_p = -\frac{\Delta G_T^0}{RT} \quad \text{or} \quad \lg K_p = -\frac{\Delta G_T^0}{4.573\, T} , \tag{69}$$

i.e., on determining the value of ΔG_T^0 one can find the value of the equilibrium constant at a given temperature using Eq. (69). One must know the technique of calculating $\Delta G = f(T)$. Under standard conditions the calculation of ΔG, ΔH and S is the net result of their algebraic summation, i.e., the sum of ΔG or ΔH or S of the initial substances is subtracted from the sum of ΔG or ΔH or S of the products of a reaction.

An interdependence exists there between ΔG, ΔH and ΔS (Gibbs-Helmholtz' equation) for chemical reactions:

$$\Delta G = \Delta H - T \Delta S , \tag{70}$$

thus, when we know ΔH and ΔS we can always determine ΔG. In order to determine ΔG, ΔH and S at any temperature we must know the temperature dependence. In accordance with Kirchhoff's equation we have:

$$\Delta H_T = \Delta H_{T_0} + \int_{T_0}^{T} \Delta C_p dT ,$$

where

$$\Delta C_p = \sum C_{\text{f.p.}} - \sum C_{\text{i.s.}} ,$$

f.p. being the final product and i.s. being the initial substance.

It is well known that:

$$S_T = S_{T_0} + \int_{T_0}^{T} \frac{C_p dT}{T} .$$

Therefore at $T = 298$ K and $p = 1$ atm the equations may be expressed as:

$$\Delta H_T^0 = \Delta H_{298}^0 + \int_{298}^{T} \Delta C_p dT , \tag{71}$$

$$S_T^0 = S_{298}^0 + \int_{298}^{T} \frac{C_p dT}{T} . \tag{72}$$

For the alteration of entropy we obtain:

$$\Delta S_T^0 = \Delta S_{298}^0 + \int_{298}^{T} \frac{\Delta C_p dT}{T} . \tag{73}$$

If we substitute Eqs (71) and (73) in Eq. (70) we obtain:

$$\Delta G_T^0 = \Delta H_{298}^0 + \int_{298}^{T} \Delta C_p dT - T \Delta S_{298}^0 - T \int_{298}^{T} \frac{\Delta C_p dT}{T}$$

or in the end

$$\Delta G_T^0 = \Delta H_{298}^0 - T \Delta S_{298}^0 + \int_{298}^{T} \Delta C_p dT - T \int_{298}^{T} \frac{\Delta C_p dT}{T} . \tag{74}$$

As we know ΔG_T^0 from Eq. (74), we can always calculate K_p at any temperature. Thus for calculating K_p and ΔG^0 at temperature T one should know the heat effect of a reaction and entropies of all the reactants at the same temperature. If however we know the values of standard entropies for 298 K calculated proceeding from the heat capacities at low temperatures, whereas the heat capacities at high temperatures may be expressed by power series, it may seem expedient to find the ΔG values at 298 K (25 °C) from Eq. (65) and substituting ΔG_{298} and $T = 298$ K in Eq. (59) to determine the integration constant y. The actual examples of the application of these methods to the technology of silicates will be given in the following chapters.

7 Phase Equilibria

As phase equilibria are of great importance in physical chemistry of silicates, one ought to make oneself familiar at least of the most essential points with the thermodynamic interpretation of phase rule and its application to silicate systems although the subject has been well covered in specific works [53, 119, 135]. Without clearly understanding the phase diagrams of equilibrium coexistence of various solid and liquid phases in silicate systems one cannot properly look into the essence of physico-chemical processes taking place in them. The phase rule enables one to determine conditions under which an equilibrium system retains the number of coexisting phases. The phase equation establishes a mathematical link between the number of degrees of freedom, that of components and that of phases in a given equilibrium system. The theory of the phase rule has been substantially developed in Palatnik's [468] and Berezhnoy's [43] works.

If a system consists of k components coexisting in n phases, according to Eq. (50) one can express $k(n-1)$ equations characterising the equilibrium state as:

$$\begin{aligned}
&(\mu_1)_1 = (\mu_1)_2 = (\mu_1)_3 = \cdots = (\mu_1)_n;\\
&(\mu_2)_1 = (\mu_2)_2 = (\mu_2)_3 = \cdots = (\mu_2)_n;\\
&(\mu_k)_1 = (\mu_k)_2 = (\mu_k)_3 = \cdots = (\mu_k)_n.
\end{aligned} \tag{75}$$

The number of external variables, such as temperature, pressure, electric field etc., necessary for establishing the state of a system will be denoted by r. The number of variables recording the constancy of the composition of each phase will make up $k-1$. Each phase will in its turn comprise the same number of components, whereby the number of variables for n phases needed to ensure the constancy of each phase will amount to $n(k-1)$. The number of degrees of freedom f will equal the number of internal variables $n(k-1)$ plus the number of external variables r minus the number of equations necessary for describing the equilibrium:

$$f = n(k-1) + r - k(n-1)$$

or

$$f = k - n + r. \tag{76}$$

In case some external variables do not alter (e.g., the external variable conditioned by the electric field etc.), the number of external variables may be taken for constant (e.g., temperature and pressure) and equal to two. Then, the phase rule may be expressed by a relation

$$f = k - n + 2. \tag{77}$$

For example, in the system consisting of $CaCO_{3(sol)}$, $CaO_{(sol)}$, $CO_{2(gas)}$ a reaction

$$CaCO_{3(sol)} = CaO_{(sol)} + CO_{2(gas)}$$

is possible.

The following indices for denoting the state of substances are used: sol = solid, gl = vitreous, am = amorphous, aq = aqueous, gas = gaseous.

In this system there are three constituents, but only two independent components, as the concentrations of substances in gas phase are found by the equation of the law of mass action

$$K_a = \frac{a_{CaO} a_{CO_2}}{a_{CaCO_3}} .$$

The activities of the pure crystalline CaO and $CaCO_3$, however, are equal to one whereas the activity of CO_2 at lower pressures can be replaced by the pressure of gas p, thus, $K_a = a_{CO_2} \approx P_{CO_2}$. So, in this system $n = 3$, $k = 2$ and $f = 2 + 2 - 3 = 1$, i.e., the system has one degree of freedom: to each temperature a specific pressure (more exact activity) of CO_2 corresponds. It should be remembered here that the system cannot consist of one component only because the equality of the number of moles of CaO and CO_2 does not mean the equality of their concentrations in gas medium. Figure 2 in Eq. (77) has been conditioned by the assumption that the equilibrium state may be effected by only two parameters: pressure and temperature. As stated above such cases also occur when the equilibrium may be influenced by other parameters (e.g., the dispersion of a substance etc.). In these cases the amount of parameters capable of altering the equilibrium of the system should be added to the two. As far as silicate systems are concerned one of the parameters, pressure, alters, so insignificantly in most practical cases that it does not substantially effect the processes (the above example is an exception). Therefore the phase rule for silicate systems will be expressed this way:

$$f = k - n + 1 . \tag{78}$$

Equations (77) and (78) remain valid in case some component does not enter the composition of some phases as well, because the number of variables and that of equations lessen in like manner and the difference between them remains unaltered. Similar cases may be viewed from the standpoint of independent states as well [227].

In addition to the phase rule, of great importance to the study of equilibria is the Clausius-Clapeyron equation that characterizes the temperature and pressure dependence of a thermodynamic equilibrium in single-component systems.

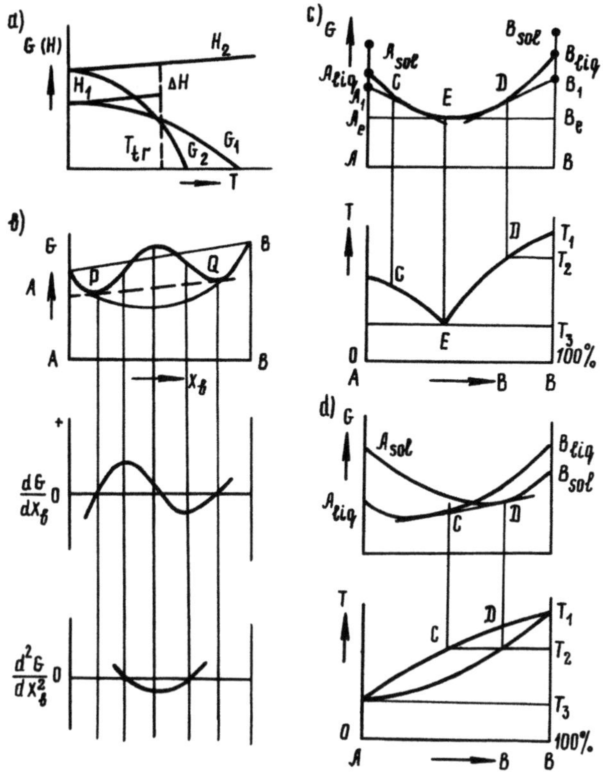

Fig. 2. The plotting of the diagrams of phase equilibria by thermodynamic data. **a** for phase transformations in a single-component system; **b** for a binary system; **c** for a binary system with eutectic; **d** for a binary system with full mutual miscibility (a continuous series of solid solutions)

In the case of two phases in a single-component system a correlation $\mu_1 = \mu_2$ should be valid.

As further on

$$dG = -SdT + vdp = d\mu_{1,2},$$

then

$$V_1 dp - S_1 dT = V_2 dp - S_2 dT.$$

Thus

$$\frac{dp}{dT} = \frac{S_2 - S_1}{V_2 - V_1} = \frac{\Delta S}{V_2 - V_1}.$$

According to Eq. (40):

$$\Delta H_p = \Delta S_p T_p ,$$

where ΔH_p, alteration of heat content; T_p, temperature of phase transformation.

The finally we obtain

$$\frac{dp}{dT} = \frac{\Delta H_p}{T_p(V_2 - V_1)} . \tag{79}$$

In an equilibrium between a liquid and a gaseous phase the volume of the liquid one may be neglected. The volume of the gaseous phase can be determined from the equation of state for an ideal gas. Hence the Clausius-Clapeyron equation may be expressed as:

$$\frac{dp}{dT} = \frac{p \Delta H_v}{RT^2} , \tag{80}$$

where ΔH_v is the molar alteration of the enthalpy at evaporation.

Similarly one can obtain an equation for the equilibrium of a solid and a liquid phases of a single-component system:

$$\frac{dp}{dT} = \frac{p \Delta H_{\mathrm{sub}}}{RT^2} , \tag{81}$$

where ΔH_{sub} is a molar alteration of enthalpy at sublimation.

Thus, if the difference between the substance volumes corresponding to a higher (V_2) and a lower (V_1) temperatures is positive, the values of temperature and latent transformation heat in the right part of the equation being always positive, the left part of the equation will always be positive as well. The volume of silicates increases in the process of melting, therefore their temperature of melting generally rises with the increase of pressure (very important for understanding geochemical processes). Moreover, Eq. (79) enables one to calculate the value of the latent heat of any phase transformation on the basis of the difference of specific volumes.

8 Phase Diagrams

Considering the relative disposition of curves of the thermodynamic potential values, one can determine the boundaries of the phase stability and use these data for building phase diagrams [330, 607].

In a single-component system (Fig. 2a) and at a low temperature phase 1 possessing the least thermodynamic potential value is stable and at temperature $>T_{tr}$ the most stable one is phase 2. At the temperature of transition $G_1 = G_2$ and the alteration of heat content $\Delta H = H_2 - H_1 = T_{tr}S_2 - T_{tr}S_1 = T_{tr}\Delta S$.

In a two-component system (Fig. 2b) the thermodynamic potential of a mechanical mixture is geometrically termed by a straight line connecting points A and B. If components A and B give a stable solution, the value of its thermodynamic potential is to be lower than that of a mechanical mixture in the whole range, i.e., the given points are to be connected by a curve bent downwards.

In case the components are not fully mutually miscible, on the curve a section bent upward appears. The compositions of the two extreme solutions to which the system dissociates can be thermodynamically determined by taking the first and the second-order derivative on a certain section x_B. The probability of a composition deviation from the mean x_B determines the composition fluctuations up to the possible limits of dissociation.

Figures 2c and d respectively illustrate clearly the deduction of the types of diagrams for the systems with simple eutectic and for the case of formation of a continuous series of solid solutions, for which only one temperature has been chosen.

More complicated diagrams are not considered here and the reader is referred to corresponding primary sources [1, 43, 109, 302, 509].

9 Methods of Obtaining Initial Thermodynamic Data

The thermodynamic research method enables one to theoretically analyze the most diverse physico-chemical processes, therefore, at present thermodynamic calculations are widely used both by scientists and industrial engineers as well as designers.

The final object of the thermodynamic analysis is the determination of the Gibbs free energy of a reaction ΔG_T^0 or of the equilibrium constant K. This problem can be solved by using the second law of thermodynamics and Kirchhoff's and Gibbs-Helmholtz' equations. In cases, however, when in the temperature range considered, the reactants undergo phase transformations – polymorphic transitions, melting, sublimation – the calculations become more sophisticated.

While conducting a thermodynamic analysis the following data for all the reactant compounds (including phase transformations if any) are necessary.

1. Standard thermodynamic constants [64, 65, 247, 311, 528]: ΔH_{298}^0, the alteration of enthalpy in forming a given substance at 298 K; S_{298}^0, the entropy of the substance at 298 K; instead of entropy the alteration of the Gibbs free energy in substance formation from the elements at 298 K may be known.

2. Temperatures of phase transformations: T_{tr}, polymorphic transformation temperature; T_m, melting temperature; T_{ev}, evaporation temperature.

3. Enthalpy alterations in phase transformations: ΔH_{tr}, enthalpy alteration in a polymorphic transformation; ΔH_m, enthalpy alteration in melting; ΔH_{ev}, enthalpy alteration in evaporation.

In phase transformations instead of an enthalpy alteration one can employ an entropy alteration in the same phase transformation, because these values, as noted above, are linked by the relationship:

$$\Delta H_{tr} = \frac{\Delta S_{tr}}{T_{tr}}.$$

4. Equations of the temperature dependence of the heat capacity in form of a power function for the entire temperature range under consideration: $C_p = a + bT + cT^2$ or $C_p = a + bT + cT^{-2}$.

In case solutions and melts are formed in a system, quite a precise thermodynamic analysis can be carried out only with the activities of constituents in view.

Of the initial data enumerated above the most important values are ΔH^0_{298}, S^0_{298} and ΔG^0_{298}. With knowledge of these values and employing heat capacity equations one can carry out a thermodynamic analysis up to the temperature of melting of an ionic low-melting compound taking part in the reaction.

Thermic constants may be obtained by several methods:

1. by calorimetric measurements enabling one to establish the temperature dependence of heat capacities, the heat of compound formation and the heat of phase and polymorphic transformations; experimental data then serve as a starting point for calculating the thermodynamic functions ΔH, ΔG, S etc.;
2. by an experimental study of equilibria at a given temperature and pressure;
3. by measuring the electromotive force of galvanic elements [336, 550, 639];
4. by thermochemical calculations based on Hess' law;
5. by calculations based on experimental spectroscopic data (by methods of statistical and quantum mechanics) [333, 623];
6. by the calculations based on semi-empirical regularities most often used in cases when there are no data obtained immediately by experimental measurements.

For compounds employed in silicate technology the most reliable results can so far be drawn from immediate calorimetric measurements.

The latter allows determining the heats of the reactions, of phase transformations and heat capacities by measuring the heat absorbed or released in corresponding processes.

Enthalpies of formation of various inorganic compounds, including silicates, are most often found by measuring heats of their dissolving in mixtures of diverse mineral acids. They can also be found by measuring the heat of a reaction (hydration etc.) with the subsequent calculation of enthalpy of formation of one of the reactants, if for the rest of the reactants, the enthalpies of formation are known.

In order to find standard entropies, the heat capacities of the substances should be determined beginning with the temperatures close to the absolute zero up to room temperature. In this case the entropies are calculated with the help of graphic integration by means of determining the area delimitated by the curve $C_p/T = f(T)$ and axes of coordinates.

The equations of the temperature dependence of the heat capacity are obtained by experimental determination of heat capacity at high temperatures. The mean heat capacities may be measured by an ordinary "mixing method", whereas the true ones, in vacuum and radiation calorimeters, it being a long-drawn and laborious work. More detailed information about experimentally determining thermodynamic constant may be obtained from special handbooks [124, 200, 249, 488, 528, 580, 600].

The structural-thermodynamic aspect of the application of low-temperature calorimetry based on the works by Tarasov [611] is of particular interest. Developing the quantum theory of heat capacity as applied to chain and layer structures, Tarasov was the first to state that the character of an alteration of heat capacity in the range of low temperatures is connected with the internal structure of the substance studied and permits understanding structural peculiarities of a lot of substances based on experimental data. This method, e.g., permits establishing the characteristic peculiarities of glass structure.

Unfortunately, the experimental data on thermodynamic constants published for silicates and other compounds widely used in silicate technology are few because of particular experimental difficulties arising while determining them. This situation necessitates the use of approximate methods for estimating the missing initial data. Therefore the authors thought it expedient to give some approximate methods of calculating thermodynamic constants relevant to the fields of their application. The application of these methods allows in most cases obtaining thermodynamic constants correct within $\pm(3-5)\%$ which is quite acceptable for practical use.

Before pursuing the examination of approximate methods of determining thermodynamic constants, one should concentrate on the formulation of monotypeness – similarity and analogy of substances and reactions – concept, which is of great importance in carrying out any comparative calculations.

While defining the concept of monotypeness of compounds – reactions – one should first mention that the monotype-compounds can be theoretically subdivided into monotype-compounds proper, and those which are not monotype-proper, but may however be used for conducting comparative calculations where the initial data for monotype-compounds-proper are lacking. Certainly, if we lessen the monotypeness of the compounds used we thereby lower the precision of the final results of estimations.

Table 1. Methods of determining ΔH^0_{298} and ΔG^0_{298}. (Compiled with the assistance of Frenkel)

Author	Necessary conditions for the calculation	Formula	Explanation to the formula	Possibility of the application to the study of silicates	References
Kapustinskiy	Data available for ΔH^0_{298} on several monotype-compounds	$\Delta H^0_{298} = A \lg N + B$	N, cation ordinal member; A, B, constants for the given group of compounds	–	[236]
Karapet'yants	1. Data available for ΔH^0_{298} on two series of monotype-compounds	$\Delta H^0_{298_{II}} = A \Delta H_{298_I} + B$	$\Delta H^0_{298_{II}}$ and $\Delta H^0_{298_I}$, alteration of enthalpy at forming monotype-compounds of two series (I and II) from elements	Valid for the equations of the dependence of ΔH^0_{298} of orthosilicates on ΔH^0_{298} of metasilicates with cations of the main subgroup of the II group of the periodic system	[244]
	2. Data available on the properties of compounds of the binary system	$\Delta H^0_{298} = AZ + B$	A and B, constants for the group of compounds in question; Z, properties of the system in question	–	
Matveyev	1. Data available on the structural peculiarities of meta- and orthosilicate structures	$Z_{Me_2SiO_4} - Z_{Me'_2SiO_4} = 2(Z_{MeSiO_3} - Z_{Me'SiO_3})$	Z, thermodynamic properties	For meta- and orthosilicates of the II group of the periodic system	[356]
	2. Constant value of ΔZ_{Me} for a compound of the binary system available	$\Delta Z^n = \Delta Z_{Me} + kn$	ΔZ^n, alteration of the thermodynamic property at the formation of a compound; ΔZ_{Me}, the part of the thermodynamic	For calculating the thermodynamic properties in binary and ternary systems	

			property value introduced by the metal cation; $k = \Delta Z_{Si}$, contribution at the formation of the silica skeleton of metasilicate; n, mole number SiO_2		
Chao-Hsia Chen	$\sum \Delta G_i^0$ available (the sum of ΔG^0 of compounds in combination for a series of the reactions of the silicate mineral formation growing subsequently complicated)	$\sum \Delta G_i^0 = a \cdot e^{bx} + c$	a and b, constants determined by regression analysis; c, extrapolated value of a mineral; x, an independent variable	For the estimation of the Gibbs free energy of the formation of silicate minerals from elements	[94]
Hoppe	Presence of the completed electronic shell of a cation, i.e., for salts of cations of the main subgroups of the I, II and III groups of the periodic system. The advantage of the method is the possibility of using well-known data for chlorides and oxides	$\nabla(Cl, O)_{Me} = A \nabla(Si, O)_{Me} + B$	$\nabla(Cl, O)_{Me} = [\Delta H(MeCl_2) - \Delta H(MeO)]/2$, where $\Delta H = \Delta H^0_{298}$ of the compounds in brackets. The equations for the I and the II groups have the same coefficients A and B	$\nabla(SiO_3, O)_{Me} = [\Delta H(MeSiO_3) - \Delta H(MeO) - \Delta H(SiO_2)]/2$ $\nabla(SiO_4, O)_{Me} = [\Delta H(Me_2SiO_4) - 2\Delta H(MeO) - \Delta H(SiO_2)]/4$ On determining A and B $(SiO_3, O)_{Me}$ (or any other similar to it) is calculated and further ΔH^0_{298}	[209]

Table 1 (continued)

Author	Necessary conditions for the calculation	Formula	Explanation to the formula	Possibility of the application to the study of silicates	References
Reznitskiy	A linear equation available linking ΔH^0_{298} of a salt with a normal electrode potential of a cation for salts of the same acid	$-\Delta H^0_{298} = A - BE$	E, a normal electrode potential of an element (cation) in V; ΔH^0_{298}, heat of salt formation in cal/g · equ^{-1}	May be used for determining the ΔH^0_{298} of salts of silicic acids (oxygen, fluorine)	[512]
Glushkova, Grebenshchikov (modified isoatom method by Shchukaryov)	ΔH^0_{298} value of mono-type-compounds available	The plotting of the dependence of a $\Delta H_{298}/n$ isocomponent value by compositions	ΔH^0_{298}, heat of formation; n, number of moles of the oxides of the compound in question	For determining the ΔH^0_{298} of silicates, titanates and zirconates	[162, 163, 164, 566]
Babushkin, Mchedlov-Petrossyan	1. Structural analogy presence manifested by the equality of the part of energy of one or several bonds of separate structural groups (oxides, ions, saturated molecules)[a]	$\Delta H^0_{298} = \sum Pe$	Pe, part of energy of separate structural units	For calcium hydroaluminates, hydrosulphoaluminates and hydrocarboaluminates	[414]

	2. Values of mean bond energies of a compound (Cottrell), the energy of the sublimation and dissociation of the components composing it are available	$\Delta H^0_{298} = -\sum \varepsilon_{bond} + \sum \Delta U_{sublimation} + \sum \Delta U_{dissociation}$	ε_{bond}, mean values of bond energies in a compound; $U_{sublimation}$, energy of the sublimation of solid elements of which a compound is formed (Si, Ca, Al, etc.); $U_{dissociation}$, energy of the dissociation of gaseous molecules of which a compound is formed (O_2, H_2 etc.)	For calculating the unknown values of heats of calcium hydrosilicate formation	[410]
Karapet'yants	1. Reliable values of ΔH^0_{298} and ΔG^0_{298} for a series of compounds-monotype to the tested one as well as the ΔH^0_{298} of the substance studied available	$\Delta G^0_{298} = \Delta H^0_{298} + B$	–	For calculating ΔG^0_{298} of metasilicates: $n/m\ Me_{(sol)} + Si_{(sol)} + \frac{3}{2}O_{2(g)} = 0.993\ \Delta H^0_{298} + 17.92$ (correct to ± 0.6 kcal/mol)	[244]
	2. When the coefficients of the preceding equation are unknown	$\Delta G^0_{298} = 0.986\,\Delta G^0_{298} - 9.4\,\Delta n$	Δn, alteration of the mole number of a gaseous substance in a reaction of the compound formation from elements	–	

[a] The combination of the structural analogy principle with the comparative calculation method resulted in the crystallochemical energetics [165]

Within the framework of thermodynamics of silicates such compounds may be considered to be monotype-proper ones as the salts of the same acid, e.g., silicates – titanates, germanates, zirconates, etc. – formed by metals belonging to the same sub-group of the periodic system. Other examples of monotype-compounds-proper are oxides of elements of the same sub-group (SiO_2, TiO_2 etc.) or salts of metasilicic, metatitanic, metazirconic acids formed by the same cation. Thus, within the framework of the thermodynamics of silicates such compounds may be viewed as monotype-proper, one part of the formula whereof (acidic or basic) is constant, and the other alters within the same sub-group of the periodic system.

A smaller but still high degree of monotypeness should be ascribed to the salts of the same metal that differ in the content of an acidic oxide, e.g., sodium salts of various silicic acids are also monotype-compounds. In some cases monotypeness of compounds may be viewed not within one subgroup but within a period of the periodic system. For silicate compounds this case is quite rare.

Sometimes due to insufficient initial data one must neglect the monotypeness-proper of compounds used in comparative calculations. One can, e.g., carry out comparative calculations for all the metasilicates of the II group of the periodic system, for all the metasilicates with bivalent cations and for any known metasilicates. The three cases rank according to the monotypeness of the compounds which will accordingly effect the final preciseness of calculations.

Limited by the extent of the book the authors can only give concise characteristics of the calculation methods based on already found practical application.

In Table 1 approximate methods of calculating ΔH^0_{298} and ΔG^0_{298} for inorganic compounds are systematized, in Table 2, the corresponding methods of calculating S^0_{298} and equations $C_p = f(T)$.

It should be noted that while using only one of the above methods of evaluating the thermodynamic data (when no experimental data is available for comparison whereby the precision of the values obtained could thus be ascertained) one cannot be certain whether the constants calculated are correct. The application of several methods when the disagreement of the values obtained is not more than 1 – 3 kcal/mol enables one to consider them as quite reliable.

10 The Calculation of $\Delta G = f(T)$ of Reactions in Silicate Systems Based on Simple Thermodynamic Data

Certain thermochemical data given, Eqs. (48) – (74) must be used in the following way.

The Calculation of ΔG by Eq. (74)

A. The Data Indispensable for the Calculation

1. The enthalpy (ΔH^0_{298}) of formation of initial substances and products from elements under standard conditions.

2. The entropies (S^0_{298}) of the initial substances and products under standard conditions.

3. The equations of the temperature dependence of the heat capacity of the initial substances and products $C_p = f(T)$. Most often tabulated approximate equations such as

$$C_p = a + bT + cT^{-2}$$

are used.

B. The Order of the Calculation

1. Determine the enthalpy of the reaction at 298 K: $\Delta H^0_{298} = \sum \Delta H^0_{298}$ of the products $- \sum \Delta H^0_{298}$ of the initial substances.

2. Determine the entropy of the reaction at 298 K: $\Delta S^0_{298} = \sum S^0_{298}$ of the products $- \sum S^0_{298}$ of the initial substances.

3. Determine the coefficients of the equation of the temperature dependence of the heat capacity of the reaction: $\Delta C_p = \sum C_p$ of the products $- \sum C_p$ of the initial substances; in other terms: $\Delta C_p = \Delta a + \Delta bT + \Delta cT^{-2}$.

4. Substituting the values of ΔH^0_{298} and $\Delta C_p = f(T)$, determine ΔH^0_T according to the formula

$$\Delta H^0_T = \Delta H^0_{298} + \int_{298}^{T} \Delta C_p dT .$$

5. Knowing ΔS^0_{298} and $\Delta C_p = f(T)$, we determine ΔS^0_T by the formula

$$\Delta S^0_T = \Delta S^0_{298} + \int_{298}^{T} \frac{\Delta C_p}{T} dT .$$

6. Knowing ΔH^0_T and ΔS^0_T we determine ΔG^0_T by Eq. (70)

$$\Delta G^0_T = \Delta H^0_T - T\Delta S^0_T .$$

The application of thermodynamic methods to the study of geochemical processes allows to make an analysis of important problems of petrogenesis and evolution of the earth crust [34, 83, 237, 263, 291 – 293, 326, 347, 438, 451 – 453, 499, 547, 581, 582, 627, 670].

The Calculation of ΔG by Integrating Eq. (48)

A. The Data Indispensable for the Calculation

1. The enthalpies of the formation of the initial substances and products from elements under standard conditions (ΔH^0_{298}).

Table 2. Methods of determining S^0_{298} and $C_p = f(T)$. (Compiled with the assistance of Frenkel)

Author	Necessary conditions for the calculation	Formula	Explanation to the formula	Possibility of the application to the study of silicates	References
Turkdogan, Pearson	Data on molar volumes for related groups of compounds (silicates, carbonates, sulphates) are available	$S^0_{298} = av^n$	v, molar volume [ml]; a and n, coefficients for the group of compounds in question	Applicable to silicates	[647]
Latimer	The values of atomic weights available	$S^0_{298} = 1.5\,R\ln A - 0.94$	A, atomic weight	–	[323]
Karapet'yants	1. Values of C^0_{p298} available for monotype-compound groups	$S^0_{298} = a + bC_p$	C_p, heat capacity at 298 K; a and b, constants for a group of monotype-compounds	–	[243]
	2. The value of the entropy for one of the monotype-compounds	$S^0_{298_\mathrm{I}} = a + bS^0_{298_\mathrm{II}}$	$S^0_{298_\mathrm{I}}$ and $S^0_{298_\mathrm{II}}$, entropies of two monotype-compounds I and II	–	
Kapustinskiy, Yatsimirskiy	The values of the atomic weight, the charge and the radius of an ion are available	$S_{\mathrm{ion}} = 1.9R\ln A - 1.5\,z^2/r$	A, atomic weight; z, ion charge; r, thermochemical radius	Applicable to the calculation of single- and two-charge ions	[238]

Kireyev	The presence of additivity of entropies of constituents of complex monotype-compounds	$S^0_{298} = \sum S_{cp} + \Delta S'$	S_{cp}, entropy of compounds; $\Delta S'$, a constant identical for a series of monotype-compounds	Applicable to hydrated forms of silicates, aluminates etc.	[120, 274, 275, 276]
Landiya	The link between the heat capacities of solids and their entropies present	The equation $C_p = a + bT + cT^2$ is calculated by a specially elaborated scheme	–	Most fruitful while determining the equations of heat capacity for silicates, aluminates and their hydrated forms	[320]
Ivanova	The temp. of the first phase transformation of compounds available	$C_p = n\,(5.283 - 1.987)\,T_{tr}$	n, number of atoms in molecules of compounds; T_{tr}, temperature of the first phase transition	–	[214]
Karapet'yants	The value of the common heat capacity available	$T_N = AT_M + B$	T_N, T_M, temps. at which the heat capacities of the substances N and M are equal	Determining heat capacities at high temperatures	[245]

2. The Gibbs free energies of the formation of the initial substances and products from elements under standard conditions (ΔG^0_{298}).

3. The equations of the temperature dependence of the heat capacity of the initial substances and products $C_p = f(T)$.

B. The Order of the Calculation

1. Determine the enthalpy of the reaction at 298 K:
$\Delta H^0_{298} = \sum \Delta H^0_{298}$ of the products $- \sum \Delta H^0_{298}$ of the initial substances.

2. Determine the alteration of the Gibbs free energy of the reaction at 298 K:
$\Delta G^0_{298} = \sum \Delta G^0_{298}$ of the products $- \sum \Delta G^0_{298}$ of the initial substances.

3. Determine the equation of the heat capacity alteration depending on the temperature of the reaction:

$\Delta C_p = \sum \Delta C_p$ of the products $- \sum \Delta C_p$ of the initial substances; in other terms

$$\Delta C_p = \Delta a + \Delta b T + \Delta c T^{-2}.$$

4. Integrating the heat capacity equation derived and substituting ΔH^0_{298} and $T = 298$ K in it, determine the constant of integration (ΔH_0):

$$\Delta H_0 = \Delta H^0_{298} - \Delta a \cdot 298 - \tfrac{1}{2}\Delta b(298)^2 + \Delta c(298)^{-1}.$$

5. Term ΔG of the reaction for any temperature by the equation $\Delta G^0_T = \Delta H_0 - \Delta a T \ln T - \frac{1}{2}\Delta b T^2 - \frac{1}{2}\Delta c T^{-1} + yT$ where y is the second constant of the integration.

After that we substitute $T = 298$ K and ΔH_0 in the equation obtained and determine constant y from the equation:

$$\Delta G^0_{298} = \Delta H_0 - \Delta a 298 \ln 298 - \tfrac{1}{2}\Delta b(298)^2 - \tfrac{1}{2}\Delta c(298)^{-1} + yT.$$

6. Knowing ΔH_0 and y we set up an equation of the temperature dependence of the Gibbs free energy:

$$\Delta G^0_T = \Delta H_0 - \Delta a T \ln T - \tfrac{1}{2}\Delta b T^2 - \tfrac{1}{2} c T^{-1} + yT.$$

7. Using the above equation we calculate the values of ΔG^0 of the reaction for the temperature or the temperature range required.

The Calculation of ΔG_T at High Temperatures by the Thermodynamic Properties of Individual Substances

While carrying out thermodynamic calculations in practice, and in case of a great deal of phase and polymorphic transformations in particular, calculating special equations for ΔH_T^0 and ΔG_T^0 of the reactions under consideration turns out to be unadvisable. It is easier to calculate first the ΔH_T^0 and ΔG_T^0 values of the individual reactants in question in the entire temperature range in order to later determine ΔH_T^0 and ΔG_T^0 of the reactions under consideration with the thus obtained values.

With this purpose in view $H_T - H_{298} = f(T)$ and $S_T - S_{298} = f(T)$ for individual substances are calculated before. The calculations are carried out by the following formulae:

$$H_T - H_{298} = \int_{298}^{T_{tr}} C_{p_1} dT + \Delta H_{tr}^0 + \int_{T_{tr}}^{T_m} C_{p_2} dT + \Delta H_m^0 + \int_{T_m}^{T} C_{p_3} dT,$$

$$S_T - S_{298} = \int_{298}^{T_{tr}} C_{p_1} \frac{dT}{T} + \Delta S_{tr}^0 + \int_{T_{tr}}^{T_m} C_{p_2} \frac{dT}{T} + \Delta S_m^0 + \int_{T_m}^{T} C_{p_3} \frac{dT}{T},$$

where C_{p_1}, C_{p_2} and C_{p_3} are heat capacities in temperature ranges of 298 to T_{tr}; T_{tr} to T_m and T_m to T.

Then ΔH_T^0 and ΔS_T^0 are determined by the formulae:

$$\Delta H_T^0 = \Delta H_{298}^0 + (H_T - H_{298});$$

$$\Delta S_T^0 = \Delta S_{298}^0 + (S_T - S_{298}) .$$

Finally, by Gibbs-Helmholtz' equation ΔG_T^0 is determined:

$$\Delta G_T^0 = \Delta H_T^0 - T \Delta S_T^0 .$$

ΔH_T^0 and ΔG_T^0 of the reactions under consideration are determined as the differences of the values of ΔH and ΔG for the products of the reactions and the initial substances.

The ΔG_T Calculation by Reduced Thermodynamic Potentials

Recently a new function, the reduced thermodynamic potential [157, 546, 623], has become popular in the practice of conducting thermodynamic calculations. The reduced potential ϕ^* is a ratio

$$\phi_T^* = -\frac{G_T^0 - H_0^0}{T},$$

where G_T^0 is the standard Gibbs free energy at a temperature of T K; H_0^0 – the enthalpy at 0 K; T – the temperature in K.

The absolute values may be calculated the same way as the values of entropy S_T^0 and that of heat capacity C_p. The application of ϕ^* is more appropriate for calculating thermodynamic functions of gases by statistic methods. Moreover, it can be easily shown that

$$G_T^0 = H_T^0 - TS_T^0 = H_0^0 + \int_0^T C_p dT - \int_0^T \frac{C_p dT}{T},$$

whence

$$-\phi_T^* = \frac{G_T^0 - H_0^0}{T} = \frac{1}{T} \int_0^T C_p dT - \int_0^T \frac{C_p dT}{T}.$$

Thus ϕ^* depends wholly on the heat capacity of the reactants.

In some works instead of function ϕ^* a somewhat different reduced potential is used, which may be denoted by a symbol ϕ'. Potential ϕ' is characterized by a ratio

$$\phi_T' = -\frac{G_T^0 - H_{298}^0}{T}.$$

The interrelation between ϕ^* and ϕ' is determined by a ratio

$$\phi' = \phi^* + \frac{H_{298}^0 - H_0^0}{T}.$$

The alterations of ϕ^* and ϕ' resulting from a chemical reaction are determined by equality $\sum \Delta\phi^* = \sum \phi^*$ of the products $- \sum \phi^*$ of initial substances.

From the definition of $\phi^* (\phi')$ it follows that

$$G_T^0 = -\phi_T^* T + H_0^0;$$
$$G_T^0 = -\phi_T' T + H_{298}^0.$$

Hence the alterations of ΔG of a reaction are found from equations:

$$\Delta G_T^0 = -\Delta\phi_T^* T + \Delta H_0^0,$$
$$\Delta G_T^0 = -\Delta\phi_T' T + \Delta H_{298}^0.$$

The values of potentials ϕ^* and ϕ' are calculated on the basis of the enthalpy and entropy values by formulae:

$$\phi_T^* = S_T^0 - \frac{H_T^0 - H_0^0}{T};$$

$$\phi'_T = S^0_T - \frac{H_T - H^0_{298}}{T}.$$ [6]

11 The Peculiarities of Applying the Methods of Chemical Thermodynamics to Silicate Systems

Chemical and physico-chemical processes taking place chiefly with solids and liquids at various temperatures are characteristic of silicate technology, solids being crystalline and vitreous silicates. Liquids embrace melts as well as water and electrolyte aqueous solutions.

With reactions in silicate systems reaching stable equilibria is highly hampered because of a small area of phase boundaries, high melt viscosity and a marked tendency of silicates to forming unstable states (vitrification, forming colloid solutions etc.).

As the laws acting in silicate systems are much different from "normal" ones, we thought it reasonable to discuss all the peculiarities of the application of the laws of chemical thermodynamics to silicates in a special paragraph.

As reactions in silicate systems involve mainly solids and liquids particular attention must be paid to the application of thermodynamics to condensed systems: solid-phase reactions – polymorphic transformations, reactions of ceramic and refractory sintering and those for producing binders – melt reactions, as well as the reactions of the interaction of solids with water and electrolyte solutions.

We have accordingly adopted the following classifications of reactions in silicate systems:

1. Solid-phase reactions:
   a) polymorphism reactions.
   b) solid-phase reactions proper.
2. Reactions with the participation of gaseous products.
3. Reactions with the participation of melts.
4. Hydration reactions.
5. Combined reactions.

The chemical interaction of substances in condensed systems was first considered by van't Hoff [203]. The foundations of the thermodynamic approach towards the study of solid-phase reactions were laid by Tammann [608, 609]. He showed the reaction of the chemical interaction of two crystalline substances to theoretically advances towards emanation of heat and the equilibrium only to be able to take place under certain conditions.

---

[6] For information on approximate and accelerated methods of calculation by Ulich, Tyomkin, Shvartsman, Vladimirov et al. refer to special handbooks [65, 258, 277, 380, 476, 652, 658, 673, 713]

In fact one can see from the equation of the alteration of the Gibbs free energy ($\Delta G = \Delta H - T\Delta S$) that ΔG can be close to zero in the following cases:

1. the heat of the synthesized compound formation is quite small (in this case ΔH is minute and the difference $\Delta H - T\Delta S \rightarrow 0$);

2. gases and liquids take part in a reaction (in this case ΔS unlike that of solids reaches great values and $T\Delta S$ can approach the value of ΔH, especially at high temperatures, their difference approaching zero);

3. the mutual solubility of phases is possible as well as the formation of mixed crystals and solid solutions [346, 454, 466], of glass etc. (All these processes result in the increase of ΔS);

4. A considerable difference exists between the sums of the heat capacities of the initial substances and those of the interaction products (the increase of ΔS; $\Delta G \rightarrow 0$).

In silicate systems the heats of formation of the reaction products in crystalline systems reach considerable values and are generally measured, while these products are obtained from elements, by hundreds of kcal/mol and while obtained from oxides, by tens of kcal/mol (with some exceptions, e.g., lead silicates). Therefore oxide interaction reactions which we come across in practice, in the solid phase, theoretically advance to the end towards emanation of heat, unless gaseous and liquid substances take part in the reactions and phases with great defects of the crystalline lattice arise. If the latter takes place, the entropy member $T\Delta S$ in the equation $\Delta G = \Delta H - T\Delta S$ will contribute to the reaction proceeding in the direction of a greater release of the liquid and gaseous phases, while the reaction deviating from van't Hoff's rule may become endothermic.

The degree to which the interaction is completed is conditioned by a proper dosage of the components and by purely kinetic factors, and varies from a practically complete one to the value less than that required by thermodynamics.

It is worth mentioning that Tammann's principle of reactions proceeding in solid phases theoretically in the direction of heat release is sometimes challenged, one argument against being: enantiotropic processes of $A_\alpha \rightleftarrows A_\beta$ type like a classical example of the transition of gray tin to white. In this case a polymorphic transformation being a simplest solid-phase reaction and a chemical interaction in solid-phase are mixed up. It is well known [203] that for a reaction between solid phases it is sufficient that heat release should be greater than 1 kcal/g-at, because $T\Delta S < 1$ kcal/g-at, as the value of ΔS is generally lower than 3 cal/mol K per 1 g-at. The table data show that polymorphic transformations usually produce a lower heat effect (fractions of a kcal/mol) than the effects of the chemical interaction of solid oxides in silicate systems.

What has been stated about the thermodynamic regularities of the solid phase chemical interaction is true in the case of normal pressure. At elevated temperatures high pressures can alter the situation because of the role played by the alteration of the volume of the phase formed.

Recently the thermodynamic analysis of subsequent solid-phase reactions is increasingly more often used for solving various problems of the stability of particular phases and compounds and for establishing the most general regularities of the course of solid-phase reactions. In spite of this, however, the main principles of the application of the thermodynamic method to such problems are still inadequately researched.

The scope of the thermodynamic method combined with experimental kinetic data as applied to the discussion of the sequence of the reactions' course in solid phases is shown below [7, 8, 202, 324, 383].

We shall at first consider the substances characterized by a certain stoichiometric composition and possessing n modifications. In this case the thermodynamic stability of these modifications will be determined by the magnitude of the corresponding values of the Gibbs free energy ΔG_1, ΔG_2, $\Delta G_3 \ldots \Delta G_n$, the most stable one being the form with the minimum ΔG. A series of these ΔG values can always be ranged so that ΔG_1 will correspond to the maximum value of ΔG and ΔG_n (a series of modifications A_1, A_2, $A_3 \ldots A_n$) to the minimum one. Thus, we obtain a so-called series of the thermodynamic sequence of formation of modifications.

Independent of the way it is obtained from the thermodynamic point of view, the modification with the smallest ΔG value, i.e., phase A_n will always be the final product (the final phase). If however, the values of ΔG_1, ΔG_2, $\Delta G_3 \ldots \Delta G_{n-1}$ are negative, the formation of these phases will be probable as well from the standpoint of thermodynamics, but the sequence of their formation cannot be determined a priori on the basis of thermodynamic data only. The accomplishment of one or the other reaction will depend not only on the value of ΔG, but on kinetic factors as well, the main ones being the energy of activation and the rate of the realization of the process in question. As solid-phase reactions in silicate systems take place at high temperatures, the effect of activation barriers becomes negligible.

The rate of a reaction in this case may be of substantial importance: the greater the rate of the reaction the greater the probability of the reaction taking place under given conditions. Apparently the rates of the formation of compounds $A_1, A_2, A_3 \ldots A_n$ either can alter following the alteration of ΔG with the passage from A_1 to A_n or there will be no such sequence. Hence two main cases ensue.

In the first case: a) with the decrease of ΔG the rate of a reaction enhances, i.e., to the compound with the minimum thermodynamic stability the minimum reaction rate corresponds and to the compound with the maximum thermodynamic stability A_n –, the maximum reaction rate. In this case immediately the most stable compound A_n with the maximum formation rate will form. Thus the thermodynamic probability will coincide with the kinetic one, and we arrive at the rule of the primary phase formation formulated by Mchedlov-Petrossyan [395, 397, 412]; b) with the decrease of ΔG the reaction rate is reduced, i.e., to the least stable compound A_1 the maximum reaction rate will correspond and to the most stable compound A_n –, the minimum reaction rate. In this case compound A_1 with the minimum thermodynamic stability and the maximum reaction rate is formed first. A further transition from a less stable compound to a more stable one proceeds subsequently with a decrease of ΔG. Thus we arrive at the well-known Ostwald stage rule [467].

In the second, principal case, when the reaction rates are not subsequently linked to the alterations of the thermodynamic stability, a compound with the maximum reaction rate forms first (the condition $\Delta G < 0$ being performed) and the sequence of the reactions will be determined by their rate. Nevertheless in this instance as well, kinetic conditions being favourable, the final compounds must always be the compounds with the minimum ΔG. Therefore, it is not unusual that earlier Stranski and Totomanow [597] stated that the sequence of the rise of various compound modifications is determined by the work of the formation of the nuclei of a new phase.[7] Later it was corroborated by other works as well [280, 423]. While studying the effect of adsorption on the processes of a new phase formation it has been found that the "neat conditions" encourage keeping to the stage rule.

In a similar way the above considerations can be applied to the reactions taking place in two- and more component systems. In this case an indispensable condition of a correct comparison of the thermodynamic sequence of reactions in the same system takes into account the stoichiometry of reactions. The latter can be performed on the basis of what we call the principle of ΔG calculation "by compositions" [20, 405, 408].

The essence of this principle consists in taking the same amount of the initial substances in the left parts of the equations for all the reactions in a given system. Thus, any possible reaction will be related to the same amount of initial substances and, consequently, the corresponding values of ΔG will be comparable.

Finally for a certain range of compounds a series of the thermodynamic sequence – stability – can be set up which may be used for determining the most stable compounds and correcting kinetic data. Examples of the application of thermodynamic calculations to the analysis of the sequence of solid-phase and other reactions are given below.

As we have already shown, the formation of silicates from oxides on account of reactions in solid phase proceeds to the end until the appearance of a melt in most cases. The reactions of binder hardening take place with the formation of new crystalline and amorphous phases precipitating from solutions supersaturated with respect to them, therefore the equilibrium in these systems is practically always shifted towards the formation of hydrated forms. In the same way in the process of glass formation from melts no equilibrium is practically achieved.

---

[7] On the problem of the energetics of a new phase formation see [536, 675]

For most chemical reactions of silicate technology we observe a complicated multi-stage process that is difficult to be studied by the known methods of classical thermodynamics. These problems may be solved with the help of the laws of thermodynamics of irreversible processes [114, 115, 116, 339, 492] the equations of which include time-factor; however, the related data is not yet available, the results of the calculation cannot be numerically expressed in all instances. In this connection we shall analyze the possibility of the application of the principles of classical thermodynamics to the study of actual processes. For chemical processes that present a set of chemical and phase transformations, mass and heat transfer thermodynamic potentials as a measure of the motive forces must be decisive in determining their relative rate [121].

Moreover, in these cases it is of importance to know the total alteration of the enthalpy in a reaction, as it shows the amount of heat to be imparted to the system in order to perform a reaction and maintain a temperature needed for the rate of the reaction to meet the requirements. The reaction cannot be viewed as impossible solely because the respective alteration of the free energy is not a large positive value. In case it appears to be feasible and expedient, one should calculate the constants of equilibrium for a series of temperatures and calculate them by the maximum degrees of transformations for a set of concentrations and pressures before drawing any conclusions [667].

Kinetic factors determine the rate of a reaction, the actual output of products and finally, the possibility or impossibility of the processes ruled by the main regularities of thermodynamics. Thus, the principal possibility of a process predicted by thermodynamics does not always hold in view the effect of kinetic factors. While studying silicates, however, thermodynamic factors more often than e.g., in organic chemistry determine the course of the reactions of silicate formation [266, 579, 620].

Among the substantial kinetic factors one should mention the degree of phase contact resulting partly from the grinding fineness and the time of the process etc. Equally important is the effect of the alteration of a modification and of the method of the preparation of components on the thermodynamic constants of the initial substances and subsequently on the expected result of the interaction. Therefore one cannot say that in principle, in all cases, the thermodynamic evaluation of a process will predetermine the character and the sequence of the reactions under consideration. In particular cases however, kinetic factors can be limited or properly taken into account thereby permitting information to be obtained about the character of the possible reactions on the basis of a thermodynamic analysis. A similar limitation of kinetic factors turns out to be performable for a greater number of reactions taking place in silicate systems. In such cases the reaction rates are very often predetermined by diffusion processes connected with the transport of structural units of the components taking part in the reaction to the corresponding phase boundaries. Therefore in case of binar silicate and similar systems the reaction mechanism and the rates of the formation of various compounds within the limits of particular systems will be approximately the same and the effect of activation barriers at high temperatures, insignificant. Under these circumstances by the results of a thermodynamic calculation one can not only determine the most stable product of the reaction but in individual cases find whether it is primary as well.

In general, however, there is absolutely no single-valued dependence between the rate of various processes and the corresponding potentials; the thermodynamic potentials can be used for quantitatively evaluating only one part of a phenomenon, the measure of its moving forces.

D'yakonov [121] has demonstrated that in the energy balance of actual chemical processes the role of the transformations which are connected with the rise of resistances in the system is quite small, and they may be neglected as compared to the transformation due to the total moving forces of the process. Thus, with an insignificant error any actual irreversible chemical process may be presented as a reversible one passing the same intermediate stages; the works of these processes are equal, but in the second case the work is linked with overcoming only external forces (since we have neglected the resistances within the system), whereas in the first it is the objective work of overcoming all actual forces of a non-equilibrium system.

In this connection for the study of reactions in silicate systems one can, with sufficient approximation, use the laws of classical thermodynamics.

In reactions in which only solids and liquids of constant composition (pure substances) taken part there is either no equilibrium state possible or it is possible only at a certain strictly determined temperature (in case of modification transformations). The ΔG value in this case equals the alteration of the Gibbs free energy in a transition from pure initial to pure final substances and the sign of ΔG at a given temperature will always indicate the direction of the reaction; the reaction will proceed to the end in the direction that corresponds to the negative value of ΔG. A reaction in the reverse direction at a given temperature is impossible without being combined with other reactions accompanied by a lessening of the Gibbs free energy.

Thus, for these reactions the equilibrium constant has no physical sense and therefore the calculation of equilibrium concentrations and of the output of products seems unfeasible. In this case the most expedient application of thermodynamics to reactions is the determination of the thermodynamic stability and of the preference of some reactions to others in the same system or in different systems, but with a common component.

Based on the above as well as on the results of the thermodynamic studies of reactions in various systems we have come to the following conclusion. In a system consisting of two simple solids in which several compounds are formed and the compound most favourable from the point of view of structure and kinetic factors is at the same time the most preferable from the point of view of thermodynamics, the compound with the most negative value of ΔG will be primary. In the contrary case the formation of intermediate products is possible, but if we view the preference of the formation of compounds in some system without taking into account the amount of the initial components in each reaction, ΔG in those reactions will always be related to different amounts of reactants and therefore ΔG in this case cannot be used as a standard for comparing the moving forces of a reaction. Moreover, while making a thermodynamic analysis of systems with solid phases taking part, one should consider that the ΔG value of the reactions directly depends on the amount of crystalline substances obtained in the reaction. In this connection taking into account the stoichiometry of reactions is an indispensable condition for a correct thermodynamic analysis of the preference of reactions in the same system which facilitates obtaining absolute energy characteristics of the formation of the final, most probable and stable compounds while using reliable initial thermochemical data. In order to obtain these conditions we have suggested using the principle of the calculation by compounds and tested it on a whole series of reactions in various systems.

Essentially the principle states that for all the reactions in a system in the left parts of the equations the same amounts of the initial substances are taken, then ΔG of each possible reaction will be related to the same amount of the initial substances and the initial thermodynamic conditions for all the reactions will be the same. In cases where the structure of the compounds most preferable thermodynamically are the least different from those of the initial compounds or those of the solution or melt from which the latter crystallize, at high temperatures in particular, they will be most preferable from the kinetic standpoint as well and at the same time primary and most stable under these conditions.

The principle suggested agrees with the consideration of reactions in any silicate system with any number of components, moreover, one can view both primary and side reactions where compounds are obtained with a surplus of one or the other component.

Other investigators have come to similar conclusions while calculating reactions in binary systems [287].

A reaction with a melt involved – type 3 – and those in solutions – type 4 – can, although slowly, reach metastable equilibria, and therefore, the equilibrium constant for them is determined by a corresponding ratio of activities. The equations of equilibrium constants of the reactions of decomposition and thermic dissociation – type 2 – will contain the fugacities of gaseous phases instead of solution activities. For these reactions one can calculate the output of products at various temperatures.

One should especially mention the effect of the exactness of the initial thermic constants on the exactness of the final results of thermodynamic calculations and in this connection the evaluation of the limits of applicability of the calculation results to the solution of concrete problems: possibility or impossibility of a process and the stability degree of a certain product etc.

The thermochemical measurements of the heats of the formation of silicate compounds from oxides are generally carried out correctly within ±200 cal/mol, those of the entropy – correctly within ±5 cal/mol K, and the Gibbs free energy, ±500 cal/mol.

In thermodynamic calculations of chemical reactions ΔH, ΔG and ΔS are arrived at by a simple algebraic addition. The errors will also add up. Moreover, the value of the total error depends on the stoichiometry of the reaction.

Thus while calculating ΔH^0_{298} and ΔG^0_{298} of the reactions the total error, if we use the data listed in the handbook by Rossini et al. [528] does not generally surpass ±1 kcal/mol.

While calculating $\Delta G = f(T)$ more errors, namely those produced by the coefficients of the equation of the temperature dependence of the heat capacity, are added to the final result besides the additional summing up operations. The exactness of these equations is generally within the limits of ±5%. According to our calculations the total error while estimating ΔG for high temperatures may reach ±(3 – 10)%.

Most reactions in silicate systems have the absolute value of ΔG as a rule more than 10 kcal/mol; for this reason in the instances considered the calculation error does not effect the reliability of possible inferences.

The relative calculation error becomes substantial only in the reactions in which the minuend – the sum of ΔH or ΔG of the reactants standing in the right part of the equation – and the subtrahend – the sum of ΔH or ΔG of the equation members standing in the left part of the equation – differ insignificantly, thus the difference is comparatively small. This refers mainly to the polymorphism reactions and when studying this reaction one should be very careful, because the error percentage can reach 60 and more.

In more detail all the applications of thermodynamics to the silicate technology processes will be expounded in special chapters.

Some words should be said about the interconnection of thermodynamics and kinetics of some solid-phase reactions.

As early as 1938 Evans and Polanyi [130] laid the foundations of the new trend in the study of the complicated phenomena of the interconnection between the kinetic and thermodynamic parameters of chemical reactions. They first gave a theoretical basis to the approximate rule linking the value of the activation barrier with the absolute value of the heat effect ΔH for a series of monotype reactions:

$$E_a = -\alpha \Delta H, \tag{82}$$

where E_a is the activation barrier the same for a direct and a reverse reaction. They were also the first to verify the correctness of relation (82) in case of the interaction of a sodium atom with various halogenated hydrocarbons. Later this was further supported by Semyonov [563] who tested it by a series of other monotype chemical reactions in solutions.

We [23] made an attempt to broaden Polanyi's concepts by establishing a link of the activation energy of chemical reactions not only with their heat effects, but also with the alteration of the Gibbs free energies and to verify at the same time the correctness of these conclusions in reactions with solids taking part that are of great importance for hightemperature chemistry.

Indeed the link between the activation energy and the Gibbs free energy can be easily established on the following premises. Since according to Eq. (53) $\Delta G = -RT \ln K_p$, where K_p is the equilibrium constant ($K_p = K_1/K_2$) and bearing in mind that the constants of the reaction rate K_1 and K_2 are in exponential dependence on activation energy E:

$$K_1 = A_1 e^{-E_1/RT} \quad K_2 = A_2 e^{-E_2/RT},$$

where A is the so-called frequency factor, we arrive at

$$\Delta G = -RT \ln \left(\frac{A_1 e^{-E_1/RT}}{A_2 e^{-E_2/RT}} \right) \tag{83}$$

or

$$\Delta E = \Delta G + RT \ln \frac{A_1}{A_2}, \tag{84}$$

where

$$\Delta E = E_1 - E_2 .$$

Thus, between ΔE and the alteration of the Gibbs free energy of monotype chemical reactions linear dependence exists that allows determining their kinetic characteristics by the known thermodynamic data. One should however bear in mind that the values of coefficients A even for monotype reactions can alter within considerable limits which will effect in the exactness of the results of the estimation by formula (84).

In order to check the correctness of the relation derived (Eq. 84) we have examined several types of solid-phase reactions including those of spinel formation ($MeO - R_2O_3$ system), of orthosilicate formation ($2MeO - SiO_2$ system) and of carbonate decomposition ($MeCO_3 - MeO - CO_2$ system). For all the reactions ΔH^0_{298} and ΔG^0_{298} have been calculated by the usual methods. The initial thermochemical data and the activation energy values have been taken from the well-known sources [20, 43, 195, 302, 323, 711].

The calculation results are assembled in Table 3, and from the table data, diagrams (Fig. 3) of the activation energy of the reactions examined as a function of ΔH^0_{298} and ΔG^0_{298} have been plotted.

On the analysis of the diagrams of equations of $E = f(\Delta H^0, \Delta G^0)$ have been derived (Table 4).

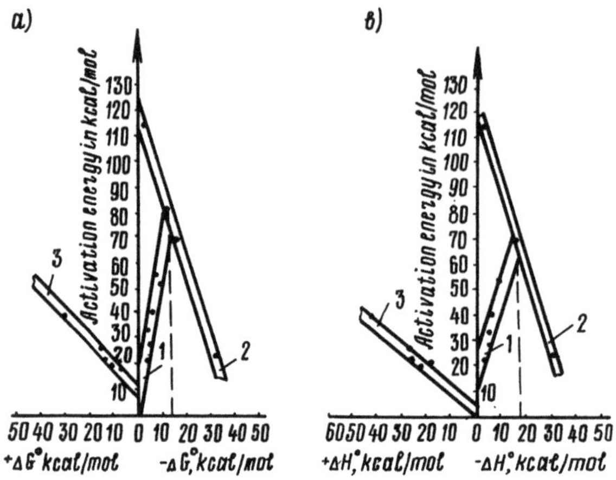

Fig. 3. The dependence between the activation energy and the alteration of the Gibbs free energy (**a**) and the heat of the reactions (**b**) in the systems *1* $MeO-R_2O_3$; *2* $2MeO-SiO_2$; *3* $MeCO_3-MeO-CO_2$

The equations derived allow to a certain degree of reliability determining the activation energy by the values of ΔH^0 and ΔG^0 for monotype reactions which we have done for several reactions (see Table 3).

An analysis of the equations listed in Table 4 shows different reaction types to have different dependence between the activation energy and the absolute values of the Gibbs free energy and heat effect alterations.

Thus, for exothermic solid-phase reactions of spinel formation the increase of the heat effect and the Gibbs free energy absolute value is directly proportional to the increase of the activation energy (Fig. 3a, straight line 1 and Fig. 3b, straight line 1).

As one can see from Table 3, the value of the heat effect of most reactions of spinel formation is within 2 – 8 kcal/mol, and, naturally, in such reactions it will not effect the kinetics of the process to any substantial degree.

In the exothermic reactions however, in which the heat effect value surpasses 15 kcal/mol, the linear dependence between the activation energy and thermodynamic parameters is retained, but the direction of the straight line changes (e.g., the reactions of orthosilicate formation). In this connection from the kinetic point of view only those monotype reactions can be considered as monotype-proper the ΔH^0 value of which changes within certain limits. Thus, e.g., the reactions of $CaO \cdot Fe_2O_3$, $BaO \cdot Al_2O_3$ and $CaO \cdot B_2O_3$ formation will not follow the same regularities as the reactions of spinel formation, because they belong to highly exothermic ones, ΔH^0_{298} of them equals -49.6; -24; -29.4 kcal/mol respectively, so we did not find it possible to determine their activation energies using the equations of Table 4.

Considering Fig. 3 for low exothermic reactions (ΔH^0 less than 10 kcal/mol) the negative ΔG^0 value at low temperatures already provides for their thermodynamic possibility and the application of energy from without is necessary to overcome the activation barrier involved by the loosening of the crystal

Table 3. Monotype reactions in the systems $MeO-R_2O_3$, $MeO-SiO_2$ and $MeCO_3-MeO-CO_2$

Reactions	ΔH^0_{298} from oxides [kcal/mol]	ΔG_{298} from oxides [kcal/mol]	E [kcal/mol]	
			Experimental	Determined by Fig. 3
$MgO+Al_2O_3=MgO\cdot Al_2O_3$	−5	−5.2	40.7	–
$MgO+Fe_2O_3=MgO\cdot Fe_2O_3$	−3.6	−3.72	33.8	–
$MgO+Cr_2O_3=MgO\cdot Cr_2O_3$	−5	−5.75	27.6	–
$FeO+Al_2O_3=FeO\cdot Al_2O_3$	−2.3	−2	–	15.0±6
$CaO+Cr_2O_3=CaO\cdot Cr_2O_3$	−8.5	−8.95	53.5	–
$CaO+Al_2O_3=CaO\cdot Al_2O_3$	−7.7	−6.9	55	–
$FeO+Fe_2O_3=FeO\cdot Fe_2O_3$	−2.5	−4.63	21.67	–
$SrO+Al_2O_3=SrO\cdot Al_2O_3$	−2	−2.12	–	14±6
$ZnO+Fe_2O_3=ZnO\cdot Fe_2O_3$	−7	−6.85	–	45.0±6
$2MgO+SiO_2=2MgO\cdot SiO_2$	−15.1	−15.07	69.2	–
$2CaO+SiO_2=2CaO\cdot SiO_2$	−30.19	−30.64	23.2	–
$2FeO+SiO_2=2FeO\cdot SiO_2$	−8.4	−7.51	46.1	–
$2ZnO+SiO_2=2ZnO\cdot SiO_2$	−7	−6.88	–	–
$2MgO+SnO=2MgO\cdot SnO_2$	–	–	85.1	–
$2PbO+SiO_2=2PbO\cdot SiO_2$	−2.5	−2.8	114	–
$CaCO_3=CaO+CO_2$	+42.5	+31.12	40	–
$Ag_2CO_3=Ag_2O+CO_2$	+28.11	+15.61	26	–
$MgCO_3=MgO+CO_2$	+19.5	+7.63	20.4	–
$CdCO_3=CdO+CO_2$	+23.79	+12.15	19.7	–
$MnCO_3=MnO+CO_2$	+27.85	+14.34	22.5	–
$BaCO_3=BaO+CO_2$	+63.85	+51.64	–	54±5
$SrCO_2=SrO+CO_2$	+56.75	+43.84	–	51±5
$Na_2CO_3=Na_2O+CO_2$	+76.85	+66.14	–	61±5
$K_2CO_3=K_2O+CO_2$	+93.48	+115.04	–	81±5
$FeCO_3=FeO+CO_2$	+20.25	+8.4	–	18±5
$ZnCO_3=ZnO+CO_2$	+16.98	+4.49	–	16±5
$PbCO_3=PbO+CO_2$	+20.85	+10.19	–	20±5

Table 4. Equations $E=f(\Delta H^0, \Delta G^0)$ for various reactions

Reaction type	Equation $E=f(\Delta H^0, \Delta G^0)$
$MeO+R_2O_3=MeO\cdot R_2O_3$	$E=6.5\Delta H^0\pm 6$; $E=-5-7.5\Delta G^0\pm 9$
$2MeO+SiO_2=2MeO\cdot SiO_2$	$E=121-3.1\Delta H^0$; $E=121-3.1\Delta G^0$
$MeCO_3=MeO+CO_2$	$E=2+0.84\cdot\Delta H^0\pm 2.5$; $E=11.7+0.83\Delta G^0\pm 2.5$

lattice of the initial compounds and the effectiveness of diffusion processes being provided for (see Fig. 3a, straight line 1 and Fig. 3b, straight line 1). The small value of the heat effect in the case under consideration cannot influence the decrease of the activation energy.

In reactions of orthosilicate formation considerable values of heat effects (>15 kcal/mol) readily influence the activation energy. For lead orthosilicate at heat effect of -2.5 kcal/mol, $E = 114$ kcal/mol, for calcium orthosilicate at $\Delta H^0_{298} = -30.64$ kcal/mol, $E = 23.2$ kcal/mol.

It is also noteworthy that straight lines 1 and 2 (Figs. 3a and b) intersect at a point corresponding to nearly 15 kcal/mol, testifying once more to the fact that on reaching the heat effect value of higher than 15 kcal/mol a great effect of exothermics on the kinetics of reactions begins. Thus, the point for ferrous orthosilicate does not fall within the straight line because ΔH^0_{298} for this reaction does not reach $-$ 15 kcal/mol (it equals -8.4 kcal/mol), therefore there are more grounds for relating it to the low exothermic reaction type. In fact the point for ferrous orthosilicate ($\Delta H^0_{298} = -8.4$ kcal/mol; $E = 46.4$ kcal/mol) is well placed upon the straight line of spinel formation (straight line 1, Fig. 3b) and the same holds for zinc orthosilicate.

As far as the endothermic reactions of carbonate decomposition are concerned all of them without exception have positive values of ΔG^0_{298} at 298 K, and they need the elevation of temperature in the first place for ΔG to become negative. In this connection in endothermic reactions of carbonate decomposition the dependence between the activation energy and heat effect must be directly proportional, i.e., the greater the endothermic effect of a reaction, the higher its activation energy, because in this case its value is the sum of the amount of heat necessary both for providing the thermodynamic possibility of the start of the reaction and for overcoming the activation barrier and the loosening of the crystal lattice. This statement has been fully corroborated (see Figs. 3a and b, straight line 3).

Thus, the interconnection between the activation energy and the thermodynamic parameters established by us has been corroborated by three types of solid-phase reactions. Further investigations of the matter may result in more important inferences and open up new possibilities in the study of chemical reactions.

In a more general form the dependence between the kinetic and thermodynamic characteristics is expressed by the following equation [698]:

$$\Delta G_A = \Delta G^0 - \alpha_1 \ln a_i + \alpha_2 , \tag{85}$$

where ΔG_A, alteration of the Gibbs free energy of activation; ΔG^0, the alteration of the Gibbs free energy of the reaction under standard conditions; α_1 and α_2, constants; a_i, product of activities.

Chapter II. Pyrosilicate Reactions

The most important processes underlying silicate technology are those of thermal treatment resulting in the synthesis of such products as calcium silicates and calcium aluminates in the manufacturing of various kinds of cements, complex sodium and calcium silicates and aluminosilicates in the manufacturing of glass, ferrites in the production of magnetic ceramics, titanates and similar substances in the manufacturing of special dielectrics; the processes of the sintering and making up of the ceramic body, connected with the strengthening of products without change of their shape etc. In all these cases chemical reactions at high temperatures with solid reactants are employed. These reactions are called pyrosilicate ones.

In spite of these processes and reactions being widespread the methods of controlling them and the properties of the substances obtained require further studies. Thermodynamic methods can play a major role in this field. These methods allow establishing the main regularities of the course of pyrosilicate and similar reactions and thus contribute to the intensification of the industrial processes.

1 General Principles

1.1 Types of Pyrosilicate Reactions

Related to the pyrosilicate reactions including all types of high temperature reactions of silicate production and partly of petrogenesis are the following most characteristic types of reactions:

1. solid-state polymorphic transformations of silicates and oxides: reversible – enantiotropic – and irreversible – monotropic – ones;

2. solid-phase reactions, to which those of the interaction of elements, oxides and silicates in solid state belong;

3. reactions with substances in gaseous state to which such reactions as those of carbonate, sulphate and oxide dissociation, those of the decomposition of clays at heating, those of hydrate, hydrosilicate and hydroaluminate dehydration, those of silicate formation with gaseous phase etc. belong;

4. reactions with liquid phase; to this group numerous reactions concerned with melt formation and those in melts belong;

5. combined reactions with solid, liquid and gaseous phases.

This classification quite suits the thermodynamic consideration of pyrosilicate reactions. Further on each group of reactions will be discussed individually.

1.2 Polymorphic Transformations of Silicates and Oxides

A single-component system undergoing polymorphic transformations may be in equilibrium established between the modifications of substances, each modification being stable at a certain temperature and pressure. Modifications, stable at high temperatures, formed immediately from a melt are generally denoted as α-modifications. All the following modifications, stable at lower temperatures, are subsequently denoted with letters of the Greek alphabet (β, γ etc.).

Allow a transformation of β-modification to α-modification to take place. Then at a constant pressure the alteration of the Gibbs free energy determining the possibility and the direction of a polymorphic transformation will be determined by the equation:

$$\Delta G_{\beta\to\alpha} = \Delta H_{\beta\to\alpha} - T\Delta S_{\beta\to\alpha}, \tag{86}$$

where $\Delta G_{\beta\to\alpha}$, $\Delta H_{\beta\to\alpha}$, $\Delta S_{\beta\to\alpha}$ are the alterations of the Gibbs free energy, enthalpy and entropy respectively accompanying the said polymorphic transformation [296].

The value of $\Delta G_{\beta\to\alpha}$ obtained allows judging the relative stability of the modifications considered and the direction of the polymorphic transformations.

1.2.1 The Case of Enantiotropic Transformations

Allow β-modification to be stable at low temperatures and α-modification to be stable at high temperatures. Then the stable state of the low-temperature modification must be characterized by positive values of $\Delta G_{\beta\to\alpha}$ and $\Delta H_{\beta\to\alpha}$. With the elevation of temperature the effect of the entropy term of Eq. (86) will grow, as a result of this with the increase of temperature in enantiotropic transformations the value of $\Delta G_{\beta\to\alpha}$ decreases. At $\Delta G_{\beta\to\alpha} = 0$ an equilibrium sets in and the transformation proves to be possible. With a further elevation of temperature $\Delta G_{\beta\to\alpha}$ becomes negative, the value of $\Delta H_{\beta\to\alpha}$ continually remaining positive.

Thus, the decrease of the Gibbs free energy due to the increase of the entropy term in absolute value will be characteristic of enantiotropic transformations. Such polymorphic transformations will always be accompanied by absorption of heat.

1.2.2 The Case of Monotropic Transformations

Allow β to be a low-temperature – unstable – modification, and α, a high-temperature – stable – one.

Then at any temperature the transformation $\beta \rightarrow \alpha$ will feature a negative value of the alteration of the Gibbs free energy and enthalpy, i.e., $\Delta G_{\beta \rightarrow \alpha}$ and $\Delta H_{\beta \rightarrow \alpha}$ will always be less than 0.

With the elevation of temperature the entropy term in Eq. (86) will alter.

At $\Delta S > 0$ the value of $\Delta G_{\beta \rightarrow \alpha}$ will also increase in absolute value remaining continually negative. With $\Delta S < 0$ the value of $\Delta G_{\beta \rightarrow \alpha}$ will decrease in absolute value up to the temperature when a given polymorphic transformation turns out to be impossible.

With the ΔS value equal or near to the zero the value of $\Delta G_{\beta \rightarrow \alpha}$ tends to $\Delta H_{\beta \rightarrow \alpha}$. In this case however, the direction of the process will be governed by the value of the transformation heat effect. In all cases monotropic transformations will be accompanied by heat release and no equilibrium in monotropic transformations ever takes place.

Examples of the evaluation of modification stabilities and of the alteration of the Gibbs free energy in polymorphic transformations are listed in Eitel [124, 305, 441]. These examples prove the calculation to be in agreement with the experiment.

Rarely, however, does the data of the thermochemical calculations of enantiotropic transformations agree so precisely with the test. This is due to the total calculation error depending on those arising while obtaining thermochemical data, mainly of heat effects and of the equations of the temperature dependence of the heat capacity for particular modifications, being comparable with the ΔG value of a polymorphic transformation. Therefore the results of the calculations for enantiotropic polymorphic transformations should be interpreted more cautiously.

The above facts encourage more thorough consideration of the initial thermochemical properties of SiO_2, accepted in further calculations.

Thermodynamic constants of quartz, cristobalite and tridymite are quite close to each other and the accuracy of the heat capacity equations of various modifications of SiO_2 still does not allow establishing reliable correlations for transformations of various SiO_2 modifications. Therefore in calculations for all the modifications of silica, in most cases only β-quartz was studied; in several cases when the calculations were conducted for high temperatures the equation of heat capacity for α-quartz was used. Crystoballite and tridymite were not studied. The bases for such an assumption, besides the lack of exhaustive data on thermodynamics of phase transitions in the system SiO_2, is that quartz is the main modification being a part of silica-containing raw materials used in silicate production.

It should be noted that the thermodynamic constants of quartz, cristobalite and tridymite are so close to each other that the above assumptions will not substantially affect the exactness of the calculations and the results obtained.

ΔH^0_{298} for β-quartz has a conderable "chronological" tendency to increase (in absolute value). Thus Richter [516] obtained the value of -205400 cal/mol which is to date adduce in various handbooks [101, 124, 184, 528].

Troitzsch [643] found for β-quartz the value of $\Delta H^0_{298} = -208300$ cal/mol.

In 1952 Humphrey and King [210] experimentally determined the value of $\Delta H^0_{298} = -210260$ cal/mol and in 1956 Golutvin [158] found the negative value to be within the limits 215000 – 219000 cal/mol. Recent information based on the experiment of burning β-quartz in fluorine [97, 101, 160, 692] upholds that this value is from -215800 to -217750 cal/mol. As the most reliable we have adopted the value of ΔH^0_{298} for β-quartz equal to -217750 cal/mol. Accordingly we have adopted the value of $\Delta G^0_{298} = -204750$ cal/mol.

All the values of ΔH^0_{298} and ΔG^0_{298} presented in the book were recalculated to this value.

1.3 Chemical Interaction in Solid Phases

General ideas concerning the determination of the direction of the reaction course and the conditions of the establishment of equilibrium adduced above may be spread to the case of idealized reactions taking place in solid phases. In effect, unless the reactants form solid solutions and if they are in condensed phases – ideally pure phases –, the activities of all the components will equal one. Then for an ideal case of an interaction in solid phases Eq. (65) can be expressed as:

$$\Delta G = \Delta G^0 . \tag{87}$$

Hence, the alteration of the Gibbs free energy in case of solid-phase reactions (polymorphic transformations included) can be derived immediately from the data for standard values of the Gibbs free energy presented in reference literature.

In case of an equilibrium $\Delta G = 0$, in an ideal solid-phase reaction the equilibrium will be reached at $\Delta G^0 = 0$. A case like this however seldom occurs in practice because ΔG^0 for solid-phase reactions is, as a rule, far from being 0 in the entire temperature range as a result of a great value of heat effects and an insignificant effect of the entropy term.

The results of the thermodynamic studies of ideal solid-phase reactions allow:

1. clearing up the principal possibility and the probable directions of reactions by the sign and value of ΔG^0 (a reaction is possible at $\Delta G^0 < 0$);

2. carrying out a comparison of the thermodynamic probability of solid-phase reactions and of the stability of particular compounds in silicate and similar systems. In this case it is most advisable to compare ΔG^0 of reactions depending on the proportions of the initial component, the reaction having the

minimum value of ΔG^0 being thermodynamically the most probable and the thus formed compound, the most stable.

As an example below, we give a thermodynamic analysis of solid-phase reactions in the system $CaO-SiO_2$.

Of great theoretical and practical interest is the study of solid-phase reactions in the system $CaO-SiO_2$ as they play an important role in the production of cement and glass. Bringing to light the sequence of these reactions and the thermodynamic probability of the formation of one or the other compounds assists in explaining the processes and in finding new ways to perfect the production technique. The knowledge of heat effects at high temperatures contributes to making up heat balances of industrial processes and enables one to develop heat plants. Moreover, it should be noted that for the system $CaO-SiO_2$ the most reliable thermochemical data have been obtained. A great deal of research has been made in the kinetics of the reactions in this system as well.

A thermodynamic analysis of the reactions in the system $CaO-SiO_2$ was first carried out by Landiya and Mchedlov-Petrossyan [321], subsequently these calculations were specified [222, 361, 405].

According to data [40, 75, 123, 221, 361] in the system $CaO-SiO_2$ the following compounds exist:

$$CaO \cdot SiO_2;\ 3CaO \cdot 2SiO_2,\ 2CaO \cdot SiO_2,\ 3CaO \cdot SiO_2\ .$$

Let us view the main reactions of the formation of these compounds from oxides:

1. $CaO + SiO_2 = CaO \cdot SiO_2$
2. $3CaO + 2SiO_2 = 3CaO \cdot 2SiO_2$
3. $2CaO + SiO_2 = 2CaO \cdot SiO_2$
4. $3CaO + SiO_2 = 3CaO \cdot SiO_2$

Here polymorphic transformations are to be taken into account:

α-quartz $\rightarrow$ α-tridymite (at 1140 K);
β-$CaO \cdot SiO_2 \rightarrow \alpha$-$CaO \cdot SiO_2$ (at 1400 K);
β-$2CaO \cdot SiO_2 \rightarrow \alpha'$-$2CaO \cdot SiO_2$ (at 970 K).

As initial data for analysis the thermochemical ones (Table 5) were taken. In the calculations instead of the data for α-tridymite those for γ-tridymite were taken which resulted in an error of approximately 0.1%.

With the above data given, the following equations of enthalpy and the Gibbs free energy alterations with temperature were set up for all the reactions:

Table 5. The initial thermochemical data for compounds in the system $CaO-SiO_2$

Compounds	$-\Delta H^0_{298(el)}$	$-\Delta G^0_{298(el)}$	$-\Delta H^0_{298(ox)}$	$-\Delta G^0_{298(ox)}$	$C_p=f(T)$		
	[cal/mol]				a	$b\cdot10^3$	$c\cdot10^{-5}$
CaO	151900	144400	–	–	11.67	1.08	−1.56
SiO_2 α-quartz	217600	–	–	–	14.41	1.94	
SiO_2 β-quartz	217750	204750	–	–	11.22	8.2	−2.7
SiO_2 γ-tridymite	217150	204250	–	–	13.64	2.64	
$CaO\cdot SiO_2$ β-wollastonite	390950	370550	21300	21400	26.64	3.6	−6.52
$CaO\cdot SiO_2$ α-wollastonite	389750	369750	20100	20600	25.85	3.94	−5.65
$3CaO\cdot 2SiO_2$	938720	890800	47520	48100	64	9.05	−16.6
β-$2CaO\cdot SiO_2$	551740	524190	30190	30640	34.87	9.74	−6.26
$3CaO\cdot SiO_2$	709450	665470	26980	27520	49.85	8.62	−10.15

Table 6. The alteration of ΔG^0 with temperature for main reactions in the system $CaO-SiO_2$

Compounds	$-\Delta G^0$ [cal/mol] at temp. in K									
	298	1000	1100	1200	1300	1400	1500	1600	1700	1800
$C+S=CS$	21400	20420	20290	20646	20517	20425	20812	20774	20732	20654
$3C+2S=C_3S_2$	48100	46728	46474	47200	46966	46747	46560	46410	46250	46107
$2C+S=C_2S$	30640	31287	31685	32241	32458	32967	33468	34159	34944	35519
$3C+S=C_3S$	27520	28340	28550	29280	29480	29900	30270	30730	31140	31580

Note: From here on the following abbreviations for oxides in writing formulae in the dualistic form are adopted: A – Al_2O_3; B – BaO; Be – BeO; C – CaO; Cs – Cs_2O; Fe – FeO; K – K_2O; Li – Li_2O; Mg – MgO; Mn – MnO; Na – Na_2O; Pb – PbO; Rb – Rb_2O; S – SiO_2; Sr – SrO; Zr – ZrO_2

1. Prior to the polymorphic transformation
   α-quartz $\rightarrow$ α-tridymite
   $\Delta H_1^0 = -23158 + 0.56\,T + 0.29 \cdot 10^{-3}T^2 + 4.96 \cdot 10^5 T^{-1}$;
   $\Delta H_2^0 = -51656 + 0.17\,T + 0.96 \cdot 10^{-3}T^2 + 11.92 \cdot 10^5 T^{-1}$;
   $\Delta H_3^0 = -31239 + 3.51\,T - 0.62 \cdot 10^{-3}T^2 - 3.12 \cdot 10^5 T^{-1}$;
   $\Delta H_4^0 = -29096 + 0.43\,T + 1.72 \cdot 10^{-3}T^2 + 5.48 \cdot 10^5 T^{-1}$;
   $\Delta G_1^0 = -23158 - 0.56\,T \ln T - 0.29 \cdot 10^{-3}T^2 + 2.48 \cdot 10^5 T^{-1} + 6.64\,T$;
   $\Delta G_2^0 = -51656 - 0.17\,T \ln T - 0.965 \cdot 10^{-3}T^2 + 5.96 \cdot 10^5 T^{-1} + 6.47\,T$;
   $\Delta G_3^0 = -31293 - 3.51\,T \ln T - 0.62 \cdot 10^{-3}T^2 - 1.56 \cdot 10^5 T^{-1} + 25.51\,T$;
   $\Delta G_4^0 = -29096 - 0.43\,T \ln T - 1.72 \cdot 10^{-3}T^2 + 2.74 \cdot 10^5 T^{-1} + 5.61\,T$.

2. With polymorphic transformations
   α-quartz $\rightarrow$ α-tridymite (1140 K)
   $\Delta H_1^0 = -23959 + 1.33\,T - 0.65 \cdot 10^{-3}T^2 + 4.96 \cdot 10^5 T^{-1}$;
   $\Delta H_2^0 = -53252 + 1.71\,T + 0.265 \cdot 10^{-3}T^2 - 3.12 \cdot 10^5 \cdot T^{-1}$;
   $\Delta H_3^0 = -32091 + 4.28\,T + 0.27 \cdot 10^{-3}T^2 - 3.12 \cdot 10^5 T^{-1}$;
   $\Delta H_4^0 = -2989 + 1.20\,T + 1.37 \cdot 10^{-3}T^2 + 5.48 \cdot 10^5 T^{-1}$;
   $\Delta G_1^0 = -23959.5 - 1.33\,T \ln T + 0.06 \cdot 10^{-3}T^2 + 2.48 \cdot 10^5 T^{-1} + 11.94\,T$;
   $\Delta G_2^0 = -53252 - 1.71\,T \ln T - 0.265 \cdot 10^{-3} \cdot T^2 + 5.96 \cdot 10^5 T^{-1} + 17.07\,T$;
   $\Delta G_3^0 = -32091.5 - 4.28\,T \ln T - 0.27 \cdot 10^{-3}T^2 - 1.56 \cdot 10^5 T^{-1} + 30.81\,T$;
   $\Delta G_4^0 = -29894.5 - 1.21\,T \ln T - 1.37 \cdot 10^{-3}T^2 + 2.74 \cdot 10^5 T^{-1} + 10.46\,T$.

3. With phase transformations
   β-$CaO \cdot SiO_2 \rightarrow \alpha$-$CaO \cdot SiO_2$ (1400 K) and
   β-$2CaO \cdot SiO_2 \rightarrow \alpha'$-$2CaO \cdot SiO_2$ (970 K)
   $\Delta H_1^0 = -25672 + 2.12\,T + 0.23 \cdot 10^{-3}T^2 + 5.84 \cdot 10^5 T^{-1}$;
   $\Delta H_2^0 = -53252 + 1.71\,T + 0.26 \cdot 10^{-3}T^2 + 11.92 \cdot 10^5 T^{-1}$;
   $\Delta H_3^0 = -35204 + 6.99\,T - 0.37 \cdot 10^{-3}T^2 + 3.14 \cdot 10^5 T^{-1}$;
   $\Delta H_4^0 = -29894 + 1.21\,T + 1.37 \cdot 10^{-3}T^2 + 5.48 \cdot 10^5 T^{-1}$;
   $\Delta G_1^0 = -25672.5 - 2.12\,T \ln T - 0.23 \cdot 10^{-3}T^2 + 2.92 \cdot 10^5 T^{-1} + 18.94\,T$;
   $\Delta G_2^0 = -23252 - 1.71\,T \ln T - 0.26 \cdot 10^{-3}T^2 + 5.96 \cdot 10^5 T^{-1} + 17.07\,T$;
   $\Delta G_3^0 = -35204.5 - 6.99\,T \ln T + 0.37 \cdot 10^{-3}T^2 + 1.57 \cdot 10^5 T^{-1} + 51.64\,T$;
   $\Delta G_4^0 = -29894.5 - 1.2\,T \ln T - 1.37 \cdot 10^{-3}T^2 + 2.74 \cdot 10^5 T^{-1} + 10.46\,T$.

The alterations of the Gibbs free energy in the temperature range 1000 – 1800 K were calculated by these equations. The results obtained are presented in Table 6 and Fig. 4.

For evaluating the thermodynamic sequence of reactions the dependence of ΔG^0 of the reactions on the compositions of the initial mixtures was examined. The authors analyzed the values at the ratios of initial mixtures of $CaO/SiO_2 = 1/1$; 3/2; 2/1 and 3/1. The results obtained are presented in Table 7, and the diagrammatic interpretation is given in Fig. 5.

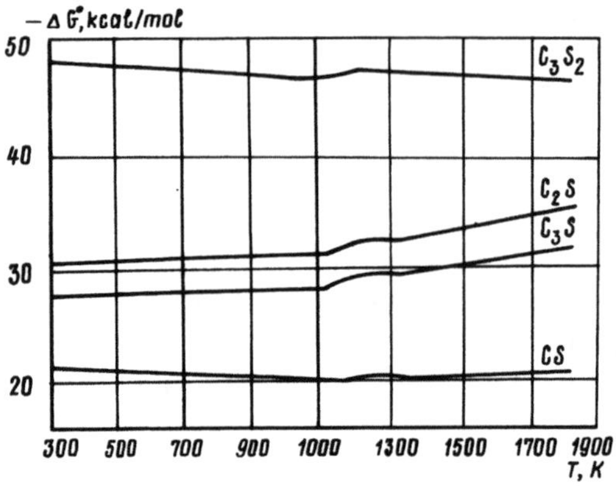

Fig. 4. The temperature dependence of ΔG^0 of reactions in the system $CaO-SiO_2$ in the temperature range of 1000 to 1800 K

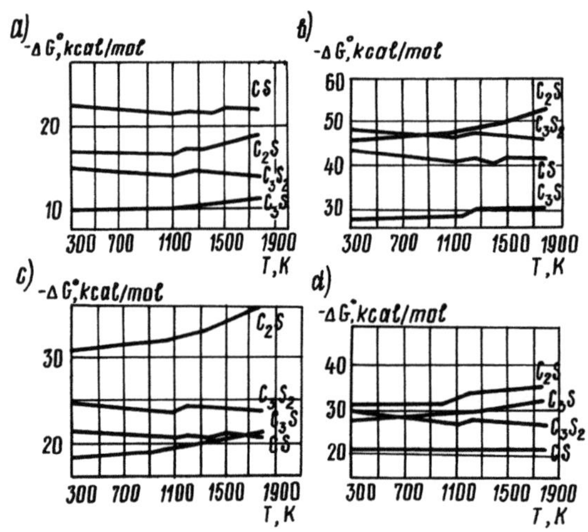

Fig. 5a–d. The temperature dependence of ΔG^0 of reactions in the system $CaO-SiO_2$ in the temperature range of 1000 to 1800 K for the ratios. **a** C/S = 1/1; **b** C/S = 3/2; **c** C/S = 2/1; **d** C/S = 3/1

On the above theoretical premises the calculation allows determining thermodynamically most stable reaction products for the given binary silicate system and permits drawing some conclusions concerning the sequence of reactions in this system.

Based on the results obtained we make up the so-called "thermodynamic stability series" of compounds with diverse ratios of initial components. These series are presented in form of a table (Table 8) in which the value of the Gibbs free energy increases from left to right (stability, accordingly, falls).

According to these calculations when the proportion of the initial components C/S = 1/1, the most stable compound (find product) will be calcium monosilicate. This is corroborated by new data for the system [50].

When the proportion of the initial components C/S = 3/2 and more – the range rich in lime –, the most stable compound will be dicalcium silicate.

Table 7. The alteration of ΔG^0 with temperature for the reactions in the system $CaO-SiO_2$ at various proportions of components

Reactions	$-\Delta G^0$ [cal/mol] at temps. in K									
	298	1000	1100	1200	1300	1400	1500	1600	1700	1800
a) $CaO/SiO_2 = 1/1$										
$C+S = CS$	21400	20420	20290	20646	20517	20425	20812	20774	20732	20654
$C+S = \frac{1}{3}C_3S_2 + \frac{1}{3}S$	18030	15576	15491	15730	15655	15583	15520	15470	15416	15369
$C+S = \frac{1}{2}C_2S + \frac{1}{2}S$	15320	15643	15842	16120	16229	16483	16734	17079	17472	17759
$C+S = \frac{1}{3}C_3S + \frac{2}{3}S$	9173	9446	9516	9760	9826	9966	10090	10243	10380	10526
b) $CaO/SiO_2 = 3/2$										
$3C+2S = 2CS+C$	42800	40840	40580	41292	41034	40850	41624	41548	41464	41303
$3C+2S = C_3S_2$	48100	46728	46474	47200	46966	46747	46560	46410	46250	46107
$3C+2S = \frac{3}{2}C_2S + \frac{1}{2}S$	45960	46929	47526	48360	48687	49449	50202	51237	52416	53277
$3C+2S = C_3S+S$	27520	28340	28550	29280	29480	29900	30270	30730	31140	31580
c) $CaO/SiO_2 = 2/1$										
$2C+S = CS+C$	21400	20420	20290	20846	20517	20425	20812	20774	20732	20654
$2C+S = \frac{1}{2}C_3S_2 + \frac{1}{2}C$	24050	23364	23237	23600	23483	23373	23280	23205	23125	23053
$2C+S = C_2S$	30640	31287	31685	32241	32458	32967	33468	34159	34944	35519
$2C+S = \frac{2}{3}C_3S + \frac{1}{3}S$	18346	18892	19032	19520	19652	19932	20180	20486	20760	21052
d) $CaO/SiO_2 = 3/1$										
$3C+S = CS+2C$	21400	20420	20290	20646	20517	20425	20812	20774	20732	20654
$3C+S = \frac{1}{2}C_3S_2 + \frac{3}{2}C$	24050	23364	23237	23600	23483	23373	23280	23205	23125	23053
$3C+S = C_2S+C$	30640	31287	31685	32241	32458	32987	33468	34159	34944	35519
$3C+S = C_3S$	27520	28340	28550	29280	29480	29900	30270	30730	31140	31580

Table 8. The thermodynamic stability sequence of compounds of the system $CaO-SiO_2$

Ratio of initial components	The thermodynamics stability series beginning with most stable compound in the temp. range of 1000–1800 K			
1/1	CS	C_2S	C_3S_2	C_3S
3/2	C_2S	CS	C_3S_2	C_3S
2/1	C_2S	CS	C_3S	C_3S_2
3/1	C_2S	C_3S	CS	C_3S_2

Through extensive investigations of the reactions in the system $CaO-SiO_2$ the primary formation of C_2S independent of the ratio C/S in initial mixtures was established, as seen in the stability diagram (see Table 8).

For this purpose we enclose C_2S in a frame. As one can see from the diagram the formation of C_3S_2 and C_3S by way of solid-phase reactions will be improbable with any proportions of C/S. The formation of CS is possible only with the ratio $C/S \leqslant 1$ by the reaction $C_2S+S=2CS$. Thus, in the system $CaO-SiO_2$, in the area rich in lime, compound C_2S will be primary and at the same time most stable.

As one can see from Table 8 obtaining tricalcium silicate by way of the addition of CaO to dicalcium silicate during interaction in pure solid phases appears impossible.

1.4 Reactions with Gaseous Substances

In the silicate production processes with gaseous phase occur very often. Among them are carbonate, sulphate and oxide dissociation, hydrate, hydrosilicate and hydroaluminate-dehydration, clay decomposition at heating, reactions of silicate formation which represent only some of the processes belonging to this type of pyrosilicate reactions.

The monotypeness of the equation describing all the above reactions permits a general introduction of the consideration of particular instances.

Allow reaction:

$$bB_{(\text{sol})}+cC_{(\text{sol})}=lL_{(\text{sol})}+mM_{(\text{gas})} \tag{88}$$

to take place.

Then bearing in mind that at comparatively low values of the total pressure $a_M=f_M$ and $a_B=a_C=a_L=1$, at a constant pressure and temperature the total alteration of Gibbs free energy will be determined by the equation:

$$\Delta G_{p,T} = \Delta G^0_{p,T} + mRT \ln p_M . \tag{89}$$

At an equilibrium $\Delta G_{pT} = 0$, for the equilibrium state we obtain:

$$\Delta G^0_{pT} = -mRT \ln p'_M , \tag{90}$$

where p'_M is an equilibrium pressure of a gaseous component formed as a result of a reaction.

Thus, when the activities of solid components are equal to 1, the equilibrium constant turns out to be immediately equal to the equilibrium pressure:

$$K = p'_M . \tag{91}$$

If in the initial equation there are two or more gaseous substances, corresponding pressures will enter the equilibrium constant term.

The above equations enable us:

1. to determine the principal possibility and the probable direction of the process;

2. to determine the equilibrium constants (equilibrium pressures), the decomposition temperatures and those of the beginning of a reaction ($p_M = 1$).

Diverse combined calculations are feasible as well. Knowing e.g., the equilibrium pressure at some temperature one can determine the temperature of the beginning of decomposition or the temperature of the beginning of a reaction.

Below we adduce some instances of the calculations of ΔG of reactions, equilibrium constants and dissociation temperatures of some substances that are of importance in the silicate production processes.

1.4.1 Carbonate Dissociation

As an example we shall present the calculation of ΔG of the decomposition reaction, of the equilibrium constant and of the temperature of the decomposition $CaCO_3$ (calcite) [182] based on the known thermochemical data.

The data needed for the estimation are presented in Table 9.

Table 9. Initial data for compounds taking part in $CaCO_3$ dissociation reaction

Compounds	$-\Delta H^0_{298}$ [cal/mol]	$-\Delta G^0_{298}$ [cal/mol]	$C_p = f(T)$ [cal/deg mol]
$CaCO_3$ (calcite)	288450	269780	$24.98 + 5.24 \cdot 10^{-3} T - 6.2 \cdot 10^5 T^{-2}$
CaO	151900	144400	$11.67 + 1.08 \cdot 10^{-3} T - 1.56 \cdot 10^5 T^{-2}$
CO_2	94050	94260	$10.55 + 2.16 \cdot 10^{-3} T - 2.04 \cdot 10^5 T^{-2}$

Below an example of calculation is given:

1. Determine the ΔH^0 of the reaction at 298 K:

$$\Delta H^0_{298} = -151900 - 94050 + 288450 = 42500 \text{ cal/mol}.$$

2. Determine the ΔG^0 of the reaction at 298 K:

$$\Delta G^0_{298} = -144400 - 94260 + 269780 = 31120 \text{ cal/mol}.$$

3. Determine constants Δa, Δb, Δc:

$$\Delta a = 11.67 + 10.55 - 24.98 = -2.76$$
$$\Delta b = (1.08 + 2.16 - 5.24) \cdot 10^{-3} = -2 \cdot 10^{-3}$$
$$\Delta c = (-1.56 - 2.04 + 6.2) \cdot 10^5 = 2.6 \cdot 10^5 .$$

The equation $C_p = f(T)$ will take form:

4. Determine ΔH_0:

$$\Delta H_0 = 42500 + 2.76 \cdot 298 + 1 \cdot 10^{-3} (298)^2 + 2.6 \cdot 10^5 (298)^{-1}$$
$$\Delta H_0 = 42500 + 823 + 89 + 873 = 44285 \text{ cal/mol}.$$

5. Formulate the equation of the dependence of the reaction enthalpy:

$$\Delta H^0_T = 44285 - 2.76\,T - 1 \cdot 10^{-3} T^2 - 2.6 \cdot 10^5 T^{-1} .$$

6. Determine y:

$$298\,y = 31120 - 44285 - 2.16 \cdot 298 \ln 298 - 1 \cdot 10^{-3} (298)^2 + 1.3 \cdot 10^5 (298)^{-1}$$
$$298\,y = 31120 - 44285 - 4685 - 89 + 436 = 17503$$
$$y = -\frac{17503}{298} = -58.7 .$$

7. Formulate the equation of the temperature dependence of the Gibbs free energy:

$$\Delta G^0_T = 44285 + 2.76\,T \ln T + 1 \cdot 10^{-3} T^2 - 1.3 \cdot 10^5 T^{-1} - 58.7\,T .$$

8. With the above equations we calculate ΔH^0_T and ΔG^0_T of the reactions of $CaCO_3$ dissociation for the temperature range of 800 to 1400 K (within each 100 K):

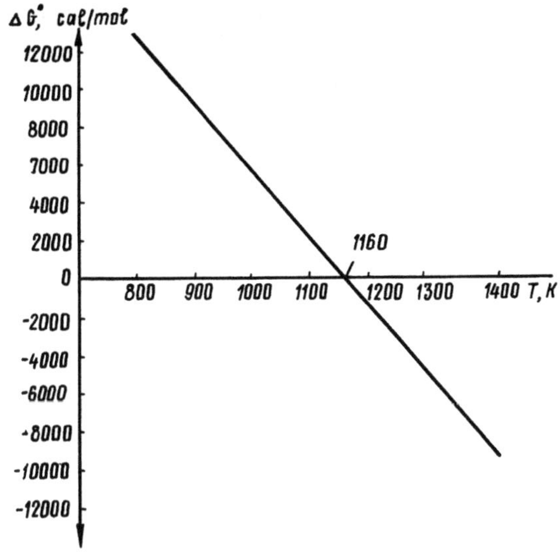

Fig. 6. The temperature dependence of ΔG^0 of the $CaCO_3$ dissociation reaction in the temperature range of 800 to 1400 K

Temperature [K]	800	900	1000	1100	1200	1300	1400
ΔG [cal/mol]	12552	9021	5535	1997	−1323	−4709	−8053
ΔH [cal/mol]	41129	40704	40264	39788	39305	38804	38278

The data obtained are presented in the diagram (Fig. 6).

The equilibrium constant according to Eqs. (90) and (91) will be determined from the ratio:

$$-\Delta G_T^0/T = 2.303\,R\,\lg K\,.$$

From which $\lg K = \dfrac{-\Delta G_T^0}{4.576\,T}$.

Thus, e.g., for the temperature of 1000 K:

$$\Delta G^0 = 5535 \text{ cal/mol and } \lg K = -\frac{5.535}{4.576 \cdot 1000} = -1.2\,.$$

From which the equilibrium constant will equal:

$$K = 0.0631 \text{ atm}\,.$$

To determine the temperature of the beginning of limestone decomposition we shall presume $p = 1$. In this case ΔG_T^0 must be equal to zero. Thus, the problem is the determination of the temperature at which $\Delta G_T^0 = 0$. In Fig. 6 this temperature corresponds to the abscissa of the points where the curve of the dependence of ΔG^0 of $CaCO_3$ decomposition on temperature intersects the zero ordinate. In our case this temperature will be equal to 1160 K agreeing with the experimentally determined temperature, i.e., 1155 K [568].

The calculation enables us also to determine the equation of the temperature dependence of the $CaCO_3$ dissociation pressure:

$$\lg p_{CO_2} = -9650\,T^{-1} - 0.6 \ln T - 0.227 \cdot 10^{-3} T + 0.28 \cdot 10^5 T^2 + 12.8 \,.$$

This example of the $CaCO_3$ dissociation shows that with initial thermochemical data exact enough, thermodynamics yields reliable solutions.

Similar calculations were also carried out for other carbonates of the second group of the periodic system that are of great importance for the silicate production processes (Table 10).

The thermodynamic conditions of the dolomite decomposition were examined by Mchedlov-Petrossyan and Babushkin [402].

The calculations presented are well in agreement with the experimental data [187, 193].

1.4.2 Sulphate Dissociation

More instances of decomposition reactions, those of sulphates [546, 697], important in binder and glass production processes may be viewed.

At high temperatures sulphates decompose according to the general equation:

$$MeSO_4 = MeO + SO_3 \,.$$

The arising SO_3, as a rule, dissociates at high temperatures according to the equation:

$$SO_3 = SO_2 + \tfrac{1}{2}O_2 \,.$$

In presence of catalysts, e.g., iron oxides, the equilibrium of the above reaction establishes quickly enough, the dissociation pressure is possible, thus we may express experimentally:

$$p = p_{SO_3} + p_{SO_2} + p_{O_2} \,,$$

as later:

$$p_{O_2} = \tfrac{1}{2} p_{SO_2} \,, \quad p = p_{SO_3} + \tfrac{3}{2} p_{SO_2} \,.$$

Table 10. The temperature of the beginning of decomposition and the equations of the ΔG^0 temperature dependence of the carbonates of the second group of the periodic system

Compounds	Temp. of the beginning of decomposition [K]	Equations $\Delta G^0 = f(T)$
$MgCO_3$	680	$\Delta G^0 = 28135 - 211\,T \ln T + 4.95 \cdot 10^{-3} T^2 - 0.32 \cdot 10^5 T^{-1} - 31\,T$
$CaCO_3$	1160	$\Delta G^0 = 44285 + 2.76\,T \ln T + 1 \cdot 10^{-3} T^2 - 1.3 \cdot 10^5 T^{-1} - 58.7\,T$
$SrCO_3$	1425	a) $\Delta G^0 = 55690 - 1.47\,T \ln T + 2.64 \cdot 10^{-3} T^2 + 0.23 \cdot 10^5 T^{-1} - 32.38\,T$ (up to 1198 K)
		b) $\Delta G^0 = 53230 + 11.0\,T \ln T - 1.64 \cdot 10^{-3} T^2 + 1.93 \cdot 10^5 T^{-1} - 115\,T$ (above 1198 K)
$BaCO_3$	1550	a) $\Delta G^0 = 63082 - 2.52\,T \ln T - 4.25 \cdot 10^{-3} T^2 + 0.58 \cdot 10^5 T^{-1} - 26\,T$ (up to 1079 K)
		b) $\Delta G^0 = 70202 + 13.71\,T \ln T - 1.60 \cdot 10^{-3} T^2 + 2.01 \cdot 10^5 T^{-1} - 143.5\,T$ (above 1079 K)
		c) $\Delta G^0 = 71984 + 15.71\,T \ln T - 1.60 \cdot 10^{-3} T^2 + 2.01 \cdot 10^5 T^{-1} - 159.7\,T$ (above 1241 K)

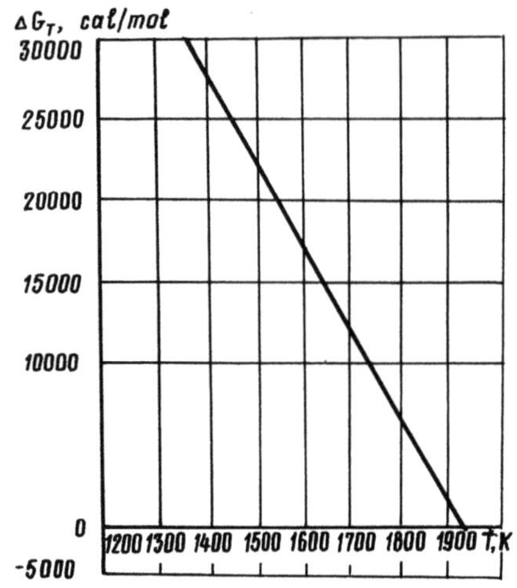

Fig. 7. The dependence of ΔG^0 of the anhydrite dissociation reaction in the temperature range of 1400 to 2000 K

Thus, for the reaction of anhydrite $CaSO_4$ dissociation:

$$CaSO_4 = CaO + SO_2 + \tfrac{1}{2}O_2 .$$

Owing to the complete dissociation of SO_3 its concentration in gaseous phase may be dispensed with. Therefore:

$$p_{SO_2} = \tfrac{2}{3}p \quad \text{and} \quad p_{O_2} = \tfrac{1}{3}p .$$

Then the equilibrium constant of the reaction of anhydrite decomposition may be expressed by an equation:

$$K = p_{SO_2} \cdot p_{O_2}^{1/2} = 0.386\, p^{3/2} .$$

On the basis of thermochemical data (Table 11) one can calculate the equilibrium constant and determine the dissociation pressure.

Table 11. The initial data for the calculation of the reaction of anhydrite decomposition

Compound	$-\Delta H^0_{298}$ [cal/mol]	$-\Delta G^0_{298}$ [cal/mol]	$C_p = f(T)$ [cal/deg mol]
$CaSO_4$	342420	315560	$18.52 + 21.97 \cdot 10^{-3}T - 1.57 \cdot 10^5 T^{-2}$
CaO	151900	144400	$11.67 + 1.08 \cdot 10^{-3}T - 1.56 \cdot 10^5 T^{-2}$
SO_2	70960	51970	$10.38 + 2.54 \cdot 10^{-3}T - 1.42 \cdot 10^5 T^{-2}$
$\tfrac{1}{2}O_2$	–	–	$4.13 + 0.13 \cdot 10^{-3}T - 0.94 \cdot 10^5 T^{-2}$

An instance of calculation:

1. Determine ΔH^0 of the reaction at 298 K:

$$\Delta H^0_{298} = -151900 - 70960 + 342420 = 119500 \text{ cal/mol} .$$

2. Determine ΔG^0 of the reaction at 298 K:

$$\Delta G^0_{298} = -144400 - 71790 + 315560 = 99370 \text{ cal/mol} .$$

3. Determine constants Δa, Δb, Δc:

$$\Delta a = 11.67 + 10.38 + 4.13 - 18.52 = 7.66$$

$$\Delta b = (1.08 + 2.54 + 0.13 - 21.97) \cdot 10^{-3} = -18.22 \cdot 10^{-3}$$

$$\Delta c = (-1.56 - 1.42 - 0.94 + 1.57) \cdot 10^5 = -2.35 \cdot 10^5 .$$

From which we have:

$$\Delta C_p = 7.66 - 18.22 \cdot 10^{-3} T - 2.35 \cdot 10^5 T^{-2} .$$

4. Determine ΔH_0:

$$\Delta H_0 = 119560 - 7.66 \cdot 298 + 9.11 \cdot 10^{-3} (298)^2 - 2.35 \cdot 10^5 (298)^{-1}$$

$$\Delta H_0 = 119560 - 2283 + 810 - 790 = 117297 .$$

5. Determine y:

$$99370 = 117297 - 7.66 \cdot 298 \ln 298 + 9.11 \cdot 10^{-3} (298)^2 + 1.175 \cdot 10^5 (298)^{-1} + 298 y .$$

$$99370 = 117297 - 13013 + 810 + 395 + 298 y$$

$$y = -\frac{6119}{298} = -20.55 .$$

6. Formulate the equation of the temperature dependence of the Gibbs free energy:

$$\Delta G^0_T = 117297 - 7.66 T \ln T + 9.11 \cdot 10^{-3} T^2 + 1.175 \cdot 10^5 T^{-1} - 20.55 T .$$

The equilibrium constant will equal:

$$\lg K = \frac{-\Delta G^0_T}{4.576}$$

and the pressure at dissociation

$$p = 1.89\,K^{2/3}\ \text{atm}\,.$$

Thus, for a temperature of 1500 K:

$$\begin{aligned}\Delta G^0_{1500} &= 117297 - 7.66 \cdot 1500 \cdot 7.31 + 9.11 \cdot 10^{-3}(1500)^2 \\ &\quad + 1.175 \cdot 10^5 (1500)^{-1} - 20.55 \cdot 1500 \\ &= 117297 - 34100 + 20500 + 78 - 30800 = 22975\,.\end{aligned}$$

Then $\lg K_p = -22975/4.576 \cdot 1500 = -3.35$.

From which

$$p = 1.89 \cdot 5.84 \cdot 10^{-3} = 0.011\ \text{atm}\,.$$

At a temperature of 1570 K according to [482] $CaSO_4$ melting takes place, ΔH^0_m constitutes 6540 cal/mol. This value was borne in mind while conducting further calculations. The heat capacity of the liquid $CaSO_4$ was assumed to be constant and equal to that of the solid $CaSO_4$ at the temperature of melting since Krestovnikov and Vigdorovich [302] stated that the heat capacities of inorganic substances in solid and liquid state coincide close to the melting point. For the liquid $CaSO_4$ $C_p = 52.95$ cal/deg mol was assumed.

For temperatures above 1570 K the following equations of the temperature dependence of ΔH^0 and ΔG^0 were taken:

$$\Delta H^0_T = 137.735 - 26.77\,T + 1.875 \cdot 10^{-3}\,T^2 - 3.92 \cdot 10^5\,T^{-1}$$

$$\Delta G^0_T = 137.735 + 26.77\,T \ln T - 1.875 \cdot 10^{-3}\,T^2 + 1.96 \cdot 10^5\,T^{-1} - 289.6\,T\,.$$

In Table 12 a comparison of equilibrium pressures of anhydrite dissociation obtained by way of calculations with the experimental data [546] for natural anhydrite is presented. Considering $CaSO_4$ melting the correlation of the calculated and experimental data improved substantially [20].

The diagrammatic determination of the temperature of anhydrite spontaneous dissociation features 1940 K (Fig. 7).

The reactions of gypsum dihydrate dehydration were investigated in the temperature range of 298 to 900 K [402]. The study showed the dehydration of $CaSO_4 \cdot 2H_2O$ to $CaSO_4 \cdot \frac{1}{2}H_2O$ and gaseous water possible, beginning with 380 K while the decomposition of β-$CaSO_4 \cdot 0.5\,H_2O$ to soluble anhydrite and water, beginning with 480 K.

The transformation of the soluble anhydrite to insoluble takes place irreversibly in the whole temperature range.

Table 12. ΔG_T^0 of dissociation and comparison of anhydrite equilibrium pressures

Temp. [K]	ΔG_T^0 [cal/mol]		Pressure [atm]		
	With no respect to melting	With respect to melting	Calculated		Experimental
			With no respect to melting	With respect to melting	
1371	30484	–	0.001	–	0.001
1400	28769	–	–	–	–
1418	27603	–	0.002	–	0.003
1472	24636	–	0.007	–	0.007
1500	22975	–	–	–	–
1503	22917	–	0.011	–	0.012
1570	19050	19050	–	–	–
1600	17363	17790	0.049	0.045	0.043
1650	14608	15196	0.097	0.088	0.080
1675	13200	13897	0.134	0.117	0.107
1700	11900	12600	0.180	0.157	0.140
1800	–	7420	–	–	–
1900	–	2090	–	–	–

It should be mentioned that kinetically the correctness of the calculation results obtained is well corroborated by a thermographical study of gypsum [260, 474]. Other transformations of $CaSO_4 \cdot 2H_2O$ and of the products of its dissociation were thoroughly examined by Kelley, Southard and Anderson [474].

1.4.3 Reactions of Silicate Formation with Gaseous Phase Release

The application of these principles to the reactions of silicates with volatile substances taking part is of considerable interest for systematically studying this type of reaction.

Thus, Danielson performed the calculation of calcite-wollastonite equilibrium, verified by improved thermochemical data [108]. This equilibrium is characterized by the equation:

$$CaCO_{3(sol)} + SiO_{2(sol)} = CaSiO_{3(sol)} + CO_{2(gas)} \cdot$$

Thus, in this instance as well the equilibrium constant, equilibrium pressure and the temperature of the beginning of the decomposition can be found by Eq. (90).

In a similar way Danielson deduced the equation of the temperature dependence of the Gibbs free energy:

$$\Delta G_T^0 = 21216 - 0.99 \ln T - 3.84 \cdot 10^{-3} T^2 - 0.17 \cdot 10^5 T^{-1} - 34.24 T.$$

Furthermore, as the dependence of ΔG on temperature is termed by Eq. (90), one can determine the equation of the temperature dependence of the reaction equilibrium pressure:

$$\lg p_{CO_2} = 7.48 - 4636\,T^{-1} + 0.498\,\lg T - 0.84 \cdot 10^{-3}\,T - 0.037 \cdot 10^{5}\,T^{-2}$$

within the temperature range from 273 to 848 K.

Above the transformation temperature, up to 1303 K, a similar result is expressed by the equation:

$$\lg p_{CO_2} = 11.61 - 4603\,T^{-1} - 1 \cdot 107\,\lg T - 0.155 \cdot 10^{-3}\,T - 0.258 \cdot 10^{5}\,T^{-2}.$$

The temperature of an equilibrium pressure $p_{CO_2} = 1$ atm, i.e., the temperature of the beginning of the reaction, equals 553 K.

The reactions of silicate formation accompanied by gaseous phase release were also studied by Repa [511] and Trzebiatowski [644].

1.5 Reactions with Substances in Molten State and Combined Reactions

For a reaction in a melt one should know the corresponding thermodynamic data of pure substances that are in a molten state. One should, moreover, know the respective activities of the reactants that are capable of mutual solution in the melt.

Should the melt be taken as an ideal solution of components it comprises, the activities of these components will be equal to their concentrations – molar fractions – in the melt. If, however, one or another pure liquid phases take part, their activities will be equal to one, and consequently, when carrying out any calculations one can use thermochemical data for pure substances in molten state directly. Practically the latter fact results in the necessity of considering the changes connected with melting processes.

The alteration of enthalpy and the Gibbs free energy of individual substances is necessary in the determination of the technique set forth in Chap. I.10.

1.5.1 On the Nature of Silicate Melts [340, 653]

As mentioned above, at present silicate melts are considered as ionic liquids [703].

By the prevailing type of bond with oxygen all the cations of such melts can be divided into two major groups. With the one this bond is mainly heteropolar, with the other covalent bonds are predominant. The first groups chiefly comprises the ions of alkali and alkali-earth metals; the second one, ions Si^{4+}, P^{5+}, B^{3+}, Ti^{4+} etc. A high proportion of rigid directed covalent bond conditions stabilize complex ions, e.g., tetrahedra SiO_4^{4-}, PO_4^{3-} etc. in a melt. These tetrahedra form a complex anion skeleton of a melt within which covalent bonds prevail, and without, ionic ones. The charge of the complex central ion depends on the force of the field of the ions surrounding the complex. By its free bonds the oxygen of the complexes is bound with the modifying ions present in a melt, the ionic character of these bonds conditioning a relatively high mobility of modifiers. The mobility of ions taking an intermediate position, such as Al^{3+}, Fe^{3+} etc., depends on their coordination [310, 342, 343].

Depending on the ratio of O/Si tetrahedra $[SiO_4]^{4-}$ may form various polytetrahedric configurations in which part of tetrahedra $[SiO_4]^{4-}$ can be replaced by $[AlO_4]^{5-}$ tetrahedra, and possibly $[FeO_4]^{5-}$ as well [335].

An interesting problem is that of the form of anions most frequently occurring in melts. Highly wide-spread are complex silicooxygen anions of $Si_xO_y^{z-}$ type of diverse complexity.

In case O/Si = 2 the lattice is an infinite three-dimensional network consisting of silicooxygen SiO_4 tetrahedra with common apices. Each atom of oxygen in an apex is connected with two atoms of silicon posited in the centres of adjacent tetrahedra. Thus, all the valences of O and Si are fully mutually saturated.

The crystal $(SiO_2)_\infty$ dissociates at melting to complexes $(SiO_2)_n$ which evidently do not bear excessive charges. Solomin [585] evaluated the mean dimension of an n complex by the temperature dependence of the silica viscosity. Since in silicate melts besides SiO_2, there is a certain amount of other oxides, the ratio of O/Si is generally greater than 2. When it reaches 2.5 in the lattice there are flat networks of tetrahedra of $(Si_2O_5^{2-})_\infty$ infinitely spreading in two dimensions (layer structure). In this structure three oxygen atoms of the tetrahedron are bound to two atoms of silicon, the fourth one belonging to one silicon atom only and because of this bears a charge.

The reaction of the formation of a complex

$$n\,SiO_2 + n\,O^{2-} = (Si_2O_5)_n^{2-}$$

turns out to be possible owing to a great affinity of SiO_2 to O^{2-}, which provides the addition of an ion of oxygen, brought by metal oxides to silica.

Complex anions consisting of two interconnected tetrahedron chains infinitely extended in the same direction correspond to O/Si ratio = 2.75. They are characteristic of amphibole minerals and have the form of $(Si_4O_{11}^{6-})_\infty$. They can be considered as pieces of an infinite two-dimensional network torn by additionally introduced O^{2-} anions.

A further increase of the concentration of oxygen particles brings about the formation of (SiO_3^{2-}) unidimensional chain complexes, in which O/Si = 3. In them each silicon atom is bound with two binding oxygen atoms, belonging to adjacent tetrahedra as well as with two single-charged oxygen atoms, procuring bonds between an anion and cations.

A case of the formation of comparatively small circle-like anions may also occur as a result of thermic disintegration of chain anion to separate links, on the one end of which is silicon and on the other oxygen. These then curve and interconnect at the end parts with oxygen and silicon.

$(Si_3O_9)^{6-}$, $(Si_4O_{12})^{8-}$ and $(Si_6O_{18})^{12-}$.

However there is no proof of such structures being present in a melt.

Addition of new portions of O^{2-} results in chemical disintegration of anions $(SiO_3)^{2-}$ to simpler complexes $(SiO_7)^{6-}$.

In these complexes $O/Si = 3.5$. They represent the simplest polymerized silicooxygen anions, arising from two linked tetrahedra. Of the seven oxygen atoms only one performs linkage functions here whereas the rest bear one negative charge each providing for the bond with cations.

The maximum breaking of silicooxygen aggregates is reached when a silicon atom is surrounded by four single-charged oxygen atoms. This takes place in a SiO_4^{4-} orthosilicate anion where $O/Si = 4$.

SiO_4^{4-} anions possessing the smallest dimensions have the greatest specific charge (per Si atom). This provides for its high potential and bond strength in diverse compounds. In this respect it is significant that the orthosilicates of alkali-earth metals have a higher melting temperature and more strongly pronounced maximums on the phase diagram than metasilicates (Table 13).

Table 13. Melting temperatures and maximum characters of ortho- and metasilicates of alkali-earth metals

System	Melting temp. [°C]		Maximum character	
	Ortho-silicate	Meta-silicate	Ortho-silicate	Meta-silicate
$MgO-SiO_2$	1890	1557	Pointed	Open
$CaO-SiO_2$	2130	1540	Pointed	Flatly sloping
$SrO-SiO_2$	1800	1580	Pointed	Flatly sloping
$BaO-SiO_2$	1800	1604	Pointed	Pointed

The existence of all the above anions in silicate melts is proven by numerous data and, in particular, by cryoscopic observations, as well as by various "composition-property" diagrams.

1.5.2 Thermodynamics of Melts

The thermodynamics of silicate melts has satisfactorily been studied only for basic slags.

The thermodynamic properties of acid melts rich in silica, to which most glasses are to be related, have been described insufficiently for lack of a comprehensive melt theory [51, 109, 315, 539, 575, 705]. In order to calculate the thermodynamic properties of silicate melts, one should work out a quantitative theory reflecting the dependence between the activity and concentration of each component. At one point the theory of ideal – perfect – and regular melts was applied to the thermodynamics of silicate melts (see the works by Yesin and his fellow workers [60, 294, 588, 589, 704–706, 711]).

Perfect melts, naturally, differ from actual ones, because such melts are theoretical: their components differ only slightly from one another, i.e., they are close by their chemical nature (e.g., Fe – Ni or FeO – MnO melts). Perfect melts, while being formed, do not change their volume and give no heat effect ($\Delta H = 0$ and $\Delta V = 0$).

For electrolytes Tyomkin [651] suggested a peculiar standard, a perfect ionic solution.

Unlike an ideal solution in a perfect ionic one the sequence of ions is restricted by their charges. Consequently, in a perfect ionic solution there are two groups of equivalent particles: cations and anions.

The heat of mixing at the formation of a perfect ionic solution equals zero, since the place of certain cations is occupied by others energically equivalent to them and certain anions are replaced by others equivalent to them in interaction.

Thus the energy and heat content of a solution are the sum of the energy and heat contents of its components. Since cations cannot replace anions, the entropy of mixing will differ from that of an ideal – diluted – solution; it will obviously be the sum of such entropies calculated for cations and anions apart. As the mixing heat equals zero, the activity will be determined as a derivative of the number of particles but with respect to the entropy of mixing.

Let us concentrate on the thermodynamic activity as a value linking the properties of actual solutions (and consequently, melts). The activity has been chosen as characteristic, by which one can predict all the remaining properties of solutions for the following reasons:

1. through the activity all thermodynamic properties of solutions are expressed;

2. in extreme instances (for ideal solutions) the activity coincides with concentration, and in actual solutions it emphasizes the degree of the deviation from ideal ones;

3. by means of the activity the form of equations expressing the laws of ideal solutions is retained.

The relationship

$$a_i = p_i/p_i^0 \tag{92}$$

allows the determination of the activity of a given component in a solution as a ratio of the vapour pressure of this component over the solution to its vapour pressure in a standard state. The activity is proportionate to the vapour pressure (p_i^0 is a constant value) thus characterizes the tendency of a component to leave the phase in question.

For solutions close to perfect ones a pure component for which $a = 1$ is assumed to be a standard state. For perfect solutions the activity is equal to molar fraction. If the behaviour of a solution [264] deviates from a perfect one, then $a_i \neq N_i$. The degree of this deviation is determined by the value of the activity coefficient (γ_i):

$$a_i/N_i = \gamma_i, \tag{93}$$

where N_i is the molar fraction of the component i.

In actual solutions the activity coefficient may be both less than one – negative deviations – and greater than one – positive deviations –.

Experimental methods of determining activity are restricted to the study of diverse equilibria with solutions taking part. Given the component activity in some phase submitted to simple laws, one can find its activity in an actual solution which is in equilibrium with this phase.

As stated above all the calculations concerning silicate melts are conducted based on the theory of perfect ionic and regular solutions. Outstanding success is achieved in calculations of liquid slags (basic in particular). The peculiarities of the entropy of the ionic melt mixing can be seen in the value of the activity as well. The ion activity in a melt turns out to be equal to the ratio of the number of ions in question to the total number of ions of the same sign. Activity, thus, coincides with the ion fraction calculated on the presumption that the melt consists of two independent ideal solutions [267].

In a perfect ionic solution unlike a diluted one the sum of all the ionic fractions equals not one, but two. Hence, the activity calculation can be carried out by the data of a chemical analysis of the melt.

The theory of perfect ionic solutions is of great value for, similar to that of ideal solutions, it enables one to establish the degree to which actual ionic melts deviate from perfect ones and to demonstrate the main causes of these deviations [707].

The deviation of the experimental data from the theory was shown to grow with the increase of silica contents in a melt.

Experience gained from the development of the electrolyte solution theory shows that one should consider not only the difference of the bonding forces but also the consequence of this difference, the peculiarities of the ion arrangement in space (e.g., ion associations). Thus, the energy equivalence of ions must result in an additional particle ordering in a melt as compared with perfect ionic solutions. Thus, in a liquid slag the Fe^{2+} and O^{2-} interaction energy considerably surpasses that of all other ion couples (e.g., Ca^{2+} and O^{2-}, Ca^{2+} and SiO_4^{4-} etc.). For this reason Fe^{2+} and O^{2-} will most often be in proximity to one another forcing weakly interacting ions to form groups of their own [148].

Thus, the interaction difference results in such an additional ordering at which in a melt comparatively large cybotaxis groups appear containing mainly highly interacting or weakly interacting ions. Within these groups temporary associations of two or several ions, i.e., quasimolecules and complex ions may exist.

In summarizing the above facts one can view silicate melts as representing a microheterogeneous electrolyte. As a measure of deviation of actual melts from perfect ionic solutions activity coefficients γ are introduced. Based upon the difference of ion interaction as well as on the microheterogeneity resulting from it, one can explain these deviations.

Of great interest is a theoretical deduction of an equation for the activity coefficients of the melt components. In order to bring it about it is necessary to formulate an expression either for the melt free enthalpy G or for its alteration ΔG when the substances constituting the melt are mixed. In their turn the two values that form ΔG (ΔH and $T\Delta S$) can be examined separately.

Earlier Yesin and Kozheurov [702, 707] considered only the second part of the term which comprises the so-called "mixing entropy" ΔS. When doing this they examined the simplest slag composed of FeO and Ca_2SiO_4. In such a microheterogeneous melt the presence of only two types of cybotaxes was assumed for the purpose of simplicity. In one of them Fe^{2+} and O^{2-} prevailed whereas in the other Ca^{2+} and SiO_4^{4-}. Therefore the heat of mixing ΔH is to be comparatively small here and it may be neglected at first approach. Attention was paid mainly to the microheterogeneity of the melt, to the uneven distribution of anions among cations and to the spatial orderliness of particles. The arrangement of anions SiO_4^{4-} predominantly around ions Ca^{2+} is conditioned not by a greater but by a smaller bond strength of SiO_4^{4-} with Ca^{2+}, than that of Fe^{2+} with O^{2-}.

To emphasize the microheterogeneity while computating ΔS not only transpositions of ions of different signs were not permitted – the theory of perfect solutions – but transpositions of ions of the same sign were restricted as well.

The equation deduced on such premises was quite complicated and contained the so-called "inequality coefficient" of the distribution of anions among cations. When a suitable value of this coefficient is chosen, one can correspond the calculated FeO activity coefficient to the observed one.

Later Kozheurov [281, 282] took a different approach formulating an expression for ΔH, but removing all additional restrictions on the particle transposition, i.e., the calculated ΔS neglecting the microheterogeneity of the melt.

Such an approach fully corresponds with the theory of regular solutions and according to the reasons forwarded by the author the values of chemical potentials of binary systems thus obtained differ little from those found with regard to the uneven distribution of particles.

The difficulty consists in revealing to what extent ΔH and $T\Delta S$ effect the ions activity coefficient in melts, as in silicate melts the differences of interaction energies and the resulting microheterogeneity of the solution are quite substantial.

1.5.3 The Application of the Theory of Regular Solutions

Regular solutions are as is known those, whose formation follows a heat effect and the ΔS value is the same as that of perfect ones.

Let us consider Kozheurov's calculation on the simplest example of a melt containing FeO and SiO_2 with mole fractions N_1 and N_2 respectively.

By E_{11} we shall denote the bond energy of a Fe atom with all the surrounding O atoms when in the next layer only Fe atoms are, and by E_{12} – the same value when the layer in question consists exclusively of Si atoms. Similar values for silicon will be E_{22} and E_{21}.

Supposing that the mean interaction energy can be calculated by the mixing rule for a Fe atom we obtain:

$$E_1 = N_1 \cdot E_{11} + N_2 \cdot E_{12},$$

and for a Si atom:

$$E_2 = N_1 \cdot E_{21} + N_2 \cdot E_{22}.$$

The total energy of a mole of a melt expresses the sum of the energies of all the Fe and Si atoms, i.e.,

$$U = N(N_1E_1 + N_2E_2),$$

where N is the Avogadro number.

Or, on appropriate mathematical transformations:

$$U = N[(N_1E_{11} + N_2E_{22}) + N_1N_2(E_{12} + E_{21} - E_{11} - E_{22})].$$

The values of $U_1 = NE_{11}$ and $U_2 = NE_{22}$ may be viewed as the energies of pure FeO and SiO_2 and the value of

$$Q_{12} = N(E_{12} + E_{21} - E_{11} - E_{22})$$

is the mixing energy. Therefore,

$$U = N_1 U_1 + N_2 U_2 + N_1 N_2 Q_{12}.$$

The entropy of a mole of melt will be

$$S = N_1 S_1 + N_2 S_2 + R(N_1 \ln N_1 + N_2 \ln N_2).$$

Substituting U and S into the free energy expression we find

$$F = N_1 F_1 + N_2 F_2 + RT(N_1 \ln N_1 + N_2 \ln N_2) + N_1 N_2 Q_{12}. \quad (94)$$

Here $F_1 = U_1 - TS_1$ and $F_2 = U_2 - TS_2$ can be viewed as molar free energies of pure components. The total free energy of a melt containing n_1 FeO moles and n_2 SiO_2 moles will be:

$$F' = (n_1 + n_2)F = n_1 F_1 + n_2 F_2 + RT(n_1 \ln N_1 + n_2 \ln N_2) + \frac{n_1 n_2}{n_1 + n_2} Q_{12}. \quad (95)$$

Consequently, the chemical potential of some component, e.g., FeO:

$$\mu_1 = \frac{\partial F_1}{\partial n_1} = F_1 + RT \ln N_1 + N_2^2 Q_{12} \quad (96)$$

and the activity coefficient:

$$\ln \gamma_1 = \frac{N_2^2 Q_{12}}{RT}. \quad (97)$$

In its form this expression is similar to the equation of the theory of regular solutions for a binary mixture.

The conclusion can be also easily applied to a slag, consisting of K components – oxides –.

In this case the formula will be expressed as:

$$RT \ln \gamma_l = \sum_{i=1}^{l-1} N_i Q_{il} + \sum_{i=l+1}^{k} N_i Q_{li} - \sum_{i=1}^{k-1} \sum_{j=i+1}^{k} N_i N_j Q_{ij}, \quad (98)$$

where γ_l, activity coefficient of the l component; Q_{ij}, Q_{il}, Q_{li}, mixing energies of the corresponding component couples; $N_l = \frac{\nu_i n_i}{\sum_{i=1}^{k} \nu_i n_i}$, an oxide mole fraction and at the same time the non-oxygen element atomic fraction calculated like an ionic fraction in a perfect solution; ν_i, number of atoms of a non-oxygen element in a chemical formula of an oxide (e.g., $\nu = 2$ for P_2O_5); n_i, number of oxide moles in a slag.

Kozheurov obtained numerical values of FeO, MnO and P_2O_5 activity coefficients fitting in the values of mixing energies. His data agree with the experimental ones for the distribution of oxygen between a metal and a slag as well as for the reactions manganese and phosphorus [282].

Interpreting his results qualitatively, Kozheurov [285] infers that situations of Si – O – Ca are more probable (mixing energy is 27 kcal), i.e., they occur more often than those of Si – O – Fe. This proves anions SiO_4^{4-} to neighbor more often cations Ca^{2+}. The great absolute value assumed for the mixing energy of CaO with P_2O_5 (48 kcal) similarly shows that PO_4^{3-} anion is drawn by ions O^{2-} from cations Fe^{2+} to those of Ca^{2+}.

In other words anions SiO_4^{4-} and PO_4^{3-} are more stable in the proximity of cations Ca^{2+} and not of Mn^{2+} and Fe^{2+}.

As a result of Kozheurov's calculations the quantitative theory of basic slag melts was advanced to the present theory of regular solutions.

Unfortunately, the relations deduced are not applied to silicate melts, although the calculations of Zhukhovitskiy, Finkel'shteyn and Kulikov [718] as well as those of Kozheurov [283] who used Frenkel's theory [140] permit elaborating a statistical theory of actual solutions and silicate melts.

Let us examine particular instances of the most thoroughly investigated systems.

1.5.4 Binary Systems

The systems MeO – SiO_2 have been studied in detail.

In the simplest binary silicate systems of MeO – SiO_2 type, sign-changing deviations from ideal solutions are observed. Besides negative deviations – formation of chemical compounds – at mean silica concentrations, separation phenomena are also observed, i.e., considerable positive deviations of the melt behaviour at high SiO_2 content.

Attempts at giving semiempirical formulae appropriate for describing systems of that kind have been made by Kulikov [316] and Kozheurov [284 – 286]. The latter neglects the short-range effect on the mixing entropy and introduces two heat effects. One of them ($Q_{12}N_1N_2$) is determined by the substitution of Me – O – Si bond for those of Si – O – Si and Me – O – Me and implies Si – O – Si groupings related to a random cation distribution. The second heat effect results from polymerization, i.e., forced rise of Si – O – Si bonds owing to the lack of oxygen ions. It is proportional to the product of the concentrations of the cation Me^{2+} (N_1) and the anions O^2 bound with two silicon ions ($3N_2 - 1$), i.e., $qN_1(3N_2 - 1)^2$. By fitting in the Q and q values one manages to describe fairly well the limits of the SiO_2 separation and solubility in MeO – SiO_2 melts.

Kozheurov's semiempirical equation for the free energy of acid silicate melts can be expressed as follows:

$$G = \sum_{i}^{k} N_i G_i + RT \sum_{i}^{k} N_i \ln N_i + \sum_{i=1}^{k=1} \sum_{j=i+1}^{k} N_i Q_{ij} + (3N_i - 1)^2 \sum_{i=1}^{k=1} N_i q_i , \qquad (99)$$

where k, number of components in a melt; N_i, ionic fractions of cations; Q_{ij}, heat of the component mixing; q_i, heat of polymerization.

This equation can easily yield expressions for chemical potentials and activities of components in acid silicate melts and can therefore be applied to calculating various equilibria in slags and glasses.

1.5.4.1 Equations of the Gibbs Free Energy and of Activity Coefficients

Let us examine a solution containing atoms A and B the molar fractions whereof are N_1 and N_2. The interaction energies of groupings AA, BB, AB, AB_2 and A_2B will equal U_{11}, U_{22}, U_{12}, U_{122}, U_{112} respectively. At first approach suppose the probability of AB_2 and A_2B combinations, unaccounted for in regular solutions, to be proportional to $N_1N_2^2$ and $N_1^2N_2$. Then the energy of a mole of the solution (now in AB interactions not N_1N_2 groupings, but fewer take part since part of AB combinations has been replaced by more complicated structures AB_2 and A_2B) is

$$E = N[U_{11}N_1^2 + U_{22}N_2^2 + 2U_{122}k_1N_1N_2^2 + 2U_{112}k_2N_1^2N_2 + 2U_{12}(N_1N_2 - k_1N_1N_2^2 - k_2N_1N_2^2 - k_2N_1^2N_2)] .$$

Further supposing

$$NU_{11} = E_1 , NU_{22} = E_2 ; N(2U_{12} - U_{11} - U_{22}) = Q_{12} ,$$

$$2Nk_1(U_{122} - U_{12}) = Q_{122} \quad \text{and}$$

$$2Nk_2(U_{112} - U_{12}) = Q_{112}$$

we derive

$$E = E_1N_1 + E_2N_2 + Q_{12}N_1N_2 + Q_{122}N_1N_2^2 + Q_{112}N_1^2N_2 . \qquad (100)$$

In accordance with Kozheurov [285] the same Eq. (100) can be shown to be valid for a binary solution with a common ion C(AC + BC) as well. The difference will show up however in the expressions for the values of Q_{12}, Q_{122}, Q_{112}.

In a silicate melt Q_{12} accounts for the energy effect when bonds Me – O – Me and Si – O – Si are replaced by bonds Me – O – Si. Coefficients Q_{122} and Q_{112} accordingly correspond to the energy released during the substitution of ortho- and disilicate bonds for metasilicate ones.

Supposing that the mixing entropy does not differ from that of ideal solutions we find the excessive Gibbs free energy $\Delta G'$

$$\Delta G' = Q_{112}N_1^2N_2 + Q_{12}N_1N_2 + Q_{122}N_1N_2^2 .$$

Substituting Q for $Q_{112}+Q_{12}$ and q for $Q_{122}+Q_{12}$ in it we obtain

$$\Delta G' = QN_1^2N_2 + qN_1N_2^2 . \tag{101}$$

Employing Eq. (101) we obtain by an ordinary thermodynamic method the following expressions for the activity coefficients:

$$RT\ln\gamma_1 = (2Q-q)N_2^2 + (2q-2Q)N_2^3 , \tag{102}$$

$$RT\ln\gamma_2 = (2q-Q)N_1^2 - (2q-2Q)N_1^3 . \tag{103}$$

Equations (102) and (103) were derived by Sryvalin, Yesin and Khlynov [589] using Kozheurov's calculation technique. Relations such as in Eqs. (102) and (103) in the form of semiempirical equations were also suggested earlier [564]. Unique to them is the definite physical sense of coefficients Q and q allowing clear interpretation of quite a number of systems that are not embraced by the equations of regular solutions, silicate in particular. The chief difference of Eqs. (102) and (103) from those derived by Kozheurov [284–286] consists in values Q and q having another physical meaning. Here parameter Q emphasizes the energic effect connected with the formation of orthosilicate groupings from pure components whereas q emphasizes the rise of disilicate structure as well as (to a certain degree) to the effect of polymerization of SiO_2.

On the basis of Eqs. (102) and (103) Sryvalin, Yesin and Khlynov described several silicate systems.

1.5.4.2 System $FeO-SiO_2$

For determining parameters Q and q phase diagrams of $FeO-SiO_2$ [315] were used. At a temperature of 1700 °C separation of the liquid was observed. The equating of the chemical potentials of each constituent in both coexistent phases yielded:

$$Q = -2700\ \text{cal} \quad \text{and} \quad q = 11000\ \text{cal} .$$

Thus for the temperature of 1600 °C:

$$\log\gamma_1 = -1.92N_2^2 + 3.2N_2^3 , \tag{104}$$

$$\log\gamma_2 = 2.88N_1^2 - 3.2N_1^3 , \tag{105}$$

$$\frac{\Delta G'}{2.3RT} = -0.32N_1^2N_2 + 1.28N_1N_2^2 . \tag{106}$$

Examining Eqs. (104) – (106) we can expect negative deviations for FeO and positive ones for SiO_2 in the homogeneous area. The excessive Gibbs free energy of a melt changes the sign; when the silica mole fraction is close to 0.2 (at $N_2 < 0.2\ \Delta G' < 0$ and at $N_2 > 0.2\ \Delta G' > 0$).

In accordance with Kozheurov's equations [284, 286, 316] which are valid for $N_2 \geqslant \frac{1}{3}$.

$$RT \ln \gamma_1 = q(3N_2 - 1 - 6N_1N_2)(3N_2 - 1)\,, \tag{107}$$

$$RT \ln \gamma_2 = 6q(3N_2 - 1)N_1^2\,. \tag{108}$$

In FeO – SiO_2 melts there are negative deviations for FeO and positive ones for SiO_2. The negative deviations of FeO at moderate concentrations of SiO_2 follow from the formula suggested by Kulikov [316]:

$$\log \gamma_1 = \frac{3217}{T} N_2^2 - 2.61 \left(1 - \frac{2805}{T} - e^{-2470/T}\right)$$
$$- 0.435 N_2 \left(1 - \frac{1}{6.55 N_2} 1.71\right). \tag{109}$$

Table 14 comprises the values of FeO and SiO_2 activities derived from Eqs. (104), (105) and (107) – (109). They demonstrate that the values obtained by Eqs. (104) and (105) agree with the results of other authors quite well.

Table 14. FeO and SiO_2 activities at 1600 °C

N_{FeO}	FeO activities derived from Eqs.			SiO_2 activities derived from Eqs.		N_{FeO}	FeO activities derived from Eqs.			SiO_2 activities derived from Eqs.	
	(104)	(107)	(109)	(105)	(108)		(104)	(107)	(109)	(105)	(108)
1.00	1.00	1.00	1.00	0.00	0.00	0.67	0.54	0.67	0.35	0.68	0.33
0.95	0.94	0.95	0.77	0.04	0.05	0.60	0.47	0.51	0.34	0.84	0.52
0.90	0.86	0.90	0.62	0.10	0.10	0.55	0.44	0.44	0.32	0.90	0.66
0.80	0.71	0.80	0.45	0.30	0.20	0.50	0.42	0.37	0.33	0.98	0.79
0.70	0.58	0.70	0.37	0.57	0.30	0.45	0.41	0.32	–	1.00	0.80

The evaluation of coefficients Q and q by means of phase diagrams is only approximate. It is more advisable to find them by two FeO or SiO_2 activities experimentally determined. However the experimental data available are contradictory [93, 129, 553, 708, 709, 710]. Based on the above we may assume that at

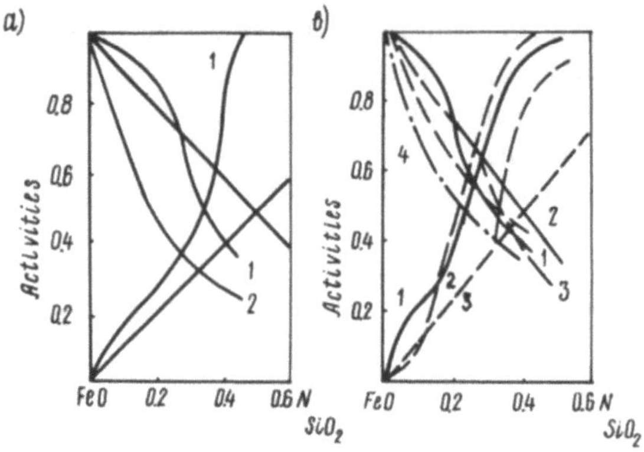

Fig. 8. The dependence of melt activity in the system $FeO-SiO_2$ on the mole fraction of SiO_2 determined by various means at temperatures: **a** from 1300 to 1400 °C: *1* by [652]; *2* by [711]; **b** 1600 °C: *1* by experimental data [129]; *2, 3* by Eqs. (104) and (105); *4* by [93, 709]

temperatures higher than FeO melting point (1380 °C) slight sign-changing deviations from ideal solutions are characteristic of the $FeO-SiO_2$ system (Figs. 8 and 9). They are connected with the change of the character of Fe, O and Si atoms ordering in a transition from high-basic melts to more acid ones. In effect the mixing heat is not large, not exceeding several kcal/mol.

1.5.4.3 System $PbO-SiO_2$

Equations (110) and (111) correspond to Eqs. (102) and (103) and may be expressed as:

$$\log \gamma_1 = -N_2^2 + 2N_2^3\,, \tag{110}$$

$$\log \gamma_2 = 2N_1^2 - 2N_2^3\,. \tag{111}$$

Equations (110) and (111), as seen from Fig. 10, describe fairly well the behaviour of the components of the mentioned melts. Therefore the excessive Gibbs free energy of mixing can be evaluated with the help of relation:

$$\frac{\Delta G'}{2.3RT} = N_1 N_2^2\,. \tag{112}$$

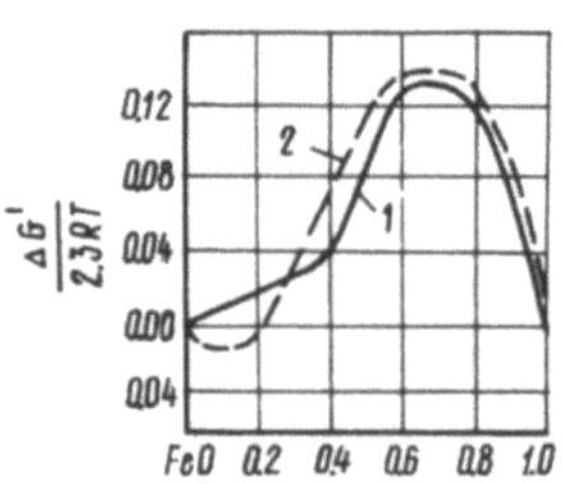

Fig. 9. The dependence of $\Delta G^0/2.3RT$ on the SiO_2 mole fraction of melts in the system $FeO-SiO_2$: *1* by experimental data [129]; *2* by Eq. (106)

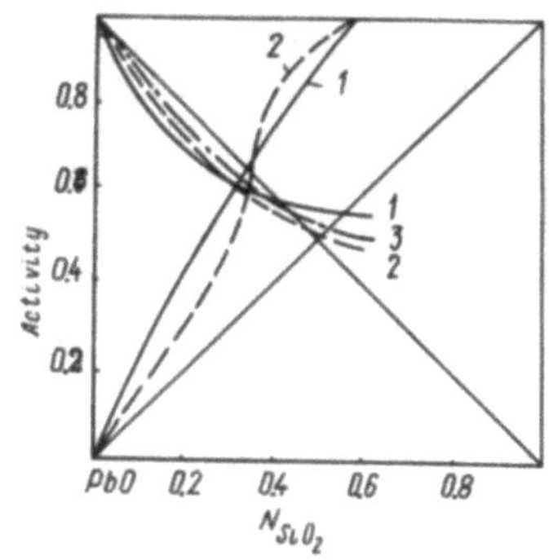

Fig. 10. The dependence of the melt activity in the system $PbO-SiO_2$ on the SiO_2 mole fraction: *1* and *3* by [710]; *2* by Eqs. (110) and (112)

In the system under consideration in compounds with a high PbO content positive and negative and in compounds with a high SiO_2 content, only positive deviations from ideal solutions are observed. Mixing heat $Q=0$ and a small value of q testifies to small deviations from an ideal state.

All this agrees with the shape of the $PbO-SiO_2$ phase diagram: the Schröder equation is applicable to the description of liquidus curves, it has flatly sloping maximums for chemical compounds, low melting temperatures and no separation in the liquid state [60, 315, 515, 710].

1.5.4.4 Systems $CaO-SiO_2$ and $MgO-SiO_2$

The thermodynamic characteristics of the system $CaO-SiO_2$ were studied experimentally by a number of authors [93, 129, 461, 708–710]. In accordance with Eqs. (110)–(112) for the present system we obtain:

$$\log \gamma_1 = -11.25N_2^2 + 11.78N_2^3, \quad (113)$$

$$\log \gamma_2 = 6.42N_1^2 - 11.78N_1^3, \quad (114)$$

$$\frac{\Delta G'}{2.3RT} = -5.36N_1^2N_2 + 0.53N_1N_2^2. \quad (115)$$

Parameters Q and q were found by the phase diagram of $CaO-SiO_2$ [315]. The comparison of the experimental data [129] as far as component activities and $\Delta G'$ are concerned with those calculated by Eqs. [283, 316, 320] proves that they satisfactorily agree. As one can see from Fig. 11, in the system $CaO-SiO_2$ negative deviations preponderate. The activities of the melts of the system $CaO-SiO_2$ were also determined by Kay and Taylor [251] on the basis of the equilibrium pressure of the reaction $SiO_2 + 3C = SiC + 2CO$.

Unfortunately reliable experimental data on the $MgO-SiO_2$ melts are lacking. Using the values of the separation limits at 1700 °C one obtains the following relations for this system at a temperature of 1600 °C:

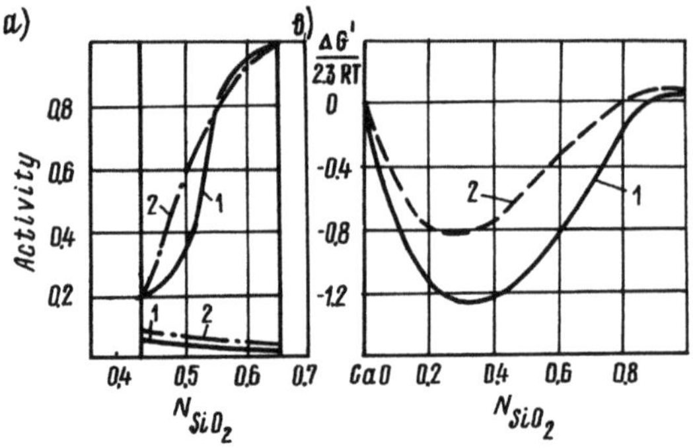

Fig. 11. The dependence of **a** the melt activity; **b** $\Delta G^0/2.3RT$ of melts: *1* according to [129]; *2* according to Eqs. (113)–(115) on the SiO_2 mole fraction in the system $CaO-SiO_2$

$$\log \gamma_1 = -8.88N_2^2+9.64N_2^3, \quad (116)$$

$$\log \gamma_2 = -5.58N_1^2-9.64N_1^3, \quad (117)$$

$$\frac{\Delta G'}{2.3RT} = -4.06N_1^2N_2+0.76N_1N_2^2. \quad (118)$$

Table 15. The separation limits for the systems $MeO-SiO_2$ at a temperature of about 1700 °C and the values of parameters Q and q

System	Separation range [molar %]		Values of parameters [kcal/mol]	
	Phase I	Phase II	Q	q
$CaO-SiO_2$	70.6	99.3	−46	4.5
$MgO-SiO_2$	62.0	98.8	−35	6.5
$FeO-SiO_2$	64.2	97.0	−2.7	11.0
$PbO-SiO_2$	–	–	0	5.8

In Table 15 the experimentally found compositions of coexistent liquid phases at 1700 °C and the calculated values of Q and q are presented.

According to the physical sense of coefficients Q and q in the systems presented to a more negative value of Q a greater stability of ortho- and metasilicates results (e.g., Ca_2SiO_4 and $CaSiO_3$ as compared to Mg_2SiO_4 and $MgSiO_3$) and a higher solubility of SiO_2 in the oxide CaO than in that of MgO (i.e., at the value of q of the same order).

For less stable silicates ($FeO-SiO_2$, $PbO-SiO_2$) the Q value is close to zero and the appearance of Me_2SiO_4 and $MeSiO_3$ at concentrations exceeding a random distribution of Me and Si ions close to oxygen ions practically does not occur.

The positive values of q show the formation of $MeSi_2O_5$ in these melts to be energetically inefficient.

The decrease of q from 11 kcal/mol for $FeO-SiO_2$ down to 4.5 kcal/mol for $CaO-SiO_2$ corresponds to the decrease of the disilicate instability.

With the parameters of Q of the same order the SiO_2 solubility in the melts of $MeO-SiO_2$ reduces with the growth of q.

1.5.5 Ternary Systems

As stated above basic slags are the most studied of silicate melts. Let us consider the systems $FeO-CaO-SiO_2$ and $CaO-Al_2O_3-SiO_2$. Their thermodynamic properties were investigated by Sryvalin and Yesin [587]. In the melts $CaO-Al_2O_3-SiO_2$ and less markedly $CaO-FeO-SiO_2$, tests show sign-reversing deviations from ideal solutions, the positive deviations that result in separation being conditioned by polymerization. Contrarily, negative deviations large enough characterize the existence of certain compounds. Positive deviations for ferrous oxide observed in the melts of $FeO-CaO-SiO_2$ also result from the energic inequivalence and microheterogeneity of a melt conditioned by it.

In order to reflect the presence of a certain structure in a melt it is necessary to pass from triatomic models over to more complicated ones. One should bear in mind the next adjacent element. Thus, Kozheurov describes the phenomenon of SiO_2 polymerization adopting two mixing energies. The one corresponds to the random distribution of Me and Si cations around oxygen atoms, the other, to a forced combination of $Si-O-Si$ caused by the lack of oxygen ions. According to the theory of regular solutions the molar heat of mixing for a ternary solution equals:

$$\Delta H = Q_{12}N_1N_2 + Q_{13}N_1N_3 + Q_{23}N_2N_3 .$$

Evaluating the effect of the energy inequivalence of particular nodes of similar atoms leads to a relation:

$$\Delta H' = Q_{12}N_1N_2 + q_{12}N_1N_2^2 + Q_{13}N_1N_3 + q_{13}N_1N_3^2 + Q_{23}N_2N_3 + q_{23}N_2N_3^2 . \tag{119}$$

Here the Q values are the same as in regular solutions whereas q deviate from them stemming from structural peculiarities.

Presuming the mixing entropy to be the same as for an ideal solution we will obtain an expression for the excessive Gibbs free energy similar to Eq. (119) from which the formulae for activity coefficient in a ternary system can be obtained. For the first component it may be expressed as:

$$\begin{aligned}\lg \gamma_1 = {} & Q_{12}N_2(1-N_1) + q_{12}N_2^2(1-2N_1) + Q_{13}N_3(1-N_1) \\ & + q_{13}N_3^2(1-2N_1) - Q_{23}N_2N_3 - 2q_{23}N_2N_3^2 .\end{aligned} \tag{120}$$

1.5.5.1 System $FeO-CaO-SiO_2$

For determining parameters Q and q the data for binary systems [93, 129, 553, 708–710] were used. $FeO-SiO_2$ and $CaO-SiO_2$ melts have been examined above. As far as $CaO-FeO$ melts are concerned, they comply fairly well with the laws for regular solutions according to Darken [109]. The experimental data obtained by the methods of distribution and electromotive force and the data found by the formulae of Eqs. (119) and (120) type satisfactorily agree.

Based on corrections to the formulae of regular solutions Sryvalin and Yesin suggested a formula of the activity coefficient of ferrous oxide at a temperature of 1600 °C.

$$\lg \gamma_1 = -0.81 N_2(1-N_1) - 0.2 N_3(1-N_1) - 0.5 N_3^2(1-2N_1) + 10 N_2 N_3 + 20 N_2 N_3^2 . \qquad (121)$$

It is noteworthy that the deviation of the calculated values from those obtained by the distribution method is regularly negative (lower than the experimental ones). It is evidently to be accounted for by the fact that Eq. (121) ignores the effect of Fe_2O_3 on the iron activity.

1.5.5.2 System $CaO-Al_2O_3-SiO_2$

Although the deviation of $CaO-Al_2O_3-SiO_2$ melts from ideal ones was experimentally determined by several authors (Chang, Derge, Yesin and Lepinskikh, Byrd, Taylor, Fulton and Chipman), the area of melts studied is rather limited and a quantitative evaluation of the component activities rarely possible. Therefore it is necessary to use the data on binary systems $CaO-SiO_2$ and $CaO-Al_2O_3$ at 1500 °C available in literature. The system $Al_2O_3-SiO_2$ has no liquid phase at the said temperature and Sryvalin and Yesin used the data for the ternary $CaO-SiO_2-Al_2O_3$ melts.

Based on the above facts they obtained the following relations for activity coefficients [8]:

$$\lg \gamma_1 = -8 N_2(1-N_1) + 7 N_2^2(1-2N_1) - 4.5 N_3(1-N_1) - 5.5 N_3^2(1-2N_1) + 4 N_2 N_3 + 0.3 ,$$

$$\lg \gamma_2 = -8 N_1(1-N_2) + 14 N_1 N_2(1-N_2) + 4.5 N_1 N_2 + 11 N_1 N_2^2 - 4 N_3(1-N_2) ,$$

$$\lg \gamma_3 = 8 N_1 N_2 - 14 N_1 N_2^3 - 4.5 N_1(1-N_3) - 11 N_1 N_3(1-N_3) - 4 N_2(1-N_3) + 2.08 .$$

[8] There are indices in the equations: $CaO-1$, SiO_2-2, Al_2O_3-3

One can thus conclude that the thermodynamic properties of the melts of the systems $CaO-Al_2O_3-SiO_2$ and $FeO-CaO-SiO_2$ can well be described by equations obtained by complicating the formulae of regular solutions. These equations as a first approximation reflect the influence of the peculiarities of the structure of melts on the mixing energy of their components.

The SiO_2 activities for the melts of the system $CaO-Al_2O_3-SiO_2$ were determined by Kay and Taylor [251] in the following range of compositions: CaO (10.8 – 49.8% by weight); Al_2O_3 (0.6 – 30.0% by weight); SiO_2 (32.0 – 69.5% by weight). The results obtained for a temperature of 1550 °C are presented in Fig. 12.

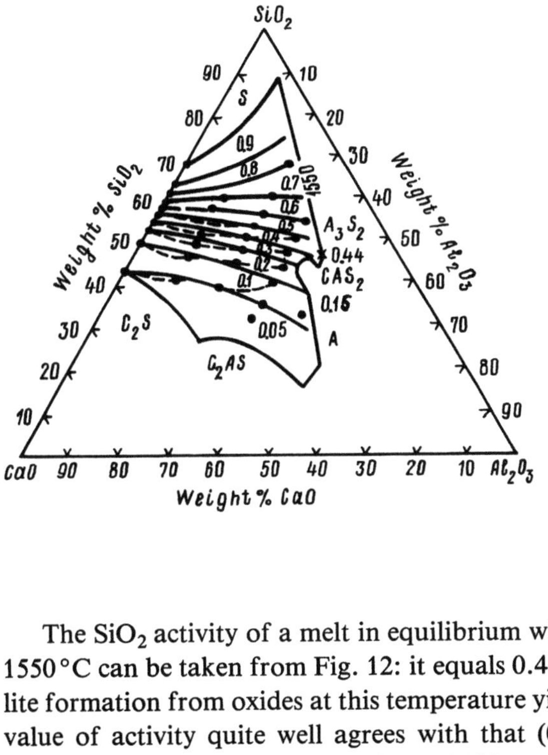

Fig. 12. Silica activities in the system $CaO-Al_2O_3-SiO_2$ at a temperature of 1550 °C

The SiO_2 activity of a melt in equilibrium with the solid mullite and Al_2O_3 at 1550 °C can be taken from Fig. 12: it equals 0.44. The calculation of ΔG of mullite formation from oxides at this temperature yields the value of −5800 cal. This value of activity quite well agrees with that (0.42) determined directly at this temperature for the reaction $3Al_2O_3 \cdot 2SiO_2 = 3Al_2O_3 + 2SiO_2$. Thus the results obtained confirm the point of view put forward by Aramaki and Roy [6], who maintain that mullite melts congruently and the mullite-alumina eutectic is shifted to the field of anorthite at the point shown on old phase diagrams of the system $CaO-Al_2O_3-SiO_2$.

The SiO_2 activities obtained were used for calculating CaO and Al_2O_3 activities at 1550 °C by Gibbs-Duhem equation.

Silica activities of melts in the systems $CaO-MgO-SiO_2$ and $MgO-Al_2O_3-SiO_2$ were determined by Henderson and Taylor [199].

2 The Thermodynamic Analysis of the Silicate Formation Reactions in Binary and Ternary Systems

The reactions of silicate formation in the systems of metallic oxides with silica play a great role in several domains of chemistry and of production of silicates. They are important, in particular in the synthesis of cements, refractories, glass ceramics and in glass manufacturing as well as in a wide variety of new industries (ceramics of pure oxides, corrosion-resistant and heat-resistant coatings etc.). Moreover, studying these reactions is no doubt of theoretical interest for establishing general rules concerning the reactions of silicate formation and their link with the position of the reactants in the periodic system. In this respect thermodynamic examination of the reactions of silicate formation in the above systems is also of concern, because it allows solving the problems of the stability of certain compounds, of mutual reactivity and of the sequence of reactions.

Below we present the results of a thermodynamic analysis of some systems of the second group of the periodic system.

2.1 System $BeO-SiO_2$

The reaction of $BeO-SiO_2$ interaction is of great importance in the technology of high-refractory materials and glass making.

A detailed investigation of the interaction of these oxides was carried out by Budnikov and Cherepanov [81].

We present a thermodynamic analysis of the reaction of Be_2SiO_4 and $BeSiO_3$ formation from oxides. The initial thermochemical data that served for the analysis were those from Table 16.

The enthalpy of $BeSiO_3$ formation from elements was calculated on the basis of the known values for Be and Mg orthosilicates and Mg metasilicate according to the previously derived Eq.:

$$\Delta H^0_{BeSiO_3} = \tfrac{1}{2}(\Delta H^0_{Be_2SiO_4} - \Delta H^0_{Mg_2SiO_4}) + \Delta H^0_{MgSiO_3}\,.$$

As one can see from Table 16, $BeCO_3$ is quite an unstable compound and its consideration is out of question. This concerns $BeSO_4$ as well which based on data of [671] dissociates at 813 K, therefore the reaction of $BeSO_4$ dissociation will not be considered either. The equations of the temperature dependence of heat capacities for $BeSiO_3$ and Be_2SiO_4 were calculated by Landiya's method.

$\Delta H^0 = f(T)$ and $\Delta G^0 = f'(T)$ for the two main reactions:

1. $BeO + SiO_2 = BeO \cdot SiO_2$
2. $2BeO + SiO_2 = 2BeO \cdot SiO_2\,,$

Table 16. The initial thermodynamic data for the analysis of solid-phase reactions in the systems $BeCO_3 - BeSO_4 - BeO - SiO_2$ and $MgCO_3 - MgSO_4 - MgO - SiO_2$

Compounds	$-\Delta H^0_{298(el)}$ [cal/mol]	$-\Delta G^0_{298(el)}$ [cal/mol]	S^0_{298} [cal/mol deg]	$-\Delta H^0_{298(ox)}$ [cal/mol]	$-\Delta G^0_{298(ox)}$ [cal/mol]	$C_p = f(T)$ [cal/mol deg]
BeO	143100	136150	3.4	–	–	$8.45 + 4.0 \cdot 10^{-3} T - 3.17 \cdot 10^5 T^{-2}$
$BeCO_3$	234600	215300	12.5	−2550	−15110	–
$BeSO_4$	286410	260200	21.5	−72350	−52310	–
$BeO \cdot SiO_2$	367980 [a]	347150 [a]	12.9	7130	6260	$22.80 + 2.08 \cdot 10^{-3} T - 10.95 \cdot 10^5 T^{-2}$
$2BeO \cdot SiO_2$	515950	488400	15.4	12000	11350	$24.63 + 5.65 \cdot 10^{-3} T - 14.05 \cdot 10^5 T^{-2}$
MgO	143840	136130	6.4	–	–	$10.18 + 1.74 \cdot 10^{-3} T - 1.48 \cdot 10^5 T^{-2}$
$MgCO_3$	266000	246000	15.7	28100	15610	$18.62 + 13.8 \cdot 10^{-3} T - 4.16 \cdot 10^5 T^{-2}$
$MgSO_4$	305500	280500	21.9	–	72580	$26.18 + 8.56 \cdot 10^{-3} T - 5.08 \cdot 10^5 T^{-2}$
$MgO \cdot SiO_2$ enstatite	370250	349480	16.2	−3660	−8670	$24.55 + 4.74 \cdot 10^{-3} T - 6.28 \cdot 10^5 T^{-2}$
$2MgO \cdot SiO_2$ forsterite	520530	492150	22.7	−15100	−15140	$35.81 + 6.54 \cdot 10^{-3} T - 8.52 \cdot 10^5 T^{-2}$

[a] Calculated by the authors

were calculated by the following equations:

$$\Delta H^0 = f(T)$$

1. $\Delta H_T^0 = -9551 - 0.06\,T - 1.93 \cdot 10^{-3} T + 7.78 \cdot 10^5 T^{-1}$,
2. $\Delta H_T^0 = 12420 + 6.68\,T + 2.145 \cdot 10^{-3} T + 7.71 \cdot 10^5 T^{-1}$.

$$\Delta G^0 = f'(T)$$

1. $\Delta G_T^0 = -9551 + 0.06\,T \ln T + 1.93 \cdot 10^{-3} T^2 + 3.89 \cdot 10^5 T^{-1} + 5.76\,T$,
2. $\Delta G_T^0 = 12420 + 6.68\,T \ln T + 2.145 \cdot 10^{-3} T + 3.57 \cdot 10^5 T^{-1} - 39.4\,T$.

The values of enthalpy and of the Gibbs free energy in the temperature range of 500 to 1800 K (within each 100 K) were calculated based on these equations. The results obtained are presented in Fig. 13.

The examination of the data obtained permits drawing the following conclusions. The formation of $BeSiO_3$ and Be_2SiO_4 is followed by a considerable heat effect, the amount of heat released increasing with the elevation of temperature.

The Gibbs free energies of the $BeSiO_3$ and Be_2SiO_4 formation increase and intersect the zero ordinate. The range of the possible formation of $BeSiO_3$ lies below 1100 K and that of Be_2SiO_4 – below 1165 K. As in this area the kinetic conditions seem to be unfavourable for the proceeding of reactions, the formation of beryllium silicates from pure oxides is essentially impossible fully agreeing with the results of Budnikov and Cherepanov's work. The calculation by compositions shows that beryllium orthosilicate will in all cases be the most stable com-

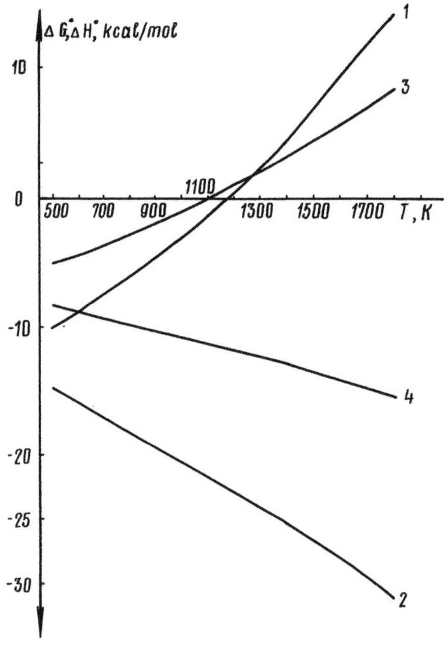

Fig. 13. The temperature dependence of ΔH^0 and ΔG^0 of the reactions in the system $BeO - SiO_2$ within a temperature range of 500 to 1800 K. *1* $\Delta G^0 = f(T)$ and *2* $\Delta H^0 = f'(T)$ for Be_2SiO_4; *3* $\Delta G^0 = f(T)$ and *4* $\Delta H^0 = f'(T)$ for $BeSiO_3$

pound. The most probable way of obtaining beryllium orthosilicate (besides the introduction of mineralizers) is the crystallization from a melt or glass, because in the range of low temperatures the ΔG value will be negative.

2.2 System $MgCO_3-MgSO_4-MgO-SiO_2$

2.2.1 System $MgO-SiO_2$

Mchedlov-Petrossyan was the first to carry out the analysis of solid-phase reactions in the system $MgO-SiO_2$ [395]; later the appearance of new data on the enthalpy of forsterite and enstatite formation resulted in our making corrections in the calculations.

$\Delta H^0 = f(T)$ and $\Delta G^0 = f'(T)$ were calculated for the main reactions of the formation of compounds $MgO \cdot SiO_2$ and $2MgO \cdot SiO_2$ existent in the system $MgO-SiO_2$ [44, 75, 219].

1. $MgO + SiO_2 = MgO \cdot SiO_2$
2. $2MgO + SiO_2 = 2MgO \cdot SiO_2$

The initial thermodynamic data for the examination we reborrowed from [242, 247, 528, 683]. The data are summarized in Table 16.

For the estimations following equations were deduced:

for $\Delta H^0 = f(T)$
1. $\Delta H_T^0 = -10318 - 0.04T + 0.53 \cdot 10^{-3}T^2 + 4.8 \cdot 10^5 T^{-1}$
2. $\Delta H_T^0 = -17330 + 1.04T + 0.56 \cdot 10^{-3}T^2 + 5.56 \cdot 10^5 T^{-1}$

for $\Delta G^0 = f'(T)$
1. $\Delta G_T^0 = -10318 + 0.4T\ln T - 0.53 \cdot 10^{-3}T^2 + 2.4 \cdot 10^5 T^{-1} + 2.8T.$
2. $\Delta G_T^0 = -17330 - 1.04T\ln T - 0.56 \cdot 10^{-3}T^2 + 2.78 \cdot 10^5 T^{-1} + 10.3T.$

By these equations the values of the enthalpy and of the Gibbs free energy over the temperature range of 298 to 1800 K (within each 100 K) were calculated. According to the data of [482] $MgO \cdot SiO_2$ melts at 1798 K; ΔH_m^0 at this temperature is 14700 cal/mol. The above effect was taken into account while calculating ΔH_T^0 for $MgO \cdot SiO_2$. The results obtained are given in Fig. 14.

For evaluating the thermodynamic sequence of the compound formation in the system $MgO-SiO_2$ the dependence of ΔG_T^0 of the reactions on the composition of the initial mixtures was examined. With this aim in view the values of ΔG_T^0 were analyzed, the ratios of the initial mixtures MgO/SiO_2 being 1/1 and 2/1. The thermodynamic stability series obtained (Table 17) permits stating that

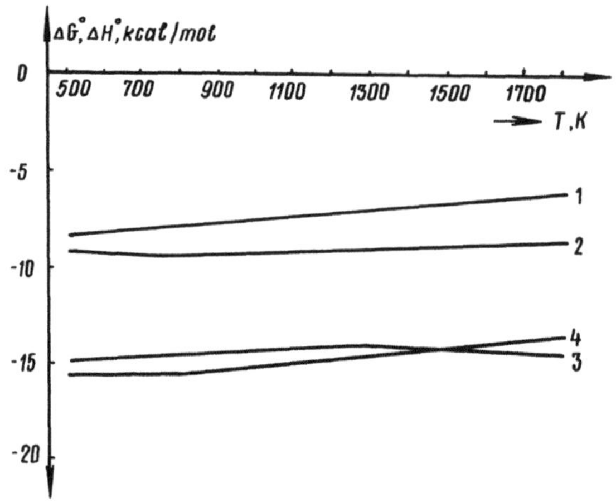

Fig. 14. The temperature dependence of ΔH^0 and ΔG^0 in the system $MgO-SiO_2$ within a temperature range of 500 to 1800 K. *1* $\Delta G^0 = f(T)$ and *2* $\Delta H^0 = f'(T)$ for $MgSiO_3$; *3* $\Delta G^0 = f(T)$ and *4* $\Delta H^0 = f'(T)$ for Mg_2SiO_4

at temperatures above 1200 K independent of MgO/SiO_2 ratio the most probable thermodynamically is the formation of forsterite. The latter fact combined with that of the primary formation of forsterite established earlier [75] allows elucidation of the results of fundamental research into the field of refractories [44].

The interaction of $MgO \cdot SiO_2$ and excess MgO, $2MgO \cdot SiO_2$ (1400 K $\Delta G^0 = -7490$ cal/mol) being formed, is quite possible in the solid phase over the entire temperature range, whereas the reaction of forsterite interaction with SiO_2 possible at low temperatures becomes thermodynamically impossible at as low a temperature as 1200 K.

Table 17. The thermodynamic stability sequence of compounds of the system $MgO-SiO_2$

Ratio of initial components	Thermodynamic stability beginning with the more stable compounds in the temp. range of 298 to 1800 K	
	from 298 to 1100 K	from 1200 to 1800 K
1/1	MgS, Mg_2S	Mg_2S, MgS
2/1	Mg_2S, MgS	Mg_2S, MgS

2.2.2 System $MgCO_3-MgO-SiO_2$

The data on ΔG_T^0 obtained for solid-phase reactions in the system $MgO-SiO_2$ combined with ΔG_T^0 of $MgCO_3$ dissociation enable us to conduct a thermodynamic study of the main reactions in the system $MgCO_3-MgO-SiO_2$.

By combining the obtained values of ΔG^0 and those calculated before for solid-phase reactions of $MgO \cdot SiO_2$ and $2MgO \cdot SiO_2$ formation we derive ΔG^0 for the following reactions:

$$MgCO_3+SiO_2=MgO\cdot SiO_2+CO_2$$

$$2MgCO_3+SiO_2=2MgO\cdot SiO_2+2CO_2\,.$$

The results of the calculations are presented in Fig. 15.

The examination of the data shows that the temperature of the beginning of the magnesite dissociation equals 678 K, which agrees well with the experimentally found value of 410 °C [37]. The temperature dependence of the $MgCO_3$ dissociation pressure can be expressed by an equation:

$$\lg p_{CO_2}=6145\,T^{-1}+0.46\ln T-1.08\cdot 10^{-3}\,T+0.07\cdot 10^5\,T^{-2}+6.77\,.$$

The reaction of the $MgO\cdot SiO_2$ formation is thermodynamically possible beginning with 465 K, that of the $2MgO\cdot SiO_2$ formation – beginning with 490 K.

Above the temperature of the magnesite decomposition, 678 K, ΔG^0 of reactions coincide with those for solid-phase reactions in the system $MgO-SiO_2$.

2.2.3 System $MgSO_4-MgO-SiO_2$

The calculation is in principle similar to that considered in the preceding paragraph.

The initial data for the thermodynamic analysis were those presented in Table 16.

The coefficients of the equation of the temperature dependence of the $MgSO_4$ heat capacity were calculated according to the entropy method suggested by Landiya [320]. The temperature of magnesium sulfate melting was accepted as equal to 1400 K in accordance with [483]. The transition to high-temperature modification formed at temperatures higher than 773 K was not taken into account for lack of thermodynamic data concerning this modification. As the heat capacity of $MgSO_4$ in molten state is unknown, above the temperature of melting the heat capacity of $MgSO_4$ for the solid phase was adopted.

On the basis of the data adduced the alteration of ΔG_T^0 of the reaction of the $MgSO_4$ decomposition over the temperature range of 300 to 1600 K was calculated.

Combining the obtained values of ΔG_T^0 and those calculated earlier for solid-phase reactions of the $MgO\cdot SiO_2$ and $2MgO\cdot SiO_2$ formation we require ΔG_T^0 for the following reactions:

$$MgSO_4+SiO_2=MgO\cdot SiO_2+SO_2+\tfrac{1}{2}O_2$$

$$2MgSO_4+SiO_2=2MgO\cdot SiO_2+2SO_2+O_2\,.$$

The results of the calculations are presented in Fig. 15.

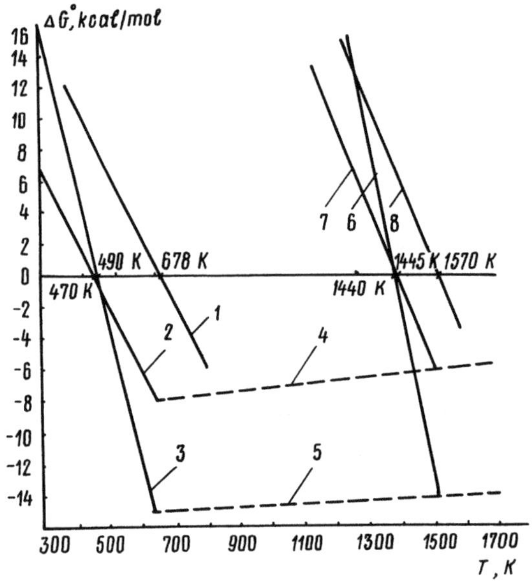

Fig. 15. The temperature dependence of ΔG^0 of reactions in the system $MgCO_3-MgSO_4-MgO-SiO_2$ within a temperature range of 298 to 1700 K for: *1* $MgCO_3 = MgO + CO_2$; *2* $MgCO_3 + SiO_2 = MgO \cdot SiO_2 + CO_2$; *3* $2MgCO_3 + SiO_2 = 2MgO_2 \cdot SiO_2 + 2CO_2$; *4* $MgO + SiO_2 = MgO \cdot SiO_2$; *5* $MgSO_4 = MgO + SO_2 + 1/2O_2$; *6* $MgSO_4 + SiO_2 = MgO \cdot SiO_2 + SO_2 + \frac{1}{2}O_2$; *7* $2MgSO_4 + SiO_2 = 2MgO \cdot SiO_2 + 2SO_2 + O_2$; *8* $2MgO + SiO_2 = 2MgO \cdot SiO_2$

The examination of the data obtained shows that the temperature of the beginning of the $MgSO_4$ dissociation equals 1570 K [102].

The reaction of the $MgO \cdot SiO_2$ formation becomes thermodynamically possible beginning with 1445 K whereas that of $2MgO \cdot SiO_2$ formation, with 1440 K, i.e., practically at the same temperature. Above the temperature of $MgSO_4$ dissociation ΔG^0 of the reactions coincide with the values of ΔG^0 for solid-phase reactions in the system $MgO-SiO_2$.

2.3 System $CaCO_3-CaSO_4-CaO-SiO_2$

2.3.1 System $CaCO_3-CaO-SiO_2$

The data previously arrived at for solid-phase reactions in the system $CaO-SiO_2$ combined with ΔG^0 of the $CaCO_3$ dissociation permit conducting a thermodynamic study of the main reactions in the system $CaCO_3-CaO-SiO_2$.

Four main reactions were considered:

1. $CaCO_3 + SiO_2 = CaO \cdot SiO_2 + CO_2$
2. $3CaCO_3 + 2SiO_2 = 3CaO \cdot 2SiO_2 + 3CO_2$
3. $2CaCO_3 + SiO_2 = 2CaO \cdot SiO_2 + 2CO_2$
4. $3CaCO_3 + SiO_2 = 3CaO \cdot SiO_2 + 3CO_2$.

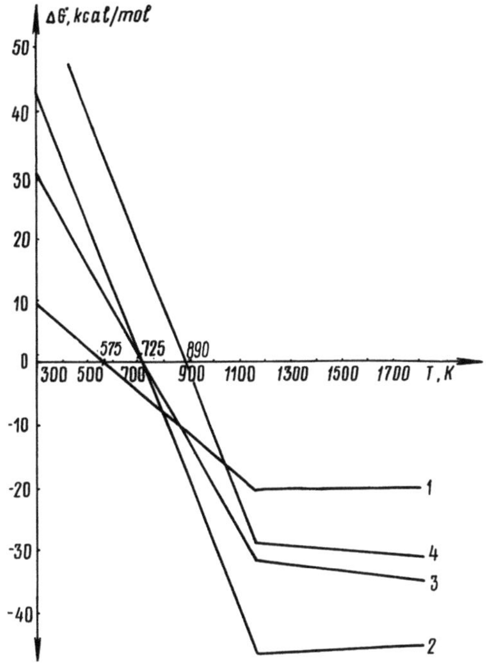

Fig. 16. The temperature dependence of ΔG^0 of reactions in the system $CaCO_3 - CaO - SiO_2$ within a temperature range of 298 to 1800 K. *1* $CaO \cdot SiO_2$; *2* $3CaO \cdot 2SiO_2$; *3* $2CaO \cdot SiO_2$; *4* $3CaO \cdot SiO_2$

For all these reactions the values of ΔG_T^0 in the temperature range of 298 to 1160 K were calculated. For temperatures above 1160 K the data for solid-phase reactions in the system $CaO - SiO_2$ were used.

The results obtained are presented in Fig. 16.

The examination of the data shows that the reaction of the $CaO \cdot SiO_2$ formation is thermodynamically possible beginning with 575 K, the reaction of $2CaO \cdot SiO_2$ formation, beginning with 715 K, that of $3CaO \cdot 2SiO_2$ formation, with 725 K and finally, that of $3CaO \cdot SiO_2$ formation, with 890 K.

For evaluating the possible sequence of reactions the ΔG^0 values were examined in which the ratios of the $CaCO_3$ and SiO_2 mixtures were taken into consideration, CaO/SiO_2 being 1/1, 3/2, 2/1, 3/1 (Fig. 17).

The consideration of the data obtained shows that, the ratio being 1/1, over the whole temperature range the formation of $CaO \cdot SiO_2$ will be the most probable. The ratio being 3/2, up to the temperature of 1000 K the most probable is the formation of $CaO \cdot SiO_2$ and above that temperature, of $2CaO \cdot SiO_2$. The ratio of CaO/SiO_2 being 2/1, in the temperature range up to 870 K the most probable is the formation of $CaO \cdot SiO_2$ and above it, of $2CaO \cdot SiO_2$. A similar situation is observed at a ratio of 3/1 as well. Thus, a tendency to the $CaO \cdot SiO_2$ formation is characteristic of compounds rich in SiO_2 ($CaO \cdot SiO_2$ features also a lower temperature of the beginning of the reaction), while for the compounds with high $CaCO_3$ content the formation of $2CaO \cdot SiO_2$ is possible at higher temperatures of the onset of the reactions.

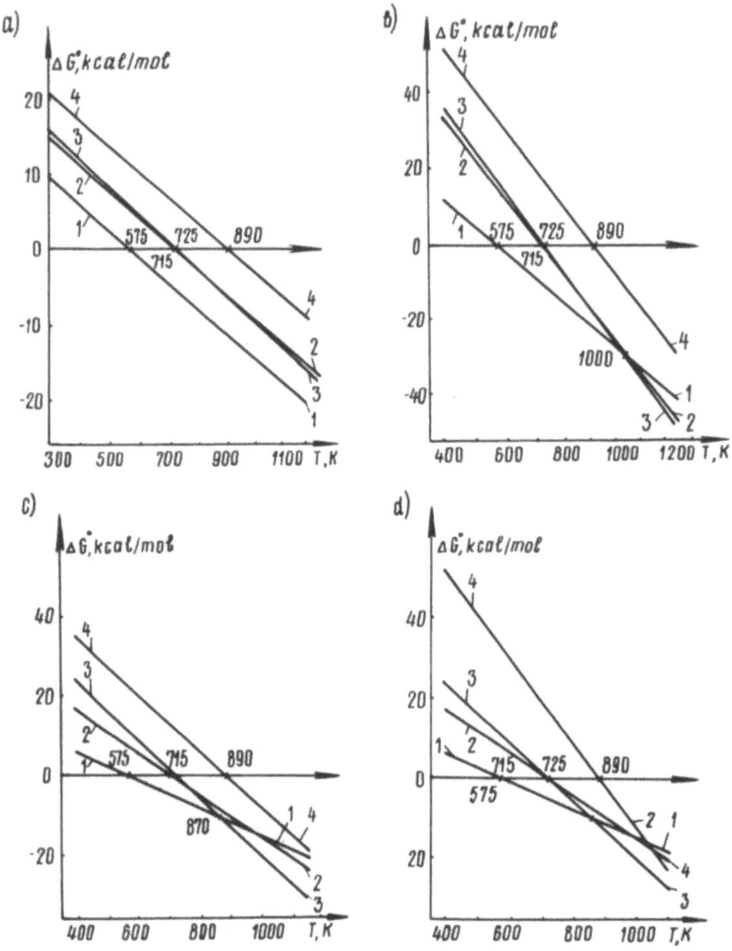

Fig. 17. The temperature dependence of ΔG^0 of reactions in the system $CaCO_3-CaO-SiO_2$ within a temperature range of 298 to 1160 K, the ratios of $CaO-SiO_2$ being: **a** 1/1; **b** 3/2; **c** 2/1; **d** 3/1

The thermodynamic sequence of reactions at temperatures higher than that of $CaCO_3$ dissociation will be similar to that of the solid-phase reactions in the system (see Chap. I).

2.3.2 System $CaSO_4-SiO_2$

An analogous technique of thermodynamic analysis was used for the system $CaSO_4-SiO_2$ as well. The main reactions of the silicate formation in this system were analyzed:

1. $CaSO_4+SiO_2=CaO\cdot SiO_2+SO_2+\frac{1}{2}O_2$
2. $3CaSO_4+2SiO_2=3CaO\cdot 2SiO_2+3SO_2+\frac{3}{2}O_2$

3. $2CaSO_4 + SiO_2 = 2CaO \cdot SiO_2 + 2SO_2 + O_2$
4. $3CaSO_4 + SiO_2 = 3CaO \cdot SiO_2 + 3SO_2 + \frac{3}{2}O_2$.

The ΔG^0 values for these reactions were determined by means of combining those of the main reactions of the system $CaO - SiO_2$ and that of the reaction of the anhydrite decomposition in the temperature range of 1200 to 1800 K.

The results obtained are presented in Fig. 18.

An analysis of the results obtained indicates that the interaction between $CaSO_4$ and SiO_2 is possible beginning with 1545 K, when the formation of $CaSiO_3$ is thermodynamically probable. The reaction of the $2CaO \cdot SiO_2$ formation is thermodynamically possible beginning with 1615 K, that of the $3CaO \cdot 2SiO_2$ formation, with 1645 K and that of the $3CaO \cdot SiO_2$ formation becomes probable only beginning with 1740 K.

For evaluation, the possible sequence of reactions were analyzed the ΔG^0 values, the ratios of the initial mixtures of $CaSO_4$ and SiO_2 having been considered in which $CaO/SiO_2 = 1/1$, 3/2, 2/1 and 3/1 (see Fig. 19).

The examination of the data obtained shows that, the ratio being 1/1, in the whole temperature range the $CaO \cdot SiO_2$ formation will be the most probable. The ratio being 3/2, in the temperature range up to 1725 K the most preferred is the $CaO \cdot SiO_2$ formation, at a higher temperature, that of $2CaO \cdot SiO_2$. The CaO/SiO_2 ratio being 2/1, at a temperature up to 1680 K the most probable is the $CaO \cdot SiO_2$ formation, at higher temperatures, that of $2CaO/SiO_2$. An analogous situation is observed, the ratio being 3/1. Thus, as in the case of the system $CaCO_3 - SiO_2$ a tendency toward $CaO \cdot SiO_2$ formation will be

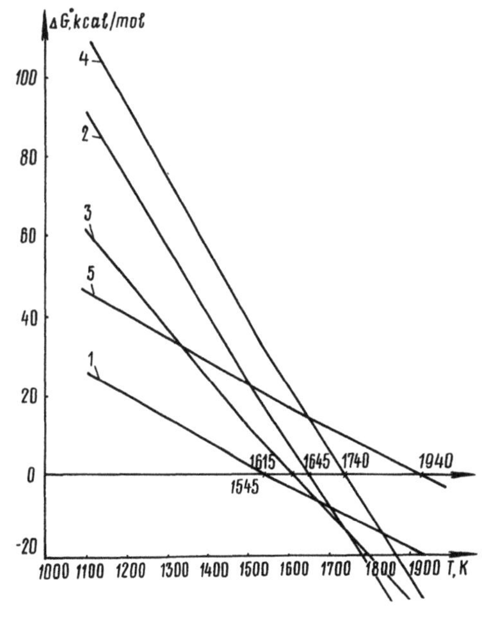

Fig. 18. The temperature dependence of ΔG^0 of reactions in the system $CaSO_4 - SiO_2$ within a temperature range of 1100 to 1940 K

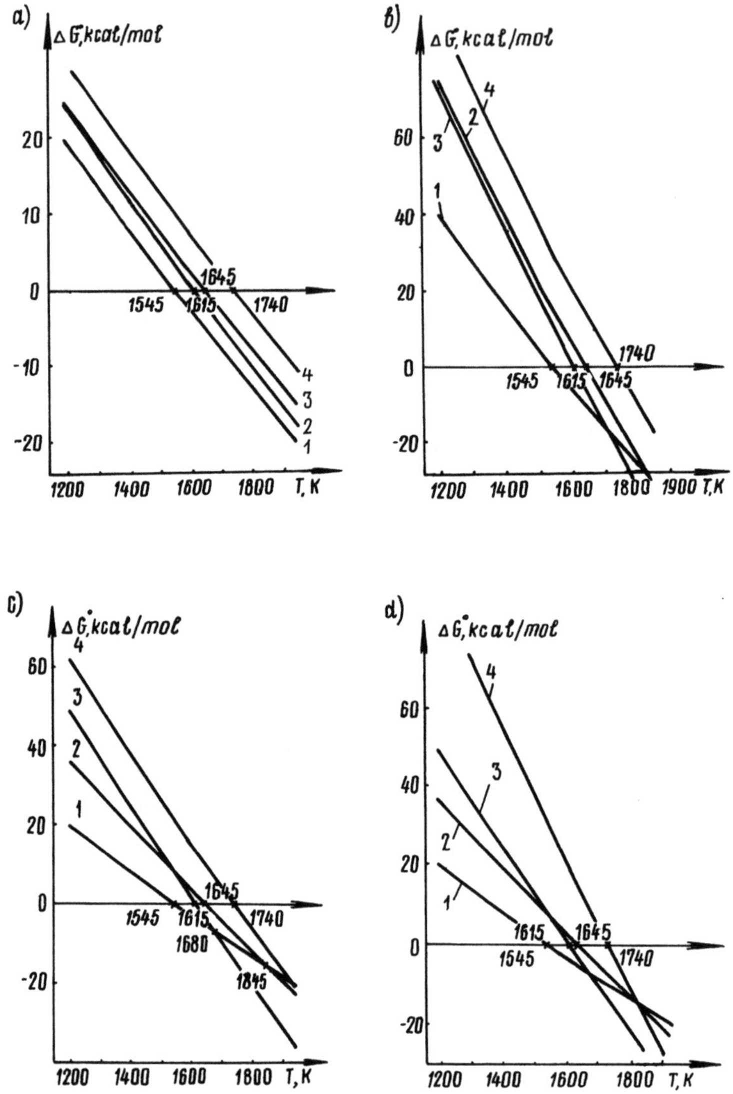

Fig. 19. The temperature dependence of ΔH^0 of reactions in the system $CaSO_4-SiO_2$ within a temperature range of 1100 to 1940 K, the CaO/SiO_2 ratios being **a** 1/1; **b** 3/2; **c** 2/1; **d** 3/1

characteristic of the compounds rich in SiO_2 at a lower temperature of the beginning of reactions. For compounds with a high $CaSO_4$ content the $2CaO \cdot SiO_2$ formation is possible at higher temperatures of the beginning of reactions.

2.4 System $SrCO_3 - SrSO_4 - SrO - SiO_2$

2.4.1 System $SrCO_3 - SrO - SiO_2$

A thermodynamic analysis of solid-phase reactions in the system $SrCO_3 - SrO - SiO_2$ was first done by Glushkova [156]. However the appearance of experimental data on the enthalpies of strontium silicates [32] as well as the modification of the technique of calculations itself required a revaluation of Glushkova's results.

In the system $SrO - SiO_2$ the existence of three strontium silicates was found:

$SrO \cdot SiO_2$, $2SrO \cdot SiO_2$ and $3SrO \cdot SiO_2$ [254].

The initial thermochemical data for the analysis were those presented in Table 18 [32, 156]. ΔG^0 of the following main reactions were calculated:

1. $SrO + SiO_2 = SrO \cdot SiO_2$
2. $2SrO + SiO_2 = 2SrO \cdot SiO_2$
3. $3SrO + SiO_2 = 3SrO \cdot SiO_2$
4. $SrCO_3 = SrO + CO_2$.

Later by combining the corresponding ΔG^0 values of the main reactions ΔG^0 for the following ones were determined:

1. $SrCO_3 + SiO_2 = SrO \cdot SiO_2 + CO_2$
2. $2SrCO_3 + SiO_2 = 2SrO \cdot SiO_2 + 2CO_2$
3. $3SrCO_3 + SiO_2 = 3SrO \cdot SiO_2 + 3CO_2$.

The following equations for calculating $\Delta G^0 = f(T)$ were obtained:[9]

1. $\Delta G_T^0 = -33131 - 0.01\, T\ln T - 0.765 \cdot 10^{-3} T^2 + 2.715 \cdot 10^5 T^{-1} + 0.725\, T$
2. $\Delta G_T^0 = -50966 + 2.29\, T\ln T - 1.26 \cdot 10^{-3} T^2 + 1.95 \cdot 10^5 T^{-1} - 8.17\, T$
3. $\Delta G_T^0 = -71554 - 5.59\, T\ln T - 1.56 \cdot 10^{-3} T^2 + 1.35 \cdot 10^5 T^{-1} + 39.8\, T$
4. $\Delta G_T^0 = -55690 - 1.47\, T\ln T + 2.64 \cdot 10^{-3} T^2 + 0.23 \cdot 10^5 T^{-1} - 32.38\, T$

(up to 1198 K)

$$\Delta G_T^0 = 59541 + 11.11\, T\ln T - 1.64 \cdot 10^{-3} T^2 + 1.93 \cdot 10^5 T^{-1} - 119.5\, T$$
$$\Delta H_T^0 = 59.948 + 11.11 \cdot T + 1.64 \cdot 10^{-3} T^2 + 3.846 \cdot 10^5 T^{-1}.$$

(above 1198 K)

Using these equations the values of the Gibbs free energy were calculated over the temperature range of 500 to 1800 K (within each 100 K) for all the above reactions.

[9] For the reaction of the $SrCO_3$ decomposition the following equations for the $\Delta H^0 = f(T)$ calculation were obtained: up to 1198 K:
$\Delta H_T^0 = 55.996 + 1.47 \cdot T - 2.64 \cdot 10^{-3} T^2 + 0.45 \cdot 10^5 T^{-1}$

Table 18. The initial thermochemical data for the analysis of solid-phase reactions in the system $SrCO_3-SrO-SiO_2$

Compounds	$-\Delta H^0_{298(el)}$ [cal/mol]	$-\Delta G^0_{298(el)}$ [cal/mol]	S^0_{298} [cal/mol deg]	$-\Delta H^0_{298(ox)}$ [cal/mol]	$-\Delta G^0_{298(ox)}$ [cal/mol]	$C_p=f(T)$ [cal/mol deg]
SrO	141100	138800	13.0	–	–	$12.34+1.12\cdot10^{-3}T-1.806\cdot10^5T^{-2}$
$SrCO_3$-α	291200	271900	23.2	−56150	−38840	$21.42+8.56\cdot10^{-3}T-3.396\cdot10^5T^{-2}$
α→β $SrCO_3$	−4700	–	–	–	–	above 1197 K 34.0
CO_2	94050	94260	51.1	–	–	$10.55+2.16\cdot10^{-3}T-2.04\cdot10^5T^{-2}$
$SrO\cdot SiO_2$	390090	374640	22.5	31240	31090	$26.758+4.593\cdot10^{-3}T-7.236\cdot10^5T^{-2}$
$2SrO\cdot SiO_2$	549990	532390	36.0	50040	50040	$36.810+6.698\cdot10^{-3}T-7.520\cdot10^5T^{-2}$
$3SrO\cdot SiO_2$ [a]	709890	689990	49.0	68840	68840	$47.019+8.430\cdot10^{-3}T-8.141\cdot10^5T^{-2}$

[a] Calculated by the authors

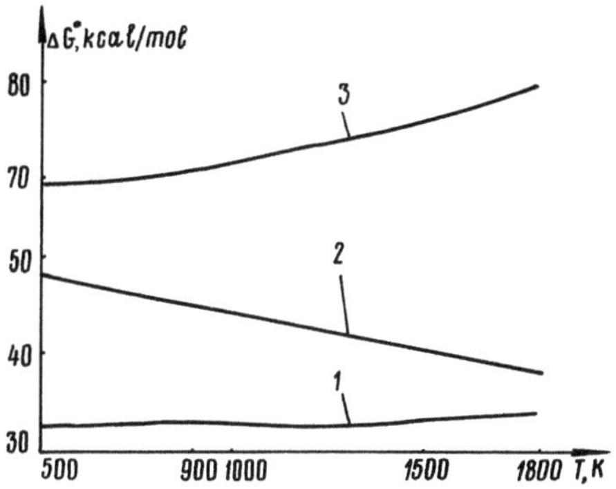

Fig. 20. The temperature dependence of ΔG^0 of reactions in the system $SrO - SiO_2$ within a temperature range of 500 to 1800 K

The results obtained are presented in Figs. 20 and 21.

For evaluating the thermodynamic sequence of the formation of the compounds of the system $SrCO_3 - SrO - SiO_2$ the dependence of ΔG^0 of the reactions on the composition of the initial mixtures was then examined. The ΔG^0 values were analyzed, the ratios of the initial SrO/SiO_2 mixtures being 1/1, 2/1, 3/1. The results obtained are presented in Table 19 and in Figs. 22 and 23. The alteration of ΔG^0 with temperature, the ratio of the initial components SrO/SiO_2 being 3/1, coincides with the data presented in Fig. 21.

The examination of all the data obtained permits drawing the following general conclusions. Theoretically the reactions of $SrCO_3$ and SiO_2 are possible be-

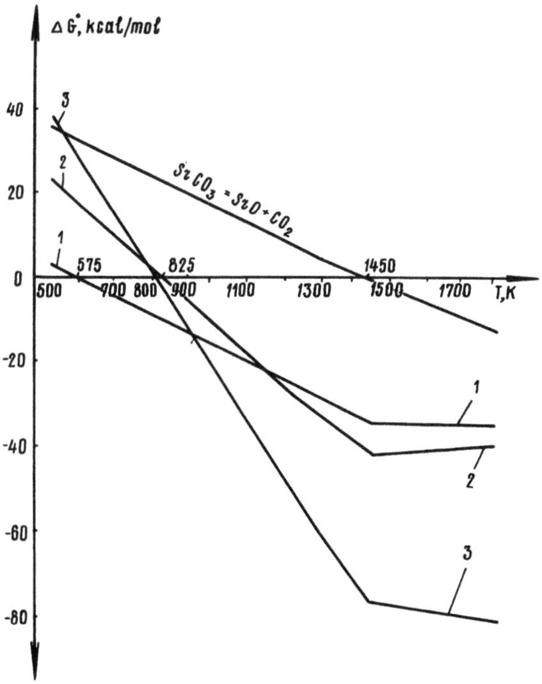

Fig. 21. The temperature dependence of ΔG^0 of reactions in the system $SrCO_3 - SrO - SiO_2$ within a temperature range of 500 to 1800 K

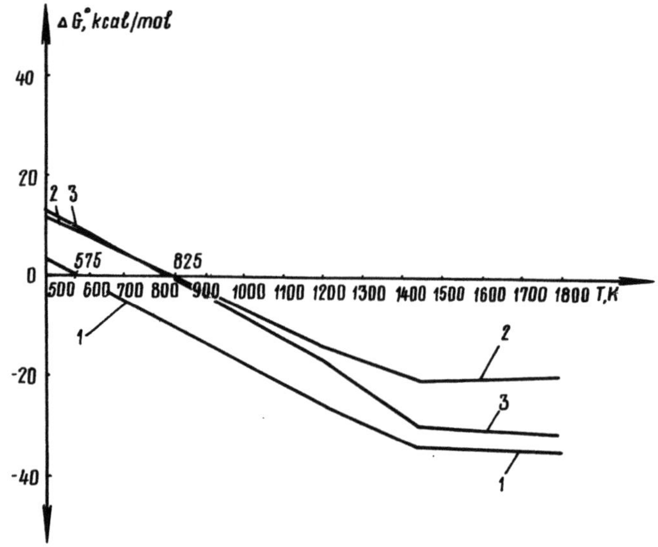

Fig. 22. The temperature dependence of ΔG^0 of reactions in the system $SrCO_3-SrO-SiO_2$ within a temperature range of 500 to 1800 K, the ratio of SrO/SiO_2 being 1/1

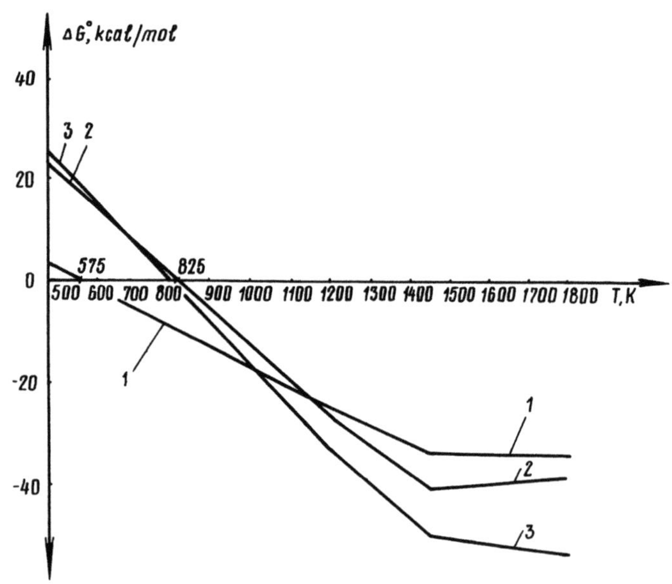

Fig. 23. The temperature dependence of ΔG^0 of reactions in the system $SrCO_3-SrO-SiO_2$ within a temperature range of 500 to 1800 K, the ratio of SrO/SiO_2 being 2/1

Table 19. The thermodynamic stability sequence of the system $SrCO_3-SrO-SiO_2$

Ratio of the initial components	The thermodynamic stability series; beginning with the most stable compound within a temp. range of 1200 to 1800 K		
1/1	SrS	Sr_3S	Sr_2S
2/1	Sr_3S	Sr_2S	SrS
3/1	Sr_3S	Sr_2S	SrS

ginning with a temperature of 575 K, when the reaction of strontium metasilicate formation is thermodynamically possible. The formation of $2SrO \cdot SiO_2$ and $3SrO \cdot SiO_2$ is possible at temperatures of 800 and 825 K respectively. The thermodynamic stability series given in Table 19 enable us to state that with the ratio $SrO/SiO_2 \geqslant 1$ the most stable (final) compound will be $SrO \cdot SiO_2$. With the ratio $SrO/SiO_2 \geqslant 2$ $3SrO \cdot SiO_2$ is revealed as the most stable compound.

Bearing in mind that the formation of $2SrO \cdot SiO_2$ in reactions in mixtures of $SrCO_3-SiO_2$ was proven to be primary independent of the initial mixtures [156], one can infer that the originally formed $2SrO \cdot SiO_2$, SiO_2 being in excess, converts into $SrO \cdot SiO_2$, and $SrCO_3$ being the excess, to $3SrO \cdot SiO_2$. Thus, at high temperatures and great holding time in mixtures rich in SiO_2 the final product will be $SrO \cdot SiO_2$, while in those rich in SrO, $3SrO \cdot SiO_2$. The above outline of the reactions course agrees well with the experimental data [76, 156].

2.4.2 System $SrSO_4-SrO-SiO_2$

The investigation of solid-phase reactions in the system $SrSO_4-SrO-SiO_2$ is of a great theoretical and practical interest, as they play an important role in the synthesis of strontium glasses, porcelain enamels and glazes [388].

The initial data for the analysis were thermochemical ones [156, 384].

Equations of $\Delta G^0 = f(T)$ were set up for the following reactions:

1. $SrSO_4 + SiO_2 = SrO \cdot SiO_2 + SO_2 + \frac{1}{2}O_2$
   $\Delta G_T^0 = 99710 - 0.61\,T\ln T + 2.985 \cdot 10^{-3}T^2 + 3.533 \cdot 10^5 T^{-1} - 81.85\,T$
2. $2SrSO_4 + SiO_2 = 2SrO \cdot SiO_2 + 2SO_2 + O_2$
   $\Delta G_T^0 = 214716 + 1.09\,T\ln T + 6.24 \cdot 10^{-3}T^2 + 3.586 \cdot 10^5 T^{-1} - 171.97\,T$
3. $3SrSO_4 + SiO_2 = 3SrO \cdot SiO_2 + 3SO_2 + \frac{3}{2}O_2$
   $\Delta G_T^0 = 326949 - 7.39\,T\ln T + 9.69 \cdot 10^{-3}T^2 + 3.804 \cdot 10^5 T^{-1} - 205.9\,T$
4. $SrSO_4 = SrO + SO_2 + \frac{1}{2}O_2$
   $\Delta G_T^0 = 132841 - 60\,T\ln T + 3.75 \cdot 10^{-3}T^2 + 0.818 \cdot 10^5 T^{-1} - 81.9\,T$.

Table 20. The alteration of ΔG^0 with temperature for the reactions in the system $SrSO_4 - SrO - SiO_2$

Reactions	ΔG^0 [kcal/mol] at temps. in K						
	400	500	600	700	800	900	1000
$Sr + S = SrS$	67.2	58.7	50.3	42.1	33.4	25.2	18.9
$2Sr + S = Sr_2S$	150.5	134.5	118.7	103.1	87.6	72.2	56.8
$3Sr + S = Sr_3S$	229.3	203.3	179.0	154.4	154.4	103.8	80.0
$SrSO_4 = SrO + SO_2 + \frac{1}{2}O_2$	99.4	91.2	32.9	74.7	66.6	58.5	50.6
Reactions	ΔG^0 [kcal/mol] at temps. in K						
	1100	1200	1300	1400	1500	1600	1700
$Sr + S = SrS$	9.0	1.1	−6.9	−5.1	−22.8	−30.6	−38.0
$2Sr + S = Sr_2S$	42.5	27.3	12.0	−23.4	−32.1	−32.1	−45.3
$3Sr + S = Sr_3S$	55.6	31.4	4.8	−17.2	−40.5	−64.8	−88.0
$SrSO_4 = SrO + SO_2 + \frac{1}{2}O_2$	42.4	35.0	27.3	19.4	11.9	3.9	−3.1

Based on these equations the ΔG_T^0 values were calculated in a temperature range of 400 to 1700 K (within each 100 K) for all of the above reactions (Table 20; Fig. 24); the curve numbers in figures correspond to the rows in the table from top to bottom.

The alteration of the Gibbs free energy with temperature enables one to determine the possibility of the reactions under consideration.

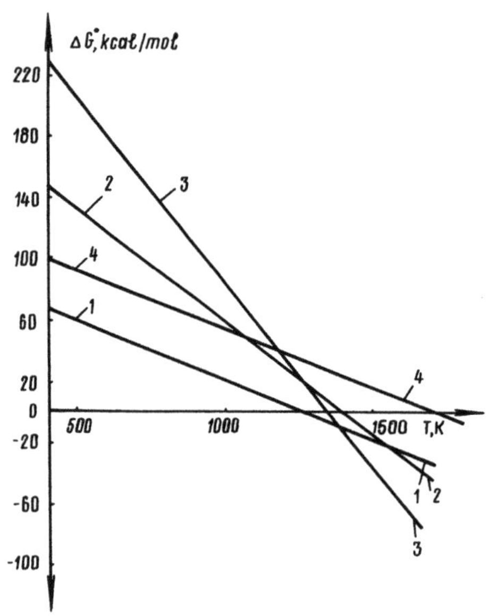

Fig. 24. The temperature dependence of ΔG^0 of reactions in the system $SrO_4 - SrO - SiO_2$ within a temperature range of 400 to 1700 K

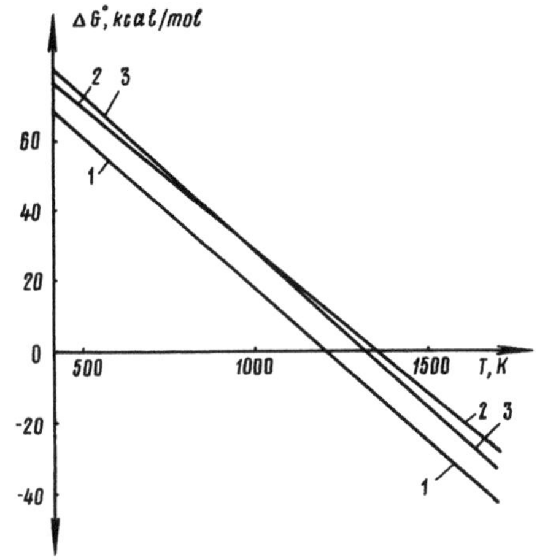

Fig. 25. The temperature dependence of ΔG^0 of reactions in the system $SrSO_4-SrO-SiO_2$ within a temperature range of 400 to 1700 K, the ratio of SrO/SiO_2 being 1/1

Further, the values of ΔG_T^0 were examined, the ratios SrO/SiO_2 being 1/1, 2/1 and 3/1 (Figs. 25 and 26). The alteration of ΔG_T^0 with temperature, SrO/SiO_2 being 3/1, coincides with the facts presented in Fig. 24.

The reaction of the $SrO \cdot SiO_2$ formation is possible at a temperature of 1225 K, that of the $2SrO \cdot SiO_2$ formation, at a temperature of 1380 K, and that of the $3SrO \cdot SiO_2$ formation, at 1340 K.

The reaction of the $SrSO_4$ thermic dissociation is possible at a temperature of 1670 K.

The series of the thermodynamic stability coinciding with those of the system $SiCO_3-SrO-SiO_2$ permit the assumption that with the ratio of $SrO/SiO_2 \leqslant 1$ the most stable – final – phase will be $SrO \cdot SiO_2$ and with the ratio of $SrO/SiO_2 \geqslant 2$ the most stable phase is $3SrO \cdot SiO_2$.

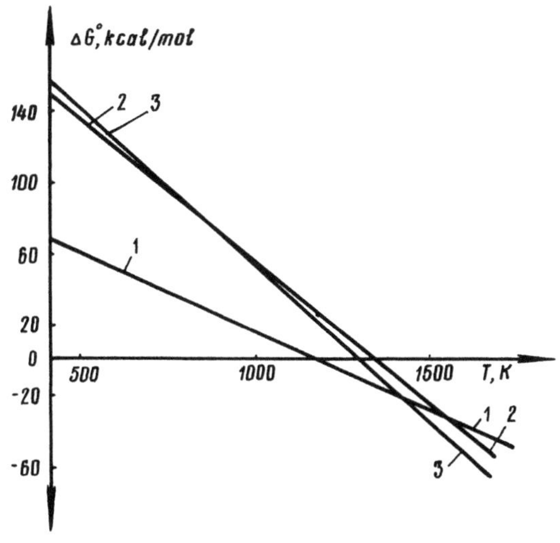

Fig. 26. The temperature dependence of ΔG^0 of reactions in the system $SrSO_4-SrO-SiO_2$ within a temperature range of 400 to 1700 K, the ratio of SrO/SiO_2 being 2/1

Consequently at high temperatures in mixtures rich in SiO_2 the final product will be $SrO \cdot SiO_2$ and in mixtures rich in SrO, $3SrO \cdot SiO_2$. This outline of the reactions in the system $SrSO_4-SrO-SiO_2$ agrees well with experimental data [76, 156, 382].

2.5 System $BaCO_3-BaSO_4-BaO-SiO_2$

An investigation of the system $BaCO_3-BaSO_4-BaO-SiO_2$ is of great interest for silicate production processes, chiefly for the manufacturing of special types of cement, glass, glass-ceramics and porcelain enamels. A thermodynamic study of the conditions of the reactions in this system permits finding the main regularities of the formation of barium silicates, important from the technical standpoint, as well as giving the energy characteristics of processes of the manufacturing of industrial materials on the basis of or using barium silicates.

2.5.1 System $BaCO_3-BaO-SiO_2$

The thermodynamics of solid-phase reactions in the system $BaCO_3-SiO_2$ was studied by several authors [156, 229, 358, 555].

The appearance of the data on the enthalpies of formation of barium silicates [32] as well as the modification of the very calculation technique illustrate the need for reevaluation of results obtained earlier.

The thermochemical data needed for the analysis are presented in Table 21 [32, 156, 229].

According to literature [43, 76, 253, 555] the system $BaO-SiO_2$ favours mostly the formation of the following compounds:

$$BaO \cdot 2SiO_2,\ 2BaO \cdot 3SiO_2,\ BaO \cdot SiO_2,\ 2BaO \cdot SiO_2 \text{ and } 3BaO \cdot SiO_2 .$$

Therefore we analyzed the main reactions of the formation of these compounds. ΔH^0 and ΔG^0 of the following main reactions were calculated:

1. $BaO+2SiO_2=BaO \cdot 2SiO_2$
2. $2BaO+3SiO_2=2BaO \cdot 3SiO_2$
3. $BaO+SiO_2=BaO \cdot SiO_2$
4. $2BaO+SiO_2=2BaO \cdot SiO_2$
5. $3BaO+SiO_2=3BaO \cdot SiO_2$.

Then combining the corresponding ΔH^0 and ΔG^0 values of the main reactions and of the reaction of the $BaCO_3$ decomposition ΔH^0 and ΔG^0 were determined for the following reactions:

Table 21. The initial thermochemical data for the analysis of solid-phase reactions in the system $BaCO_3 - BaO - SiO_2$

Compounds	$-\Delta H^0_{298(el)}$ [cal/mol]	$-\Delta G^0_{298(el)}$ [cal/mol]	S^0_{298} [cal/mol deg]	$-\Delta H^0_{298(ox)}$ [cal/mol]	$-\Delta G^0_{298(ox)}$ [cal/mol]	$C_p = f(T)$ [cal/mol deg]
BaO	133400	126300	16.8	–	–	$12.74 + 1.04 \cdot 10^{-2} T - 1.984 \cdot 10^5 T^{-2}$
$BaCO_3$-α	291300	272200	26.8	–	–	$20.77 + 11.70 \cdot 10^{-5} T - 2.86 \cdot 10^5 T^{-2}$ (298 – 1079 K)
$BaCO_3$-β	–	–	–	–	–	37.0 (1079 – 1241 K)
$BaCO_3$-γ	–	–	–	–	–	39.0 (above 1241 K)
$\alpha \rightarrow \beta$ (1083 K)	−3830	–	–	–	–	–
$\beta \leftarrow \gamma$ (1240 K)	−700	–	–	–	–	–
$BaSO_4$	350200	323380	31.6	–	–	$29.90 + 1.37 \cdot 10^{-3} T$ (up to 1551 K) $34.93 + 6.7 \cdot 10^{-3} T$ (up to 1853 K)
$BaO \cdot 2SiO_2$	610180	577080	36.8	41280	41280	$41.28 + 8.125 \cdot 10^{-3} T - 6.918 \cdot 10^5 T^{-2}$
$2BaO \cdot 3SiO_2$	1002470	949270	63.6	82420	82420	$50.53 + 9.18 \cdot 10^{-3} T - 19.05 \cdot 10^5 T^{-2}$ [a]
$BaO \cdot SiO_2$	389180	369080	26.8	38030	38030	$24.5 + 5.9 \cdot 10^{-3} T$
$2BaO \cdot SiO_2$	549030	521830	43.5	64480	64480	$34.5 + 8.0 \cdot 10^{-3} T$
$3BaO \cdot SiO_2$	708880	674580	60.4	90930 [b]	90930	$44.46 + 10.9 \cdot 10^{-3} T$

[a] Calculated by Landiya's method
[b] Calculated by [357]

1a. $BaCO_3 + 2SiO_2 = BaO \cdot 2SiO_2 + Co_2$
2a. $2BaCO_3 + 3SiO_2 = 2BaO \cdot 3SiO_2 + 2CO_2$
3a. $BaCO_3 + SiO_2 = BaO \cdot SiO_2 + CO_2$
4a. $2BaCO_3 + SiO_2 = 2BaO \cdot SiO_2 + 2CO_2$
5a. $3BaCO_3 + SiO_2 = 3BaO \cdot SiO_2 + 3CO_2$.

The Calculation of ΔH^0 of Reactions

The following equations for calculating $\Delta H^0 = f(T)$ of the main reactions were obtained.

1. $\Delta H_T^0 = -42994 - 0.28\,T + 1.60 \cdot 10^{-3}T^2 - 4.934 \cdot 10^5 T^{-1}$
2. $\Delta H_T^0 = -82107 - 18.18\,T + 0.645 \cdot 10^{-3}T^2 - 4.036 \cdot 10^5 T^{-1}$
3. $\Delta H_T^0 = -36706 - 2.65\,T - 1.46 \cdot 10^{-3}T^2 - 1.984 \cdot 10^5 T^{-1}$
4. $\Delta H_T^0 = -61728 - 5.39\,T + 1.99 \cdot 10^{-3}T^2 - 3.968 \cdot 10^5 T^{-1}$
5. $\Delta H_T^0 = -86759 - 8.17\,T + 2.92 \cdot 10^{-3}T^2 - 5.954 \cdot 10^5 T^{-1}$.

Further, for calculating the enthalpy of the reaction $BaCO_3 = BaO + CO_2$ equations were determined:

up to 1083 K
6. $\Delta H_T^0 = 63088 + 2.52\,T - 4.25 \cdot 10^{-3}T^2 + 1.164 \cdot 10^5 T^{-1}$

from 1083 K to 1240 K
7. $\Delta H_T^0 = 70202 - 13.71\,T + 1.60 \cdot 10^{-3}T^2 + 4.02 \cdot 10^5 T^{-1}$

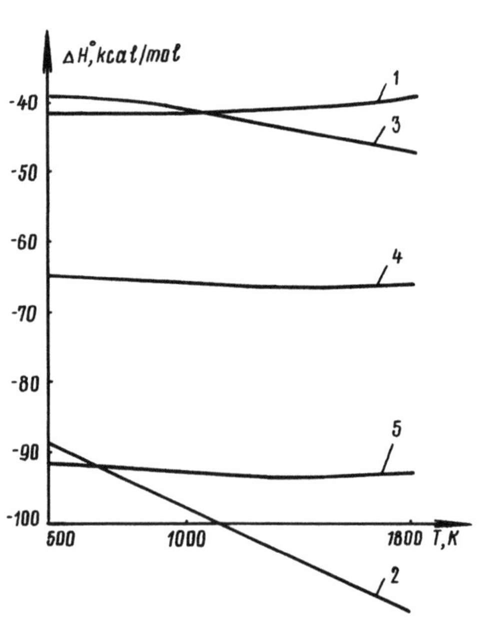

Fig. 27. The temperature dependence of ΔH^0 of reactions in the system $BaO-SiO_2$ within a temperature range of 500 to 1800 K

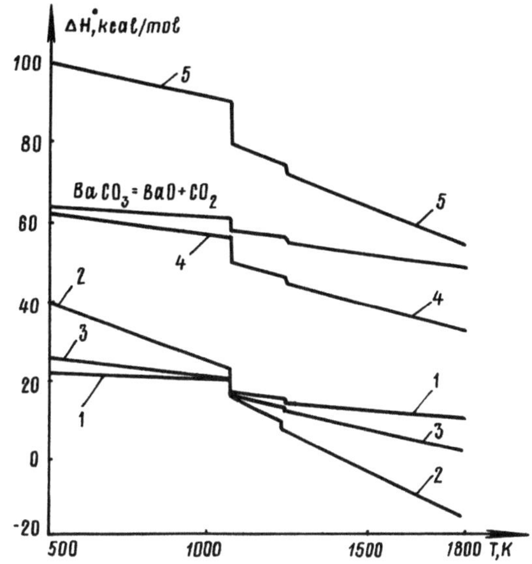

Fig. 28. The temperature dependence of ΔH^0 of reactions in the system $BaCO_3 - BaO - SiO_2$ within a temperature range of 500 to 1800 K

above 1240 K

8. $\Delta H_T^0 = 71984 - 15.71\,T + 1.60 \cdot 10^{-3} T^2 + 4.02 \cdot 10^5 T^{-1}$.

According to these equations the values of the reaction enthalpies in the temperature range of 500 to 1800 K (within each 100 K) were calculated for all of the above reactions. The results obtained are presented in Figs. 27 and 28.

The Calculation of ΔG^0 of Reactions

For determining $\Delta G^0 = f(T)$ the following equations for the main reactions were deduced:

1. $\Delta G_T^0 = -42994 + 0.28\,T \ln T - 1.60 \cdot 10^{-3} T^2 + 2.467 \cdot 10^5 T^{-1} + 1.35\,T$
2. $\Delta G_T^0 = -76824 + 13.76\,T \ln T - 1.645 \cdot 10^{-3} T^2 - 2.018 \cdot 10^5 T^{-1} - 96.5\,T$
3. $\Delta G_T^0 = -36706 + 2.65\,T \ln T - 1.46 \cdot 10^{-3} T^2 - 0.992 \cdot 10^5 T^{-1} - 14.0\,T$
4. $\Delta G_T^0 = -61728 + 5.39\,T \ln T - 1.99 \cdot 10^{-3} T^2 - 1.984 \cdot 10^5 T^{-1} - 29.3\,T$
5. $\Delta G_T^0 = -86759 + 8.17\,T \ln T - 2.92 \cdot 10^{-3} T^2 - 2.976 \cdot 10^5 T^{-1} - 56.3\,T$.

For calculating $\Delta G^0 = f(T)$ of the reaction $BaCO_3 = BaO + CO_2$ the following equations were determined:

up to 1083 K:

6. $\Delta G_T^0 = 63088 - 2.52\,T \ln T - 4.25 \cdot 10^{-3} T^2 + 0.582 \cdot 10^5 T^{-1} - 26.0\,T$

from 1083 to 1240 K:

7. $\Delta G_T^0 = 70202 + 13.71\,T \ln T - 1.60 \cdot 10^{-3} T^2 + 2.01 \cdot 10^5 T^{-1} - 143.5\,T$

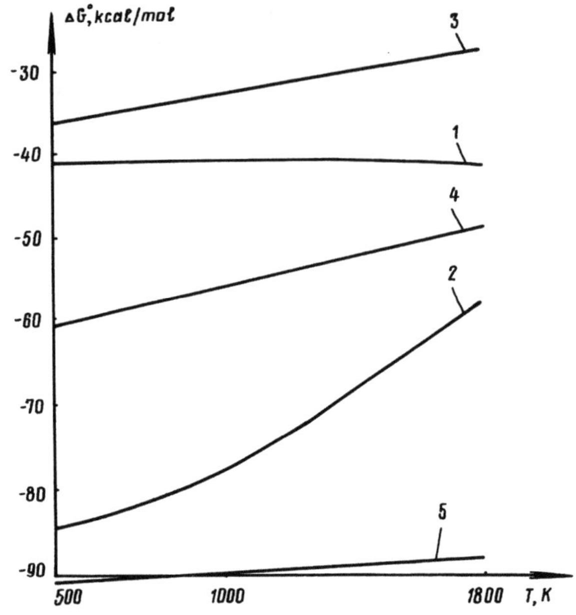

Fig. 29. The temperature dependence of ΔG^0 of reactions in the system $BaO-SiO_2$ within a temperature range of 500 to 1800 K

above 1241 K:

8. $\Delta G_T^0 = 71884 + 15.7\,T \ln T - 1.60 \cdot 10^{-3}\,T^2 + 2.01 \cdot 10^5\,T^{-1} - 159.7\,T$.

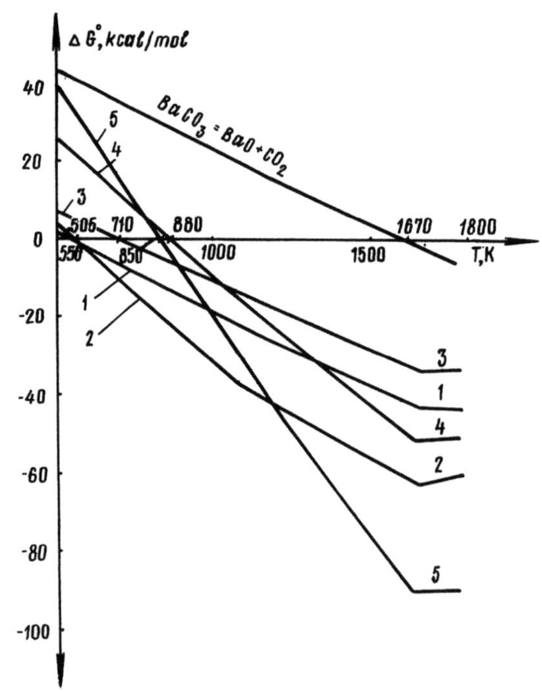

Fig. 30. The temperature dependence of ΔG^0 of reactions in the system $BaCO_3-BaO-SiO_2$ within a temperature range of 500 to 1800 K

By these equations the values of the Gibbs free energy over the temperature range of 500 to 1800 K (within each 100 K) were calculated for all of the above reactions. The results obtained are presented in Figs. 29 and 30.

For evaluating the thermodynamic sequence of the formation of compounds in the system $BaCO_3 - BaO - SiO_2$ we examined the dependence of ΔG^0 of reactions on the composition of the initial mixtures in view of the ratio of BaO/SiO_2. We analyzed the values for the ratios of initial mixtures of $BaO/SiO_2 = 1/2, 2/3, 1/1, 2/1$ and $3/1$ and the results obtained are presented in Fig. 31.

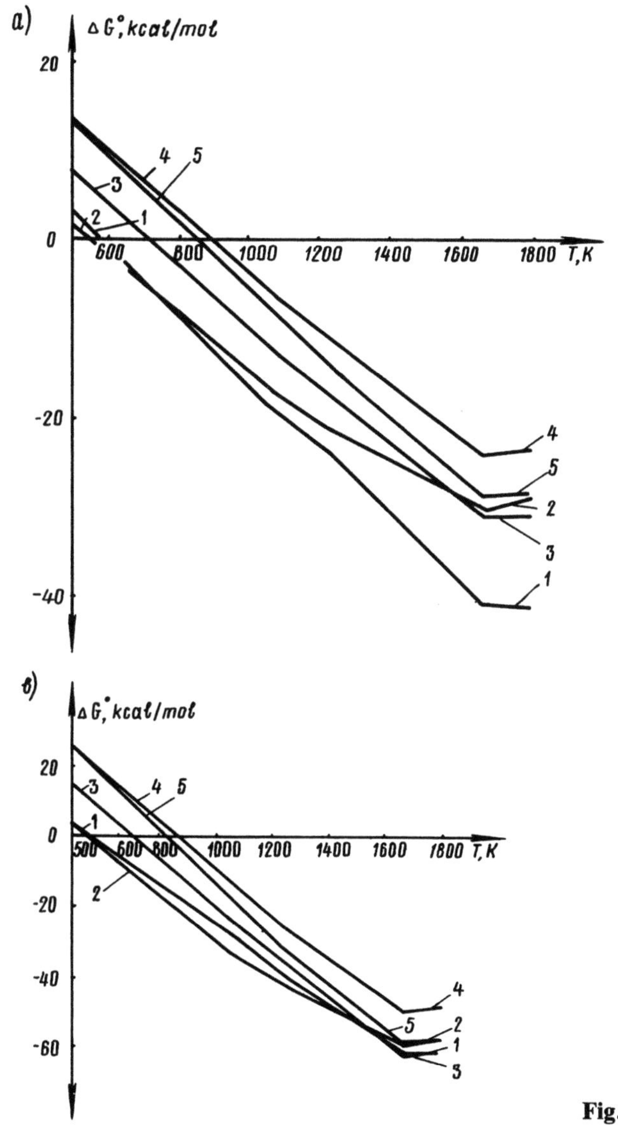

Fig. 31a – b (Legend see page 117)

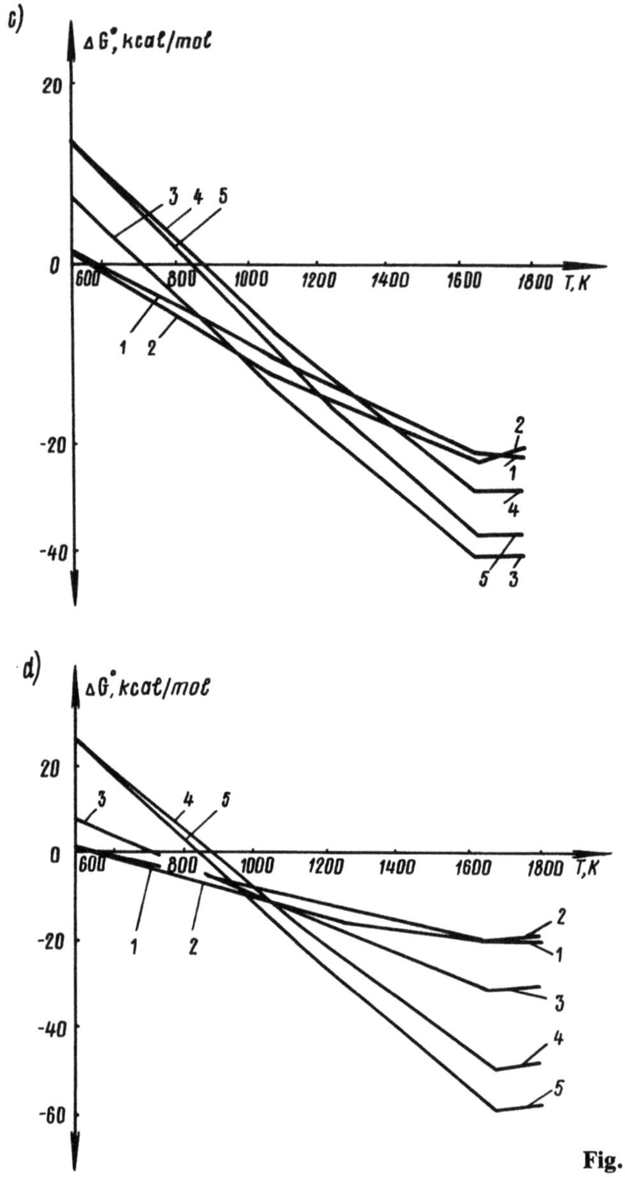

Fig. 31c–d (Legend see page 117)

An examination of the data obtained enables us to draw the following general conclusions.

Theoretically the beginning of the reactions between $BaCO_3$ and SiO_2 is possible starting with a temperature of 550 K, when the reaction of the $2BaO \cdot 3SiO_2$ formation is thermodynamically possible. Quite close to it is also the temperature of the beginning of the reaction of the $BaO \cdot 2SiO_2$ formation,

e)

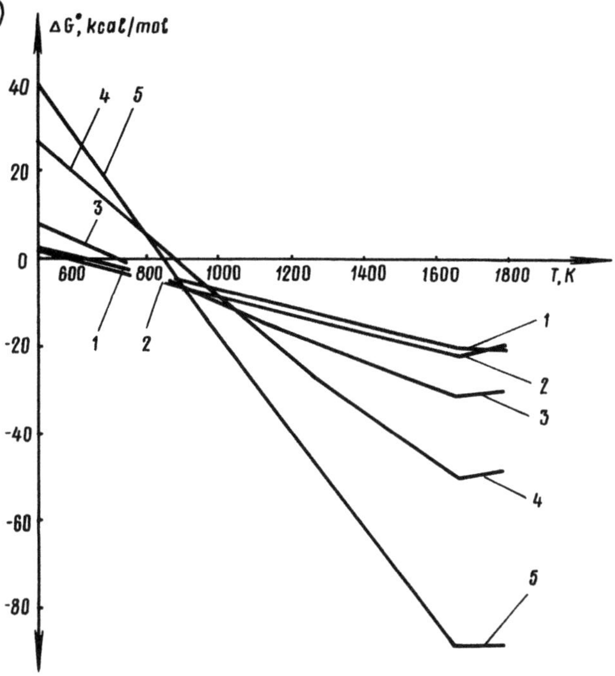

Fig. 31. The temperature dependence of ΔG^0 of reactions in the system $BaCO_3 - BaO - SiO_2$ within a temperature range of 500 to 1800 K, the ratios of BaO/SiO_2 being **a** 1/2; **b** 2/3; **c** 1/1; **d** 2/1; **e** 3/1

565 K. A reaction of the $BaO \cdot SiO_2$ formation appears to be possible beginning with 710 K. A practical study of the reaction of silicate formation for the mixtures of $BaCO_3 - SiO_2$ of different compositions has shown compounds $BaO \cdot 2SiO_2$ and $2BaO \cdot 3SiO_2$ not to be formed as primary phases whereas the formation of $BaO \cdot SiO_2$ as a result of an immediate interaction of the initial components is possible only in the narrow temperature range of 700 – 800 °C [156, 253]. The formation of $3BaO \cdot SiO_2$ and $2BaO \cdot SiO_2$ becomes theoretically possible at temperatures of 850 and 880 K respectively. Thus, the character of the primary interaction of $BaCO_3$ and SiO_2 will be determined by kinetic factors – structure simplicity, the measure of phase contact, reactant activity etc.

The results of calculations of $\Delta G^0 = f(T)$ by compositions permitted formulation of the series of the thermodynamic stability sequence of compounds in the system $BaCO_3 - BaO - SiO_2$ (Table 22). These series enable us to state that the most stable compound for a given proportion of initial constituents will be the one corresponding to that proportion by its stoichiometric composition. The only exception is $2BaO \cdot SiO_2$ which, the ratio being $BeO/SiO_2 = 2/1$, is less stable than $3BaO \cdot SiO_2$.

Table 22. The thermodynamic stability sequence of the compounds of the system $BaCO_3-BaO-SiO_2$

Ratio of initial components	The thermodynamic stability series beginning with the most stable compound within the temp. range from 1000 to 1800 K				
1/2	BS_2	BS	B_2S_3	B_3S	B_2S
2/3	BS	BS_2	B_2S_3	B_3S	B_2S
1/1	BS	B_3S	B_2S	BS_2	B_2S_3
2/1	B_3S	B_2S	BS	BS_2	B_2S_3
3/1	B_3S	B_2S	BS	BS_2	B_2S_3

Bearing in mind the primary formation of $2BaO \cdot SiO_2$ while the substances interact in the mixtures of $BaCO_3-SiO_2$ independent of the proportions of the initial mixtures [75, 156, 218] one can say that the originally formed $2BaO \cdot SiO_2$, SiO_2 being in excess, converts consequently, to $BaO \cdot SiO_2$ and $BaO \cdot 2SiO_2$ (the final compound is determined by the stoichiometric composition) and $BaCO_3$ being in excess, to $3BaO \cdot SiO_2$.

2.5.2 System $BaSO_4-SiO_2$ [366]

The study of the system $BaO-SiO_2$ combined with the knowledge of ΔH^0 and ΔG^0 of the reaction of $BaSO_4$ decomposition allowed the determination of ΔH^0 and ΔG^0 for the following reactions:

1. $BaSO_4+2SiO_2=BaO \cdot 2SiO_2+SO_2+\frac{1}{2}O_2$
2. $2BaSO_4+3SiO_2=2BaO \cdot 3SiO_2+2SO_2+O_2$
3. $BaSO_4+SiO_2=BaO \cdot SiO_2+SO_2+\frac{1}{2}O_2$
4. $2BaSO_4+SiO_2=2BaO \cdot SiO_2+2SO_2+O_2$
5. $3BaSO_4+SiO_2=3BaO \cdot SiO_2+3SO_2+\frac{3}{2}O_2$.

For the possible decomposition of barium sulfate at high temperatures the following equation is most likely to fit:

$$BaSO_4=BaO+SO_2+\tfrac{1}{2}O_2\,.$$

On the basis of the thermochemical data presented in Table 21 ΔH^0 and ΔG^0 of decomposition were determined for this equation. The equation of the temperature dependence of the barium sulfate heat capacity was determined according to Landiya's method [320] at $T_{tr}=1451$ K; $T_m=1853$ K and $S^0_{298}=31.6$ cal/mol deg.

The following equations were obtained for calculating ΔH^0 and ΔG^0 of $BaSO_4$ up to the temperature of melting:

within a temperature range of 298 to 1551 K:

$$\Delta H_T^0 = 145076 - 2.67\,T + 1.17 \cdot 10^{-3} T^2 + 4.34 \cdot 10^5 T^{-1}$$

$$\Delta G_T^0 = 145076 + 2.67\,T \ln T - 1.17 \cdot 10^{-3} T^2 + 2.17 \cdot 10^5 T^{-1} - 62.8\,T$$

within a temperature range of 1551 – 1853 K:

$$\Delta H_T^0 = 146948 - 7.71\,T - 1.50 \cdot 10^{-3} T^2 + 4.34 \cdot 10^5 T^{-1}$$

$$\Delta G_T^0 = 146948 + 7.71\,T \ln T + 1.50 \cdot 10^{-3} T^2 + 2.17 \cdot 10^5 T^{-1} - 98.75\,T .$$

The calculation results are presented in Table 23.

Table 23. The alteration of ΔH^0 and ΔG^0 with temperature for the reaction of the $BaSO_4$ decomposition

Temp. [K]	500	600	700	800	900	1000	1100
ΔH_T^0 [kcal/mol]	144.90	144.62	144.40	144.23	144.10	144.01	143.95
ΔG_T^0 [kcal/mol]	122.22	117.57	113.08	108.62	104.18	99.77	95.35
Temp. [K]	1200	1300	1400	1500	1600	1700	1800
ΔH_T^0 [kcal/mol]	143.92	143.92	143.94	143.99	131.04	139.76	128.45
ΔG_T^0 [kcal/mol]	90.94	86.52	82.11	77.69	83.94	82.05	78.23

The data obtained allow the determination of the equilibrium constant and the equation of the temperature dependence of the $BaSO_4$ dissociation pressure. Thus, for $T = 1600$ K:

$$\lg K = \frac{83941}{4.576 \cdot 1600} = -11.3 \quad \text{or} \quad K = 5 \cdot 10^{-12}\,\text{atm} .$$

The equation of the temperature dependence of the $BaSO_4$ dissociation pressure is for the temperature range of 298 to 1551 K:

$$\lg p = -31.702\,T^{-1} - 0.583 \ln T + 0.256 \cdot 10^{-3} T - 0.474 \cdot 10^5 T^{-2} + 13.72;$$

for the temperature range of 1551 to 1853 K:

$$\lg p = -32.113\,T^{-1} - 1.685 \ln T - 0.329 \cdot 10^{-3} T - 0.474 \cdot 10^5 T^{-2} + 21.58 .$$

The results obtained show the $BaSO_4$ decomposition up to the temperature of melting to be thermodynamically improbable.

The ΔH^0 and ΔG^0 values of the reaction of barium silicate formation were determined by combining the corresponding ΔH^0 and ΔG^0 values of the main reactions of the system $BaO-SiO_2$ and of those of $BaSO_4$ decomposition in a temperature range of 500 to 1800 K (within each 100 K).

The analysis of the data obtained permits drawing a general conclusion that all possible reactions of the interaction of $BaSO_4$ with silica are thermodynamically almost impossible, since for all of them the ΔG^0 values in the temperature range covered are markedly positive.

The reactions of $BaSO_4$ with Al_2O_3 were also studied [13].

2.6 General Regularities of the Reactions of Silicate Formation in the Systems Under Consideration

From the above results of the thermodynamic analysis general conclusions can be drawn concerning regularities for reactions of silicate formation in the systems and hypotheses can be set forth concerning the connection of the thermodynamic characteristics of different silicates with their position in the periodic system and with their reactivity.

2.6.1 The Interaction of Quartz with Oxides of Alkali-Earth Metals [371]

In the system $BeO-SiO_2$ the most stable compound is beryllium orthosilicate $2BeO \cdot SiO_2$, the existence range of this compound being restricted to relatively low temperatures when the kinetic conditions do not favour its formation from pure oxides.

In the system $MgO-SiO_2$ the most stable compound at high temperatures is magnesium orthosilicate, but SiO_2 being in excess at relatively low temperatures, $MgO \cdot SiO_2$ is stable as well.

In the system $CaO-SiO_2$ with the ratio $CaO/SiO_2 > 1$, $2CaO \cdot SiO_2$ is stable, and with the ratio $CaO/SiO_2 < 1$, $CaO \cdot SiO_2$.

In the system $SrO-SiO_2$ with the ratio $SrO/SiO_2 > 1$ the most stable compound is $3SrO \cdot SiO_2$, and with the ratio $SrO/SiO_2 < 1$ strontium metasilicate $SrO \cdot SiO_2$ is most stable.

In the system $BaO-SiO_2$ with the ratio $BaO/SiO_2 > 1$ $3BaO \cdot SiO_2$ is most stable, while with the ratio $BaO/SiO_2 < 1$ barium silicates with high silica content $BaO \cdot SiO_2$ and $BaO \cdot 2SiO_2$ may be stable depending on the composition of the initial mixtures.

The comparison of the reactions of silicate formation in the above systems witnesses to the growth of the number and the broadening of the area of existence of various silicates as one moves ahead by the periodic system from top to bottom.

While for the system $BeO-SiO_2$ only one compound has been obtained, $2BeO \cdot SiO_2$, and in the system $MgO-SiO_2$ already two, $MgO \cdot SiO_2$ and

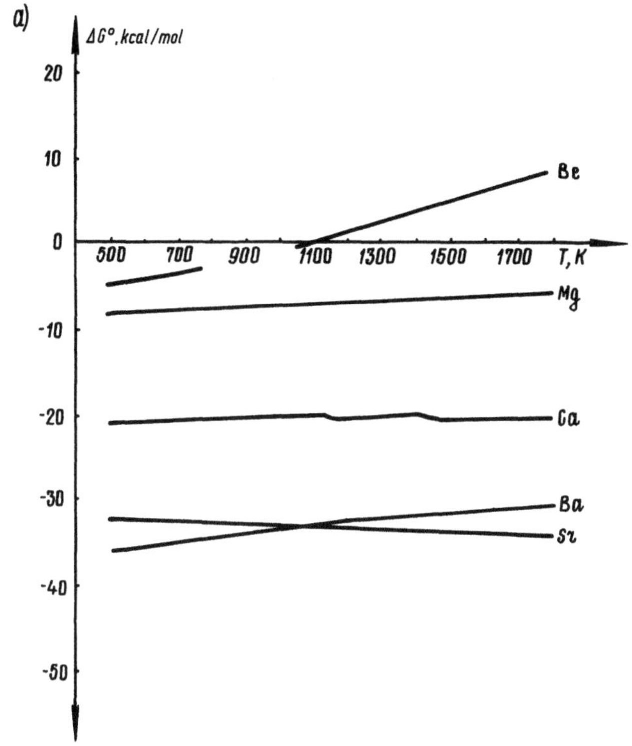

Fig. 32. $\Delta G^0 = f(T)$ of the formation of the compounds under consideration from oxides: **a** metasilicates; **b** orthosilicates

$2MgO \cdot SiO_2$, exist, $MgO \cdot SiO_2$ being however unstable at high temperatures, in the system $CaO - SiO_2$, contrarily, the metasilicate with high SiO_2 content becomes quite stable, and in the system $SrO - SiO_2$ the stability of $3SrO \cdot 2SiO_2$ intensifies as well.

In the system $BaO - SiO_2$ an outer broadening of the area of silicate formation is observed as well as a growth of the number of possible compounds (stable $3BaO \cdot SiO_2$ and $BaO \cdot 2SiO_2$).

The comparison of the Gibbs free energies of the formation of meta- and ortho-silicates from oxides of metals of the second group of the periodic system (Fig. 32) also witnesses to the growth of the stability of meta- and orthosilicates as advances down the periodic system. This circumstance allows the establishment of a peculiar series of the thermodynamic reactivity (as the latter grows) coinciding with the element position in the second group of the periodic system.

Thus, the oxides standing below in the periodic system will replace the corresponding oxides in the silicates placed above in the periodic system.

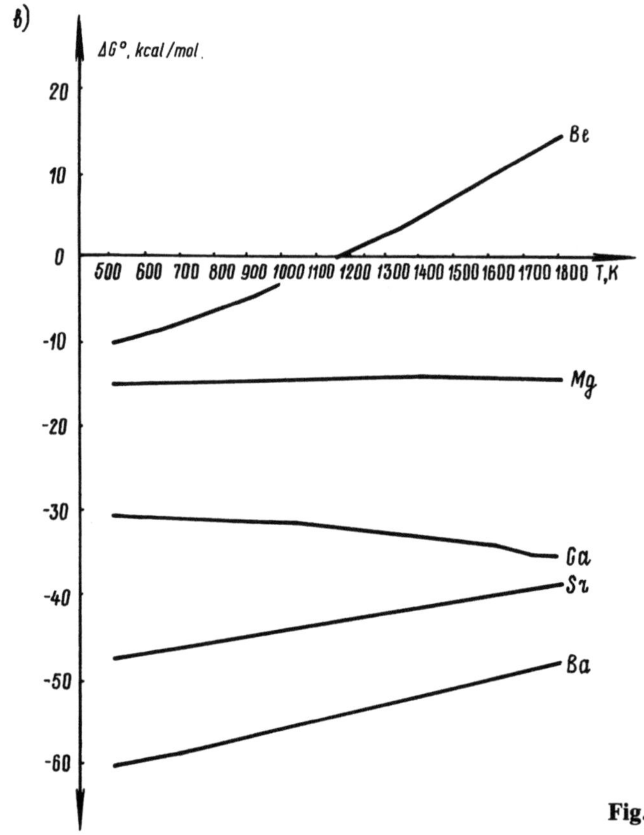

Fig. 32b (Legend see page 121)

2.6.2 The Interaction of Quartz with Carbonates of Alkali-Earth Metals

The formation of metasilicates of all the alkali-earth metals, except beryllium the carbonate of which is unstable, appears thermodynamically possible within the temperature range from 465 to 710 K (see Table 24).

The thermodynamic sequence can be expressed as the following series of silicates:

Mg (465 K); Ca, Sr (575 K) and Ba (710 K).

The formation of orthosilicates of all the alkali-earth metals is possible within a temperature range of 490 – 880 K. The sequence of the beginning of the reactions of orthosilicate formation can be expressed in the following series:

Mg (490 K); Ca (715 K); Sr (800 K) and Ba (880 K) .

At high temperatures (1800 K) the thermodynamic sequence can be characterized by a series similar to that of the interaction of quartz with the oxides of alkali-earth metals.

The formation of the silicates of Me_3SiO_5 (Ca, Sr, Ba) type becomes thermodynamically possible within a temperature range of 825 to 890 K and is characterized by a following sequence:

Sr (825 K); Ba (850 K) and Ca (890 K).

The formation of silicates of $Me_2Si_3O_8$ (550 K) and $MeSi_2O_5$ (565 K) type is possible only with Ba whereas that of the silicates of $Me_3Si_2O_7$ type, only with Ca (725 K). In general a tendency is observed towards the lowering of a temperature of the beginning of reactions ($p_r = 1$ atm) with the increase of SiO_2 content.

As a rule in the case of the interaction of carbonates of alkali-earth metals with quartz a sequence is observed corresponding to the position of the metals in question in the periodic system. It is true, however, that this sequence is considerably weaker than in the case of the interaction of quartz with oxides of alkali-earth metals.

2.6.3 The Interaction of Quartz with Sulphates of Alkali-Earth Metals [366, 388]

No formation of barium metasilicate is thermodynamically possible in the entire temperature range considered. The formation of Mg, Ca and Sr metasilicates is thermodynamically possible within a temperature range of 1225 to 1545 K in the

Table 24. Temperature of the main reactions in the silicate systems under consideration at $p_r = 1$ atm

Carbonates	Temp. [K] for silicates			
	Mg	Ca	Sr	Ba
1. $3MeCO_3 + SiO_2 = Me_3SiO_5 + 3CO_2$	–	890	825	850
2. $2MeCO_3 + SiO_2 = Me_2SiO_4 + 2CO_2$	490	715	800	880
3. $3MeCO_3 + 2SiO_2 = Me_3Si_2O_7 + 3CO_2$	–	725	–	–
4. $MeCO_3 + SiO_2 = MeSiO_3 + CO_2$	465	575	575	710
5. $2MeCO_3 + 3SiO_2 = Me_2Si_3O_8 + 3CO_2$	–	–	–	550
6. $MeCO_3 + 2SiO_2 = MeSi_2O_5 + CO_2$	–	–	–	565
Sulphates				
1. $3MeSO_4 + SiO_2 = Me_3SiO_5 + 3SO_2 + \frac{3}{2}O_2$	–	1740	1340	Impossible
2. $2MeSO_4 + SiO_2 = Me_2SiO_4 + 2SO_2 + O_2$	1440	1615	1380	Impossible
3. $3MeSO_4 + 2SiO_2 = Me_3Si_2O_7 + 3SO_2 + \frac{3}{2}O_2$	–	1645	–	–
4. $MeSO_4 + SiO_2 = MeSiO_3 + SO_2 + \frac{1}{2}O_2$	1445	1545	1225	Impossible
5. $2MeSO_4 + 3SiO_2 = Me_2Si_3O_8 + 2SO_2 + O_2$	–	–	–	Impossible
6. $MeSO_4 + 2SiO_2 = MeSi_2O_5 + SO_2 + \frac{1}{2}O_2$	–	–	–	Impossible

following sequence: strontium metasilicate (1225 K), magnesium metasilicate (1445 K) and calcium metasilicate (1545 K) (see Table 24).

The formation of barium orthosilicate is thermodynamically impossible in the entire temperature range under consideration. The formation of orthosilicates of the rest of the alkali-earth metals is possible within a temperature range of 1380 to 1615 K. The thermodynamic sequence can be expressed by the following series:

Sr (1380 K); Mg (1440 K) and Ca (1615 K).

No formation of the silicate Ba_3SiO_5 is thermodynamically possible within the entire temperature range considered. The Ca_3SiO_5 formation is possible at 1740 K and that of Sr_3SiO_5, at 1340 K. Especially noteworthy is the low temperature of reactions of the formation of strontium silicates. Strontium sulphate may be introduced as a refiner for glassmelting (in form of a natural mineral of celestine) and as a component for the melting of strontium-containing glasses and enamels.

The formation of the silicate $Ca_3Si_2O_7$ is possible at 1645 K and that of $Ba_2Si_3O_8$ and $BaSi_2O_5$ is impossible in the entire temperature range under consideration. As in the previous case a tendency towards the lowering of the reaction temperature ($p_r = 1$ atm) with the increase of SiO_2 content is observed.

2.7 System Iron Oxides – Silica

The system iron oxides-silica is of importance in mineralogy, petrography, metallurgy and in silicate production processes.

Owing to the difficulties arising in the study of this system – sensitivity to oxidation and reduction – of great interest is a thermodynamic calculation of solid-phase reactions in the area of compounds corresponding to $FeO \cdot SiO_2$ and $2FeO \cdot SiO_2$ revealed in literature [636]. Since the valence of iron in the given system is variable, it is necessary to compare the reactions of the formation of bivalent iron silicates from oxides FeO, Fe_2O_3, Fe_3O_4 and from silica (Fig. 33). For calculating the alteration of the free energy and enthalpy of the reaction of formation of iron silicates the thermochemical data given in Table 25 [248, 311, 623, 624] were used. The following reactions were considered:

I-1. $Fe_{0.947}O + 0.053Fe + SiO_2 = FeO \cdot SiO_2$

I-2. $2Fe_{0.947}O + 0.106Fe + SiO_2 = 2FeO \cdot SiO_2$

II-1. $\frac{1}{2}Fe_2O_3 + SiO_2 = FeO \cdot SiO_2 + \frac{1}{4}O_2$

II-2. $Fe_2O_3 + SiO_2 = 2FeO \cdot SiO_2 + \frac{1}{2}O_2$

III-1. $\frac{1}{3}Fe_3O_4 + SiO_2 = FeO \cdot SiO_2 + \frac{1}{6}O_2$

III-2. $\frac{2}{3}Fe_3O_4 + SiO_2 = 2FeO \cdot SiO_2 + \frac{1}{3}O_2$.

Table 25. The initial thermochemical data for calculating $\Delta H^0 = f(T)$ and $\Delta G^0 = f'(T)$ of the reactions of the iron silicate formation from oxides

Substance	$-\Delta H^0_{298(el)}$ [kcal/mol]	$-\Delta G^0_{298(el)}$ [kcal/mol]	S^0_{298} [cal/mol deg]	Temp. of melting, transformation [K]	ΔH_m (transform.) [kcal/mol]	$C_p = a + bT - cT^{-2}$			Temp. range [K]
						a	$b \cdot 10^3$	$c \cdot 10^{-5}$	
$Fe_{0.947}O$	63.4	58.4	13.76	1651	7.49	11.66	2.00	−0.67	298–1651
						14.91	–	–	1651–1800
Fe_3O_4-α	267.9	242.32	35.0	900	–	21.88	48.20	–	298–900
Fe_3O_4-β	–	–	–	1870	33.0	48.0	–	–	900–1800
Fe_2O_3-α	196.50	177.14	21.50	950	–	23.48	18.60	−3.55	298–950
Fe_2O_3-β	–	–	–	1050	–	36.00	–	–	950–1050
Fe_2O_3-δ	–	–	–	1750	–	31.7	1.76	–	1050–1750
						34.78	–	–	1750–1800
SiO_2(β-quartz)	217.75	204.75	10.0	848	0.29	11.22	8.2	−2.7	298–848
SiO_2(α-quartz)	–	–	–			14.41	1.94		848–1800
$FeO \cdot SiO_2$	285.63	267.10	23.0	1413	14.6	25.88	5.12	−4.23	298–1413
						33.44	–	–	1413–1800
$2FeO \cdot SiO_2$	353.85	330.11	34.7	1493	22.0	36.51	9.36	−6.70	298–1493
						57.50	–	–	1493–1800
Fe-α	–	–	–	–	–	4.18	5.92	–	298–1033
Fe-β	–	–	–	–	–	9.0	–	–	1033–1181
Fe-γ	–	–	–	–	–	1.84	4.66	–	1181–1674
Fe-δ	–	–	–	–	–	10.50	–	–	1674–1800
O_2	–	–	49.0	–	–	8.20	0.26	−1.88	298–1800

Note: Equation $C_p = f(T)$ for $FeO \cdot SiO_2$ was calculated by Landiya's method

Table 26. Equations for calculating $\Delta H^0 = f(T)$ of the reactions in the system iron oxides-silica

Nos. of reactions	Reactions	Temp. range [K]	Equations $\Delta H^0 = f(T)$
I-1	$Fe_{0.947}O + SiO_2 + 0.053Fe = FeO \cdot SiO_2$	298 – 848	$-5342 + 2.73T - 2.69 \cdot 10^{-3}T^2 + 0.86 \cdot 10^5 T^{-1}$
		848 – 1033	$-5496 - 0.46T + 0.43 \cdot 10^{-3}T^2 + 3.56 \cdot 10^5 T^{-1}$
		1033 – 1181	$-5399 - 0.72T + 0.59 \cdot 10^{-3}T^2 + 3.56 \cdot 10^5 T^{-1}$
		1181 – 1413	$-5675 - 0.34T + 0.47 \cdot 10^{-3}T^2 + 3.56 \cdot 10^5 T^{-1}$
		1413 – 1651	$3582 + 7.27T - 2.09 \cdot 10^{-3}T^2 - 0.67 \cdot 10^5 T^{-1}$
		1651 – 1674	$-1275 + 4.001 - 1.09 \cdot 10^{-3}T^2$
		1674 – 1800	$-853 + 3.54T - 0.97 \cdot 10^{-3}T^2$
I-2	$2Fe_{0.947}O + 0.106Fe + SiO_2 = 2FeO \cdot SiO_2$	298 – 848	$-10493 + 1.52T - 1.73 \cdot 10^{-3}T^2 + 2.66 \cdot 10^5 T^{-1}$
		848 – 1033	$-10644 - 1.66T + 1.39 \cdot 10^{-3}T^2 + 5.36 \cdot 10^5 T^{-1}$
		1033 – 1181	$-10455 - 2.17T + 1.71 \cdot 10^{-3}T^2 + 5.96 \cdot 10^5 T^{-1}$
		1181 – 1493	$-11007 - 1.41T + 1.46 \cdot 10^{-3}T^2 + 5.36 \cdot 10^5 T^{-1}$
		1493 – 1651	$-94946 + 19.57T - 3.21 \cdot 10^{-3}T^2 + 1.34 \cdot 10^5 T^{-1}$
		1651 – 1674	$-42003 + 13.03T - 1.21 \cdot 10^{-3}T^2$
		1674 – 1800	$-13050 + 12.12T - 0.97 \cdot 10^{-3}T^2$
II-1	$\frac{1}{2}Fe_2O_3 + SiO_2 = FeO \cdot SiO_2 + \frac{1}{4}O_2$	298 – 848	$29251 + 4.97T - 4.90 \cdot 10^{-3}T^2 + 0.22 \cdot 10^5 T^{-1}$
		848 – 950	$29103 + 1.78T - 1.77 \cdot 10^{-3}T^2 + 2.93 \cdot 10^5 T^{-1}$
		950 – 1050	$31782 - 4.48T + 1.62 \cdot 10^{-3}T^2 + 4.7 \cdot 10^5 T^{-1}$
		1050 – 1413	$30071 - 2.33T + 1.18 \cdot 10^{-3}T^2 + 4.7 \cdot 10^5 T^{-1}$
		1413 – 1750	$39345 + 5.23T - 1.37 \cdot 10^{-3}T^2 + 0.47 \cdot 10^5 T^{-1}$
		1750 – 1800	$40707 + 3.69T - 0.94 \cdot 10^{-3}T^2 + 0.47 \cdot 10^5 T^{-1}$

Table 26 (continued)

Nos. of reactions	Reactions	Temp. range [K]	Equations $\Delta H^0 = f(T)$
II-2	$Fe_2O_3 + SiO_2 = 2FeO \cdot SiO_2 + \frac{1}{2}O_2$	298 – 848	$58720 + 5.91\,T - 6.15 \cdot 10^{-3}T^2 + 1.39 \cdot 10^5 T^{-1}$
		848 – 950	$57560 + 2.72\,T - 3.03 \cdot 10^{-3}T^2 + 4.09 \cdot 10^5 T^{-1}$
		950 – 1050	$63950 + 9.80\,T + 3.77 \cdot 10^{-3}T^2 + 7.64 \cdot 10^5 T^{-1}$
		1050 – 1413	$60401 - 5.50\,T + 2.89 \cdot 10^{-3}T^2 + 7.64 \cdot 10^5 T^{-1}$
		1413 – 1750	$54552 + 15.49\,T - 1.78 \cdot 10^{-3}T^2 + 9.4 \cdot 10^5 T^{-1}$
		1750 – 1800	$57243 + 12.41\,T - 0.90 \cdot 10^{-3}T^2 + 9.4 \cdot 10^5 T^{-1}$
III-1	$\frac{1}{3}Fe_3O_4 + SiO_2 = FeO \cdot SiO_2 + \frac{1}{6}O_2$	298 – 848	$18962 + 8.73\,T - 9.54 \cdot 10^{-3}T^2 + 1.84 \cdot 10^5 T^{-1}$
		848 – 900	$18823 + 5.54\,T - 6.41 \cdot 10^{-3}T^2 + 4.54 \cdot 10^5 T^{-1}$
		900 – 1413	$20131 - 3.15\,T + 1.61 \cdot 10^{-3}T^2 + 4.54 \cdot 10^5 T^{-1}$
		1413 – 1800	$29466 + 4.40\,T - 0.94 \cdot 10^{-3}T^2 + 12.00 \cdot 10^5 T^{-1}$
III-2	$\frac{2}{3}Fe_3O_4 + SiO_2 = 2FeO \cdot SiO_2 + \frac{1}{3}O_3$	298 – 848	$38131 + 13.45\,T - 15.4 \cdot 10^{-3}T^2 + 4.62 \cdot 10^5 T^{-1}$
		848 – 900	$37983 + 10.26\,T - 12.3 \cdot 10^{-3}T^2 + 7.38 \cdot 10^5 T^{-1}$
		900 – 1493	$40635 - 7.14\,T + 3.75 \cdot 10^{-3}T^2 + 7.33 \cdot 10^5 T^{-1}$
		1493 – 1800	$42174 + 13.85\,T - 0.98 \cdot 10^{-3}T^2 + 6.26 \cdot 10^5 T^{-1}$

Table 27. Equations for calculating $\Delta G^0 = f(T)$ of the reactions in the system of iron oxides-silica

Nos. of reactions	Reactions	Temp. range [K]	Equations $\Delta G^0 = f(T)$
I-1	$Fe_{0.947}O + 0.053Fe + SiO_2 = FeO \cdot SiO_2$	298 – 848	$-5342 - 2.73\,T\ln T + 2.69 \cdot 10^{-3}T^2 + 0.43 \cdot 10^5 T^{-1} + 19.09\,T$
		848 – 1033	$-5496 + 0.46\,T\ln T - 0.43 \cdot 10^{-3}T^2 + 1.78 \cdot 10^5 T^{-1} + 0.23\,T$
		1033 – 1181	$-5399 + 0.72\,T\ln T - 0.59 \cdot 10^{-3}T^2 + 1.78 \cdot 10^5 T^{-1} - 1.47\,T$
		1181 – 1413	$-5675 + 0.34\,T\ln T - 0.47 \cdot 10^{-3}T^2 + 1.78 \cdot 10^5 T^{-1} + 1.30\,T$
		1413 – 1651	$3582 - 7.27\,T\ln T - 2.09 \cdot 10^{-3}T^2 - 0.34 \cdot 10^5 T^{-1} + 46.44\,T$
		1651 – 1674	$-1275 - 4.00\,T\ln T - 1.09 \cdot 10^{-3}T^2 + 26.79\,T$
		1674 – 1800	$-853 - 3.54\,T\ln T - 0.97 \cdot 10^{-3}T^2 + 23.34\,T$
I-2	$2Fe_{0.947}O + 0.106\,Fe + SiO_2 = 2FeO \cdot SiO_2$	298 – 848	$-10493 - 1.52\,T\ln T + 1.73 \cdot 10^{-3}T^2 + 1.33 \cdot 10^5 T^{-1} + 13.3\,T$
		848 – 1033	$-10644 + 1.66\,T\ln T - 1.39 \cdot 10^{-3}T^2 + 2.68 \cdot 10^5 T^{-1} - 5.52\,T$
		1033 – 1181	$-10455 + 2.17\,T\ln T - 1.71 \cdot 10^{-3}T^2 + 2.68 \cdot 10^5 T^{-1} + 8.93\,T$
		1181 – 1493	$-11007 + 1.41\,T\ln T - 1.46 \cdot 10^{-3}T^2 + 2.68 \cdot 10^5 T^{-1} - 3.38\,T$
		1493 – 1651	$-94646 - 19.57\,T\ln T + 3.21 \cdot 10^{-3}T^2 - 0.67 \cdot 10^5 T^{-1} + 142.15$
		1651 – 1674	$-42003 - 13.03\,T\ln T + 1.21 \cdot 10^{-3}T^2 + 93.76\,T$
		1674 – 1800	$-13050 - 12.12\,T\ln T + 0.97 \cdot 10^{-3}T^2 + 7.52\,T$
II-1	$\frac{1}{2}Fe_2O_3 + SiO_2 = FeO \cdot SiO_2 + \frac{1}{4}O_2$	298 – 848	$29251 - 4.97\,T\ln T + 4.9 \cdot 10^{-3}T^2 + 0.11 \cdot 10^5 T^{-1} + 16.72\,T$
		848 – 950	$29103 - 1.78\,T\ln T + 1.77 \cdot 10^{-3}T^2 + 1.46 \cdot 10^5 T^{-1} + 2.13\,T$
		950 – 1050	$31782 + 4.48\,T\ln T + 1.62 \cdot 10^{-3}T^2 + 2.35 \cdot 10^5 T^{-1} - 44.75\,T$
		1050 – 1413	$30014 + 2.33\,T\ln T - 1.18 \cdot 10^{-3}T^2 + 2.35 \cdot 10^5 T^{-1} - 28.57\,T$
		1413 – 1750	$3945 - 5.23\,T\ln T + 1.37 \cdot 10^{-3}T^2 + 0.23 \cdot 10^5 T^{-1} + 16.15\,T$
		1750 – 1800	$40707 - 3.69\,T\ln T + 0.94 \cdot 10^{-3}T^2 + 0.23 \cdot 10^5 T^{-1} + 4.65\,T$

Table 27 (continued)

Nos. of reactions	Reactions	Temp. range [K]	Equations $\Delta G^0 = f(T)$
II-2	$Fe_2O_3 + SiO_2 = 2FeO \cdot SiO_2 + \frac{1}{2}O_2$	298 – 848	$58720 - 5.91\,T\ln T + 6.15 \cdot 10^{-3}T^2 + 0.69 \cdot 10^5 T^{-1} + 7.94\,T$
		848 – 950	$58560 - 2.72\,T\ln T + 3.02 \cdot 10^{-3}T^2 + 2.05 \cdot 10^5 T^{-1} - 10.92\,T$
		950 – 1050	$63950 + 9.80\,T\ln T - 3.77 \cdot 10^{-3}T^2 + 3.82 \cdot 10^5 T^{-1} - 96.17\,T$
		1050 – 1413	$60401 + 5.50\,T\ln T - 2.89 \cdot 10^{-3}T^2 + 3.82 \cdot 10^5 T^{-1} - 63.81\,T$
		1413 – 1750	$54552 - 15.49\,T\ln T + 1.78 \cdot 10^{-3}T^2 + 4.7 \cdot 10^5 T^{-1} + 86.68\,T$
		1750 – 1800	$57243 - 12.41\,T\ln T + 0.9 \cdot 10^{-3}T^2 + 4.7 \cdot 10^5 T^{-1} + 63.69\,T$
III-1	$\frac{1}{3}Fe_3O_4 + SiO_2 = FeO \cdot SiO_2 + \frac{1}{6}O_2$	298 – 848	$18962 - 6.73\,T\ln T + 4.77 \cdot 10^{-3}T^2 + 0.92 \cdot 10^5 T^{-1} + 43.99\,T$
		848 – 900	$18823 - 5.54\,T\ln T + 3.20 \cdot 10^{-3}T^2 + 2.27 \cdot 10^5 T^{-1} + 25.13\,T$
		900 – 1413	$20131 + 3.15\,T\ln T - 0.80 \cdot 10^{-3}T^2 + 2.27 \cdot 10^5 T^{-1} - 28.29\,T$
		1413 – 1800	$29466 - 4.40\,T\ln T + 0.47 \cdot 10^{-3}T^2 + 1.56 \cdot 10^5 T^{-1} + 16.43\,T$
III-2	$\frac{3}{4}Fe_3O_4 + SiO_2 = 2FeO \cdot SiO_2 + \frac{1}{3}O_2$	298 – 848	$38131 - 13.45\,T\ln T + 15.4 \cdot 10^{-3}T^2 + 2.31 \cdot 10^5 T^{-1} + 62.51\,T$
		848 – 900	$37983 - 10.26\,T\ln T + 12.3 \cdot 10^{-3}T^2 + 3.66 \cdot 10^5 T^{-1} + 43.65\,T$
		900 – 1493	$40635 + 7.14\,T\ln T - 3.75 \cdot 10^{-3}T^2 + 3.66 \cdot 10^5 T^{-1} - 63.18\,T$
		1493 – 1800	$42174 - 13.85\,T\ln T + 0.92 \cdot 10^{-3}T^2 + 3.13 \cdot 10^5 T^{-1} + 82.35\,T$

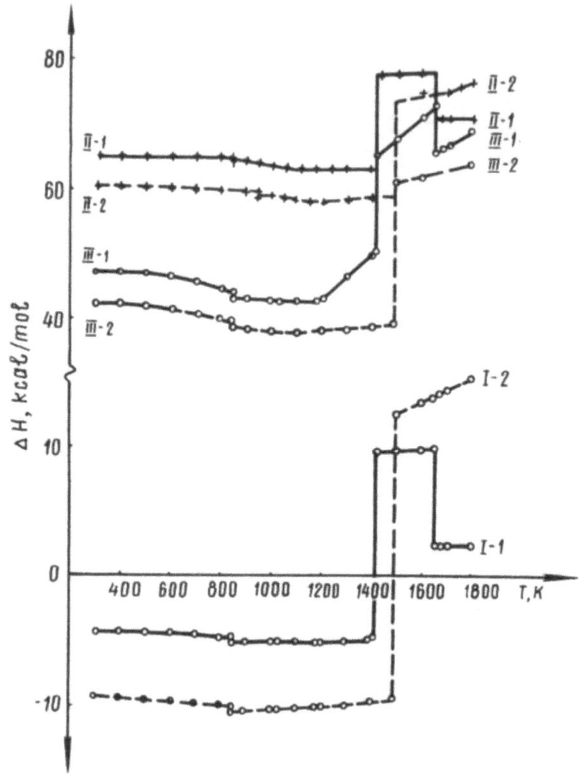

Fig. 33. The enthalpy of the reactions of formation of ferric meta- and orthosilicate as a function of temperature within the range of 298 to 1800 K

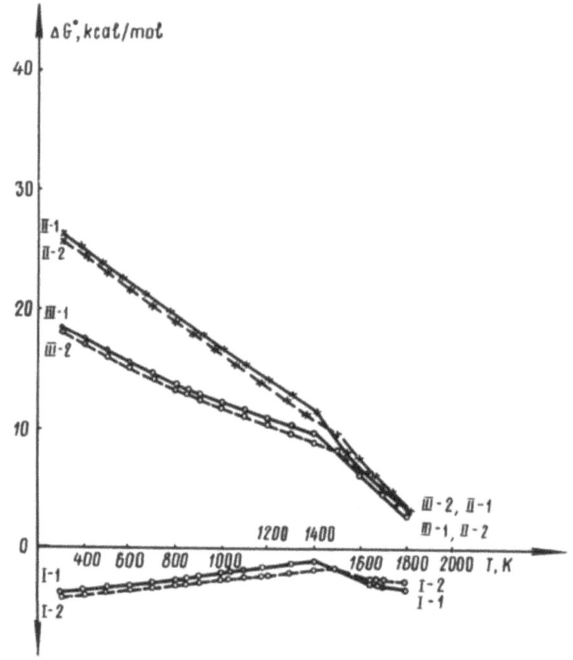

Fig. 34. The temperature dependence of ΔG^0 of ferric silicate formation from oxides, the ratio of Fe^{n+}/Si^{4+} being 1/1

Equations of $\Delta H^0 = f(T)$ and $\Delta G^0 = f'(T)$ (Tables 26 and 27) were calculated for these reactions. The dependence of the enthalpy and the Gibbs free energy alteration of the formation of iron silicates from oxides is presented in Figs. 34 and 35.

The ΔH^0 and ΔG^0 calculations were carried out within a temperature range of 298 to 1800 K.

For evaluating the thermodynamic sequence of reactions the dependence of ΔG_T^0 of the reactions on the composition of the initial mixtures was examined. The ΔG_T^0 values were examined the ratios of the initial mixtures of iron oxides: silica being 1/1 and 2/1.

The data obtained permit drawing the following conclusions:

The reactions of the formation of silicates of bivalent iron from oxides are endothermic apart from those between wüstite and quartz at low temperatures.

From the thermodynamic point of view iron silicates are more likely to be formed from wüstite; hematite and magnetite being used, no formation of iron silicates is possible up to the temperature of the order of 1700 – 2000 K.

Up to 1500 K independent of the ratio of iron and silicium cations the $2FeO \cdot SiO_2$ formation is more probable. At temperatures above 1500 K, if the proportion of ferrous oxide silica is 1/1, the formation of iron metasilicate is more probable.

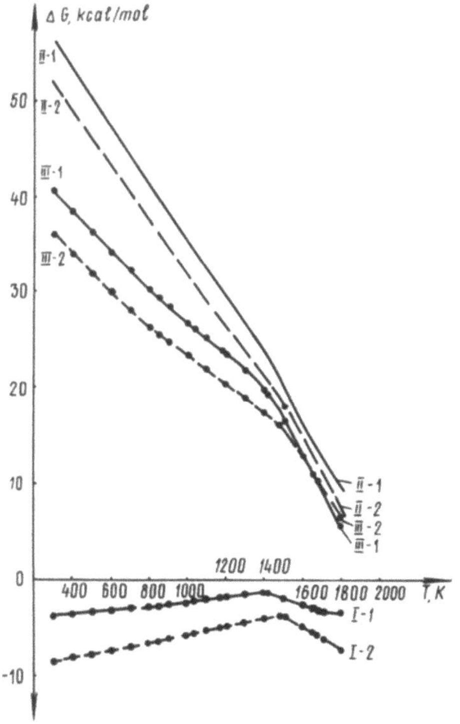

Fig. 35. The temperature dependence of ΔG^0 of ferric silicate formation from oxides, the ratio of Fe^{n+}/Si^{4+} being 2/1

2.8 System Manganese Oxides – Silica

The study of solid-phase reactions in the system manganese oxides-silica is of great theoretical and practical interest as they play an important role in a number of processes of silicate production. In the system manganese oxides-silica compounds $MnO \cdot SiO_2$ and $2MnO \cdot SiO_2$, [636] exist. In this connection it is advisable to conduct a thermodynamic analysis of reactions in this system with the formation of silicates of bivalent manganese. We have examined the following main reactions of the formation of manganese silicates from oxides. See Tables 25, 26 and 27.

I-1. $MnO + SiO_2 = MnO \cdot SiO_2$
I-2. $2MnO + SiO_2 = 2MnO \cdot SiO_2$
II-1. $\frac{1}{3}Mn_3O_4 + SiO_2 = MnO \cdot SiO_2 + \frac{1}{6}O_2$
II-2. $\frac{2}{3}Mn_3O_4 + SiO_2 = 2MnO \cdot SiO_2 + \frac{1}{3}O_2$
III-1. $\frac{1}{2}Mn_2O_3 + SiO_2 = MnO \cdot SiO_2 + \frac{1}{4}O_2$
III-2. $Mn_2O_3 + SiO_2 = 2MnO \cdot SiO_2 + \frac{1}{2}O_2$.

For calculating the alteration of enthalpy and the Gibbs free energy of the formation of manganese silicates the initial data presented in Table 28 [248, 311, 624] were used.

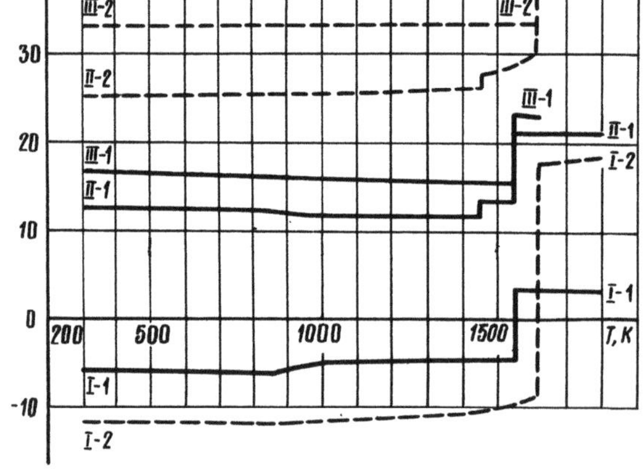

Fig. 36. The temperature dependence of ΔH of reactions of manganese silicate formation from oxides within a temperature range of 298 to 1800 K

Table 28. Initial thermochemical data for calculating ΔH_T^0 and ΔG_T^0 of the reactions of formation of manganese silicates from oxides

Compounds	T_m (transform.) [K]	$-\Delta H_{293}^0$ [kcal/mol]	$-\Delta G_{298}^0$ [kcal/mol]	S_{298}^0 [cal/mol deg]	ΔH_m (transform.) [kcal/mol]	Temp. dependence of heat capacity $C_p = f(T)$
MnO	2058	92.0	86.8	14.4	13.0	$11.1 + 1.94 \cdot 10^{-3} T - 0.88 \cdot 10^5 T^{-2}$
Mn_3O_4	1445	331.4	306.0	35.5	4.97	$34.64 + 10.82 \cdot 10^{-3} T - 2.2 \cdot 10^5 T^{-2}$
Mn_2O_3	1620	229.4	212.7	26.4	–	$24.73 + 8.38 \cdot 10^{-3} T - 3.23 \cdot 10^5 T^{-2}$
$MnSiO_3$	1545	315.65	296.55	21.3	8.0	$26.42 + 3.88 \cdot 10^{-3} T - 6.16 \cdot 10^5 T^{-2}$
Mn_2SiO_4	1618	413.55	389.58	36.9	26.15	$34.38 + 11.25 \cdot 10^{-3} T - 3.92 \cdot 10^5 T^{-2}$ [a]

[a] Calculated by Landiya's method [320]

For the above reactions equations of $\Delta H^0 = f(T)$ and $\Delta G^0 = f'(T)$ (Table 29) were calculated. Dependences of the alteration of the enthalpy and of the Gibbs free energy of the formation of manganese silicates from oxides are presented in Figs. 36–38.

For evaluating the thermodynamic sequence of reactions the dependence ΔG_T^0 system on the composition of the initial mixtures was examined. We analyzed the ΔG_T^0 values within a temperature range of 500 to 1800 K, the ratio of the initial mixtures of manganese oxides: silica being 1/1 and 2/1. The results of the thermodynamic analysis of these reactions are presented in Figs. 37 and 38.

The relations obtained allow drawing the following conclusions:

1. The ratio of manganese oxides: silica being 1/1, from the thermodynamic point of view, independent of the type of the initial raw material, more preferable are reactions of the formation of manganese orthosilicate: Mn_2O_3 and SiO_2 do not react within the temperature range under consideration, and Mn_3O_4 forms silicates of bivalent manganese only in the high-temperature range; a reaction between MnO and SiO_2 is possible at any temperature.

2. The ratio of manganese oxides: silica being 2/1, the same regularities are kept as at that of 1/1, in the low-temperature range, however, reactions II-2 and III-2 are characterized by the most positive values of ΔG^0 which is probably due to a greater contribution of ΔG^0 of oxides Mn_2O_3 and Mn_3O_4 in these reactions as compared to that of reactions II-1 and III-1. It should be mentioned that at both ratios reactions II-1 and II-2 become thermodynamically probable, beginning with temperatures 1230 and 1595 K respectively.

3. The stability of manganese silicates increases at temperatures higher than 1500 K as well; this proves that manganese silicates also retain their stability on melting.

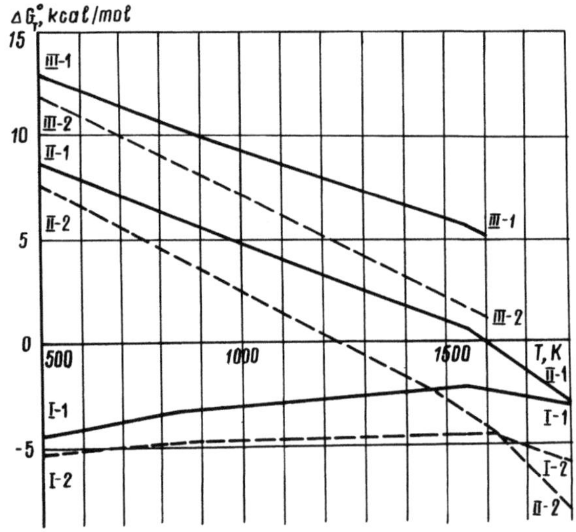

Fig. 37. The temperature dependence of ΔG^0 of the reactions of formation of manganese silicates from oxides within a temperature range of 500 to 1800 K, the ratio of manganese and silicon cations being 1/1

Table 29. Equations for calculating $\Delta H^0 = f(T)$ and $\Delta G^0 = f'(T)$ of reactions in the system manganese oxides-silica

Nos. of reactions	Temp. range [K]	Temp. dependence of ΔH_T^0 [cal/mol]	Temp. dependence of ΔG_T^0 [cal/mol]
I-2	298 – 848	$-7617+4.09T-3.13\cdot10^{-3}T^2+2.58\cdot10^5T^{-1}$	$-7617-4.09T\ln T+3.13\cdot10^{-3}T^2+1.29\cdot10^5T^{-1}+29.8T$
	848 – 1545	$-6237+0.9T+5.28\cdot10^5T^{-1}$	$-6237-0.9T\ln T+2.69\cdot10^5T^{-1}+91.5T$
	1545 – 1800	$-1782+6.45T-1.94\cdot10^{-3}T^2-0.88\cdot10^5T^{-1}$	$-1782-6.45T\ln T+1.94\cdot10^{-3}T^2-0.44\cdot10^5T^{-1}+44.12T$
I-2	298 – 848	$-12224+0.94T-0.415\cdot10^{-3}T^2-0.54\cdot10^5T^{-1}$	$-12224-0.94T\ln T+0.415\cdot10^{-3}T^2-0.27\cdot10^5T^{-1}+11.6T$
	848 – 1610	$-12083-2.25T+2.715\cdot10^{-3}T^2+2.16\cdot10^5T^{-1}$	$-12083+2.25T\ln T-2.715\cdot10^{-3}T^2+1.08\cdot10^5T^{-1}-10.34T$
	1610 – 1800	$-200+15.82T-2.91\cdot10^{-3}T^2-1.76\cdot10^5T^{-1}$	$-200-15.82T\ln T+2.91\cdot10^{-3}T^2-0.88\cdot10^5T^{-1}+106.7T$
II-1	298 – 848	$10404+5.02T-3.94\cdot10^{-3}T^2+3.04\cdot10^5T^{-1}$	$10404-5.02T\ln T+3.94\cdot10^{-3}T^2+1.52\cdot10^5T^{-1}+25T$
	848 – 1445	$10539+1.83T-0.81\cdot10^{-3}T^2+5.74\cdot10^5T^{-1}$	$10539-1.83T\ln T+0.81\cdot10^{-3}T^2+2.87\cdot10^5T^{-1}+5.73T$
	1445 – 1545	$15836-3.36T+0.99\cdot10^{-3}T^2+16.46\cdot10^5T^{-1}$	$15886+3.36T\ln T-0.99\cdot10^{-3}T^2+3.23\cdot10^5T^{-1}-33.07T$
	1545 – 1800	$20349+2.19T-0.95\cdot10^{-3}T^2+0.30\cdot10^5T^{-1}$	$20349-2.19T\ln T+0.95\cdot10^{-3}T^2+0.15\cdot10^5T^{-1}+1.87T$
II-2	298 – 848	$24360+2.8T-2.04\cdot10^{-3}T^2+0.38\cdot10^5T^{-1}$	$24360-2.8T\ln T+2.04\cdot10^{-3}T^2+0.19\cdot10^5T^{-1}-2.27T$
	848 – 1445	$24493-0.39T+1.09\cdot10^{-3}T^2+3.08\cdot10^5T^{-1}$	$24493+0.39T\ln T-1.09\cdot10^{-3}T^2+1.54\cdot10^5T^{-1}-21.44T$
	1445 – 1618	$35164-10.76T+4.7\cdot10^{-3}T^2+4.54\cdot10^5T^{-1}$	$35164+10.76T\ln T-4.7\cdot10^{-3}T^2+2.27\cdot10^5T^{-1}-99.04T$
	1618 – 1800	$47049+7.31-0.92\cdot10^{-3}T^2+0.62\cdot10^5T^{-1}$	$47049-7.31T\ln T+0.92\cdot10^{-3}T^2+0.31\cdot10^5T^{-1}+17.90T$
III-1	298 – 848	$14945+4.88T-4.22\cdot10^{-3}T^2+2.30\cdot10^5T^{-1}$	$14945-4.88T\ln T+4.22\cdot10^{-3}T^2+1.15\cdot10^5T^{-1}+23.76T$
	848 – 1545	$15080+1.69T-1.09\cdot10^{-3}T^2+0.5\cdot10^5T^{-1}$	$15080-1.69T\ln T+1.09\cdot10^{-3}T^2+2.50\cdot10^5T^{-1}+4.55T$
	1545 – 1620	$19535+1.24T-3.03\cdot10^{-3}T^2-1.14\cdot10^5T^{-1}$	$19535-7.24T\ln T+3.03\cdot10^{-3}T^2-0.57\cdot10^5T^{-1}+39.5T$
III-2	298 – 848	$33436+2.53T-2.60\cdot10^{-3}T^2-1.06\cdot10^5T^{-1}$	$33436-2.53T\ln T+2.60\cdot10^{-3}T^2-0.53\cdot10^5T^{-1}-4.46T$
	848 – 1620	$33741-0.76T+0.53\cdot10^{-3}T^2+1.62\cdot10^5T^{-1}$	$33741+0.86T\ln T-0.53\cdot10^{-3}T^2+0.81\cdot10^5T^{-1}-25.13T$

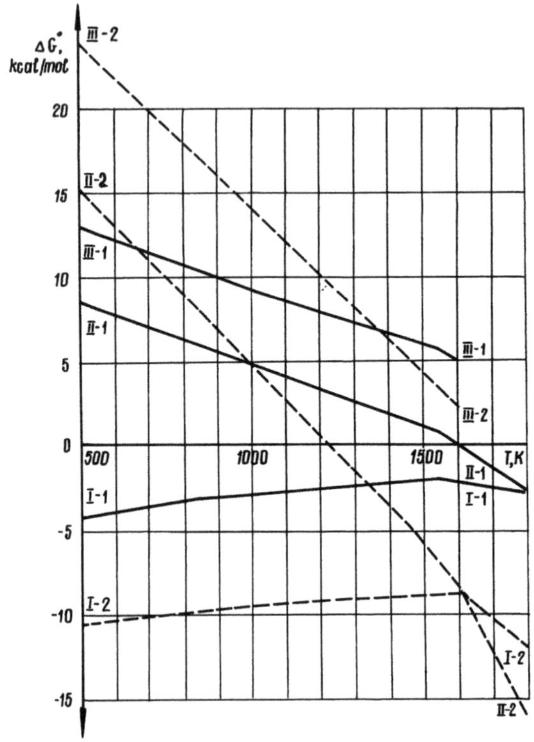

Fig. 38. The temperature dependence of ΔG^0 of reactions of formation of manganese silicates from oxides within a temperature range of 500 to 1800 K, the ratio of manganese and silicon cations being 2/1

4. The alteration of the Gibbs free energy of the reactions of formation of manganese meta- and orthosilicates from oxides regularly increases as a function of the manganese oxide type in the series:

$$MnO > Mn_3O_4 > Mn_2O_3 ,$$

in the latter case reactions being thermodynamically impossible up to the temperature 1620 K, when Mn_2O_3 decomposes.

As already shown above, a thermodynamic analysis of solid-phase reactions combined with a study of kinetic factors is of great importance in adequately explaining the formation processes and the stability areas of compounds significant in commercial respect. The use of calculation methods based on reliable experimental data permits, besides determining temperature conditions of the stability of particular compounds in ternary systems, establishing the ranges of stable existence of one or the other ternary compounds at various temperatures up to the liquids. The use of these calculations will be elucidated by means of ternary systems $MgO-SrO-SiO_2$, $Na_2O-CaO-SiO_2$ and $CaO-Al_2O_3-SiO_2$ for which reliable thermodynamic data are available.

2.9 System MgO – SrO – SiO_2

The study of solid-phase reactions of the formation of ternary compounds in the system MgO – SrO – SiO_2 is of great theoretical and practical significance in the synthesis of strontium glasses, porcelain enamels and glazes. As revealed in sources [113, 350–352, 389, 571], in the system MgO – SrO – SiO_2 a number of researchers have found the existence of following ternary compounds: $2SrO \cdot MgO \cdot 2SiO_2$, $3SrO \cdot MgO \cdot 2SiO_2$, $SrO \cdot MgO \cdot 2SiO_2$, $3SrO \cdot MgO \cdot 4SiO_2$.

In order to conduct a thermodynamic analysis of solid-phase reactions in the system MgO – SrO – SiO_2 it is imperative to know the ΔG^0 value at high temperatures for binary compounds of the system MgO – SrO – SiO_2. The function $\Delta G^0_T = f(T)$ for binary systems MgO – SiO_2 and SrO – SiO_2 was considered while analyzing these binary systems.

Using these values and employing the known thermochemical data the enthalpies ΔH^0 and the Gibbs free energies ΔG^0 of the formation from oxides at 298 K (Table 30) were calculated. Equations $C_p = f(T)$ for all compounds were calculated by Landiya's method [320].

On the basis of the data from Table 30 equations of $\Delta G^0_T = f(T)$ for the following reactions were set up:

1. $2SrO + MgO + 2SiO_2 = 2SrO \cdot MgO \cdot 2SiO_2$
   $\Delta G^0_T = -61355 + 2.28\,T \ln T - 3.82 \cdot 10^{-3} T^2 + 2.75 \cdot 10^5 T^{-1} - 16.92\,T$
2. $3SrO + MgO + 2SiO_2 = 3SrO \cdot MgO \cdot 2SiO_2$
   $\Delta G^0_T = -58666 - 0.8\,T \ln T - 1.25 \cdot 10^{-3} T^2 + 3.0 \cdot 10^5 T^{-1} - 1.13\,T$
3. $SrO + MgO + 2SiO_2 = SrO \cdot MgO \cdot 2SiO_2$
   $\Delta G^0_T = -45958 - 71.31\,T \ln T + 2.18 \cdot 10^{-3} T^2 + 6.017 \cdot 10^5 T^{-1} - 155.4\,T$
4. $3SrO + MgO + 4SiO_2 = 3SrO \cdot MgO \cdot 4SiO_2$
   $\Delta G^0_T = -107770 - 1.38\,T \ln T - 4.23 \cdot 10^{-3} T^2 + 7.84 \cdot 10^5 T^{-1} + 3.78\,T$.

Based on these equations the values of ΔG^0_T within a temperature range of 400 to 1900 K (within each 100 K) were calculated for all the above reactions (Table 31, Fig. 39); the curve numbers correspond to the reaction numbers.

In order to evaluate the preference of the formation for one or the other ternary compound in the system MgO – SrO – SiO_2 the dependence of ΔG^0_T, the ratios SrO/MgO/SiO_2 being 1/1/2, 2/1/2, 3/1/2 and 3/1/4 (Fig. 39) was considered.

The ratio of the initial components SrO/MgO/SiO_2 being 1/1/2, the thermodynamically stable compound up to a temperature of 1450 K will be $SrO \cdot MgO \cdot 2SiO_2$, which at higher temperatures converts to $2SrO \cdot MgO \cdot 2SiO_2$. This compound is the most stable one at high temperatures.

The ratio of initial components SrO/MgO/SiO_2 being 2/1/2, the compound $2SrO \cdot MgO \cdot 2SiO_2$ has the least value of ΔG^0_T. This compound at the given ratio of initial components is the most stable final phase (strontium akermanite).

Table 30. Initial thermochemical data for the system $MgO-SrO-SiO_2$

Compounds	$-\Delta H^0_{298(el)}$ [kcal/mol]	$-\Delta G^0_{298(el)}$ [kcal/mol]	$-\Delta H^0_{298(ox)}$ [kcal/mol]	$-\Delta G^0_{298(ox)}$ [kcal/mol]	S^0_{298} [cal/mol deg]	$C_p=f(T)$ [cal/mol deg]		
						a	$b\cdot10^3$	$c\cdot10^{-5}$
$2SrO\cdot MgO\cdot 2SiO_2$	921.74	885.11[b]	60.20	61.88	56.62[b]	61.4[a]	15.5[a]	−10.59[a]
$3SrO\cdot MgO\cdot 2SiO_2$	1058.94[b]	1021.43[b]	56.3	59.4	75.84[b]	76.82[a]	11.48[a]	−12.90[a]
$SrO\cdot MgO\cdot 2SiO_2$	760.34[b]	724.12[b]	39.90	39.69	38.70[b]	58.65[a]	2.37[a]	−15.32[a]
$3SrO\cdot MgO\cdot 4SiO_2$	1540.54[b]	1473.36[b]	102.40	101.83	83.68[b]	103.51[a]	21.41[a]	−22.58[a]

[a] Calculated by Landiya's method [320]
[b] Calculated by the authors' method [357]

Table 31. The temperature dependence of ΔG^0 for ternary compounds in the system $MgO-SrO-SiO_2$

Nos. of reactions	ΔG [kcal/mol] at temp. [K]														
	500	600	700	800	900	1000	1100	1200	1300	1400	1500	1600	1700	1800	1900
1	64.84	65.53	66.53	67.21	68.19	69.09	70.08	71.09	72.25	73.74	75.24	75.98	77.33	79.01	80.12
2	61.42	62.38	63.32	64.26	65.25	66.27	67.31	68.37	69.46	70.59	71.69	72.93	74.13	75.39	76.33
3	39.12	39.03	38.71	33.69	38.28	38.18	38.07	37.92	37.82	37.50	37.40	37.30	37.30	37.86	36.54
4	101.15	100.54	99.91	99.31	98.73	98.05	97.82	97.32	96.83	96.61	96.42	96.00	95.87	95.34	95.14

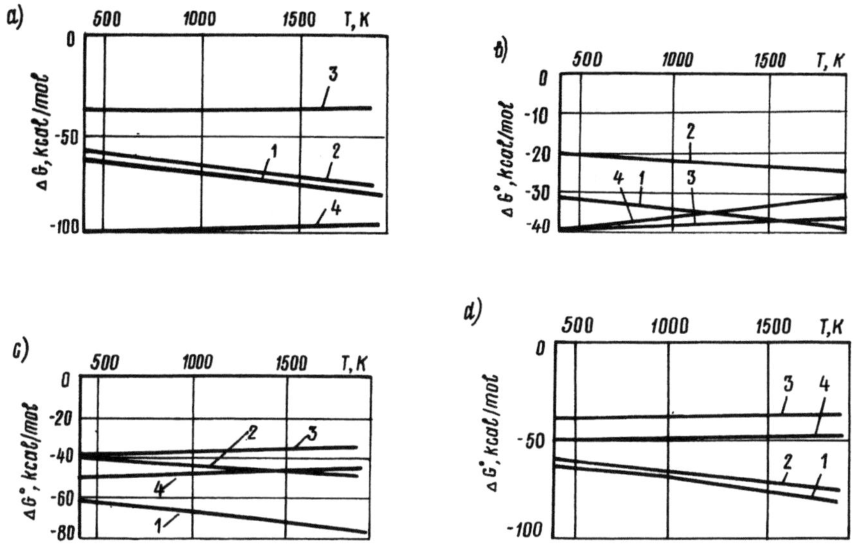

Fig. 39. The temperature dependence of ΔG^0 of compounds in the system MgO – SrO – SiO_2, the ratio of SrO/MgO/SiO_2 being: **a** 3/1/4; **b** 1/1/2; **c** 2/1/2; **d** 3/1/2

The ratio of initial components SrO/MgO/SiO_2 being 3/1/2, the most stable compound up to the temperature of 1870 K will be 3SrO · MgO · $2SiO_2$ which above this temperature converts to 2SrO · MgO · $2SiO_2$, the most stable one at high temperatures.

The ratio of initial components SrO/MgO/SiO_2 being 3/1/4, the most stable compound is 3SrO · MgO · $4SiO_2$ as synthesized by Shteynberg [571].

Each of these ternary compounds has at a certain temperature, a certain value of the Gibbs free energy ΔG_T^0 by which one can judge the possibility of the formation of one or the other ternary compound (crystalline phase), as the most stable crystalline phase has the most negative ΔG^0 value.

Based on these considerations one can show any of the said ternary compounds (crystalline phases) in the ternary diagram of the system MgO – SrO – SiO_2 as presenting the composition of each crystalline phase in mole percentage.

Research into the possibility of the existence of crystalline phases in the ternary system MgO – SrO – SiO_2 at temperatures of 1000 and 1900 K is carried out below. In Fig. 40 a diagram is presented of the stable existence of crystalline phases in the system MgO – SrO – SiO_2 at a temperature of 1000 K. At high temperatures (1900 K) the formation of only two crystalline phases: 2SrO · MgO · $2SiO_2$ and 3SrO · MgO · $4SiO_2$, is observed in the diagram (Fig. 41).

2.10 System $Na_2O-CaO-SiO_2$ [363]

In the system $Na_2O-CaO-SiO_2$ underlying glass-making (the production of lime-soda glasses) the following ternary compounds: $Na_2O \cdot 2CaO \cdot 3SiO_2$, $2Na_2O \cdot CaO \cdot 3SiO_2$ and $Na_2O \cdot 3CaO \cdot 6SiO_2$[10] have been found.

The heats of formation of these compounds were calculated on the basis of the calorimetric determination of the heat of dissolution of soda ash, calcium oxide, sodium disilicate and of the said compounds [135].

Entropies S^0_{298} of these compounds were calculated according to [357] using the previously obtained data for the systems $CaO-SiO_2$ and Na_2O-SiO_2 as

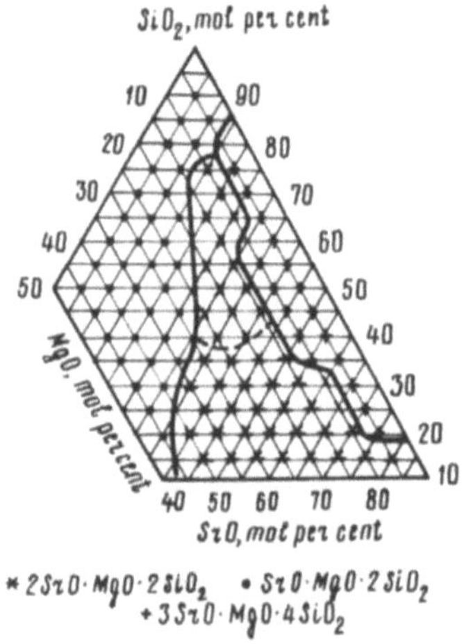

Fig. 40. Stability of crystalline phases in the system $MgO-SrO-SiO_2$ at a temperature of 1000 K

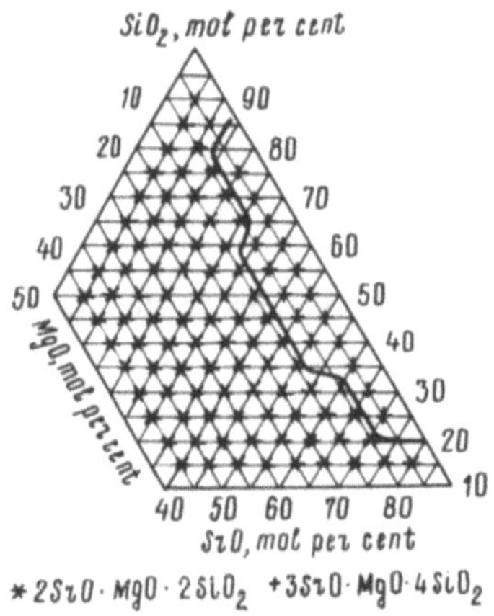

Fig. 41. Stability of crystalline phases in the system $MgO-SrO-SiO_2$ at a temperature of 1900 K

[10] Initial thermochemical data for ternary compounds in the system $Na_2O-CaO-SiO_2$ are adduced in Table 33. From now on these compounds will be abbreviated as NC_2S_3, N_2CS_3 and NC_3S_6

Table 32. Initial data for calculating $C_p=f(T)$ of ternary compounds in the system $Na_2O-CaO-SiO_2$

Compound	Melting temp. [K]	S^0_{298} [cal/mol deg]
NC_2S_3	1557	64.4
N_2CS_3	1414	74.0
NC_3S_6	1320	110.4

Table 33. Initial thermochemical data for compounds in the system $Na_2O-CaO-SiO_2$ [307–309]

Compound	$-\Delta H^0_{298\,(ox)}$ [kcal/mol]	$-\Delta G^0_{298\,(ox)}$ [kcal/mol]	$C_p=f(T)$		
			a	$b \cdot 10^3$	$c \cdot 10^{-5}$
NC_2S_3	110.75	110.75[a]	83.87	14.08	−16.31
N_2CS_3	134.44	134.35[a]	88.46	16.24	−15.05
NC_3S_6	137.39	138.73[a]	110.29	−42.13	−11.57

[a] Calculated by the authors

well. Equations of the temperature dependence of heat capacity of ternary compounds were calculated according to Landiya's methods [320] using the data on the temperatures of melting of the above ternary compounds presented in Table 32 as well.

On the basis of the initial data presented in Table 33 equations of $\Delta G^0=f(T)$ for the following reactions were set up:

1. $Na_2O+2CaO+3SiO_2=Na_2O \cdot 2CaO \cdot 3SiO_2$
2. $2Na_2O+CaO+3SiO_2=2Na_2O \cdot CaO \cdot 3SiO_2$
3. $Na_2O+3CaO+6SiO_2=Na_2O \cdot 3CaO \cdot 6SiO_2$.

These equations and the ΔG^0 values calculated according to them for 400, 700, 1000 and 1300 K are presented in Table 34 and Fig. 42. The results obtained demonstrate that with the elevation of temperature ΔG^0 of the reactions of formation of the above ternary compounds alters but slightly. A thermodynamic analysis of solid-phase reactions of compound formation at different ratios of initial constituents showed the thermodynamic sequence of compound formation essentially not to alter with the alteration of temperature.

The entire range of ratios of Na_2O, CaO and SiO_2 examined is presented in Fig. 43.

Table 34. The temperature dependence of ΔG^0 for main reactions in the system $Na_2O - CaO - SiO_2$

Reactions	Equations $\Delta G^0 = f(T)$	$-\Delta G^0$ [kcal/mol] at temp. [K]				
		298	400	700	1000	1300
$N + 2C + 3S = NC_2S_3$	$-115073 - 11.47\,T\ln T + 9.04 \cdot 10^{-3}T^2 + 2.545 \cdot 10^5 T^{-1} + 74.3\,T$ (for 298 – 848 K) $-114661 - 1.9\,T\ln T - 0.35 \cdot 10^{-3}T^2 + 6.595 \cdot 10^5 T^{-1} + 16.676\,T$ (for 848 – 1557 K)	110.75	110.77	110.88	110.80	110.78
$2N + C + 3S = N_2CS_3$	$-139024 - 12.33\,T\ln T + 10.12 \cdot 10^{-3}T^2 + 2.695 \cdot 10^5 T^{-1} + 79.88\,T$ (for 298 – 848 K) $-138615 - 2.76\,T\ln T + 0.73 \cdot 10^{-3}T^2 + 6.745 \cdot 10^5 T^{-1} + 22.273\,T$ (for 848 – 1414 K)	134.35	134.34	134.32	134.01	133.64
$N + 3C + 6S = NC_3S_6$	$-131352 + 7.44\,T\ln T + 7.855 \cdot 10^{-3}T^2 - 4.655 \cdot 10^5 T^{-1} - 64.25\,T$ (for 298 – 848 K) $-130536 + 26.58\,T\ln T - 10.925 \cdot 10^{-5}T^2 + 3.445 \cdot 10^5 T^{-1} - 179.647\,T$ (for 848 – 1320 K)	138.73	139.12	139.02	137.12	134.48

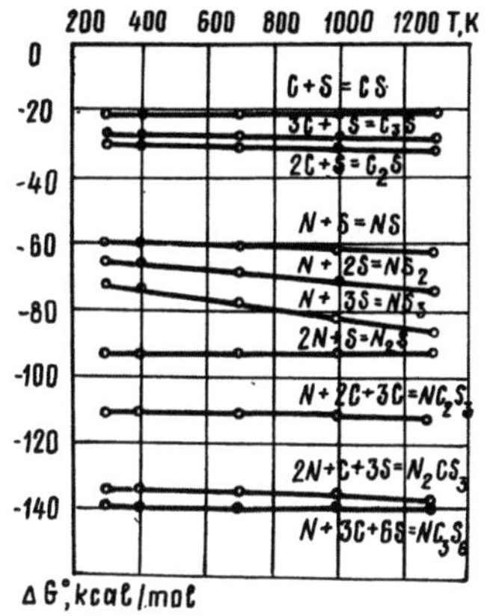

Fig. 42. The temperature dependence of ΔG^0 of reactions of formation from oxides in the system $Na_2O-CaO-SiO_2$

In the high-silica area the stable crystalline compound is devitrite NC_3S_6. At SiO_2 content less than 50 mol% and Na_2O/CaO ratio near to 1, the preferable is the compound N_2CS_3. At points with high CaO on Na_2O content, generally, the most stable is NC_2S_3.

2.11 System $CaO-Al_2O_3-SiO_2$

A study of solid-phase reactions in this system has a great theoretical and practical importance in the manufacturing of cement, aluminosilicate refractories and recently in the production of slag glass-ceramics. Two ternary compounds are most likely to exist: gehlenite $2CaO \cdot Al_2O_3 \cdot SiO_2$ and anorthite,

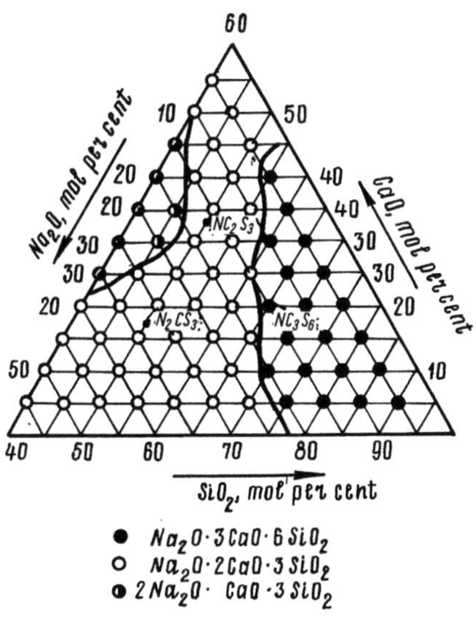

Fig. 43. Stability of crystalline ternary compounds in the system $Na_2O-CaO-SiO_2$ within temperature range of 298 to 1000 K

$CaO \cdot Al_2O_3 \cdot 2SiO_2$ [40]. Initial data for these ternary compounds are adduced in Table 35 [103, 251].

Table 35. Initial thermodynamic data for compounds in the system $CaO-Al_2O_3-SiO_2$

Compounds	$-\Delta H^0_{298}$ [kcal/mol]	$-\Delta G^0_{298}$ [kcal/mol]	S^0_{298} [cal/mol deg]	$C_p=f(T)$ a	$b \cdot 10^3$	$c \cdot 10^{-5}$	Temp. range [K]
$2CaO \cdot Al_2O_3 \cdot SiO_2$	945.55	898.74	52.9	53.73	17.68	−0.89	298−1600
$CaO \cdot Al_2O_3 \cdot 2SiO_2$	1012.45	921.83	48.4	64.42	13.70	−16.89	298−1700

Using these values and the known thermochemical data the $\Delta G^0=f(T)$ functions were calculated for the temperature range of 800 to 1800 K by the Tyomkin-Shvartsman method [652]. The data obtained are presented in Table 36.

For evaluating a greater probability of the formation of one or the other ternary or binary compound the function was examined at various ratios of initial substances. For $\Delta G^0_T=f(T)$ of binary compounds values for particular binary systems calculated earlier were used.

As the analysis of reactions at temperatures approaching the range of the appearance of the liquid phase is a subject of great interest, we have confined ourselves to calculating ΔG^0_T of the reactions at 1000, 1300 and 1600 K.

For each ratio of initial components several possible variants of formation of binary and ternary compounds were examined.

While calculating, the overall composition of components was reduced to 1 mol of substance.

Table 36. The $\Delta G^0=f(T)$ function for reactions of the formation of ternary compounds in a system

Reactions	$-\Delta G^0$ [kcal/mol] at temp. [K]					
	800	900	1000	1100	1200	
$2C+A+S=C_2AS$	34.3	35.4	36.4	37.0	38.5	
$C+A+2S=CAS_2$	29.9	30.1	30.4	31.7	31.0	
Reactions	$-\Delta G^0$ [kcal/mol] at temp. [K]					
	1300	1400	1500	1600	1700	1800
$2C+A+S=C_2AS$	39.5	40.8	41.8	43.1	44.4	45.7
$C+A+2S=CAS_2$	31.4	31.8	32.2	32.4	33.0	33.7

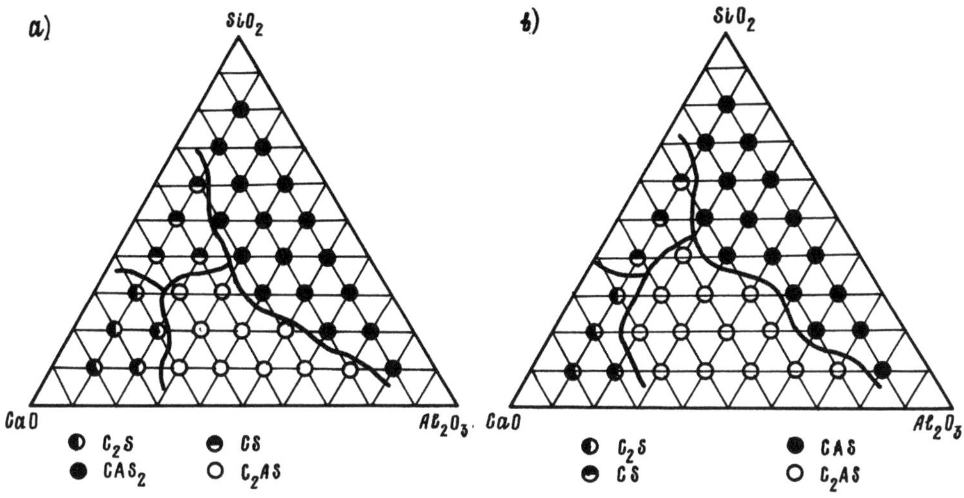

Fig. 44. The fields of the stability of crystalline compounds in the system $CaO-Al_2O_3-SiO_2$; **a** at 1000 K; **b** at 1600 K

One should bear in mind that the ΔG^0 contribution of the formation of various compounds to the overall ΔG^0 value of the reactions is different. Therefore in the thermodynamic analysis of solid-phase reactions we took into account only the compounds whose contribution to the total ΔG^0 value is the greatest. On this assumption the stable compound ranges in the system $CaO-Al_2O_3-SiO_2$ appear as shown in Fig. 44.

At temperatures 1000 and 1600 K by the most extensive ranges, as thermodynamically preferable compounds, anorthite and gehlenite are presented; of binary compounds, only wollastonite and dicalcium silicate. It results from relatively low ΔG^0 values of calcium aluminates and binary aluminosilicates at the given temperatures. As the temperature increases, one notices a certain tendency towards the extension of the ranges of the most stable existence of ternary compounds.

3 Reactions in the Firing of Ceramics and Refractories

Ceramic products are a result of the firing of diverse raw mixtures. The processes of firing are based upon reactions in the systems: $Al_2O_3-SiO_2$, $MgO-SiO_2$, $BeO-SiO_2$, ZrO_2-SiO_2 etc. as well as upon reactions of transformation of argillaceous minerals at high temperatures. Moreover, in practice and especially in the production of ceramics of pure oxides, of great importance are the problems of stability of oxides and particular compounds at high temperatures with regard to redox parameters of the environment and their interaction with melted metals. Since reactions in the systems $MgO-SiO_2$ and $BeO-SiO_2$ were examined in detail while analyzing binary and ternary systems (see Chapter II. 2), in this part of the book attention will be paid mainly to the stability and interaction of oxides, reactions in the systems $Al_2O_3-SiO_2$, ZrO_2-SiO_2 and the transformation of kaolinite at heating [73, 79].

3.1 Stability of Oxides

Extensive literature exists concerning oxide stability [5, 278, 353, 354, 546, 623, 689, 697].

An examination of the data at hand shows that every oxide displays a tendency towards dissociation at a certain temperature. Oxides of noble metals, such as Ag, Au, Pt, start dissociating at low temperatures. Contrarily, refractory oxides, even at relatively high temperatures have such insignificant dissociation pressure that it appears impossible to be measured experimentally. Nevertheless dissociation pressure can be calculated by means of Eq. (101).

Thus for the reaction

$$MeO_{(sol)} = Me_{(sol)} + \tfrac{1}{2}O_{2(gas)}: \Delta G_T^0 = -RT\frac{a_{Me}\,a_{O_2}^{1/2}}{a_{MeO}}.$$

As the activities of an oxide and a metal, at standard conditions, can be accepted being equal to one, the equilibrium of the above reaction will be expressed by the equation

$$\Delta G_T^0 = -RT\ln p_{O_2}^{1/2}. \qquad (122)$$

Therefore ΔG_T^0 or p_{O_2} can serve as a criterion for the oxide stability. A thorough analysis of oxide dissociation was carried out by Kulikov [317]. It stands to reason that for such a comparison one must relate the ΔG_T^0 value to a constant amount of oxygen. Substituting the tabulated data of the standard ΔG_{298}^0 values for diverse oxides in Eq. (122) the corresponding equilibrium pressures at a temperature of 298 K can be determined.

Since the publication of the second edition of our work Soviet literature has been enriched by a fundamental handbook on the "Thermodynamic Properties of Individual Substances" [623] in which the behaviour of several oxides which interested scientists of silicate was examined up to high temperatures. In order to determine the oxide stability at high temperatures it is necessary to calculate the ΔG^0 values at these temperatures. The calculation is conducted in a common way (see the scheme of the ΔG calculation for high temperatures), phase and polymorphic transformations both of oxides themselves and of the products of their dissociation being taken into account.

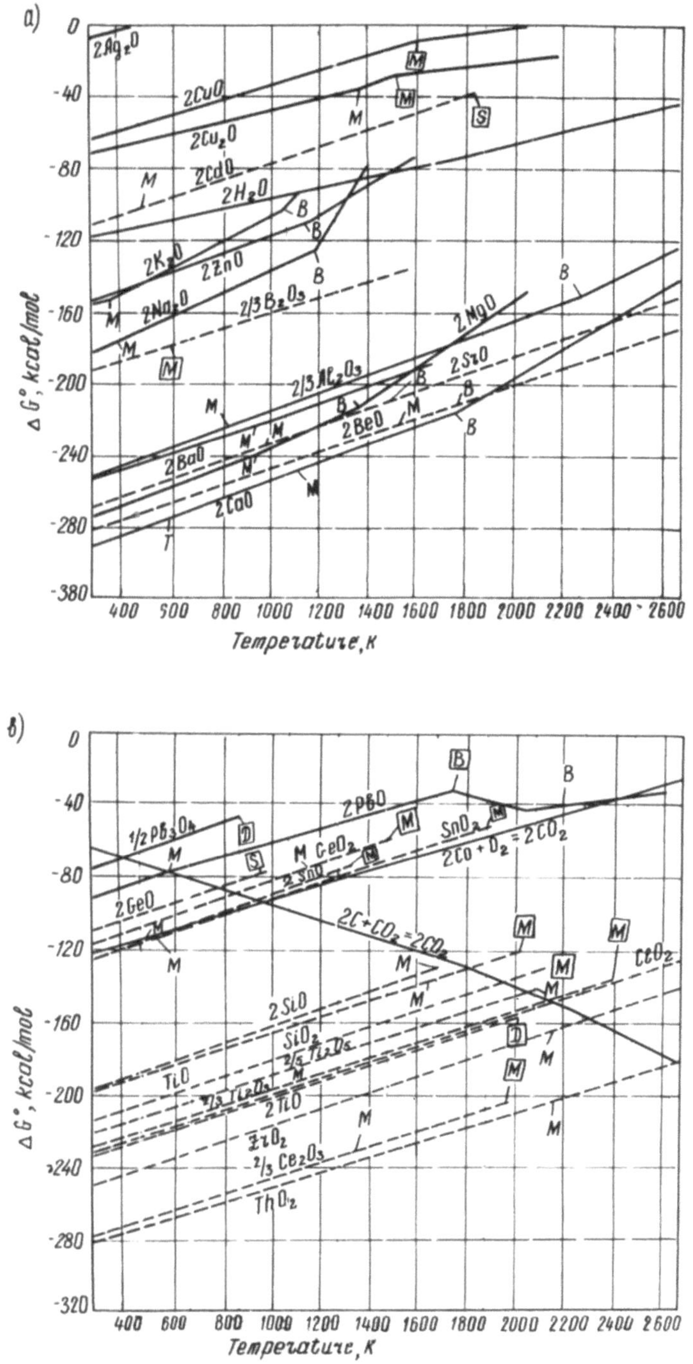

Fig. 45 a, b. The Gibbs free energies of the formation of oxides of metals of some groups of the periodic system per g-mol of oxygen: **a** group I, II and III; **b** group IV; **c** group V; **d** groups VI, VII and VIII

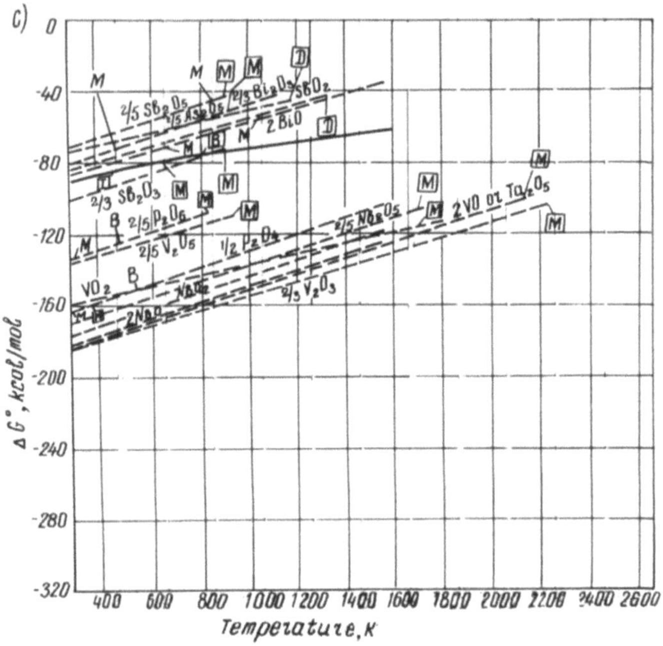

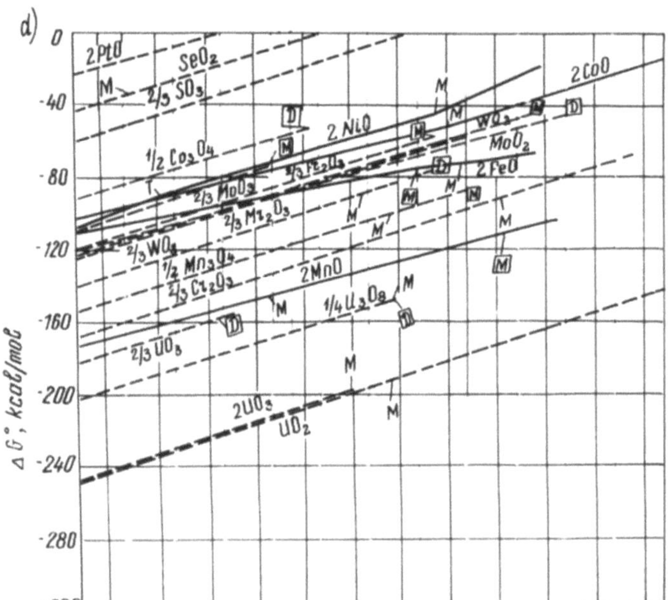

Fig. 45c, d (Legend see page 147)

In order to facilitate the calculations of this sort and mainly to increase their efficiency Tripp and King [642] have plotted diagrams of ΔG of oxide formation from elements as a function of temperature (from 0 to 2400 °C), all the corresponding ΔG values being related to a mole of oxygen at a pressure of 1 atm.

The alterations of the Gibbs free energy standard values as a function of temperature are presented in Fig. 45a. The character of the ΔG^0 alteration for all oxides is presented approximately in form of straight lines, their breaks correspond to phase and polymorphic transformations of both initial elements and resulting oxides. The following symbols are accepted: M, melting, B, boiling, T, polymorphic transformation, S, sublimation, D, dissociation. Letters in a frame mean that the products of a reaction undergo the transformations in question, letters without frames correspond to the substances having entered a reaction. By use of the diagrams it is easy to find the ΔG^0 value for a temperature under consideration and then with the help of Eq. (122) to calculate the equilibrium pressure of oxygen. Dissociation temperature, as in the previous cases, is a temperature at which ΔG^0 of an oxide is equal to 0. The use of these diagrams permits a rapid solution of several practical important problems. Thus, e.g., the comparison of ΔG^0 of various oxides enables one to elucidate the reasons for a high stability of refractory oxides.

As is clearly seen from the diagrams, the noble metal oxides are characterized by low (in the absolute value) ΔG^0 of formation, which accounts for their relatively high disposition to dissociation at comparatively low temperatures.

The diagram also permits the determination of ΔG^0 of the following reaction taking place at 1150 K:

$$3CoO + 2Cr = Cr_2O_3 + 3Co\,. \tag{123}$$

From the diagram in Fig. 45d we determine $\Delta G^0_{1150\,K}$ for $\frac{2}{3}Cr_2O_3$; it equals -130 kcal per mol O_2. Then $\Delta G^0_{1150\,K}$ for a mole of Cr_2O_3 will be $\frac{3}{2}$ $(-130) = -195$ kcal. Similarly $\Delta G^0_{1150\,K}$ for CoO will be $\frac{1}{2}$ $(-74) = -37$ kcal. On the basis of these values we obtain for reaction (123)

$$\Delta G^0_{1150\,K} = \tfrac{3}{2}(-130) - \tfrac{3}{2}(-74) = -195 + 111 = -84 \text{ kcal per } \tfrac{3}{2} \text{ mol } O_2\,.$$

As can be seen from the example presented, such calculations can be successfully employed to determine the thermodynamic values of the reactions of the interaction between oxides and elements for evaluating the ability of some oxides to be engaged in redox reactions with other oxides. Thus, oxide AO will reduce oxide BO in the temperature range where the straight line of the temperature dependence of ΔG^0_{BO} is situated above that of the temperature dependence of ΔG^0_{AO}. At the intersection point of both straight lines the reaction equilibrium is established:

$$A + BO = AO + B\,. \tag{124}$$

Above this equilibrium temperature, i.e., above the intersection point of the straight lines, element A will replace element B in its oxide. Below the equilibrium temperature the opposite picture is observed.

In general the behaviour of oxides in a melt will be characterized by the same equation as any reaction under any conditions. Naturally, it is necessary to know the corresponding thermochemical data for the pure substances that are in melted state. Moreover, one must recognize the related activities of the components reacting in a melt.

In case a melt can be viewed as an ideal solution of components, the activities of the given components will be equal to their concentrations (mole fractions) in the melt. If, however, in reactions certain pure liquid phases take part, their activities will be equal to one, and hence, conducting various calculations it is possible to use thermochemical data immediately for the substances in a melted state. Practically the latter circumstance necessitates taking into account the changes connected with the melting processes.

The enthalpy and Gibbs free energy alteration of individual substances in this case is necessary for the determination of the technique according to the formulae adduced in Chap. I.10.

On the basis of the thus calculated thermodynamic properties of individual substances, in a given temperature range the enthalpy and Gibbs free energy alteration values for the reactions in questions are determined.

The following reaction illustrates such a reaction [697]:

$$\underset{(1800\,\mathrm{K})}{Ni_{(liq)} + \tfrac{1}{3}Al_2O_{3(sol)}} = NiO_{(sol)} + \tfrac{2}{3}Al_{(liq)}\,. \tag{125}$$

The examination of Eq. (125) permits evaluating the possibility of the melting of pure Ni in a pure Al_2O_3 refractory crucible.

A standard alteration of the Gibbs free energy of this reaction can be determined on the basis of the equation

$$\Delta G^0_{1800} = \Delta H^0_{1800} - T\Delta S^0_{1800}\,.$$

ΔH^0_{1800} and ΔS^0_{1800} can be found from the corresponding values of ΔH^0_{298} and ΔS^0_{298}, ΔH^0 of melting, ΔS^0 of melting, and heat capacities of reactants. Let us take as an example the calculation for $Al_{(liq)}$:

a) $\Delta H^0_{298} = 0$; $C_{p(sol)} = 4.94 + 2.95 \cdot 10^{-3}T$;

$\Delta H^0_{m\ 931\ K} = 2570$ cal/mol

$C_{p(liq)} = 7$ cal/mol deg

$$\Delta H^0_{1800} = \Delta H^0_{298} + \int\limits_{298}^{931} (4.94 + 2.96 \cdot 10^{-3}\, T)dT$$

$$+ 2570 + \int\limits_{931}^{1800} 7dT = 4.94\,(931 - 298)$$

$$+ 1.48 \cdot 10^{-3}\,(931^2 - 298^2) + 2570 + 7(1800 - 931) = 12910 \text{ cal/mol};$$

b) $S^0_{298} = 6.77$ cal/mol deg; $\Delta S^0_{m\ 931\ K} = 2.4$ cal/mol deg;

$$S^0_{1800} = S^0_{298} + \int\limits_{298}^{931} (4.94 + 2.96 \cdot 10^{-3}\, T)\frac{dT}{T} + 2.4 + \int\limits_{931}^{1800} 7\frac{dT}{T}$$

$$= 6.77 + 4.94\,(\ln 931 - \ln 298) + 2.96 \cdot 10^{-3}\,(931 - 298) + 2.4$$

$$+ 7\,(\ln 1800 - \ln 931) = 6.77 + 14.5 = 21.27 \text{ cal/mol deg}.$$

In an analogous way we calculate ΔH^0_{1800} and S^0_{1800} for the rest of reactants. The data obtained are summarized in Table 37.

Table 37. Initial data of the compounds taking part in Eq. (125)

Compounds	S^0_{1800} [cal/mol deg]	ΔH^0_{1800} [cal/mol]
$\frac{2}{3}Al_{(liq)}$	14.2	8600
$\frac{1}{3}Al_2O_{3\,(sol)}$	20.5	−112300
$Ni_{(liq)}$	23.4	16200
$NiO_{(sol)}$	43.2	−35700

For the alteration of the reaction entropy and enthalpy we obtain:

$$\Delta S^0_{1800} = 4.5 \text{ cal/mol deg}; \quad \Delta H^0_{1800} = 69000 \text{ cal/mol}.$$

Thus,

$$\Delta G^0_{1800} = 69000 - 1800 \cdot 4.5 = 69000 - 8100 = 60900 \text{ cal/mol}.$$

As the ΔG^0 value is a positive, the rection will not take place if the materials are in a standard state, nor will the reaction take place where pure materials are formed. Nevertheless, the melt formed by aluminium can solve in nickel. The calculation of the equilibrium constant according to Eq. (69) gives:

$$\lg K = \frac{-60900}{2.3 \cdot 1.99 \cdot 1800} = -7.4;$$

$$K=\frac{[\mathrm{Al}]^{2/3}}{[\mathrm{Ni}]}=10^{-7.4}.$$

Since Ni, as a first approximation, may be considered pure, the molar fraction of Al will constitute about

$$10^{-3/2\cdot 7.4}=10^{-11.1}$$

(if Raoult's law acts here). Thus, crucibles of pure Al_2O_3 will be appropriate for melting high purity nickel.

Redox reactions are of great practical significance [546, 697]. In this case the principles of heterogeneous equilibrium can be applied as well to the reactions of oxide interaction with simple substances being in the solid, liquid, and gaseous state. These rections are of great significance in choosing proper refractories for pure metal melting, in determining the resistance of refractory oxides and heat-resistant metals towards temperature and aggressive liquid and gaseous media, in determining diverse parameters connected with the performance of the synthesis of glasses, enamels and glazes for calculating thermal balances of heat plants, as well as for other instances of practical use of high-temperature heterogeneous reactions.

A reaction of zirconium oxide interaction with hydrogen within a temperature range of 1400 to 2000 K may serve as a good example:

$$\mathrm{ZrO}_{2(\mathrm{sol})}+2\mathrm{H}_{2(\mathrm{gas})}=\mathrm{Zr}_{(\mathrm{sol})}+2\mathrm{H}_2\mathrm{O}_{(\mathrm{gas})}.$$

A thermodynamic analysis of the conditions of this reaction provides for the determination of the conditions of the retention of ZrO_2 stability in the reducing atmosphere and those of the corrosion of metallic zirconium in the water-vapour atmosphere.

The equilibrium conditions are determined from the equation

$$\Delta G_T^0=-RT\ln K_p=-RT\ln\frac{[p_{\mathrm{H_2O}}]^2}{[p_{\mathrm{H_2}}]^2},$$

from which the correlation between the equilibrium partial pressure of water steam and hydrogen can be found.

The initial data are produced in Table 38.

1. ΔH^0 of the reaction is determined:
   $\Delta H_{298}^0=-2(57798)+261500=146004$
2. ΔG^0 of the reaction is determined:
   $\Delta G_{298}^0=-2(54635)+247700=138430$

Table 38. Initial thermochemical data for compounds taking part in the reaction of the interaction of zirconia with hydrogen

Compound	$-\Delta H^0_{298}$ [cal/mol]	$-\Delta G^0_{298}$ [cal/mol]	$C_p = f(T)$
$Zr\text{-}\beta_{(sol)}$	0	0	$6.83 + 1.12 \cdot 10^{-3} T + 0.91 \cdot 10^5 T^{-2}$ (up to 1135 K)
$Zr\text{-}\alpha_{(sol)}$	–	–	7.27 (up to 1400 K)
$H_2O_{(gas)}$	57798	54635	$7.17 + 2.56 \cdot 10^{-3} T + 0.08 \cdot 10^5 T^{-2}$
$H_{2(gas)}$	0	0	$6.88 + 0.066 \cdot 10^{-3} T + 0.279 \cdot 10^{-6} T^2$
$ZrO_2\text{-}\alpha$	–	–	17.8
$ZrO_2\text{-}\beta_{(sol)}$	261500	247700	$16.64 + 1.8 \cdot 10^{-3} T - 3.36 \cdot 10^5 T^{-2}$

3. $C_p = f(T)$ is determined:

$$\Delta a = 0.44 + 14.34 - 13.76 - 16.64 = -15.62;$$
$$\Delta b = (-1.12 + 5.12 - 0.132 - 1.8) \cdot 10^{-3} = 2.068 \cdot 10^{-3};$$
$$\Delta c = (+0.91 + 0.16 + 3.36) \cdot 10^5 = 4.43 \cdot 10^5;$$
$$\Delta d = -2 \cdot 0.279 \cdot 10^{-6} = -0.558 \cdot 10^{-6};$$
$$\Delta C_p = -15.62 + 2.068 \cdot 10^{-3} T + 4.43 \cdot 10^5 T^{-2} - 0.558 \cdot 10^{-6} T^2$$

4. ΔH_0 is determined:

$$\Delta H_0 = 146004 + 15.62 \cdot 298 - \tfrac{1}{2} 2.068 \cdot 10^{-3} (298)^2 + 4.43 \cdot 10^5 (298)^{-1} + \tfrac{1}{3} 0.558 \cdot 10^{-6} (298)^3;$$
$$\Delta H_0 = 146004 + 4655 - 92 + 1490 + 5 = 152154\,.$$

5. y is determined:

$$138430 = 152154 + 15.62 \cdot 298 \cdot 5.7 - \tfrac{1}{2} \cdot 2.068 \cdot 10^{-3} (298)^2 - \tfrac{1}{2} 4.43 \cdot 10^5 \cdot (298)^{-1} + \tfrac{1}{6} \cdot 0.558 \cdot 10^{-6} (298) + 298y;$$
$$298y = -39421;\ y = \frac{-39421}{298} = -132.28\,.$$

6. An equation is set up of the temperature dependence of the Gibbs free energy: up to the polymorphic transformation $\beta \rightarrow \alpha$ ZrO_2 (1478 K)

$$\Delta G^0_T = 152154 + 15.62\, T \ln T - 1.034 \cdot 10^{-3} T^2 - 2.215 \cdot 10^5 T^{-1} + 0.093 \cdot 10^{-6} T^3 - 132.28\, T\,,$$

with a glance to the polymorphic transformation

$$\Delta G^0_T = 129844 + 0.14\, T \ln T - 1.984 \cdot 10^{-3} T^2 + 0.535 \cdot 10^5 T^{-1} + 0.093 \cdot 10^{-6} T^3 - 11.27\, T\,.$$

7. The ΔG^0 value of the reaction is calculated at 1400 and 2000 K (in the letter case it is supposed that the character of the temperature dependence of heat capacity will not change):

$$\Delta G^0_{1400} = 152154 + 15.62 \cdot 1400 \cdot 7.24 - 1.034 \cdot 10^{-3} \cdot 196 \cdot 10^4$$
$$-2.215 \cdot 10^5 \cdot 14^{-1} \cdot 10^{-2} + 0.093 \cdot 10^{-6} \cdot 2774 \cdot 10^6 - 132.28 \cdot 1400;$$

$$\Delta G^0_{1400} = 152154 + 158324 - 2030 - 161 - 256 - 185192 = 123351 \text{ cal/mol};$$

$$\Delta G^0_{2000} = 129844 + 0.14 \cdot 2000 \cdot 7.59 - 1.984 \cdot 10^{-3} \cdot 4 \cdot 10^6$$
$$+0.535 \cdot 10^5 \cdot 2^{-1} \cdot 10^{-3} + 0.093 \cdot 10^{-6} \cdot 8 \cdot 10^9 - 11.27 \cdot 2000;$$

$$\Delta G^0_{2000} = 129844 + 2125 + 7936 - 27 + 744 - 22540 = 117182 \text{ cal/mol}.$$

Given ΔG^0_{1400} and ΔG^0_{2000}, the equilibrium constant of the reaction can be determined.

$$\Delta G^0_T = -RT \ln K = -4.575\, T \lg \frac{[p_{H_2O}]^2}{[p_{H_2}]^2}$$

from which

$$\lg\left(\frac{p_{H_2O}}{p_{H_2}}\right) = -\frac{\Delta G^0_T}{2 \cdot 4.575\, T}.$$

For the temperature $T = 1400$ K we obtain

$$\lg\left(\frac{p_{H_2O}}{p_{H_2}}\right) = -9.63$$

from which

$$p_{H_2O} = 2.34 \cdot 10^{-10} \cdot p_{H_2}.$$

For the temperature $T = 2000$ K

$$\lg\left(\frac{p_{H_2O}}{p_{H_2}}\right) = -6.40$$

from which

$$p_{H_2O} = 3.98 \cdot 10^{-7} p_{H_2}.$$

At 1 atm hydrogen pressure the equilibrium partial pressure of water vapour will be $2.34 \cdot 10^{-10}$ atm at 1400 K and consequently, $3.98 \cdot 10^{-7}$ atm at 2000 K; thus, there will be no reduction if an above calculated (or greater) amount of H_2O is present. Contrarily, metallic zirconium being heated in hydrogen atmosphere, the partial pressure of water vapour higher than $2.34 \cdot 10^{-10}$ atm for 1400 K and $3.98 \cdot 10^{-7}$ atm for 2000 K will result in the reaction course from right to left, i.e., to the oxidation of metallic zirconium.

3.2 System $Al_2O_3-SiO_2$

The study of the system $Al_2O_3-SiO_2$ is theoretically and practically important as this system underlies most processes of the silicate industry (ceramics, refractories, glassmaking etc.). The conditions of the formation of mullite and sillimanite group minerals and their stability are of particular interest and have been a subject of considerable research. Thus, according to [103] sillimanite minerals turn into mullite and SiO_2 at 1200 °C, and in this connection the free energy of mullite formation at this temperature and above must be a negative value.

Avgustinik and Mchedlov-Petrossyan were the first to have carried out a thermodynamic analysis of solid-phase reactions in the system $Al_2O_3-SiO_2$ [14]. Later the calculations of reactions in the system were more accurate as new data appeared [92, 399, 412].

From the data available concerning the free enthalpy values in the systems $CaO-Al_2O_3-SiO_2$, $MgO-Al_2O_3-SiO_2$, and $Fe_2O_3-Al_2O_3-SiO_2$ [139] a conclusion was drawn that the free enthalpies of the formation of sillimanite group minerals from oxides must be from 0 to −10 kcal/mol. In [100] the equilibrium was studied between the sillimanite group minerals at high pressures. In the course of this work an inference was drawn that the possible ΔH^0_{298} values of the formation of these minerals from oxides are from −500 to −1000 cal/mol and, as the ΔS^0_{298} values are small, the ΔG^0_{298} values must be of the same order.

For the equilibrium of SiO_2 in a melt with mullite and corundum at 1550 °C the silica activity value equal to 0.44 [251] was obtained. The determination of the free enthalpy of mullite formation by the technique used for determining the SiO_2 activities in melts was carried out by Cooper, Kay and Taylor [103]. The activities were determined by equilibrium pressures in the reaction:

$$3Al_2O_3 \cdot 2SiO_2 + 6C = 3Al_2O_3 + 2SiC + 4CO\,.$$

The a_{SiO_2} value obtained in a reaction with the solid mullite at 1550 °C equals 0.42 which well agrees with the data of [251].

The free enthalpy value obtained according to the determinations can be expressed by the equation:

$$\Delta G^0_T = -2650 - 18\,T\,(1673-1873\text{ K})\,.$$

The equation shows the stability of mullite to decrease with the lowering of temperature, and the extrapolation of the results obtained indicates the free

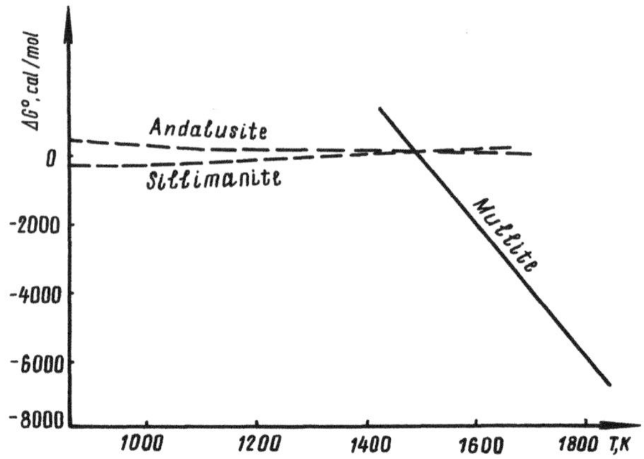

Fig. 46. The free enthalpies of the formation of compounds of the system $Al_2O_3-SiO_2$ according to Cooper, Kay and Taylor (*dotted lines* correspond to the ΔG^0 minimum values which determine the transition temperature 1200 °C)

enthalpy of mullite formation to equal zero at 1473 K. Below this temperature mullite is thermodynamically unstable as compared to quartz and corundum. The extrapolation of this kind is not, however, exact enough.

The dependence between mullite and the sillimanite group minerals can be examined with the help of a free enthalpy diagram (Fig. 46). Clark, Robertson and Birch [100] have found andalusite and sillimanite to be more stable than kyanite at a normal pressure. The relative stability of andalusite and sillimanite is less reliably specified. In all probability andalusite is more stable; the free enthalpies of the formation of these minerals are, however, near to one another.

On the basis of experimental data concerning the determination of the triple equilibrium point of andalusite-kyanite-sillimanite obtained by Khitarov and coworkers [265], Bell [39], and Clark [99], using thermodynamic data on enthalpies, heat capacities, and molar volumes Waldbaum calculated the heats and free enthalpies of Al_2SiO_5 polymorphic varieties. Applying then Kay's and Taylor's data [251] concerning the SiO_2 activity in the melts of the system $CaO-Al_2O_3-SiO_2$ combined with the data available on mullite heat capacity and enthalpy. Waldbaum determined the mullite heat and free enthalpy. Based on the values presented by Waldbaum we calculated the heats and the Gibbs free energies of the formation of all the above compounds from oxides within the temperature range from 298 to 2000 K. A temperature dependence of ΔG^0 of the reactions of compound formation from oxides in the system $Al_2O_3-SiO_2$ is presented in Fig. 47.

An analysis of the data obtained permits drawing the conclusion that among the sillimanite group compounds thermodynamically, the most stable up to the temperature of 1600 K, will be andalusite and above that temperature, sillimanite. Kyanite is unstable throughout the temperature range under consideration.

We attempted [364] to calculate, using thermodynamic constants available, the enthalpies and Gibbs free energies of mullite formation from oxides within the temperature range of 298 to 2000 K and to determine the mullite stability.

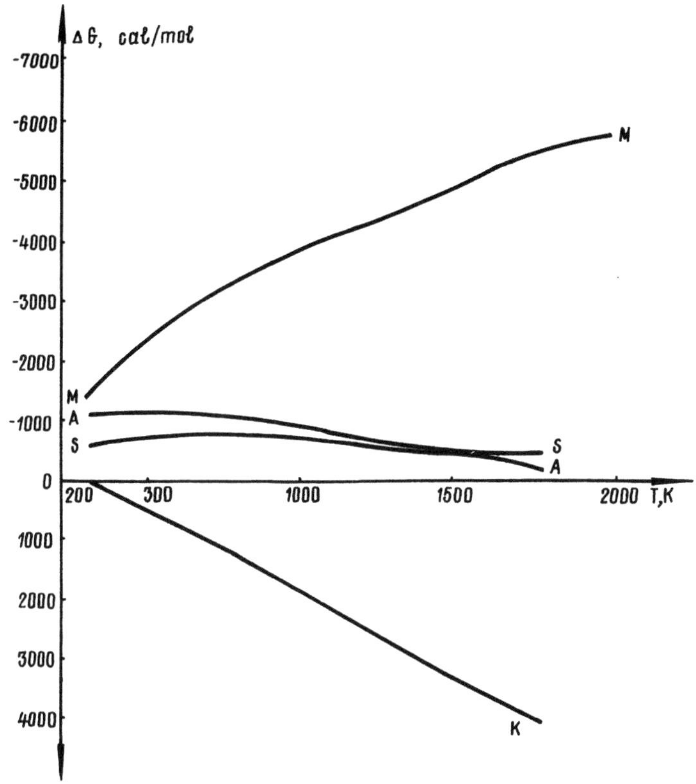

Fig. 47. The temperature dependence of ΔG^0 of the reactions of compound formation from oxides in the system $Al_2O_3-SiO_2$ within a temperature range of 298 to 2000 K according to Waldbaum

A starting point for the calculation is the knowledge of the temperature dependence of mullite heat capacity. An examination of the now existing heat capacity equations [242, 402] shows (see Fig. 48) that all of them are much understated as a result of the understated C_{p298} value formerly taken as a basis of the $C_p=f(T)$ calculation. For this reason we conducted a calculation of the equation $C_p=f(T)$ according to Landiya's entropy method [320]. The following data for mullite were taken as a basis:

$$S^0_{298}=60.8\ \text{cal/mol deg};\quad T_m=1850\,°\text{C}\ [273]\ .$$

Bearing in mind that cations Al^{3+} can take part in the anion framework creation or play the role of cations-modificators, it was accepted [36] that out of 6 ions Al^{3+} three play the role of cations-modificators and three take part in the anion framework creation.

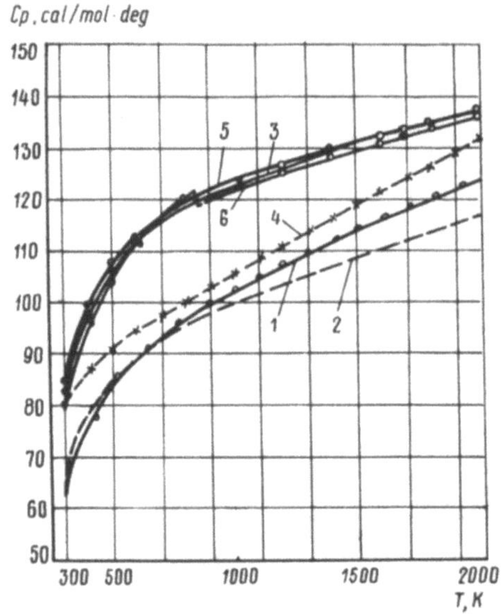

Fig. 48. The temperature dependence of the mullite heat capacity at constant pressure within a temperature range of 298 to 2000 K: *1, 2, 4* taken from literature; *3, 5* calculated by the authors; *6* obtained by summing up the data of oxides

For this instance according to Landiya's method [320] the equation of the temperature dependence of mullite isobaric heat capacity within a range of 298 K to T_m was calculated:

$$C_p = 115.9 + 11.2 \cdot 10^{-3} T - 37.0 \cdot 10^5 T^{-2}.$$

A similar $C_p = f(T)$ calculation for the instance when all the ions Al^{3+} in mullite crystalline lattice are cations-modificators resulted in the following equation of temperature dependence:

$$C_p = 112.85 + 12.61 \cdot 10^{-3} T - 29.29 \cdot 10^5 T^{-2}.$$

A comparison of the results obtained by the two equations shows that the structural factor does not effect essentially the equation of the temperature dependence of heat capacity.

The thermodynamic analysis of mullite formation from oxides is based upon the initial thermodynamic data presented in Table 39. Of all the silica modifications only β and α-quartz with the transition temperature of 848 K was considered. ΔH^0_{tr} was taken equal to -0.15 kcal/mol.

$\Delta H^0 = f(T)$ and $\Delta G^0 = f(T)$ for mullite formation from oxides were calculated, using the experimental value of $\Delta G^0_{1823} = -5.95$ kcal/mol [103, 251, 684], in the following way:

$$\Delta S^0_{1823} = \Delta S^0_{298} + \int_{298}^{848} \frac{C_{p_1} dT}{T} + \Delta S^0_{tr} + \int_{848}^{1823} \frac{C_{p_2} dT}{T}.$$

According to Gibbs-Helmholtz' equation the enthalpy of the reaction of mullite formation from oxides was determined by the equation:

$$\Delta H^0_{1823} = \Delta G^0_{1823} + 1823\,\Delta S^0_{1823} = 6866 \text{ cal/mol}\,.$$

From which

$$\Delta H^0_{298} = \Delta H^0_{1823} - \int_{298}^{848} C_{p_1} dT - \Delta H^0_{\text{tr}} - \int_{848}^{1823} C_{p_2} dT = 5290 \text{ cal/mol}\,,$$

where C_{p_1}, the β-quartz heat capacity; C_{p_2}, the α-quartz heat capacity; ΔS^0_{tr}, alteration of the entropy at transformation; ΔH^0_{tr}, alteration of the enthalpy at transformation.

Using Gibbs-Helmholtz' equations in the integrated form, the following dependences of $\Delta H^0 = f(T)$ and $\Delta G^0 = f'(T)$ in the temperature range of 298 to 848 K were determined:

$$\Delta H^0_T = 3956 + 8.12\,T - 6.485 \cdot 10^{-3} T^2 - 1.52 \cdot 10^5 T^{-1};$$

$$\Delta G^0_T = 3956 - 8.12\,T \ln T + 6.485 \cdot 10^{-3} T^2 - 0.76 \cdot 10^5 T^{-1} + 45.44\,T;$$

in the range of 848 to 2000 K:

$$\Delta H^0_T = 4229 + 1.74\,T - 0.235 \cdot 10^{-3} T^2 + 3.88 \cdot 10^5 T^{-1};$$

$$\Delta G^0_T = 4229 - 1.74\,T \ln T + 0.225 \cdot 10^{-3} T^2 + 1.94 \cdot 10^5 T^{-1} + 6.99\,T\,.$$

According to the above equations ΔH^0_T and ΔG^0_T of mullite formation from oxides in the temperature range of 298 to 2000 K were calculated. The results are presented in Fig. 49.

Function $\Delta G^0 = f(T)$ is a practically straight line, at high temperatures in particular. An intersection of this line with the zero ordinate ($\Delta G^0 = 0$) occurs at a temperature of about 944 K. At temperatures below that mullite cannot be thermodynamically formed in the reaction:

$$3Al_2O_3 + 2SiO_2 \rightarrow 3Al_2O_3 \cdot 2SiO_2\,.$$

Figure 49 also shows quite clearly that mullite has no incongruent point in the diagram of the phase equilibrium of the system.

Holm and Kleppa [206] determined the heats of kyanite, andalusite, sillimanite, and mullite formation from oxides by measuring dissolution heats at 968 K in the solvent $9PbO \cdot 3CdO \cdot 4B_2O_3$ (Table 39).

In this connection a thermodynamic analysis was conducted of the reactions in the system $Al_2O_3-SiO_2$ on the basis of experimental data concerning the enthalpies of sillimanite, andalusite, and kyanite formation. For the reaction of

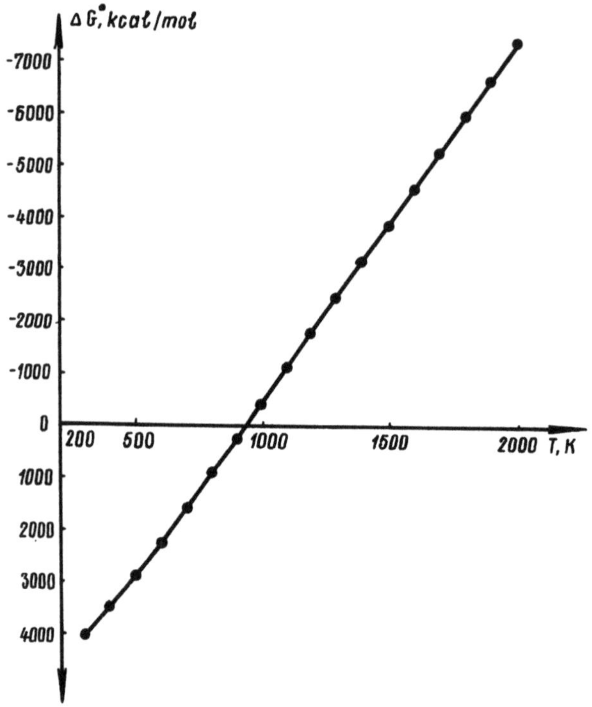

Fig. 49. The temperature dependence of ΔG^0 of the reactions of mullite formation in the system $Al_2O_3-SiO_2$ within a temperature range of 298 to 2000 K

mullite formation the equation of heat capacity previously obtained was used, because the ΔH^0_{968} value of mullite formation from oxides from our results practically coincides with the experimental one.

Equations $\Delta H^0 = f(T)$ and $\Delta G^0 = f'(T)$ of all the reactions are presented in Table 40.

Table 39. Initial thermodynamic data for compounds of the system $Al_2O_3-SiO_2$

Substances	ΔH^0_{298} from oxides [kcal/mol]	S^0_{298} [cal/mol deg]	$C_p = f(T)$	Temp. range [K]
$Al_2O_3 \cdot SiO_2$ sillimanite	-0.60	23.0	$40.09 + 5.86 \cdot 10^{-3}T - 10.13 \cdot 10^5 T^{-2}$	298–1600
$Al_2O_3 \cdot SiO_2$ andalusite	-1.34	22.3	$34.68 + 10.83 \cdot 10^{-3}T - 6.38 \cdot 10^5 T^{-2}$ [a]	298–1600
$Al_2O_3 \cdot SiO_2$ kyanite	-1.88	22.03	$45.32 + 2.34 \cdot 10^{-3}T - 16.0 \cdot 10^5 T^{-2}$	298–1700
$3Al_2O_3 \cdot 2SiO_2$	6.96	60.8	$115.9 + 11.2 \cdot 10^{-3}T - 37.0 \cdot 10^5 T^{-2}$ [a]	298–2000

[a] Calculated according to Landiya's method

Table 40. Dependences of the enthalpy and Gibbs free energy alteration of the reaction of aluminium silicate formation from oxides

Reactions	Temp. range [K]	Temp. dependence of the enthalpy alteration [cal/mol]	Temp. dependence of the Gibbs free energy alteration [cal/mol]
$3\alpha\text{-}Al_2O_3+2SiO_2 = 3Al_2O_3\cdot 2SiO_2$	298 – 848	$3293+3.12T-6.485\cdot 10^{-3}T^2 -1.52\cdot 10^5T^{-1}$	$3293-8.12T\ln T+6.485\cdot 10^{-3}T^2-0.76\cdot 10^5T^{-1}+45.44T$
mullite	848 – 2000	$3566+1.74T-0.225\cdot 10^{-3}T^2 +3.88\cdot 10^5T^{-1}$	$3566-1.74T\ln T+0.225\cdot 10^{-3}T^2+1.94\cdot 10^5T^{-1}+6.99T$
$\alpha\text{-}Al_2O_3+SiO_2 = Al_2O_3\cdot SiO_2$	298 – 848	$-526+1.44T-2.7\cdot 10^{-3}T^2 -1.04\cdot 10^5T^{-1}$	$-526-1.44T\ln T+2.7\cdot 10^{-3}T^2-0.52\cdot 10^5T^{-1}+6.64T$
sillimanite	848 – 1600	$-390-1.75T+0.43\cdot 10^{-3}T^2 +1.65\cdot 10^5T^{-1}$	$-390+1.75T\ln T-0.43\cdot 10^{-3}T^2+0.83\cdot 10^5T^{-1}-12662T$
$\alpha\text{-}Al_2O_3+SiO_2 = Al_2O_3\cdot SiO_2$	298 – 848	$2990-3.97T-0.215\cdot 10^{-3}T^2 -4.70\cdot 10^5T^{-1}$	$2290+3.97T\ln T+0.215\cdot 10^{-3}T^2-2.395\cdot 10^5T^{-1}-29.63T$
andalusite	848 – 1600	$2426-7.16T+2.915\cdot 10^{-3}T^2 -2.09\cdot 10^5T^{-1}$	$2426+7.16T\ln T-2.915\cdot 10^{-3}T^2-1.045\cdot 10^5T^{-1}-48.71T$
$\alpha\text{-}Al_2O_3+SiO_2 = Al_2O_3\cdot SiO_2$	298 – 848	$-5407+6.67T-4.46\cdot 10^{-3}T^2 +4.83\cdot 10^5T^{-1}$	$-5407-6.67T\ln T+4.46\cdot 10^{-3}T^2+2.415\cdot 10^5T^{-1}+46.94T$
kyanite	848 – 1700	$-5271+3.48T-1.33\cdot 10^{-3}T^2 +7.53\cdot 10^5T^{-1}$	$-5271-3.48T\ln T+1.33\cdot 10^{-3}T^2+3.765\cdot 10^5T^{-1}+27.71T$

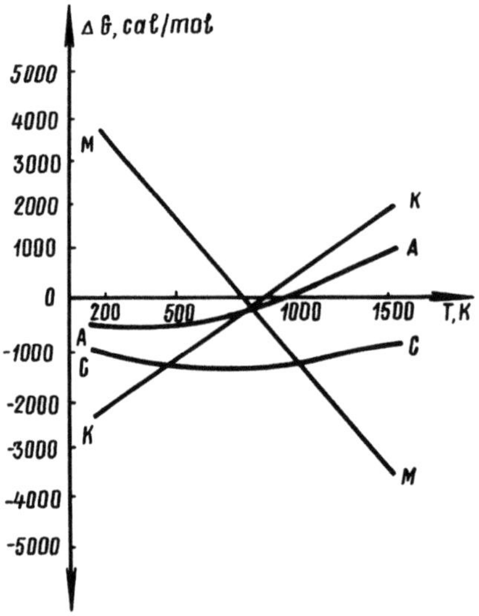

Fig. 50. The temperature dependence of ΔG^0 of the compound formation from oxides in the system $Al_2O_3-SiO_2$ within a temperature range of 298 to 1800 K

According to these equations the ΔH_T^0 and ΔG_T^0 values within the temperature range of 298 to 1700 K were calculated with an interval of 100 K. The results are presented in Fig. 50.

The calculation results for the ratios of $Al_2O_3/SiO_2 = 1/1$ and 3/2 are presented in Fig. 51, the thermodynamic stability sequence is presented in Table 41.

An analysis of the data obtained brings us to the following conclusions. Aluminium silicates of $Al_2O_3 \cdot SiO_2$ type throughout the temperature range under consideration are formed with heat release, the reaction of mullite formation being endothermic. The dependences presented are in agreement with the opinion shared by most researchers that kyanite and andalusite dissociate with the elevation of temperature [44, 92, 169, 471, 583], and that mullite is energically more stable in the range of high temperatures, up to its melting. The tendency towards the increase of ΔG_T^0 with the elevation of temperature is unlike andalusite and kyanite less characteristic of sillimanite.

Table 41. Thermodynamic stability sequence of compounds in the system $Al_2O_3-SiO_2$ [a]

Ratio of initial components	Thermodynamic stability series beginning with the most stable compound at temp. [K]			
	298 – 500	500 – 900	1000 – 1200	above 1350
1/1	KSAM	SKAM	SMAK	MSAK
3/2	KSAM	SKAM	SMAK	MSAK

[a] A – andalusite, K – kyanite, M – mullite, S – sillimanite

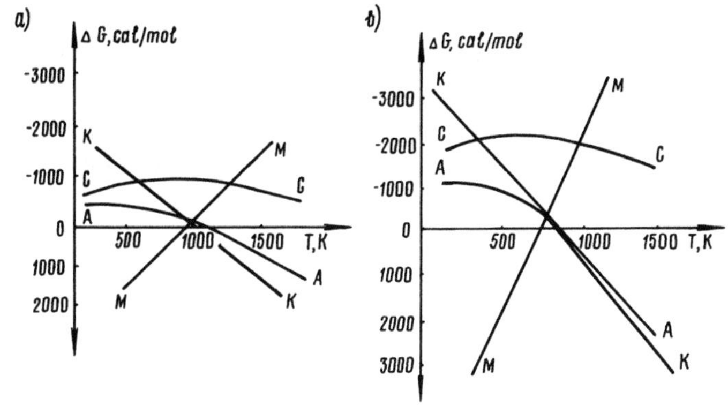

Fig. 51. The stability boundaries of compounds of the system $Al_2O_3-SiO_2$; **a** $Al_2O_3/SiO_2 = 1/1$; **b** $Al_2O_3/SiO_2 = 3/2$

The ratio Al_2O_3/SiO_2 being 1/1, up to a temperature of 1350 K the most stable compound is sillimanite, and above that temperature, mullite. At a temperature of 875 K a transformation of andalusite and kyanite into sillimanite and mullite is possible [3, 583]. The stability of the former compounds differs at low temperatures when kyanite is more stable. The ratio Al_2O_3/SiO_2 being 3/2, up to the temperature of 1215 K the most stable compound is sillimanite, above that temperature, mullite. Above 875 K a transformation of andalusite and kyanite into mullite is possible as well. At low temperatures kyanite is the most stable compound.

3.3 System ZrO_2-SiO_2 [365]

The previous systematic research into the melting equilibria of the mixture ZrO_2+SiO_2 resulted in the conclusion that zircon melts congruently [300, 685, 677, 717] at temperatures of 2175 to 2550 °C.

Observation of zircon sintering [540] however and repeated research into the equilibrium in the system ZrO_2-SiO_2 [73, 107, 330] make it evident that $ZrO_2 \cdot SiO_2$ dissociates at temperatures above 1700 °C. From the viewpoint of crystal chemistry the incongruent melting of zircon seems to be quite probable, because ions Zr^{4+} produce an intense counterpolarizing action weakening the coordination group $[SiO_4]^{4-}$ and even destroying it [125].

In this connection, of substantial interest is the thermodynamic analysis of solid-phase reactions in the system ZrO_2-SiO_2 within a temperature range of 298 to 1800 K. The extrapolation of the curve of the temperature dependence of the standard free energy of zircon formation to higher temperatures may elucidate the $ZrSiO_4$ stability in the high-temperature range.

An experimental method of determining the standard free energy of zircon formation from zirconium dioxide and silica [526] has been recently elaborated. The study of the equilibrium between Co, the gaseous phase ($CO + CO_2$) with a certain partial oxygen pressure, and some phases of the systems $CoO - ZrO_2 - SiO_2$ and $CoO - SiO_2$ enabled taking the values (in kcal/mol) of the reaction $ZrO_2 + SiO_2 = ZrSiO_4$ correct to ± 0.1 kcal at temperatures:

1180 °C – 2.38
1242 °C – 1.98
1300 °C – 1.64
1366 °C – 1.29 .

For calculating equations $\Delta H^0 = f(T)$ and $\Delta G^0 = f'(T)$ of the reaction $ZrO_2 + SiO_2 = ZrO_2 \cdot SiO_2$ the initial thermochemical data presented in Table 42 [20, 272, 623] were used.

The alteration of the enthalpy and Gibbs free energy of the reaction $ZrO_2 + SiO_2 = ZrSiO_4$ (within each 100 K) was calculated according to the following equations:

Table 42. Initial thermodynamic data for the system $ZrO_2 - SiO_2$

Substances	$-\Delta H^0_{298}$ [kcal/mol]	$-\Delta G^0_{298}$ [kcal/mol]	S^0_{298} [cal/mol deg]	$C_p = f(T)$
ZrO_2 monoclinic	261.5	247.7	12.12	$16.64 + 1.8 \cdot 10^{-3} T - 3.36 \cdot 10^5 T^{-2}$ (298 – 1478 K)
ZrO_2 tetragonal	–	–	–	17.8 (1478 – 1850 K)
$ZrSiO_4$	483.3 [a]	456.9 [a]	20.2	$31.48 + 3.92 \cdot 10^{-3} T - 8.08 \cdot 10^5 T^{-2}$ (298 – 1800 K)

[a] Calculated by the authors

298 – 848 K
$$\Delta H^0_T = -6565 + 3.62 \cdot T - 3.04 \cdot 10^{-3} T^2 + 2.02 \cdot 10^5 T^{-1}$$
$$\Delta G^0_T = -6565 - 3.62\, T \ln T + 3.04 \cdot 10^{-3} T^2 + 1.01 \cdot 10^5 T^{-1} + 25.5\, T$$

848 – 1478 K
$$\Delta H^0_T = -6429 + 0.43\, T + 0.09 \cdot 10^{-3} T^2 + 4.72 \cdot 10^5 T^{-1}$$
$$\Delta G^0_T = -6429 - 0.43\, T \ln T - 0.09 \cdot 10^{-3} T^2 + 2.36 \cdot 10^5 T^{-1} + 5.94\, T$$

1478 – 1800 K
$$\Delta H^0_T = -8328 - 0.73\, T + 0.99 \cdot 10^{-3} T^2 + 8.08 \cdot 10^5 T^{-1}$$
$$\Delta G^0_T = -8328 + 0.73\, T \ln T - 0.99 \cdot 10^{-3} T^2 + 4.04 \cdot 10^5 T^{-1} + 0.37\, T .$$

The values of ΔH^0_T and ΔG^0_T obtained are seen in Fig. 52.

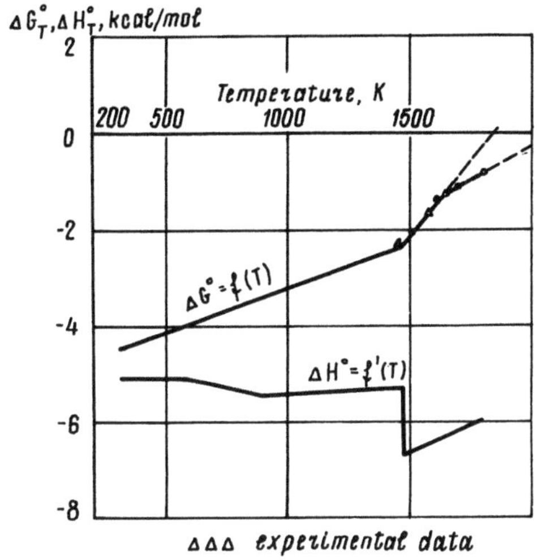

Fig. 52. The temperature dependence of ΔH^0 and ΔG^0 of zircon formation from oxides within a temperature range of 298 to 1800 K

The extrapolation of the curve drawn through the experimentally found ΔG_T^0 points up to the line $\Delta G^0 = 0$ yields a temperature of about 1800 K. On the other hand, a variety of calculation by the equation $\Delta G_T^0 = f(T)$ taking into consideration the point $\Delta G_{1639\,K}^0$ yields $\Delta G^0 = 0$ at the temperature of 2100 K, lower nonetheless, than that of zircon melting which, according to Kreidl, lies within the limits of 2500 to 2800 K.

It should be taken into account that the calculation greatly depends on the equation of zircon heat capacity. This equation must be more exact before final conclusions concerning the temperature of zircon dissociation can be drawn.

Thus, the analysis conducted shows that with the elevation of temperature the stability of zircon decreases and its dissociation seems quite probable. At any rate the temperature range of zircon dissociation may be determined as from 1800 to 2100 K. The experimental data concerning the temperature of zircon dissociation [353, 596] proves the same.

As no data written is available concerning such important characteristics of zircon as the heat of formation and the Gibbs free energy of formation from elements, we have calculated these values for the standard state. For the calculation the data obtained in the present work and those taken from literature [312, 623] were used.

The following values have been obtained: $H_{298}^0 = -483.3$ and $\Delta G_{298}^0 = -456.95$ kcal/mol.

3.4 Kaolinite Dissociation at Heating

The reactions of kaolinite transformation at heating underlie the manufacturing process of the most important kinds of ceramic products and refractories.

Although several works by Soviet scientists and by those of other countries deal with the behaviour of clays at heating, a great number of problems still require research. Thus, we do not know exactly what phases arise in argillaceous minerals at heating and what processes cause heat effects observed in the thermic analysis curves.

In this book we shall discuss the behaviour of one of the most important argillaceous minerals, kaolinite, at heating.

The reactions of the kaolinite dissociation at heating [62, 141, 397, 426, 430] to the following products were considered:

1. $Al_2O_3 \cdot 2SiO_2 + 2H_2O_{gas}$;
2. $Al_2O_3 \cdot SiO_{2\ sill} + SiO_{2\ \beta\text{-quartz}} + 2H_2O_{gas}$;
3. $\alpha\text{-}Al_2O_3 + 2SiO_{2\ \beta\text{-quartz}} + 2H_2O_{gas}$;
4. $\frac{1}{3}(3Al_2O_3 \cdot 2SiO_2) + \frac{4}{3}SiO_{2\ \beta\text{-quartz}} + 2H_2O_{gas}$.

The calculation results are presented in Table 43.

Table 43. The results of the ΔH^0_{298} and $\Delta G^0 = f(T)$ calculation for the reactions of the kaolinite dissociation

Nos. of reactions	Equations $\Delta G^0 = f(T)$	ΔH^0_{298} [kcal/mol]	ΔG^0 [kcal/mol] at temp. [K]			
			298	573	973	1273
1	$\Delta G^0_T = 57132 - 8.7\,T\ln T + 11.14 \cdot 10^{-3}T^2 - 3.65 \cdot 10^5 T^{-1} - 22.6\,T$	+56.38	+35.24	+15.31	−13.42	−33.46
2	$\Delta G^0_T = 27004 - 3.84\,T\ln T + 6.66 \cdot 10^{-3}T^2 + 0.33 \cdot 10^5 T^{-1} - 52.2\,T$	+27.78	+5.62	−14.64	−43.1	−63.53
3	$\Delta G^0_T = 27734 - 1.59\,T\ln T + 3.96 \cdot 20^{-3}T^2 + 0.85 \cdot 10^5 T^{-1} - 65.7\,T$	+28.14	+6.11	−14.21	−43.04	−63.95
4	$\Delta G^0_T = 30138 + 5.25\,T\ln T + 4.89 \cdot 10^{-3}T^2 - 0.12 \cdot 10^5 T^{-1} - 113.3\,T$	+28.06	+5.67	−14.1	−40.32	−58.76

An analysis of the equations $\Delta G = f(T)$ presented in the table shows that all the reactions under consideration are possible beginning with 800 K. Based on the results of the calculations it is possible to state that if there were no kinetic obstacles to direct transitions of kaolinite to sillimanite, mullite, and oxides, there would be equal probability for either a mixture of oxides or sillimanite to be the most stable products in this system. However, from the crystal chemistry viewpoint the transition of kaolinite to metakaolin is least impeded, and this transformation becomes thermodynamically possible at 800 K, apparently, reaction I is to be regarded as the most preferable at the first stage of the kaolinite transformation.

These facts involve the consideration of the reaction of metakaolinite transformation to mullite:

$$Al_2O_3 \cdot 2SiO_2 = \tfrac{1}{3}(3Al_2O_3 \cdot 2SiO_2) + \tfrac{4}{3}SiO_2 . \quad (126)$$

Equation $\Delta G^0 = f(T)$ for reaction (126) appears as follows:

$$\Delta G_T^0 = -31.435 + 3.33\,T \ln T - 4.76 \cdot 10^{-3} T^2 + 5.5 \cdot 10^5 T^{-1} - 16.55\,T . \quad (127)$$

In final form the $\Delta G^0 = f(T)$ diagram of the reactions of the kaolinite dissociation to metakaolin and its transformation to mullite is presented in Fig. 53 [422].

As one can see, thermodynamically the transformation of kaolinite to metakaolin is possible beginning with 800 K which is close to the results of previous calculations [394, 399, 426] and to the known experimental data, and the reaction of metakaolin transformation to mullite, at a temperature above 1235 K which is also in agreement with the experimental data.

Similar results have been obtained by other authors in their calculations [543] as well.

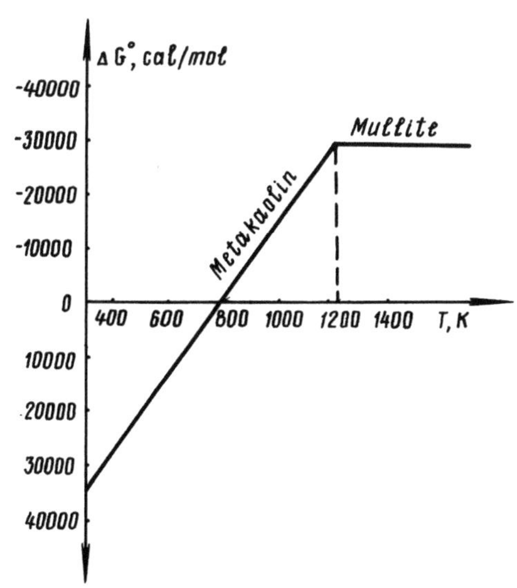

Fig. 53. The diagram of $\Delta G^0 = f(T)$ of kaolinite transformations at heating

4 Reactions of Glass Formation and Devitrification

4.1 General[11]

Reactions of glass formation and devitrification are of great importance in silicate science and, in particular, in such fields as the production of glass, glass ceramics and enamels – the melting of most glasses, water glasses etc. –, the production and life of refractories – attack of refractories by volatile components of the batch, as well as in the chemical analysis of silicates and of the products based on them.

As the theoretical aspect of glass formation and devitrification has not been sufficiently studied, we found it possible to examine the general problems of the thermodynamics of these specific processes and of the stability of the vitreous substances formed as well.

While melting glass polymorphic transformations, solid-phase chemical interaction, reactions with gases evolving and melts being formed all take place. The main reactions here will be those of the interaction of carbonates and sulfates of the metals of the first and second groups of the periodic system with quartz, as well as the reactions of lead silicate formation from oxides. As the interaction of carbonates and sulfates of the metals of the second group of the periodic system with silica has been considered in detail in the chapter dealing with the thermodynamics of binary and ternary systems (see Chap. II.2), following are examinations of the reactions of the interaction of carbonates and sulfates of the metals of the first group with quartz and the reactions of silicate formation in the system $PbO - SiO_2$.

Glasses, as is commonly known, are amorphous bodies obtained by the supercooling of melts, independent of their chemical composition and the temperature range of solidification, and possessing, as a result of a gradual increase of viscosity, properties of bodies solid in mechanical sense; the process of the transition from the liquid state to the vitreous must be reversible [439, 447, 472, 493, 569, 688].

Thus, at heating glasses do not melt like crystals, but gradually soften passing from a fragile state to a liquid-drop one with a continuous and even alteration of all their properties. Additionally, the reversibility of the processes and properties indicate that glass forming melts and the solidified glass can be viewed as true solutions [4]. Glass, as a supercooled liquid is thermodynamically unstable. Therefore glass crystallization at temperatures below liquidus, i.e., in the vitrification range, is a spontaneous process.

The range known as vitrification proper is the temperature range where at a set rate of temperature alteration properties depend not only on the temperature, but on the direction of its alteration as well [392]. At temperatures surpassing the

[11] Compiled with the assistance of A. K. Zhuravlyov

vitrification range glass is in a metastable state. Below that range the glass structure practically does not change with the alteration of temperature: this state is a frozen – stabilized – one. Reversibility is conditioned only by the temperatures above the vitrification range – metastable state – and below that range – frozen state.

For calculating the thermodynamic parameters in the vitrification range one can use the method of extrapolating the heat content curves beginning with temperatures below and above the vitrification range to their meeting in some middle point of the range corresponding to a certain structural temperature T_s of the glass under consideration. The glass produced by the industry does not reach the stabilized state for it does not undergo special continuous heat treatment in high-viscosity state. Therefore a glass of the same composition may have diverse properties and diverse structures depending on its approaching the equilibrium.

The idea of the equilibrium state of glass does not contradict the definition of the vitreous state as that characterized by instability. Any system may be believed to be in several equilibrium states of different stability degrees. Thus, outlining the Gibbs energy ΔG of a vitreous system depending on a certain generalized structural parameter ξ that characterizes the degree of the orderliness of the glass structure (Fig. 54), at a temperature T a particular non-equilibrium glass structure is characterized by the value of parameter $\xi = \xi_T$ [134]. At $\xi = \xi_{gl}$ the structure corresponds to the stabilized state of a glass formed with long annealing and ξ_{cr} corresponds to the ξ value after the transformation of glass to crystal. The structures are separated by potential barriers. The highest barrier separates the vitreous and crystalline states. There may be no energy barrier between the vitreous structures ξ_T and ξ_{gl} at all. The approach of a glass to the equilibrium state is connected with a change in its structure, but it cannot be identified with the processes of crystallization and even liquation. In order to obtain the equilibrium state glass must be conditioned in a softened state when chemical processes finish before nuclei of crystallization or liquation surpassing the critical size appear. At a temperature below T_s glass will remain in the equilibrium state. To characterize this phenomenon the term "stabilization" of glass was introduced [4].

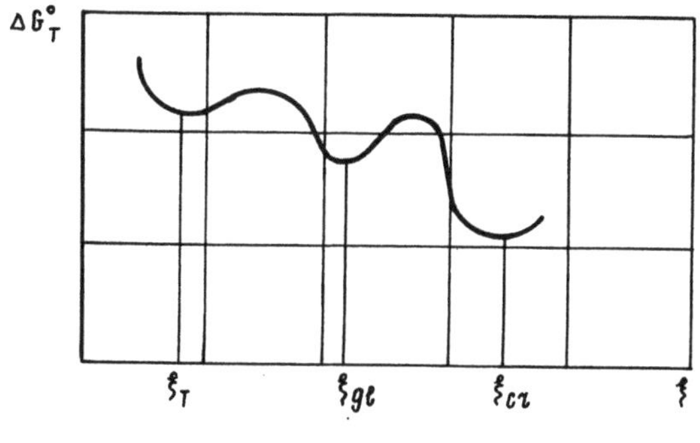

Fig. 54. The scheme of the alteration of the Gibbs free energy of the system microvolume while converting from the vitreous to the crystalline state

In this connection we shall consider the applicability of the classical thermodynamics to the vitreous state. The limits of the applicability of the principles of the classical thermodynamics to the study of silicate systems have been discussed above (see Chap. I. 11).

A peculiarity of vitreous silicate systems, unlike crystalline ones, is their lesser thermodynamic stability. However, the vitreous state can be also viewed as quasi-stable and hence the laws of the classical thermodynamics can be applied to it. A possibility of a reversible transition of the liquid state to the vitreous and crystalline one allows determining thermodynamically the conditions and the direction of these transitions.

The temperatures of glass formation are high enough and exceed the activation barriers of the reaction; the thermodynamic analysis in this case clearly determines the sequence of vitreous compound formation. The initial reactions of silicate formation in solid state should also be taken into account.

One of the difficulties of the application of thermodynamics to the vitreous state is that of determining the zero entropy.

In the works by Gutzov an attempt was made at determining the zero entropy of substances in vitreous state [175, 177]. He determines the glass entropy at the absolute zero by the following formula:

$$\Delta S_0^0 = \Delta S_m^0 + \Delta S_f^0 + \Delta S_c^0 + \Delta S_s^0 \tag{128}$$

where ΔS_m^0, entropy conditioned by the presence of molecules with different polymerization degree determined by the mean polymerization degree $\bar{P}$; ΔS_f^0, entropy conditioned by the free volume as a result of the difference of the glass molar volume and of that of the corresponding crystal; ΔS_c^0, entropy conditioned by the diversity of the configuration possibilities of long chains of glass structures which can be determined by the lattice-cell model of polymeric solutions [674].

$$\Delta S_{c\,\max}^0 \simeq 2.8 \text{ cal/mol deg} . \tag{129}$$

ΔS_s^0, entropy conditioned by the relative position and symmetry of molecules constituting the glass.

$$\Delta S_{s\,\max}^0 = R \ln 2 \quad \text{when} \quad y = 0.5 ,$$

because

$$\Delta S_s^0 = -R[y \ln y + (1-y) \ln (1-y)] , \tag{130}$$

where y, number of frozen orientations.

In particular instances with the glass constituted by symmetrical structural elements $\Delta S_s^0 = 0$.

For the glass in which $\bar{P}=1$ or $\bar{P}=\infty$ $\Delta S_{\mathrm{m}}^{0}=0$, because

$$\Delta S_{\mathrm{m}}^{0}=R\frac{1}{\bar{P}}[\bar{P}\ln\bar{P}=(\bar{P}-1)\ln(\bar{P}-1)] \tag{131}$$

$$\Delta S_{\mathrm{m\,max}}^{0}=R\ln 2,\quad \text{when}\quad \bar{P}=2\,.$$

If $\bar{P}=1$, then $\Delta S_{\mathrm{c}}^{0}=0$.

Gutzov supposes that if the values of the zero entropy, $\bar{P}$, free volume V_{f} and other values (see above) are known, one can draw conclusions concerning the glass structure or at least rule out the given structure as impossible. Oxide glasses in which chain structural elements are statistically indiscernible have the zero entropy value depending only on $\Delta S_{\mathrm{f}}^{0}$.

In the general form the configuration entropy equals:

$$\Delta S_{\mathrm{f}}^{0}=-R\left[\ln\beta_1+\frac{1-\beta_1}{\beta_1}\ln(1-\beta_1)\right] \tag{132}$$

where $\beta_1=N_0/(N_0+n_{\mathrm{h}})$. Here N_0, number of occupied places in the glass lattice; n_{h}, number of holes in the glass lattice.

$$\beta_1=\frac{N_0 v_0}{(N_0+n_{\mathrm{h}})v_0}=\frac{V_{M\,\mathrm{cr}}}{V_{M\,\mathrm{gl}}}=\frac{M}{\rho_{\mathrm{cr}}}\bigg/\frac{M}{\rho_{\mathrm{gl}}}, \tag{133}$$

where v_0, volume of a hole or, as a first approximation, volume of a monomeric element.

Given $\rho_{\mathrm{cr}}=2.65\ \mathrm{g/cm^3}$ (β-quartz) [20] and $\rho_{\mathrm{gl}}=2.203\ \mathrm{g/cm^3}$ (quartz glass) [183], we determine $\beta=\dfrac{2.203}{2.65}=0.8313$.

$$\Delta S^{0}=1.08937\ \text{cal/mol deg}\approx 1.08\ \text{cal/mol deg}\,.$$

Experimentally the zero entropy for quartz glass was determined by Simon and Lange [577] at the temperature of 9 K which is close enough to the absolute zero. According to their data S_0^0 (quartz glass) $=0.9\pm0.3$ cal/mol deg. The entropy at 0 K was found by means of extrapolation from the temperature of 9 K to 0 K.

Thus, the calculated data for vitreous quartz coincide to a good extent with the experimental data concerning the zero entropy.

However, if we apply the calculation method suggested in [177, 626] to the vitreous B_2O_3 and P_2O_5 we shall obtain results widely differing from the experimental ones.

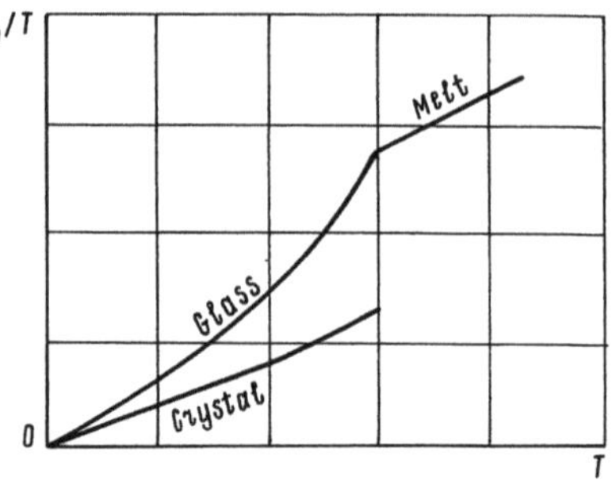

Fig. 55. The temperature dependence of C_p/T of a substance in crystalline and vitreous state

In [433] on the basis of a supposed heat capacity alteration of the same substance in crystalline and vitreous state (Fig. 55) a formula for calculating the zero entropy of a substance in vitreous state is deduced.

At point I the entropies of glass and of a crystal coincide.

$$S^0_{(cr)} = \frac{Q_{m\,(cr)}}{T_m} + \int\limits_0^{T_m} \frac{C_{p\,(cr)}}{T}\, dT\,, \tag{134}$$

$$S^0_{(gl)} = S^0_{0(gl)} + \int\limits_0^{T_m} \frac{C_{p(gl)}}{T}\, dT\,. \tag{135}$$

At point I $S^0_{(cr)} = S^0_{(gl)}$. Consequently,

$$S^0_{0(gl)} = \frac{Q_{m(cr)}}{T_m} + \int\limits_0^{T_m} \frac{C_{p(cr)} - C_{p(gl)}}{T}\, dT\,. \tag{136}$$

The calculation according to the formula is, however, hampered by the lack of data concerning heat capacity equations for crystalline and vitreous substances in particular in as wide a temperature range as from 0 K to T_m. For determining the glass zero entropy the following formula has been suggested [449]:

$$\frac{\Delta S_c T_m}{\mu} = \frac{2}{3}\left[\frac{\Delta S^0_0 + \Delta S_c T_m}{\mu} - 0.9\right], \tag{137}$$

where $\Delta S_c T_m$, configuration entropy corresponding to the melting temperature; ΔS^0_0, zero entropy of glass; μ, number of connecting links per formal unit of monomer.

This formula is, however, applicable chiefly to organic substances; cases exist where it cannot be applied to inorganic glasses.

For calculating the entropy of vitreous compounds with respect to the zero entropy data were used on the low-temperature heat capacity of alkaline silicates in vitreous state [611]. Given function $C_p = f(T)$, one can determine the entropy diagrammatically. This is how the entropy of the vitreous $Na_2O \cdot SiO_2$ and $Na_2O \cdot 2SiO_2$ at a temperature of 298 K was found. From the entropy of the crystalline $Na_2O \cdot SiO_2$ and $Na_2O \cdot 2SiO_2$ [20] at 298 K the entropy of the vitreous $Na_2O \cdot SiO_2$ and $Na_2O \cdot 2SiO_2$ at the absolute zero was deduced. It is 2.35 cal/mol deg and 3.15 cal/mol deg respectively.

For silicates and vitreous substances the calorimetric method of measuring heat capacity first used in [693] is employed. While measuring the mean heat capacities of three container glasses in the range of 353 to 1573 K it was found that above 1173 K the steep curve of the heat capacity dependence smoothens, the mean heat capacity only increasing slightly [558].

In [54] a sharp change of the heat capacity of some window and container glasses in the temperature range T_g was found a fact which was corroborated by [524].

In [191, 469, 626, 630, 631, 632] sharp changes of glass heat capacities in the range of 773 – 973 K were traced as well. It is noteworthy, however, than in several works [190, 201] no sharp change of heat capacity in the temperature range T_g was found.

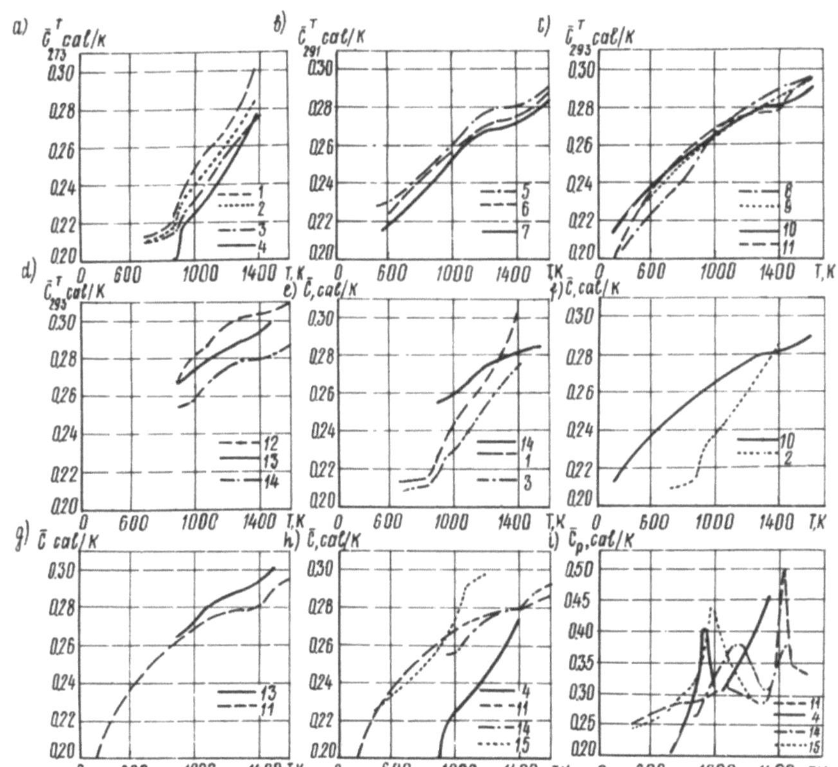

Fig. 56a – i. The temperature dependence of the mean heat capacity of glasses of different composition

Table 44. Mean heat capacities of diverse glasses

Glasses	1	2	3	4	5	6	7	8	9	10	11	12	13	14	15
References	[54]	[54]	[54]	[54]	[558]	[558]	[558]	[191]	[190]	[190]	[190]	[650]	[650]	[650]	[699]
SiO_2	71.84	70.97	72.86	70.95	66.10	64.32	62.84	72.26	74.9	71.12	71.78	70.0	72.3	72.0	72.51
TiO_2	0.05	0.07	0.08	0.05	–	–	–	–	–	–	–	–	–	–	–
Al_2O_3	1.59	3.43	1.63	1.86	6.03	7.52	7.08	1.99	1.59	3.67	0.47	–	0.7	2.0	0.89
Fe_2O_3	0.17	0.16	0.14	0.14	1.27	1.90	1.22	0.12	0.046	0.53	0.13	–	0.07	0.25	0.12
CaO	9.11	8.57	9.61	10.14	12.20	12.85	12.76	9.77	8.34	10.42	9.56	9.6	8.2	8.3	8.74
MgO	2.99	2.80	1.7	2.73	0.35	0.53	0.40	–	–	0.81	2.53	–	3.6	3.0	3.48
Na_2O	14.00	13.82	14.21	13.88	–	–	–	14.68	14.2	12.42	14.82	20.4	15.2	13.7	13.77
K_2O	–	–	–	–	12.59	12.44	8.64	1.08	0.61	0.89	0.46	–	–	–	–
MnO	–	–	–	–	1.40	0.44	7.06	–	–	–	–	–	–	–	–
SO_3	0.03	0.02	–	0.04	–	–	–	–	–	–	–	–	–	0.1	0.52

In [699] results of the examination of mean heat capacities of some silicate glasses are adduced (see Fig. 56 and Table 44). The numbering of glasses in the table and in the diagrams is the same. The glass composition is presented in weight percent. As one can see from Fig. 56 from a to d and from e to h the values of mean specific heat capacities of glasses differing in composition obtained by the same investigator differ as much or even less than those of similar glasses obtained by different scientists. The results of differentiating the values of the enthalpy alteration for glasses 4, 11, 14, and 15 the values of the mean specific heat capacities of which are presented in Fig. 56h are visualized in Fig. 56i. The figure illustrates clearly that on the basis of the experimental material available it is impossible to choose some optimum value of the true specific heat capacity.

It is of interest to consider the calculation methods of determining heat capacity that mainly amount to examining the contribution of particular oxides making up glass to its heat capacity. As shown in [690], the mean specific heat capacity of glasses within the temperature range of 16° to 100 °C is calculated by adding the empirical coefficients multiplied by the percentage of the content of constituents. A calculation of the actual heat capacity from 0° to 1300 °C was carried out in [439, 559, 565] according to the following formula:

$$C_p = \frac{at + C_m}{0.00146\,t + 1}, \qquad (138)$$

where C_m, mean heat capacity; t, temperature in °C.

$$C_m = \frac{at + C_0}{0.00146\,t + 1}. \qquad (139)$$

Here a and C_0 are constant coefficients for the glass components (presented in Table 1).

The data of [469] permit stating that lead oxide introduced instead of silica or instead of oxides of light elements, substantially decreases the heat capacity of glass. Moreover, the calculated and measured values of the heat capacities of lead glasses do not coincide, which is accounted for [191, 192] by different values of the specific factor of the lead oxide heat capacity differing with bi- and tetravalent lead in compounds as well as by the character of the compounds it enters. The effect of the chemical composition on the heat capacity of glasses was studied in [499, 522, 523].

For lack of experimental data in [214] the calculation formula is suggested

$$C_p = 5.283 + 1.987\frac{T}{T_{tr}} \qquad (140)$$

Table 45

Oxide	a	C_0
SiO_2	0.000468	0.1657
Al_2O_3	0.000453	0.1765
CaO	0.000410	0.1709
MgO	0.000514	0.2142
K_2O	0.000445	0.1756
Na_2O	0.000829	0.2229
B_2O_3	0.000598	0.1935
PbO	0.000013	0.0490
SO_3	0.000830	0.1890
Mn_3O_4	0.000294	0.1498
Fe_2O_3 [a]	0.000380	0.1449

[a] Valid only up to 600 °C

converted into gram atom heat capacity. It is noted that at an equal ratio of T/T_{tr} in two substances where T_{tr} is the temperature of the phase transition (in glasses T_g may be taken for this temperature) the heat capacities of the substances are equal. The formula is applicable to the range of $T = 0.3\,T_{tr}$ to $T = T_{tr}$. For compounds:

$$C_p = n(5.283 + 1.987\,T/T_{tr}) \tag{141}$$

where n, number of atoms in a compound.

An experimental study of the heat capacity of some glasses brought the authors to the conclusion that for practical purposes the data obtained by additivity equations [439, 559] for all the glasses studied will suffice. Empirical temperature dependence of heat capacities of some vitreous substances of the type of $C_p = a + bT - cT^{-2}$ are adduced in [257].

We have compared some calculation methods for quartz glass, vitreous sodium silicates $Na_2O \cdot SiO_2$ and $Na_2O \cdot 2SiO_2$, Lilie and Pyrex 7740 glasses [181].

For these glasses a calculation of the true heat capacity up to the temperature of liquid state (equal to the temperature of the substance melting in crystalline state) was conducted according to the formulae suggested in works [214, 439, 559] and simultaneously according to the empirical formulae suggested in works [181, 257].

The calculation results show the heat capacity values presented in [257] to be the nearest to experimental ones. [214] is of interest only in calculating heat capacities at the melting temperature of crystralline substances, whereas the course of the heat capacity curve, generally does not correspond to the actual course of the heat capacity at all. Thus, while calculating thermodynamic functions of vitreous substances, if no experimental data concerning the course of heat capacity depending on temperature are available, it is better to use the empirical formulae from [257]. Unfortunately, the number of vitreous substances in work [257] for which empirical formulae of heat capacity are adduced is insignificant. Therefore from now on for calculating the thermodynamic parameters of diverse vitreous substances we will use the comparative method of calculating heat capacities of vitreous substances.

Considering the thermodynamic stability of a substance in vitreous state it should be taken into account that a liquid can solidify in two ways: it may form crystals of a solid where particles are regularly disposed in a lattice or it may form an amorphous solid – vitreous state – where the particle disposition is characterized by an absence of long-range order. Let us consider the conditions of the formation of a stable nucleus in the crystallization processes and those of non-crystallizable glasses in the glass formation processes.

We shall examine the process of homogeneous nucleus formation from liquid at a constant temperature and atmospheric pressure, supposing that the nuclei being formed have a spherical shape and the specific crystal-liquid boundary energy is isotropic.

With these assumptions the free energy of the crystalline nucleus formation ΔG^* can be expressed [98, 648, 653, 654]

$$\Delta G^* = \frac{16\pi\sigma^3}{3(\Delta G_v)^2} \tag{142}$$

and the number of molecules in a critical nucleus equals

$$n^* = \frac{32\pi}{3V_m}\left(\frac{\sigma}{\Delta G_v}\right)^3, \tag{143}$$

where σ, surface tension on the crystal-liquid boundary; V_m, volume of a molecule; ΔG_v, difference of the standard Gibbs free energy per unit of the volume of the nucleus being formed.

At a little supercooling

$$\Delta G_v \approx \Delta H_{f(v)}(\Delta T/T_E), \tag{144}$$

where $\Delta H_{f(v)}$, latent melting heat per unit of the nucleus volume; T_E, equilibrium temperature (melting temperature); $\Delta T = T_E - T$, supercooling.

At a great supercooling [205]

$$\Delta G_v \approx \Delta H_{f(v)}\frac{\Delta T}{T_E}\cdot\frac{T}{T_E}. \tag{145}$$

The nucleation rate in glass-forming systems can be expressed by the following formula [656]:

$$I_v = \frac{N_v^0 D'}{a_0^2} e^{-(\Delta G^*)/kT}, \tag{146}$$

where D', kinetic coefficient for the molecular transfer through the nucleus-matrix surface; a_0, diameter of a molecule.

$$D' = a_0^2 \nu_0 e^{-(\Delta G'_m)/kT}, \tag{147}$$

where ν_0, frequency with which molecules try to surpass the energy barrier of the nucleus-matrix surface; $\Delta G'_m$, activation free energy for the motion through the nucleus-matrix surface.

At a nucleation temperature T^* the free energy of the formation of a crystalline nucleus ΔG^* is generally $50\,kT$ and less [655]. The T^* value is from 0.75 to $0.85\,T_E$.

It should be noted that the nucleation rate for vitreous liquids is relatively low as a result of high viscosity of such liquids.

While studying the crystallization of glass of the composition $Li_2O \cdot 2SiO_2$ [198], the authors of the same book came to the conclusion that the nucleation rate is expressed by the following formula:

$$I = \sum_{j=0}^{j=\max} \frac{N_j}{V} \nu_j e^{-(\Delta G_j^* + \Delta G'_m)/kT} \tag{148}$$

where $\frac{N_j}{V}$, number of nuclei per unit of volume; ΔG_j^*, free energy of the formation of a stable nucleus of size j.

The Gibbs free energy of the glass transition into crystalline state [605] equals:

$$\Delta G = \Delta H_f \frac{\Delta T}{T_m}, \tag{149}$$

where T_m, melting temperature; ΔT, supercooling; ΔH_f, melting enthalpy.

In [391] the crystallization of glasses $Li_2O \cdot 2SiO_2$, $Na_2O \cdot 2SiO_2$ and $K_2O \cdot 2SiO_2$ was considered and ΔG was shown to have the greatest effect on the nucleus formation rate. The highest crystallization free energy is that of the $Li_2O \cdot 2SiO_2$ glass, the lowest one is that of the $K_2O \cdot 2SiO_2$ glass, therefore the $Li_2O \cdot 2SiO_2$ glasses have the highest nucleus formation rate.

In order to determine the conditions of the non-crystallizable glass formation one must find the minimum values of the corresponding energies and ΔG^*, and $\Delta G'_m$ necessary for the glass formation from the equations determining the formation of crystallization nuclei. With this purpose in view one must first choose the values of the rate of the formation of crystallization nuclei I. In case $I = 1\ \text{cm}^{-3}\,\text{s}^{-1}$, glass will be formed if at no temperature T^*, the frequency of the formation of crystallization nuclei reaches a value greater than $1\ \text{cm}^{-3}\,\text{s}^{-1}$. If a nucleus is spherical, $N_j/V = 1611/3$; $\nu_j = 10^{12}\,\text{s}^{-1}$, the glass formation is possible when

$$(\Delta G^* + \Delta G'_{\text{m}}) \geqslant 30 R T_{\text{m}}\,, \tag{150}$$

where T_{m}, melting temperature.

Pure substances in which there are no external nuclei and the free energy of the activation of the rate of the formation of crystallization nuclei and of the rate of the crystal growth is greater than $30\,R\,T_{\text{m}}$ were shown not to crystallize.

The lower the melting temperature, the greater the probability of the glass formation.

Since

$$T_{\text{m}} = \Delta H_{\text{m}} / \Delta S_{\text{m}}\,, \tag{151}$$

where ΔH_{m}, crystallization heat; ΔS_{m}, crystallization entropy, one can say that the greater the tendency towards vitrification at a constant bond energy, the less energy is spent on creating a certain number of disordering. Asymmetrization increases the entropy of a liquid, therefore the specific energy expenditure on the creation of disorder decreases. However, while applying the above theory to actual substances it is difficult to answer the questions: which factors favour the increase of the free energy of the activation of the crystallization nucleus formation and of the growth of crystals and which at the same time lower the solidus and liquidus temperatures.

To answer the first question one must know the melt structure of the crystalline lattice and the energy of the molecular bonds broken. At present data concerning the first two factors are extremely limited and hard to be obtained. Therefore one has to replace these fundamental factors by the data concerning the melt viscosity as a measure of the difficulty of the liquid restructuring. This procedure yields good results for simple substances, but one cannot expect them to be universal for the structural factors determining viscous flow and crystallization are different. The lack of knowledge about the structure of melts also greatly hampers the study of factors determining the melting temperature.

As the motive force of crystallization the alteration of the chemical potential may be considered.

The authors of [176, 178, 179, 180] deduced the temperature dependence of $\Delta\mu$ for vitreous supercooled melts.

$$\Delta\mu = \Delta S_m \Delta T\left[1 - \frac{a}{2T_m}\Delta T + \frac{2}{3}bT - \frac{1}{3}bT_m\left(1 + \frac{T}{T_m}\right)\frac{T}{T_m}\right], \tag{152}$$

where

$$a = \frac{\Delta S_m - \Delta S_g}{T_m - T_g} \cdot \frac{T_m}{\Delta S_m} + b(T_m + T_g), \tag{153}$$

$$b = \frac{1}{2}\,\frac{1}{T_m}\left(\frac{Z}{x-y} - \frac{1}{1-y}\right)\frac{1}{1-x}, \tag{154}$$

where

$$x = \frac{T_g}{T_m}; \quad y = \frac{T_0}{T_m}; \quad z = \frac{\Delta S_g}{\Delta S_m}, \tag{155}$$

$$\Delta S_g \cong (1.1 \div 1.115)\Delta S_0^0. \tag{156}$$

On the other hand, it has been demonstrated that

$$\Delta S_0^0 \cong \tfrac{1}{3}\Delta S_m, \tag{157}$$

where ΔS_0^0, zero entropy of the vitreous state at $T = 0$ K; ΔS_m, melting entropy at melting temperature T_m; ΔS_g, glass formation entropy at vitrification temperature T_g.

The authors of [176, 180] also suggest a simplified formula for $\Delta\mu$:

$$\Delta\mu \approx \Delta S_m \Delta T \frac{T}{T_m}, \tag{158}$$

where T, supercooling temperature.

After having considered the general thermodynamic regularities of glass formation it is necessary to view the thermodynamics of the main reactions of glass formation. We shall examine primarily the reactions of the interaction of alkali metal carbonates and sulfates with silica, as they are of paramount importance in silicate science and, in particular, in such fields as glass and enamel manufacturing – the melting of most glasses, of water glasses etc. –, production and life of refractories – the attack of refractories by volatile components of the batch –, as well as for the chemical analysis of silicates and their products.

4.2 Silicate Formation in Glass Production

A thermodynamic analysis of these reactions was conducted by M. A. Matveyev, B. N. Frenkel' and G. M. Matveyev [377]. The data on the formation of liquid and solid solutions in the systems under consideration being scarce, the activities of condensed phases in all cases were taken for equal to one. In all cases the standard thermodynamic constants were taken without regard to the effect of the comminution degree and of the heat treatment of components. The presence of vitreous phases wide-spread in the systems in question was also not taken into consideration. While determining the top of the temperature range it is necessary to stop at a temperature of 1800 K which corresponds to the temperature maximum of most up-to-date glassmaking furnaces.

4.2.1 System $Na_2CO_3-Na_2SO_4-SiO_2$ [369, 377]

In accordance with reference-books it was accepted that for soda ash $\Delta H^0_{298} = -270300$ cal/mol and $\Delta G^0_{298} = -250400$ cal/mol [183, 185, 528]. The enthalpy alteration at melting was taken equal to 7303 cal/mol [489] at $T_m = 1123$ K. Although soda ash has three modifications only the phase transition at 723 K was taken into account. The enthalpy alteration in a phase transition was taken equal to 703 cal/mol which corresponds to the sum of the enthalpy alterations of two phase transitions equal to 255 and 448 cal/mol respectively [489]. Such an assumption was necessary for lack of data on the soda ash heat capacity within the range between the two phase transitions as well as for the value of the enthalpy alteration being relatively small.

For the soda ash heat capacity [311] suggests an equation $C_p = 13.98 + 54.4 \cdot 10^{-3} T - 3.125 \cdot 10^5 T^{-2}$ (from 298 to 450 K).

G. M. Matveyev and Hanikiř [369] offer an equation calculated on the basis of the technique suggested by Landiya [320]:

$$C_p = 29.29 + 9.55 \cdot 10^{-3} T - 5.18 \cdot 10^5 T^{-2}.$$

An equation from [256] is quite close to it:

$$C_p = 27.13 + 15.62 \cdot 10^{-3} T - 4.78 \cdot 10^5 T^{-2}. \text{ [541]}$$

Popov and Ginzburg [489] adduce the following equation based on their experiments: $C_p = 14.48 + 3.902 \cdot 10^{-2} T$ (298 – 723 K); $C_p = 32.197 + 8.43 \cdot 10^{-3} T$ (723 – 1123 K) and $C_p = -312.63 + 0.56 T - 21.94 \cdot 10^{-5} T^2$ (1123 – 1500 K).

Owing to the great difference between the equations suggested by [311, 369, 489] calculations of the soda ash thermodynamic properties within the range of 298 to 1800 K were carried out on the basis of the three variants of the heat capacity equations. As the main variant that calculated according to Popov and Ginzburg's data was taken. C_p calculated according to the other variants were compared with the main one and the results of the comparison are presented below.

For Na_2SO_4 [183, 184, 528] $\Delta H^0_{298} = -330900$ and $\Delta G^0_{298} = -302780$ cal/mol were assumed. The enthalpy alteration at melting ($T_m = 1157$ K) was assumed equal to 5750 cal/mol [311].

The enthalpy alteration at the phase transition ($T = 518$ K) is taken for equal to 2770 cal/mol which corresponds to the sum of two phase transitions according to [311].

The heat capacity of sodium sulfate according to [104, 311] is determined by equations $C_p = 23.5 + 31.74 \cdot 10^{-3} T$(298 – 450 K) and $C_p = 29.06 + 19.34 \cdot 10^{-3} T$ (514 – 1157 K).

Popov and Ginzburg [489] adduce the following experimental equations: $C_p = -13.94 + 0.0554\,T$ (from 298 – 518 K); $C_p = 31.75 + 1.548 \cdot 10^{-2}$ (from 518 – 1157 K) and $C_p = 136.09 + 0.072394\,T$ (from 1157 to 1500 K). As the main variant the one calculated according to Popov and Ginzburg's data is taken. The calculation according to [311] was carried out for comparison.

The ΔH^0_{298} and ΔG^0_{298} values for diverse sodium silicates taken on the basis of [554, 595] are given in [377, 510].

The values of the enthalpy alteration in the process of the melting of meta- and disilicates is taken for equal to 12.470 and 8460 cal/mol respectively according to Kelley [256]. Hummel and Schwiete [554] using more complicated technique found $\Delta H^0_m = 10500$ cal/mol for the disilicate; nevertheless, the values obtained by Kelley by a direct method should be used. For the ortho- and trisilicate the ΔH^0_m values are lacking. As a first approximation one may assume the melting entropy to depend linearly on the Na_2O molar fraction within the range of compounds from sodium ortho- to trisilicate. Thus, the entropy alterations in the process of melting: for Na_4SiO_4 $\Delta S^0_m = 10.69$ cal/mol deg and for $Na_2SiO_3O_7$ $\Delta S^0_m = 6.47$ cal/mol deg were obtained. Thus with regard to the melting temperatures of the ortho- and trisilicate (1291 and 1029 K) we obtain $\Delta H^0_m = 13800$ cal/mol for the orthosilicate and 6619 cal/mol for the trisilicate.

The assumption made is confirmed to a certain extent by the fact that the entropy of quartz melting is 3400 : 2016 = 1.68 cal/mol deg.

The heat capacities of Na_2SiO_3 according to Kelley [256] are $C_p = 31.14 + 9.6 \cdot 10^{-3} T - 6.47 \cdot 10^5 T^{-2}$ (from 298 to 1361 K) and $C_p = 42.8$ (from 1361 to 1800 K), and for the disilicate (according to the same data) $C_p = 44.38 + 16.86 \cdot 10^{-3} T - 10.67 \cdot 10^5 T^{-2}$ (from 298 to 1147 K) and $C_p = 62.35$ (from 1147 to 1800 K).

For the solid Na_4SiO_4 and $Na_2Si_3O_7$ the calculated heat capacity equations [369] were accepted: for the orthosilicate $C_p = 38.86 + 17.74 \cdot 10^{-3} T$ and for the trisilicate $C_p = 48.25 + 26.69 \cdot 10^{-3} T - 4.98 \cdot 10^5 T^{-2}$ (from 298 K to T_m).

For determining the heat capacities of the liquid Na_4SiO_4 and $Na_2Si_3O_7$ it was taken into consideration that the heat capacities of inorganic substances in solid and liquid state near the melting point coincide (more exactly, almost coincide), which was found by Krestovnikov and Vigdorovich [302]. Control calculations for meta- and disilicate proved to hold well. The heat capacities of the liquid Na_4SiO_4 and $Na_2Si_3O_7$ found by this method are 62 and 75 respectively.

In the system $Na_2CO_3 - SiO_2$ the following reactions were considered as the main ones:

1. $2Na_2CO_3 + SiO_2 = Na_4SiO_4 + 2CO_2$
2. $Na_2CO_3 + SiO_2 = Na_2SiO_3 + CO_2$
3. $Na_2CO_3 + 2SiO_2 = Na_2Si_2O_5 + CO_2$
4. $Na_2CO_3 + 3SiO_2 = Na_2Si_3O_7 + CO_2$.

In the system $Na_2SO_4 - SiO_2$ the following main reactions were considered:

5. $2Na_2SO_4 + SiO_2 = Na_4SiO_4 + 2SO_2 + O_2$
6. $Na_2SO_4 + SiO_2 = Na_2SiO_3 + SO_2 + \frac{1}{2}O_2$
7. $Na_2SO_4 + 2SiO_2 = Na_2Si_2O_5 + SO_2 + \frac{1}{2}O_2$
8. $Na_2SO_4 + 3SiO_2 = Na_2Si_3O_7 + SO_2 + \frac{1}{2}O_2$.

Reactions of the dissociation of soda ash and sodium sulfate to oxides:

9. $Na_2CO_3 = Na_2O + CO_2$
10. $Na_2SO_4 = Na_2O + SO_2 + \frac{1}{2}O_2$

were also examined in order to understand whether such a dissociation effects in any way the course of the reactions of silicate formation.

The results of all the calculations are presented in Fig. 57.

In Fig. 57a the results of the calculations of the main reactions of the system $Na_2CO_3 - SiO_2$ are given. Viewing the illustration one sees that the reaction of $Na_2O \cdot 3SiO_2$ formation is thermodynamically possible within the whole temperature range under consideration, that of $Na_2O \cdot 2SiO_2$ formation, beginning with 420 K, that of $Na_2O \cdot SiO_2$ formation, beginning with 550 K whereas that of $2Na_2O \cdot SiO_2$ formation can take place beginning only with 920 K.

For evaluating the thermodynamic sequence of the compound formation in the system $Na_2CO_3 - SiO_2$ the dependence of ΔG^0 of the main reactions on the composition of the initial mixtures was considered. The ΔG^0 values for the ratios

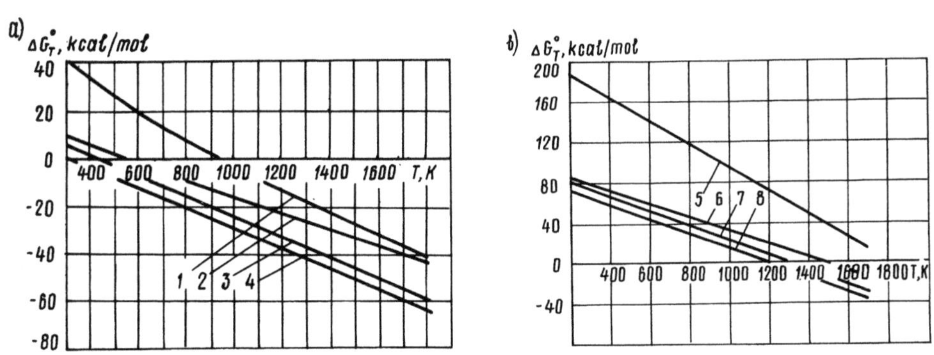

Fig. 57. The function $\Delta G^0 = f(T)$ for the main reactions in the systems: **a** $Na_2CO_3 - SiO_2$ and **b** $Na_2SO_4 - SiO_2$

of the initial mixtures Na_2O/SiO_2 being 2/1; 1/1, 1/2, 1/3 were analyzed. The results obtained are presented in Table 46.

Table 46. Thermodynamic stability sequence of compounds in the system $Na_2CO_3-SiO_2$

Ratio of initial components	Thermodynamic stability series beginning with the most stable compound at a temp. of:			
	a) 1000 K			
2/1	NS	NS_3	Ns_2	N_2S
1/1	NS	NS_3	Ns_2	N_2S
1/2	NS_3	NS_2	NS	N_2S
1/3	NS_3	NS_2	NS	N_2S
	b) 1800 K			
2/1	NS	N_2S	NS_2	NS_3
1/1	NS	NS_2	NS_3	N_2S
1/2	NS_2	NS_3	NS	N_2S
1/3	NS_3	NS_2	NS	N_2S

Abbreviations accepted: N, Na_2O and S, SiO_2

Thus, within the temperature range in question the most probable will be the formation of compounds with a higher SiO_2 content than that of $2Na_2O \cdot SiO_2$. The most stable will be the silicates with a higher SiO_2 content. The formation of sodium orthosilicate is improbable. Bearing in mind the primary formation of sodium metasilicate $Na_2O \cdot SiO_2$ in mixtures $Na_2CO_3-SiO_2$ [75] independent of the ratio of the initial mixtures, $Na_2O \cdot SiO_2$ primarily formed, SiO_2 being in excess, is believed to change into $Na_2O \cdot 2SiO$ and $Na_2O \cdot 3SiO_2$ respectively, (the final compound is determined by the stoichiometric composition).

While considering the main reactions in the system $Na_2SO_4-SiO_2$ (see Fig. 57b) the same principal regularities can be easily observed as in the system $Na_2CO_3-SiO_2$, namely a greater thermodynamic stability of the silicates with a high SiO_2 content.

4.2.2 System $K_2CO_3-K_2SO_4-SiO_2$

In this system by M. A. Matveyev, Frenkel' and G. M. Matveyev [379] ΔH^0 and ΔG^0 were in a similar manner calculated for 21 reactions in mixtures of quartz with potassium carbonate and sulphate in the temperature range of 298 to 1800 K. Standard thermodynamic values and equations adduced at the end of the book were accepted for the reactants.

In the system $K_2CO_3-SiO_2$ the following main reactions were considered:

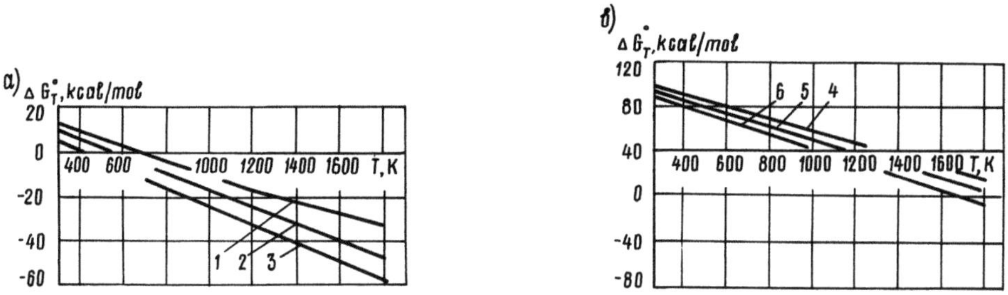

Fig. 58. The function $\Delta G^0 = f(T)$ for the main reactions in the systems: **a** $K_2CO_3-SiO_2$ and **b** $K_2SO_4-SiO_2$

1. $K_2CO_3 + SiO_2 = K_2SiO_3 + CO_2$
2. $K_2CO_3 + 2SiO_2 = K_2Si_2O_3 + CO_2$
3. $K_2CO_3 + 4SiO_2 = K_2Si_4O_3 + CO_2$.

In the system $K_2SO_4-SiO_2$ the following main reactions were considered:

4. $K_2SO_4 + SiO_2 = K_2SiO_3 + SO_2 + \frac{1}{2}O_2$
5. $K_2SO_4 + 2SiO_2 = K_2Si_2O_5 + \frac{1}{2}O_2$
6. $K_2SO_4 + 4SiO_2 = K_2Si_4O_9 + SO_2 + \frac{1}{2}O_2$.

For all the reactions ΔG_T^0 and ΔH_T^0 for the temperature range of 298 to 1800 K were calculated with an interval of 100 K.

In Fig. 58a a diagrammatic representation of the function $\Delta G^0 = f(T)$ for the main reactions in the system $K_2CO_3-SiO_2$ is given. As one can see, the reaction of $K_2O \cdot 4SiO_2$ formation is thermodynamically possible beginning with 400 K, that of $K_2O \cdot 2SiO_2$ formation, with 560 K, and that of $K_2O \cdot SiO_2$ formation, with 690 K.

For evaluating the thermodynamic sequence of the compound formation in the system $K_2CO_3-SiO_2$ the ΔG^0 dependence of the main reactions on the composition of the initial mixtures was considered. The ΔG^0 values for the ratios of the initial mixtures of K_2O/SiO_2 being 1/1, 1/2, and 1/4 were analyzed and the results obtained are presented in Table 47.

Assuming that as a primary compound in the system $K_2CO_3-SiO_2$ potassium metasilicate $K_2O \cdot SiO_2$ is formed one can draw the conclusion that $K_2O \cdot SiO_2$ primarily formed, SiO_2 being in excess, turns into $K_2O \cdot 2SiO_2$ and $K_2O \cdot 4SiO_2$ respectively (the final compound being determined by the stoichiometric composition of the mixture). As in the case of the system $Na_2CO_3-SiO_2$, the most stable will be the silicates with the higher SiO_2 content.

An analysis of the main reactions in the system $K_2SO_4-SiO_2$ (Fig. 58b) proves the reactions in sulfate-containing mixtures to be practically impossible within the range in question. Thus, the reaction of $K_2O \cdot 4SiO_2$ formation be-

Table 47. Thermodynamic stability sequence of compounds in the system $K_2CO_3-SiO_2$

Ratio of the initial components	Thermodynamic stability series beginning with the most stable compound within the range of 1000 to 1800 K		
1/1	KS	KS_2	KS_4
1/2	KS_2	KS	KS_4
1/4	KS_4	KS_2	KS

Abbreviations accepted: K, K_2O and S, SiO_2

comes thermodynamically possible beginning only with 1620 K, that of $K_2O \cdot 2SiO_2$ formation, with 1700 K, and that of $K_2O \cdot SiO_2$ is thermodynamically impossible in the entire temperature range under consideration.

All these conclusions are in agreement with the practice of the production of potassium silicates [661].

4.2.3 System $Li_2CO_3-Li_2SO_4-Li_2O-SiO_2$ [378, 380, 381]

As was the case with the two previous systems, the thermodynamic analysis was conducted in the temperature range of 298 to 1800 K. Three perfectly known compounds [125], lithium orthosilicate, metasilicate and disilicate, were examined.

The ΔH^0_{298} and ΔG^0_{298} values were determined by calculations using Hoppe's and Karapet'yants' methods [209, 246]. The heat capacity equations for the three silicates were calculated according to Landiya's technique [320]. The heat capacities of the silicates in liquid state were taken equal to those of the corresponding solid ones at the melting temperature. The melting heats of lithium ortho- and metasilicate were taken from Eitel's work [124]. The melting heat of lithium disilicate was determined based on the assumption that within the range of the compounds from Li_4SiO_4 to $Li_2Si_2O_5$ the entropy alteration while melting depends linearly on the Li_2O molar content. ΔH^0_{298}, ΔH^0_{298} and the heat capacity equations for Li_2CO_3 and Li_2SO_4 were chosen using an examination of literary data as a base [220].

The ΔH^0_{298} value of lithium oxide was taken from the latest work by Kolesov, Skuratov and Zaykin [288] and the entropy, from [528]. The heat capacity equation was borrowed from Schomate and Cohen's work [552] whose data were confirmed by Rodigina and Gomel'skiy [521]. Next ΔG^0_T for the main and derivative reactions of lithium oxide, carbonate and sulphate with silica were calculated. ΔG^0_T of the decomposition reactions were determined as well in order to bring to light whether such a decomposition effects in any way the course of the reactions of silicate formation.

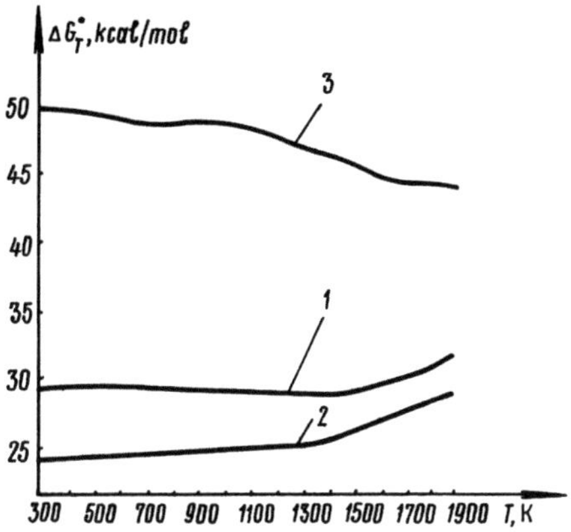

Fig. 59. The function $\Delta G^0 = f(T)$ for the main reactions in the system $Li_2O - SiO_2$

The following main reactions in the system $Li_2O - SiO_2$ were considered:

1. $Li_2O + SiO_2 = Li_2SiO_3$
2. $Li_2O + 2SiO_2 = Li_2Si_2O_5$
3. $2Li_2O + SiO_2 = Li_4SiO_4$.

The calculation results are presented in Fig. 59.

For evaluating the thermodynamic sequence of the compound formation in the system $Li_2O - SiO_2$ the dependence of ΔG^0 of the main reactions on the composition of the initial mixtures was considered. The ΔG^0 values were analyzed for the ratios of mixtures of Li_2O/SiO_2 being 2/1, 1/1 and 1/2. The results of the analysis are presented in Table 48.

An analysis of all the data obtained shows silicates with a higher content of lithium oxide in the system $Li_2O - SiO_2$ to possess a higher thermodynamic stability. The ratio of Li_2O/SiO_2 being 2/1 and more (the range rich in Li_2O), the most stable compound will be lithium orthosilicate. At a ratio of $Li_2O/SiO_2 \leqslant 1$ the most stable is lithium metasilicate. The latter fact makes the system $Li_2O - SiO_2$ substantially different from similar systems with sodium and potassium where sodium trisilicate and potassium tetrasilicate are more stable respectively.

Table 48. Thermodynamic stability sequence of compounds in the system $Li_2O - SiO_2$

Ratio of the initial components	Thermodynamic stability series beginning with the most stable compound within the temp. range of 300 to 1800 K		
2/1	Li_2S	LiS	LiS_2
1/1	LiS	Li_2S	LiS_2
1/2	LiS	Li_2S	LiS_2

Abbreviations accepted: Li, Li_2O and S, SiO_2

In the system $Li_2CO_3-SiO_2$ the following reactions were considered:

1. $Li_2CO_3+SiO_2=Li_2SiO_3+CO_2$
2. $Li_2CO_3+2SiO_2=Li_2Si_2O_5+CO_2$
3. $2Li_2CO_3+SiO_2=Li_4SiO_4+2CO_2$.

In the system $Li_2SO_4-SiO_2$ the following reactions were studied:

4. $Li_2SO_4+SiO_2=Li_2SiO_3+SO_2+\frac{1}{2}O_2$
5. $Li_2SO_4+2SiO_2=Li_2Si_2O_5+SO_2+\frac{1}{3}O_2$
6. $2Li_2SO_4+SiO_2=Li_4SiO_4+2SO_2+O_2$.

In order to clarify the effect of the lithium carbonate and sulphate dissociation on the course of the reactions of silicate formation ΔH_T^0 and ΔG_T^0 of the reactions of the dissociation of the given substances to oxides were determined:

7. $Li_2CO_3=Li_2O+CO_2$
8. $Li_2SO_4=Li_2O+SO_2+\frac{1}{2}O_2$.

For all the reactions ΔH_T^0 and ΔG_T^0 for a temperature range of 298 to 1800 K were calculated with an interval of 100 K. The calculation results are presented in Fig. 60.

An analysis of the main reactions in the system $Li_2CO_3-SiO_2$ (Fig. 60a) shows the reaction of $Li_2O \cdot SiO_2$ formation to be thermodynamically possible beginning with 600 K, that of the $Li_2O \cdot 2SiO_2$ formation, with 700 K and that of the $2Li_2O \cdot SiO_2$ formation, with 740 K.

An analysis of the main reactions in the system $Li_2SO_4-SiO_2$ (Fig. 60b) shows no reaction with lithium sulphate to be at all practically possible up to 1800 K. On the whole these inferences agree with those obtained earlier for the

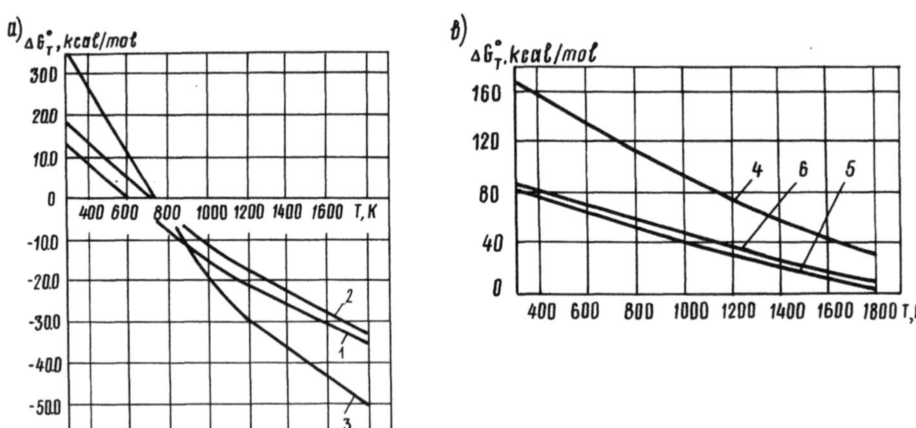

Fig. 60. The function $\Delta G^0=f(T)$ for the main reactions in the systems: **a** $Li_2CO_3-SiO_2$ and **b** $Li_2SO_4-SiO_2$

Table 49. Thermodynamic stability sequence of the compounds in the system $Li_2CO_3-SiO_2$

Ratio of the initial components	Thermodynamic stability series beginning with the most stable compound within a temp. range of 298 to 1800 K		
1/1	LiS	Li_2S	LiS_2
1/2	LiS	LiS_2	Li_2S
2/1	a) up to 800 K		
	LiS	Li_2S	LiS_2
	b) above 800 K		
	Li_2S	LiS	LiS_2

Abbreviations accepted: Li, Li_2O and S, SiO_2

reactions in mixtures of quartz with potassium and sodium carbonate and sulphate.

For evaluating the thermodynamic sequence of the compound formation in the system $Li_2CO_3-SiO_2$ the dependence of ΔG^0 of the main reactions on the composition of the initial mixtures was considered. The ΔG^0 values for the ratios Li_2O/SiO_2 being 1/1, 1/2 and 2/1 were examined (Table 49).

The calculations carried out showed lithium metasilicate to be formed at the lowest temperature. Moreover with all compositions at low temperatures – up to 800 K – this is the most stable lithium silicate. Among compounds formed at ratios $Li_2O/SiO_2 \geqslant 2/1$ at temperatures above 800 K the most stable is still lithium orthosilicate.

The examination of ΔG_T^0 of the reactions of the Li_2CO_3 and Li_2SO_4 dissociation shows that lithium sulfate does not dissociate within the temperature range under consideration; lithium carbonate, on the contrary, has a negative ΔG_T^0 value while dissociating at 1700 K. Below, equilibrium pressures of CO_2 as a function of temperature are adduced. These data are in agreement with the results of Rode's [480, 520] and Rassonskaya and Semendyayeva's [500] tests.

Temperature [K]	CO_2 equilibrium pressure [atm]
900	$3.44 \cdot 10^{-5}$
1000	$2.75 \cdot 10^{-4}$
1100	$3.50 \cdot 10^{-3}$
1200	$1.60 \cdot 10^{-2}$
1300	$5.08 \cdot 10^{-2}$
1400	$2.14 \cdot 10^{-1}$
1500	$3.26 \cdot 10^{-1}$
1600	$8.01 \cdot 10^{-1}$
1700	1.08
1800	1.83

Thus, unlike the system $Na_2CO_3-SiO_2$, the dissociation of lithium carbonate may effect the reactions in the system $Li_2CO_3-SiO_2$, particularily at high temperatures. Nonetheless, definite conclusions about the extent of this effect can be drawn only after obtaining necessary kinetic data.

4.2.4 System $Rb_2CO_3-Rb_2SO_4-SiO_2$ [371, 381]

In accordance with the phase diagram of the binary system Rb_2O-SiO_2, the reactions of the formation of Rb_2SiO_3, $Rb_2Si_2O_5$ and $Rb_2Si_4O_9$ were studied.

In the system $Rb_2CO_3-SiO_2$ the following main reactions were considered:

1. $Rb_2CO_3+SiO_2=Rb_2SiO_3+CO_2$
2. $Rb_2CO_3+2SiO_2=Rb_2Si_2O_5+CO_2$
3. $Rb_2CO_3+4SiO_2=Rb_2Si_4O_9+CO_2$.

In the system $Rb_2SO_4-SiO_2$ the following main reactions were considered:

4. $Rb_2SO_4+SiO_2=Rb_2SiO_3+SO_2+\frac{1}{2}O_2$
5. $Rb_2SO_4+2SiO_2=Rb_2Si_2O_5+SO_2+\frac{1}{2}O_2$
6. $Rb_2SO_4+4SiO_2=Rb_2Si_4O_9+SO_2+\frac{1}{2}O_2$.

For all the reactions ΔG_T^0 in a temperature range of 298 to 1800 K were calculated with an interval of 100 K. The calculation results are presented in Fig. 61.

An analysis of the results obtained shows the reaction of $Rb_2O \cdot 4SiO_2$ formation to be thermodynamically possible beginning with 320 K, that of $Rb_2O \cdot 2SiO_2$, with 600 K, and of $Rb_2O \cdot SiO_2$, with 700 K.

For evaluating the thermodynamic sequence of the compound formation in the system $Rb_2CO_3-SiO_2$ ΔG^0 of the main reactions as a function of the composition of the initial mixtures was studied. The ΔG_T^0 values for the ratios of the initial mixtures Rb_2O/SiO_2 being 1/1, 1/2 and 1/4 were analyzed. The results obtained are presented in Table 50.

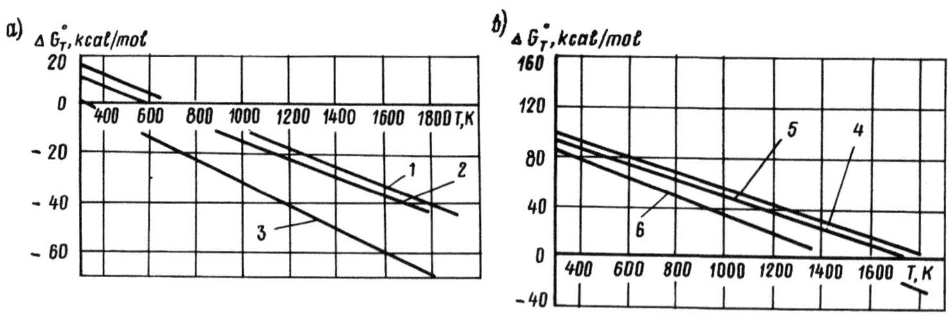

Fig. 61. The function $\Delta G^0=f(T)$ for the main reactions in the systems: **a** $Rb_2CO_3-SiO_2$ and **b** $Rb_2SO_4-SiO_2$

Table 50. Thermodynamic stability sequence of compounds in the system $Rb_2CO_3 - SiO_2$

Ratio of the initial components	Thermodynamic stability series beginning with the most stable compound within a temp. range of 1000 to 1800 K		
1/1	RbS	RbS_2	RbS_4
1/2	RbS_2	RbS	RbS_4
1/4	RbS_4	RbS_2	RbS

Abbreviations accepted: Rb, Rb_2O and S, SiO_2

When one next supposes that as the primary compound in the system $Rb_2CO_3 - SiO_2$ rubidium metasilicate $Rb_2O \cdot SiO_2$ is formed [189], one can draw the conclusion that the primarily formed $Rb_2O \cdot SiO_2$, SiO_2 being in excess, is transformed, depending on the stoichiometric composition of the mixture, into $Rb_2O \cdot 2SiO_2$ and $Rb_2O \cdot 4SiO_2$. As in the case of the sodium and potassium systems, the most stable will be the silicates with a higher SiO_2 content.

An analysis of the main reactions in the system $Rb_2SO_4 - SiO_2$ shows the reactions in sulphate-containing mixtures to proceed at essentially higher temperatures than in similar mixtures with the carbonate. Thus, the reaction of the $Rb_2O \cdot 4SiO_2$ formation becomes thermodynamically possible only beginning with a temperature of 1460 K, that of the $Rb_2O \cdot 2SiO_2$, with 1720 K, and that of the $Rb_2O \cdot SiO_2$, with 1800 K.

4.2.5 System $Cs_2CO_3 - Cs_2SO_4 - SiO_2$ [378, 381]

It is well known that at present there is no complete fusibility diagram for the system $Cs_2O - SiO_2$. As, however, the properties of rubidium and caesium compounds are close to each other, we assumed the existence of silicates Cs_2SiO_3, $Cs_2Si_2O_5$ and $Cs_2Si_4O_9$ in the system $Cs_2O - SiO_2$. The melting temperatures of caesium silicates were determined using the known melting temperatures of the corresponding silicates of the other alkali metals.

In the system $Cs_2CO_3 - SiO_2$ the following main reactions were considered:

1. $Cs_2CO_3 + SiO_2 = Cs_2SiO_3 + CO_2$
2. $Cs_2CO_3 + 2SiO_2 = Cs_2Si_2O_5 + CO_2$
3. $Cs_2CO_3 + 4SiO_2 = Cs_2Si_4O_9 + CO_2$.

In the system $Cs_2SO_4 - SiO_2$ the following main reactions were considered:

4. $Cs_2SO_4 + SiO_2 = Cs_2SiO_3 + SO_2 + \frac{1}{2}O_2$
5. $Cs_2SO_4 + 2SiO_2 = Cs_2Si_2O_5 + SO_2 + \frac{1}{2}O_2$
6. $Cs_2SO_4 + 4SiO_2 = Cs_2Si_4O_9 + SO_2 + \frac{1}{2}O_2$.

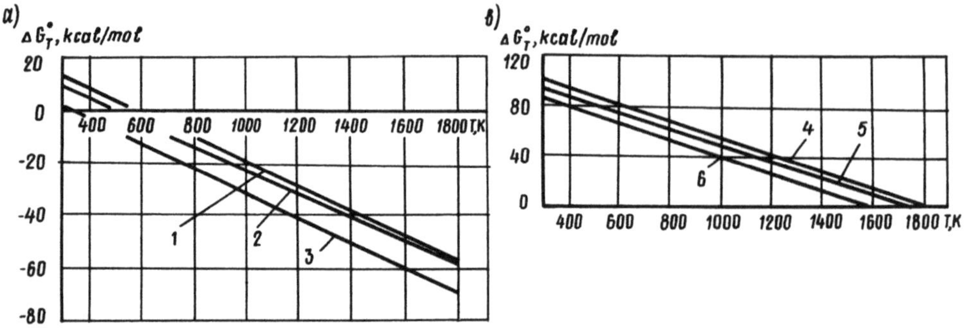

Fig. 62. The function $\Delta G^0 = f(T)$ for the main reactions in the systems: **a** $Cs_2CO_3-SiO_2$ and **b** $Cs_2SO_4-SiO_2$

For all the reactions ΔG^0 for a temperature range of 298 to 1800 K were calculated with an interval of 100 K. The calculation results are presented in Fig. 62.

In Fig. 62a, a diagrammatic representation of $\Delta G^0 = f(T)$ for the main reactions in the system $Cs_2CO_3-SiO_2$ is given. An analysis of the results obtained shows the reaction of the $Cs_2O \cdot 4SiO_2$ formation to be thermodynamically possible beginning with 340 K, that of the $Cs_2O \cdot 2SiO_2$, with 500 K, and that of the $Cs_2O \cdot SiO_2$, with 590 K.

For evaluating the thermodynamic sequence of compound formation in the system $Cs_2CO_3-SiO_2$ ΔG^0 of the main reactions as a function of the composition of the initial mixtures were considered. The values for the ratios of the initial mixtures Cs_2O/SiO_2 being 1/1, 1/2 and 1/4 were analyzed. The results obtained are presented in Table 51.

When one supposes that the primary compound caesium metasilicate is formed in the system $Cs_2CO_3-SiO_2$ [376], one can draw the conclusion that the primarily formed $Cs_2O \cdot SiO_2$, SiO_2 being in excess, changes into $Cs_2O \cdot 2SiO_2$ or $Cs_2O \cdot 4SiO_2$ depending on the composition of the mixture. As in the case of the sodium, potassium and rubidium systems, the most stable will be the silicates with the higher SiO_2 content.

Table 51. Thermodynamic stability sequence of compounds in the system $Cs_2CO_3-SiO_2$

Ratio of the initial components	Thermodynamic stability series beginning with the most stable compound within a temp. range of 100 to 1800 K		
1/1	CsS	CsS_2	CsS_4
1/2	CsS_2	CsS	CsS_4
1/4	CsS_4	CsS_2	CsS

Abbreviations accepted: Cs, Cs_2O and S, SiO_2

An analysis of the main reactions in the system $Cs_2SO_4 - SiO_2$ (see Fig. 62b) proves that reactions in sulphate-containing mixtures will proceed at essentially higher temperatures than in similar mixtures with a carbonate. Thus, the reaction of the $Cs_2O \cdot 4SiO_2$ formation becomes thermodynamically possible beginning only with a temperature of 1600 K, that of the $Cs_2O \cdot 2SiO_2$, with 1720 K, and that of the $Cs_2O \cdot SiO_2$, with 1800 K.

In Table 52 the temperatures of the start of all the reactions between quartz and alkaline metal carbonates and sulphates are presented:

Table 52. Temperatures of the beginning of the reactions studied

Reactions	Temp. of the beginning of the reactions with silicates [K]				
	Li	Na	K	Rb	Cs
Carbonates					
1. $2Me_2CO_3 + SiO_2 = Me_4SiO_2 + 2CO_2$	740	920	–	–	–
2. $Me_2CO_3 + SiO_2 = Me_2SiO_3 + CO_2$	600	550	690	700	590
3. $Me_2CO_3 + 2SiO_2 = Me_2Si_2O_5 + CO_2$	700	440	570	600	500
4. $Me_2CO_3 + 3SiO_2 = Me_2Si_3O_7 + CO_2$	–	400	–	–	–
5. $Me_2CO_3 + 4SiO_2 = Me_2Si_4O_9 + CO_2$	–	–	400	320	240
Sulphates					
1. $2Me_2SO_4 + SiO_2 = Me_4SiO_4 + 2SO_2 + O_2$	no reaction	–	–	–	–
2. $Me_2SO_4 + SiO_2 = Me_2SiO_3 + SO_2 + \frac{1}{2}O_2$	no reaction	1550	no reaction possible	1800	1800
3. $Me_2SO_4 + 2SiO_2 = Me_2Si_2O_5 + SO_2 + \frac{1}{2}O_2$	no-reaction	1400	1800	1720	1750
4. $Me_2SO_4 + 3SiO_2 = Me_2Si_3O_7 + SO_2 + \frac{1}{2}O_2$	–	1340	–	–	–
5. $Me_2SO_4 + 4SiO_2 = Me_2Si_4O_9 + SO_2 + \frac{1}{2}O_2$	–	–	1620	1460	1600

4.2.6 System PbO – SiO_2 [367]

The reactions of silicate-formation in this system were studied within the temperature range of 298 to 1800 K (the upper limit corresponds to the maximum temperature of the modern glass-making furnaces).

An analysis of the literary data [41, 125, 147, 636, 713] showed that for carrying out calculations in the system PbO – SiO_2 it is most reasonable to consider the following lead silicates: $PbO \cdot SiO_2$, $2PbO \cdot SiO_2$ and $4PbO \cdot SiO_2$. The initial thermodynamic constants for the calculation were taken from the summarizing table adduced at the end of the book after a detailed analysis of them [367, 434, 435, 657].

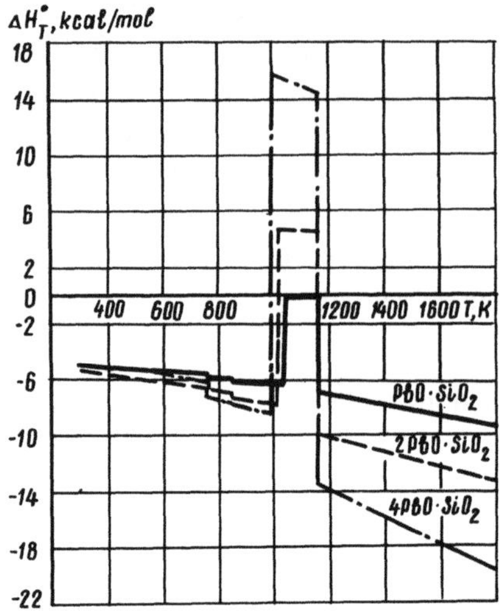

Fig. 63. The temperature dependence of the enthalpy of the reactions of the lead silicate formation

For determining $\Delta H_{\mathrm{m}}^{0}$ the entropy criterion suggested in the works of Petit, Darmois et al. [477–479] was used. For lead metasilicate the entropy criterion is $\nu = 2$, whence $\Delta S_{\mathrm{m}}^{0} = 6$ cal/mol deg and $\Delta H_{\mathrm{m}}^{0} = 6.1037 = 6222$ cal/mol.

For the approximate determination of $\Delta H_{\mathrm{m}}^{0}$ of $2PbO \cdot SiO_2$ and $4PbO \cdot SiO_2$ the iso-atom method was used [157, 163, 566]. The $\Delta H_{\mathrm{m}}^{0}$ values obtained for $2PbO \cdot SiO_2$ and $4PbO \cdot SiO_2$ are 12200 and 24000 cal/mol respectively. The ΔG^{0} values were calculated by the thermodynamic properties of individual substances.

Starting from the ΔH_{T}^{0} and ΔG_{T}^{0} values obtained for individual substances, the enthalpies and the Gibbs free energies of the reaction of lead silicate formation were determined. A diagrammatic representation of the results obtained is given in Fig. 63 and 64.

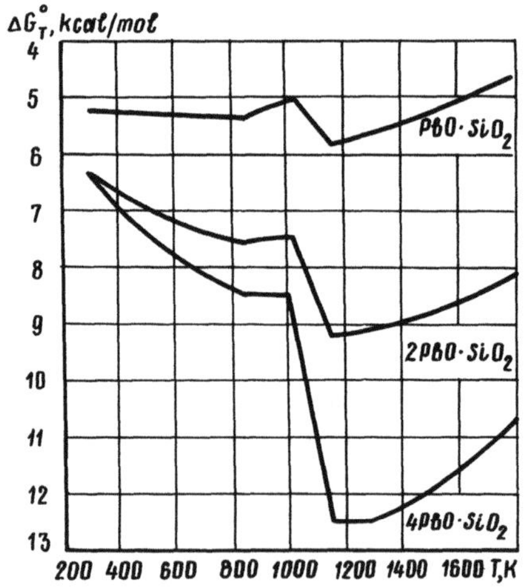

Fig. 64. The temperature dependence of the Gibbs free energies of the lead silicate formation

An analysis of equations $\Delta G^0 = f(T)$ shows the formation of the three lead silicates to be possible throughout the temperature range studied. These data also allow a comparative evaluation of the lead silicate stability at various ratios of lead oxide and silica to be made.

In accordance with the results of the calculations conducted a thermodynamic stability series of the compounds in the system $PbO - SiO_2$ was set up (Table 53).

The calculations carried out show that at any ratio of components in the system $PbO - SiO_2$ the most stable are the compounds of the stoichiometric composition corresponding to the given ratio.

4.3 Peculiarities of Some Vitreous Systems

In [610] a semiempirical method of calculating the entropy per mole of the group $[SiO_4]^{4-}$ for the main structural elements of crystalline and vitreous silicates was described. These entropies can be calculated by means of dividing the molar entropy of each structural element termed in form of a rational formula by the number of the groups $[SiO_4]^{4-}$ present in the structural element. The entropies of structural elements were found starting from those of the silicate constituents.

An alteration of the free energy in the process of the silicate glass formation was determined on the basis of the standard alterations of enthalpies per mole of each group $[SiO_4]^{4-}$ taking part in a reaction which in their turn can be calculated with the help of additive functions found out in [494].

At elevated temperatures the standard alteration of the free energy can be calculated knowing the heat capacity values of substances at these temperatures. These free energy calculations enable us to forecast the equilibrium conditions for the reactions connected with the formation of silicate glasses.

Next, in [610] the author concludes that the method of calculating the equilibrium constants reported in the article in question permits the assumption that the silicate glass has a microheterogeneous structure containing various types of groups of silicium tetroxide $[SiO_4]^{4-}$.

In [173] a thermodynamic analysis of reactions appearing in the process of glass formation in the system $SiO_2 - B_2O_3 - Al_2O_3 - ZnO - CaO - Na_2O$ of a certain composition was conducted. Based on this analysis the author of [173] cal-

Table 53. Thermodynamic stability sequence of compounds in the system $PbO - SiO_2$

Ratio of the initial components	Thermodynamic stability series beginning with the most stable compound within a temp. range of 298 to 1800 K		
1/1	$PbO \cdot SiO_2$	$2PbO \cdot SiO_2$	$4PbO \cdot SiO_2$
2/1	$2PbO \cdot SiO_2$	$4PbO \cdot SiO_2$	$PbO \cdot SiO_2$
4/1	$4PbO \cdot SiO_2$	$2PbO \cdot SiO_2$	$PbO \cdot SiO_2$

culated the points of the theoretical curves relating to the initial stage of the glass formation and to the finished glass melt which characterize the alteration in time of the chemical affinity of metal oxides formed to silica in the process of glass melting as well as the alteration of oxide concentration in time.

For simplifying and shortening the calculations it was assumed that during glass formation only primary reactions of silicate formation take place, namely:

1. $Na_2O + SiO_2 = Na_2SiO_3$
2. $Al_2O_3 + SiO_2 = Al_2SiO_5$
3. $CaO + SiO_2 = CaSiO_3$
4. $ZnO + SiO_2 = ZnSiO_3$
5. $B_2O_3 + SiO_2 = B_2SiO_5$.

The Gibbs free energy of the reactions was considered according to the following formula at a working temperature of 1693 K:

$$\Delta G_i = \Delta G_i^0 + 2.3 \lg \left(\frac{X_{\mathrm{MeSiO_3}}}{X_{\mathrm{MeO}} \cdot X_{\mathrm{SiO_2}}} \right)$$

$$\Delta G_i^0 = \Delta H_i^0 - T \Delta S_i^0 .$$

In [334] the authors carried out a calculation of the thermodynamic properties of glass of the composition (weight percent): SiO_2: 73.7; CaO: 10.1; Na_2O: 16.2.

The batch prepared at a temperature of 298 K and at atmospheric pressure consisting of a mixture of quartz, soda ash and calcium carbonate was melted at a temperature of 1733 K and then again cooled down to room temperature. Starting from the physical and chemical processes taking place in course of heating and cooling – phase transformations, silicate formation – the Gibbs free energy, entropy, enthalpy and the viscosity energy alteration were calculated.

The authors of [334] believe that in the glass under consideration chemical reactions begin at a temperature above 848 K.

Any reaction below that temperature then should be held to be only a physical transformation (in the case in question phase transformations).

At temperatures above 1483 K all the reactions finalize, and the melt consists of Na_2SiO_3, $CaSiO_3$, CaO and Na_2O.

The authors in question seemingly do not take into account the excess of SiO_2 and, on the contrary, CaO and Na_2O cannot be present in free state, as they must interact with the excessive SiO_2.

In [660], based on the calculation of ΔG_T^0 of the reactions in the system $Al(PO_3)_3 - NaF$, the authors draw a conclusion about the probability of the presumed processes and about the preference of compounds for producing glass.

For calculating ΔG_T^0 the authors of [660] took the initial thermodynamic data for crystalline compounds and allowed for an insignificant error in the calculation of ΔG_T^0 (to the order of 1.5%) of the glass formation reactions.

The Gibbs free energy for all the reactions was calculated within a temperature range of 623 to 1173 K, i.e., from the annealing temperatures to the maximum temperature of melting with an interval 100 K.

By a thermodynamic analysis of the reactions taking place in the process of glass formation in the system $Al(PO_3)_3-NaF$ the reactions of aluminium metaphosphate dissociation and those of sodium phosphate formation were shown to have a positive value of the Gibbs free energy. The most probable thermodynamically are the reactions of the interaction of aluminium metaphosphate with sodium fluoride resulting in the formation of sodium phosphates, cryolite, aluminium orthophosphate and gaseous phosphorus oxidetrifluoride and phosphorus pentoxide.

We calculated S_{298}^0, ΔH_{298}^0 and ΔG_{298}^0 for some vitreous silicates.

The entropy for the vitreous $Na_2O \cdot SiO_2$ and $Na_2O \cdot 2SiO_2$ at 298 K was calculated using the data of [611] concerning the low-temperature heat capacity. The entropies of the rest of the vitreous sodium silicates for which the corresponding crystalline compounds are known [638], were determined starting from the constancy of the dependences between the entropies of the same substance in crystalline and vitreous state and by the direct dependence between the entropy and the SiO_2/Na_2O ratio ([357] (Fig. 65)).

We determined the entropies of the vitreous potassium and lithium silicates on the basis of the structural similarity of potassium, lithium and sodium silicates and knowing the direct dependence between the entropy and the SiO_2/Me_2O ratio where Me: Li, Na, K (Fig. 65).

For calculating the enthalpies of the vitreous alkali silicates data of works [604] were used in which the heats of dissolution of a glass and a crystal of the same composition were determined. The difference between the dissolution heats of a glass and a crystal, the sample dissolving completely, may be viewed as the difference between the heats of the formation of a glass and a crystal of the same

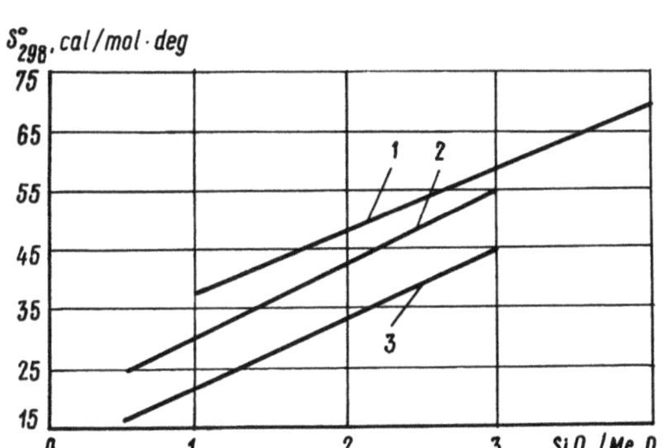

Fig. 65. Entropies of formation of potassium, sodium and lithium silicates in vitreous state related to a mole of Me_2O; *1* system K_2O-SiO_2; *2* system Na_2O-SiO_2; *3* system Li_2O-SiO_2

composition. The data concerning the heats of the crystalline alkali silicate formation were taken from [20]. Thus, the enthalpies of the formation of vitreous alkali silicates were calculated according to the formula:

$$\Delta H^0_{298\,(gl)} = \Delta H^0_{298\,(cr)} - \Delta H^0_{298\,(gl\text{-}cr)}\,, \tag{159}$$

where $\Delta H^0_{298\,(gl)}$, enthalpy of a vitreous alkali silicate formation; $\Delta H^0_{298\,(cr)}$, enthalpy of a crystalline alkali silicate formation; $\Delta H^0_{298\,(gl\text{-}cr)}$, difference between the heats of formation of a glass and a crystal of the same alkali silicate.

The calculation of the Gibbs free energy of vitreous alkali silicates was carried out according to the formula:

$$\Delta G^0_{298} = \Delta H^0_{298} - 298\,\Delta S^0_{298}\,.$$

The results of calculating the thermodynamic properties of vitreous alkali silicates are presented in Table 54. For comparison the thermodynamic proper-

Table 54. Calculation of the thermodynamic properties of vitreous alkali

Compound	S^0_{298} [cal/mol deg]	$-\Delta H^0_{298\,(el)}$ [kcal/mol]	$-\Delta G^0_{298\,(el)}$ [kcal/mol]	$-\Delta H^0_{298\,(ox)}$ [kcal/mol]	$-\Delta G^0_{298\,(ox)}$ [kcal/mol]
$2Li_2O \cdot SiO_2$ gl	32.30	530.00	501.09	28.92	29.82
$2Li_2O \cdot SiO_2$ cr	30.00	552.50	522.91	49.15	49.71
$Li_2O \cdot SiO_2$ gl	21.70	371.00	350.23	12.72	11.23
$Li_2O \cdot SiO_2$ cr	20.00	390.00	368.72	29.45	29.72
$Li_2O \cdot 2SiO_2$ gl	33.80	589.50	556.39	15.74	16.45
$Li_2O \cdot 2SiO_2$ cr	30.00	602.50	568.26	24.20	24.46
$Li_2O \cdot 3SiO_2$ gl	45.00	793.00	745.80	3.76	2.99
$Li_2O \cdot 3SiO_2$ cr	40.00	807.00	759.80	10.95	11.20
$2Na_2O \cdot SiO_2$ gl	50.50	500.25	470.21	86.49	87.84
$2Na_2O \cdot SiO_2$ cr	46.80	507.25	476.11	91.22	91.86
$Na_2O \cdot SiO_2$ gl	29.55	363.06	341.35	48.44	48.73
$Na_2O \cdot SiO_2$ cr	27.20	372.06	349.65	55.17	55.15
$Na_2O \cdot 2SiO_2$ gl	42.55	584.35	550.57	54.25	55.08
$Na_2O \cdot 2SiO_2$ cr	39.40	591.35	556.63	56.71	57.28
$Na_2O \cdot 3SiO_2$ gl	55.08	801.75	755.76	56.17	57.40
$Na_2O \cdot 3SiO_2$ cr	51.60	809.65	762.63	57.26	53.48
$K_2O \cdot SiO_2$ gl	35.55	370.55	348.79	68.62	99.72
$K_2O \cdot SiO_2$ cr	33.00	372.50	350.03	68.35	99.03
$K_2O \cdot 2SiO_2$ gl	49.22	590.50	556.92	73.14	104.98
$K_2O \cdot 2SiO_2$ cr	43.50	593.50	558.22	71.60	102.42
$K_2O \cdot 4SiO_2$ gl	68.58	1010.00	950.30	61.68	92.62
$K_2O \cdot 4SiO_2$ cr	63.50	1035.00	973.79	77.60	108.39

ties of the corresponding alkali silicates in crystalline state from [20] are also adduced.

For the verification of the calculated ΔH^0_{298} and ΔG^0_{298} values the isoatom method of the enthalpies and free energies of formation from oxides [157, 163, 566] was used. In Fig. 66 curve I represents the isocomponent of the free energy of formation for the system $Me_2O - SiO_2$ in vitreous state, curves II and III represent the isooxides of the free energy of formation for an alkali metal and silicon respectively. The even property alteration (Fig. 66) confirms the ΔH^0_{298} and ΔG^0_{298} calculation to be correct. As one can see from Fig. 66a, curve I has the minimum at the highest absolute value of $\Delta G/\sum n$ which falls within sodium orthosilicate $2Na_2O \cdot SiO_2$.

Curve II in Fig. 66a shows the glass structural network to alter only slightly as the amount of SiO_2 increases. As this takes place, $\Delta G/\sum n$ grows insignificantly in the absolute value as the amount of silica increases, contrarily, as the amount of sodium oxide increases (curve III, Fig. 66a) $\Delta G/\sum n$ soars, testifying to greater structural alterations when Na_2O is added than when SiO_2 is. A similar picture is observed while considering vitreous compounds in the systems $K_2O - SiO_2$ and $Li_2O - SiO_2$ (Fig. 66b, 66c) as well. The minimum free energy falls within the reaction of potassium metasilicate $K_2O \cdot SiO_2$ and lithium orthosilicate $2Li_2O \cdot SiO_2$ formation.

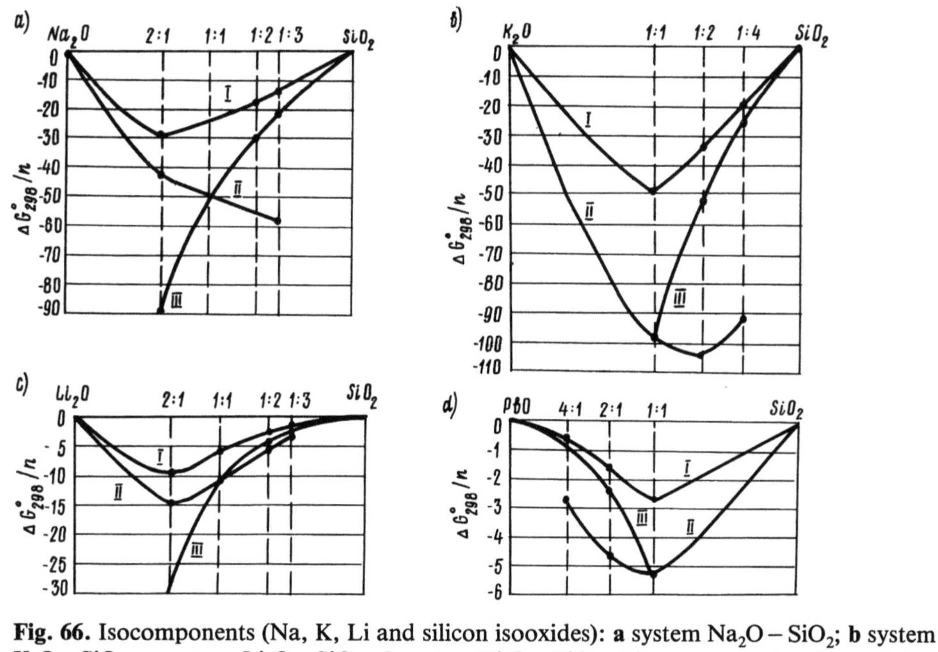

Fig. 66. Isocomponents (Na, K, Li and silicon isooxides): **a** system $Na_2O - SiO_2$; **b** system $K_2O - SiO_2$; **c** system $Li_2O - SiO_2$; **d** system $PbO - SiO_2$; *I* isocomponents; *II* metal isooxide; *III* silicon isooxide

Table 55. Values of thermodynamic functions for crystalline and glass lead silicates

Compound	S^0_{298} [cal/mol deg]	$-\Delta H^0_{298(el)}$ [kcal/mol]	$-\Delta G^0_{298(el)}$ [kcal/mol]	$-\Delta H^0_{298(ox)}$ [kcal/mol]	$-\Delta G^0_{298(ox)}$ [kcal/mol]
$PbO \cdot SiO_2$ gl	27.26	273.04	253.30	5.11	5.25
$PbO \cdot SiO_2$ cr	26.20	274.97	254.91	4.77	4.93
$2PbO \cdot SiO_2$ gl	46.40	323.79	297.83	3.41	4.60
$2PbO \cdot SiO_2$ cr	44.60	327.39	300.90	4.74	5.74
$4PbO \cdot SiO_2$ gl	80.73	426.13	386.55	0.85	2.96
$4PbO \cdot SiO_2$ cr	77.60	431.42	390.42	3.87	5.39

For calculating the thermodynamic properties of lead silicates in vitreous state [517] was used in which ΔS^0_{298}, ΔH^0_{298} and ΔG^0_{298} for $2PbO \cdot SiO_2$ transforming from glass into crystal were determined. These values with regard to the data of [681] are equal to 1.8 kcal/mol deg, 3.6 kcal/mol and 3.1 kcal/mol respectively.

The calculation results are presented in Table 55 and in Fig. 66d. For comparison the values of thermodynamic functions for crystalline lead silicates studied in [386] are given in Table 55.

It follows from Fig. 66d that the minimum free energy fall within the reaction of lead metasilicate $PbO \cdot SiO_2$ formation. The shape of isooxide curves observed here differs from those of alkali silicates which seems to be connected with a particular role of lead oxide that appears to be a glassformer forming groups $[PbO_4]^{6-}$ in the glass network [297, 298].

The calculation of the thermodynamic properties of ternary vitreous systems at $T = 298$ K was carried out with regard to [369] and based on the data on binary vitreous systems.

Let us consider the system $Na_2O - CaO - SiO_2$. In this system three compounds are known: $Na_2O \cdot 2CaO \cdot 3SiO_2$, $2Na_2O \cdot CaO \cdot 3SiO_2$ and $Na_2O \cdot 3CaO \cdot 6SiO_2$ [638]. The thermodynamic properties of the above ternary compounds in crystalline state are adduced in [363]. The entropies and enthalpies of ternary compounds in vitreous state were calculated according to the procedure adduced in [369].

$\Delta G^0_{298(el)}$ was calculated with the following formula:

$$\Delta G^0_{298(el)} = \Delta H^0_{298(el)} - 298 \Delta S^0_{298} .$$

For vitreous compounds in the ternary systems $Na_2O - MgO - SiO_2$ and $Na_2O - Al_2O_3 - SiO_2$ thermodynamic properties were calculated in the same manner which based on data of [82] concerning the formation of the corresponding crystalline silicates.

ΔH^0_{298}, ΔG^0_{298} and S^0_{298} for compounds in ternary systems in vitreous and crystalline state are presented in Table 56.

Table 56. Compounds in ternary systems in vitreous and crystalline state

Compound	S^0_{298} [cal/mol deg]	$-\Delta H^0_{298(el)}$ [kcal/mol]	$-\Delta G^0_{298(el)}$ [kcal/mol]	$-\Delta H^0_{298(ox)}$ [kcal/mol]	$-\Delta G^0_{298(ox)}$ [kcal/mol]
$Na_2O \cdot 2CaO \cdot 3SiO_2$ gl	68.75	1153.40	1090.95	104.02	103.79
$Na_2O \cdot 2CaO \cdot 3SiO_2$ cr	66.40	1166.94	1103.79	110.75	110.99
$2Na_2O \cdot CaO \cdot 3SiO_2$ gl	78.70	1117.60	1053.81	120.98	121.30
$2Na_2O \cdot CaO \cdot 3SiO_2$ cr	74.00	1137.87	1072.68	134.44	137.50
$Na_2O \cdot 3CaO \cdot 6SiO_2$ gl	113.88	1980.52	1873.43	132.80	133.26
$Na_2O \cdot 3CaO \cdot 6SiO_2$ cr	110.40	1998.73	1890.60	137.39	139.15
$Na_2O \cdot Al_2O_3 \cdot 2SiO_2$ gl	61.10	994.97	940.78	65.77	68.50
$Na_2O \cdot Al_2O_3 \cdot 2SiO_2$ cr	58.75	1006.24	951.35	72.50	75.21
$Na_2O \cdot Al_2O_3 \cdot 4SiO_2$ gl	67.55	1441.20	1357.05	81.04	79.03
$Na_2O \cdot Al_2O_3 \cdot 4SiO_2$ cr	64.07	1452.74	1367.55	83.50	81.81
$Na_2O \cdot Al_2O_3 \cdot 6SiO_2$ gl	107.48	1869.03	1764.89	77.91	81.13
$Na_2O \cdot Al_2O_3 \cdot 6SiO_2$ cr	101.60	1880.84	1774.95	76.10	79.61
$Na_2O \cdot 2MgO \cdot 6SiO_2$ gl	98.68	1755.71	1670.34	76.01	75.92
$Na_2O \cdot 2MgO \cdot 6SiO_2$ cr	94.00	1771.79	1655.15	78.47	79.53

For calculating the thermodynamic properties of binary and ternary compounds at elevated temperatures one must know the temperature dependence of the heat capacity of vitreous substances. While calculating heat capacity we used a method [357] with which we arrived at the contribution of particular oxides in vitreous and crystalline state to the heat capacity. This contribution was used for calculating the heat capacities of binary and ternary silicates in vitreous state.

Before calculating the heat capacity of vitreous alkali silicates the contribution of $SiO_{2\ gl}$, Na_2O, K_2O and Li_2O to the heat capacity was determined.

The contribution of $SiO_{2\ gl}$ was determined by means of the equation of the temperature dependence of heat capacity for the vitreous $Na_2O \cdot 2SiO_2$ taken from [257].

For calculating the heat capacity as a function of temperature for compounds in the system $PbO-SiO_2$, C_p was determined starting from the heat capacity of lead silicates in vitreous state. While calculating the heat capacity for ternary compounds the contribution of binary compounds in vitreous state to the heat capacity was taken into account.

For calculating the heat capacities of vitreous compounds in the systems $Na_2O-MgO-SiO_2$ and $Na_2O-Al_2O_3-SiO_2$ the heat capacities of these compounds in crystalline state were first calculated by Landiya's method [320].

The results of the calculations of the temperature dependence of heat capacity for vitreous and crystalline compounds are adduced in Table 57.

The calculation of ΔH^0_T, ΔS^0_T and ΔG^0_T at elevated temperatures for some vitreous silicates was carried out with the following formulae:

Table 57. Temperature dependence of heat capacity for vitreous and crystalline compounds

Compounds	$C_p = a + bT - cT^{-2}$			Temp. range [K]
	a	$b \cdot 10^3$	$c \cdot 10^{-5}$	
$2Na_2O \cdot SiO_2$ gl	43.35	20.17	4.23	298 – 1393
$2Na_2O \cdot SiO_2$ cr	38.86	17.74	–	
$Na_2O \cdot SiO_2$ gl	25.45	17.83	1.96	298 – 1361
$Na_2O \cdot SiO_2$ cr	31.14	9.60	6.47	
$Na_2O \cdot 2SiO_2$ gl	32.91	33.32	1.64	298 – 1147
$Na_2O \cdot 2SiO_2$ cr	44.38	16.86	10.67	
$Na_2O \cdot 3SiO_2$ gl	40.55	44.20	1.34	298 – 1023
$Na_2O \cdot 3SiO_2$ cr	48.25	26.69	4.98	
$K_2O \cdot SiO_2$ gl	26.93	14.08	0.64	298 – 1249
$K_2O \cdot SiO_2$ cr	32.62	5.85	5.15	
$K_2O \cdot 2SiO_2$ gl	21.31	41.63	–9.02	298 – 1318
$K_2O \cdot 2SiO_2$ cr	32.69	35.17	–	
$K_2O \cdot 4SiO_2$ gl	10.07	126.73	–66.98	298 – 1038
$K_2O \cdot 4SiO_2$ cr	60.62	38.09	–	
$2Li_2O \cdot SiO_2$ gl	38.69	15.23	6.41	298 – 1528
$2Li_2O \cdot SiO_2$ cr	38.82	16.33	2.35	
$Li_2O \cdot SiO_2$ gl	20.82	16.60	–0.87	298 – 1474
$Li_2O \cdot SiO_2$ cr	26.91	8.37	3.64	
$Li_2O \cdot 2SiO_2$ gl	24.97	34.17	–9.02	298 – 1306
$Li_2O \cdot 2SiO_2$ cr	36.35	18.11	–	
$4PbO \cdot SiO_2$ gl	43.29	42.00	3.32	298 – 1797
$4PbO \cdot SiO_2$ cr	45.30	33.80	–6.24	
$2PbO \cdot SiO_2$ gl	37.81	15.32	8.06	298 – 1016
$2PbO \cdot SiO_2$ cr	42.51	1.70	7.93	
$PbO \cdot SiO_2$ gl	35.07	1.98	10.43	298 – 1037
$PbO \cdot SiO_2$ cr	38.06	9.70	13.91	
$Na_2O \cdot 2CaO \cdot 3SiO_2$ gl	66.81	38.76	2.70	298 – 1557
$Na_2O \cdot 2CaO \cdot 3SiO_2$ cr	83.87	14.08	16.31	
$2Na_2O \cdot CaO \cdot 3SiO_2$ gl	71.40	40.91	1.54	298 – 1414
$2Na_2O \cdot CaO \cdot 3SiO_2$ cr	88.46	16.24	15.05	
$Na_2O \cdot 3CaO \cdot 6SiO_2$ gl	76.17	91.48	–15.46	298 – 1320
$Na_2O \cdot 3CaO \cdot 6SiO_2$ cr	110.29	42.13	11.57	
$Na_2O \cdot Al_2O_3 \cdot 2SiO_2$ gl	44.12	33.21	–11.02	
$Na_2O \cdot Al_2O_3 \cdot 2SiO_2$ cr	53.74	25.55	–2.12	
$Na_2O \cdot Al_2O_3 \cdot 4SiO_2$ gl	83.36	34.61	10.64	
$Na_2O \cdot Al_2O_3 \cdot 4SiO_2$ cr	77.00	29.10	2.04	
$Na_2O \cdot Al_2O_3 \cdot 6SiO_2$ gl	122.62	36.00	32.32	
$Na_2O \cdot Al_2O_3 \cdot 6SiO_2$ cr	130.98	20.96	37.88	
$Na_2O \cdot 2MgO \cdot 6SiO_2$ gl	69.63	73.25	–20.09	
$Na_2O \cdot 2MgO \cdot 6SiO_2$ cr	97.72	40.67	1.53	

$$\Delta H_T^0 = \Delta H_{298}^0 + \int_{298}^{T} C_p dT$$

$$\Delta S_T^0 = \Delta S_{298}^0 + \int_{298}^{T} \frac{C_p dT}{T}$$

$$\Delta G_T^0 = \Delta H_T^0 - T\Delta S_T^0.$$

The results of the calculations of ΔH_T^0 and ΔG_T^0 for some vitreous silicates are presented in Tables 58 and 59.

4.4 Devitrification of Glasses

The study of the glass devitrification – crystallisation – processes, as stated earlier, is also highly important theoretically and vital practically.

The investigations of the possibility of endowing glass and glass-crystalline materials with needed properties are mostly carried out by studying kinetic factors exerting influence on the direction of the process. In addition to that, it would be of value to determine through calculations the direction of the process most probable from the thermodynamic viewpoint, subsequently determining the necessary corrections required by kinetics. In any glass-crystalline system knowing the composition of the final, most stable crystalline phase is also important, because in a lot of cases it predetermines the main properties of products. In this respect the thermodynamic investigation method may prove to be highly useful: while dealing with reliable thermochemical data it allows to finding the right location of these materials on the energy scale glass-crystal and the final crystallization products for any vitreous system [20, 385, 415].

As glass features a lesser structural orderliness than a crystalline substance of related composition, the crystallization process will be characterized by a negative alteration of the Gibbs free energy the value of which can be directly calculated by the formula:

$$\Delta G_{(devitr)} = \sum \Delta G_{(cr)} - \sum \Delta G_{(gl)} \quad (160)$$

where $\Delta G_{(devitr)}$, the Gibbs free energy alteration at devitrification; $\Delta G_{(cr)}$, sum of the Gibbs free energies of crystallization products; $\Delta G_{(gl)}$, sum of the Gibbs free energies of initial glasses.

The minimum value of the Gibbs free energy calculated by the above equation for diverse crystallization products will testify to the greatest stability of the crystalline products which have given that minimum value. Thus, for conducting the calculations it is necessary to know the Gibbs free energies both of probable crystallization products and of initial glasses.

Table 58. Calculations of ΔH_T^0 for some vitreous silicates

Compounds in vitreous state	$-\Delta H^0_{(el)}$ [kcal/mol]										
	Temperature [K]										
	400	500	600	700	800	900	1000	1100	1200	1300	1400
$2Na_2O \cdot SiO_2$	495.47	490.43	485.13	479.59	473.82	467.82	461.63	455.21	448.59	441.74	434.81
$Na_2O \cdot SiO_2$	360.00	356.75	353.28	349.64	345.78	341.75	337.53	337.14	328.56	323.80	318.86
$Na_2O \cdot 2SiO_2$	579.94	575.17	570.17	564.76	558.99	552.89	547.46	539.68	532.57	525.11	517.34
$Na_2O \cdot 3SiO_2$	796.15	790.18	783.74	776.84	769.49	761.70	753.46	744.78	735.65	726.08	716.06
$K_2O \cdot SiO_2$	367.30	364.01	360.56	356.97	353.23	349.35	345.34	341.16	336.86	332.52	328.54
$K_2O \cdot 2SiO_2$	586.08	581.61	576.88	571.84	566.42	560.62	554.44	547.86	540.87	533.47	525.67
$K_2O \cdot 4SiO_2$	998.73	988.68	978.45	967.62	955.90	943.20	929.41	914.48	898.39	881.11	862.62
$2Li_2O \cdot SiO_2$	526.06	521.82	517.34	512.63	507.74	502.66	497.42	492.01	486.44	480.70	474.81
$Li_2O \cdot SiO_2$	368.22	365.33	362.32	359.13	356.00	352.29	348.62	344.78	340.79	336.63	332.30
$Li_2O \cdot 2SiO_2$	584.96	580.48	575.80	570.86	565.65	560.11	554.27	548.10	541.61	534.78	528.21
$4PbO \cdot SiO_2$	420.49	419.45	407.92	400.94	393.51	386.66	377.36	368.67	359.53	349.97	339.99
$2PbO \cdot SiO_2$	320.20	316.13	311.78	307.20	320.42	297.44	292.29	286.98	281.50	275.86	270.05
$PbO \cdot SiO_2$	270.50	267.42	264.16	260.77	257.30	253.78	250.20	246.57	242.92	239.23	235.51
$Na_2O \cdot 2CaO \cdot 3SiO_2$	1145.45	1137.16	1128.43	1119.32	1109.77	1099.83	1089.50	1078.78	1067.66	1056.16	1044.26
$2Na_2O \cdot CaO \cdot 3SiO_2$	1108.99	1100.09	1090.75	1081.00	1070.82	1060.22	1049.22	1037.72	1025.96	1013.72	1001.06
$Na_2O \cdot 3CaO \cdot 6SiO_2$	1968.17	1955.66	1942.51	1928.57	1913.81	1898.21	1881.73	1864.38	1846.11	1826.96	1823.01

Table 59. Calculations of ΔG_T^0 for some vitreous silicates

Compounds in vitreous state	$-\Delta G^0_{(el)}$ [kcal/mol]										
	Temperature [K]										
	400	500	600	700	800	900	1000	1100	1200	1300	1400
$2Na_2O \cdot SiO_2$	460.44	452.52	445.42	439.21	433.85	429.24	425.24	421.89	419.17	416.95	415.35
$Na_2O \cdot SiO_2$	334.38	328.36	322.97	318.22	313.99	310.25	306.98	304.13	301.69	299.65	297.97
$Na_2O \cdot 2SiO_2$	539.66	530.09	521.59	513.90	507.03	500.89	495.46	490.68	486.53	482.98	480.03
$Na_2O \cdot 3SiO_2$	740.86	727.75	715.23	705.03	694.82	686.45	678.53	671.34	665.19	659.68	654.94
$K_2O \cdot SiO_2$	341.84	335.86	330.54	325.81	321.61	317.89	314.61	311.72	309.24	306.95	304.33
$K_2O \cdot 2SiO_2$	546.12	536.63	528.05	520.31	513.32	507.01	501.38	496.40	492.02	488.22	485.04
$K_2O \cdot 4SiO_2$	931.67	916.67	902.49	890.69	880.47	871.80	864.59	858.81	854.45	851.48	849.88
$2Li_2O \cdot SiO_2$	491.78	483.72	476.50	470.04	464.32	459.17	454.62	450.57	447.08	444.00	441.45
$Li_2O \cdot SiO_2$	343.55	337.71	332.47	327.72	323.68	319.65	316.21	313.13	310.44	308.07	306.04
$Li_2O \cdot 2SiO_2$	545.75	536.48	528.10	520.52	513.69	507.50	501.98	497.03	492.69	488.86	485.13
$4PbO \cdot SiO_2$	373.83	362.89	353.16	344.55	337.00	330.39	324.70	319.83	315.80	312.52	310.00
$2PbO \cdot SiO_2$	289.61	282.44	276.10	270.10	265.69	261.26	257.51	254.29	251.56	249.28	247.45
$PbO \cdot SiO_2$	247.16	242.68	236.83	232.52	228.72	225.37	222.40	219.77	217.52	215.53	213.85
$Na_2O \cdot 2CaO \cdot 3SiO_2$	1070.73	1063.10	1037.04	1022.51	1009.34	997.35	986.52	976.72	967.94	960.08	953.12
$2Na_2O \cdot CaO \cdot 3SiO_2$	1033.27	1025.41	999.29	984.80	971.76	959.99	949.45	939.98	931.62	924.24	917.82
$Na_2O \cdot 3CaO \cdot 6SiO_2$	1838.65	1807.09	1779.35	1753.35	1729.15	1707.00	1686.63	1667.93	1650.86	1635.34	1621.34

The Gibbs free energies of crystalline silicates may be taken from the previous calculations. For vitreous substances little related data concerning the Gibbs free energies are available thus complicating the situation. Therefore it often seems impossible to calculate the absolute alterations of the Gibbs free energy at devitrification.

The application of the electrochemical method to solid and melted electrolytes while studying solid-phase reactions and the ones in melts [639, 715] has been recently developing successfully. This method using the interconnection of thermodynamics and electrochemistry allows the immediate determination of the alteration of the Gibbs free energy in the reactions and processes in condensed phases. It is apparently the very basis for experimentally determining the reliable ΔG values of vitreous substances.

As stated above, however, we are interested in the crystallization products bringing about the minimum value of the Gibbs free energy of devitrification. As in this case the composition and properties of the initial glass continue being the same, we may thus believe $\sum \Delta G_{(gl)}$ to be constant and, consequently, the value of the Gibbs free energy of devitrification for various crystallization products to depend only on the sum of the Gibbs free energies of the crystallization products. Thus, the calculation amounts to finding the difference of the Gibbs free energies of various crystallization products.

The applicability of these calculation methods will be elucidated by the instance of the devitrification of glasses of the system Na_2O-SiO_2.

The calculation by the absolute method was conducted for glasses composed of sodium di- and trisilicate. The values of the enthalpy of the formation of the glass composed of sodium disilicate were taken from [554]. On the basis of an analysis of the courses of the solution heat curves for crystalline and vitreous substances as a function of the molar content of sodium oxide [554] the enthalpy values for the glass composed of sodium trisilicate were calculated.

The data at hand concerning the entropy of the crystalline sodium di- and trisilicate and its alteration at melting and a transition into a new state [127] structurally close to the vitreous one permitted calculating the entropy value for vitreous substances.

For calculating the data concerning the alteration of the heat capacities of vitreous substances Tarasov's results [611] were used. By the equations suggested the shapes of the heat capacity curves up to the temperatures needed were calculated and the equation coefficients $C_p = f(T)$ determined.

The reactions examined for the glasses composed of sodium di- and trisilicate are adduced below:

Glass composed of $Na_2O \cdot 2SiO_2$

1. $Na_2O \cdot 2SiO_2\,gl \rightarrow Na_2O \cdot SiO_2\,cr + SiO_2\,cr$
2. $Na_2O \cdot 2SiO_2\,gl \rightarrow Na_2O \cdot 2SiO_2\,cr$
3. $Na_2O \cdot 2SiO_2\,gl \rightarrow \frac{2}{3}(Na_2O \cdot 3SiO_2\,cr) + \frac{1}{3}Na_2O\,cr$

4. $Na_2O \cdot 2SiO_2\,gl \rightarrow \frac{1}{2}(Na_2O \cdot SiO_2\,cr) + \frac{1}{2}(Na_2O \cdot 3SiO_2\,cr)$
5. $Na_2O \cdot 2SiO_2\,gl \rightarrow \frac{1}{2}(2Na_2O \cdot SiO_2\,cr) + \frac{3}{2}SiO_2\,cr$

Glass composed of $Na_2O \cdot 3SiO_2$
1. $Na_2O \cdot 3SiO_2\,gl \rightarrow Na_2O \cdot SiO_2\,cr + 2SiO_2\,cr$
2. $Na_2O \cdot 3SiO_2\,gl \rightarrow Na_2O \cdot 2SiO_2\,cr + SiO_2\,cr$
3. $Na_2O \cdot 3SiO_2\,gl \rightarrow Na_2O \cdot 3SiO_2\,cr$
4. $Na_2O \cdot 3SiO_2\,gl \rightarrow \frac{1}{2}(Na_2O \cdot SiO_2\,cr) + \frac{3}{2}SiO_2\,cr + \frac{1}{2}(Na_2O \cdot 2SiO_2\,cr)$
5. $Na_2O \cdot 3SiO_2\,gl \rightarrow \frac{1}{2}(2Na_2O \cdot SiO_2\,cr) + \frac{5}{2}SiO_2\,cr$

The calculation results are presented in Fig. 67.

An analysis of calculations carried out shows the formation of crystalline sodium disilicate as well as of a mixture of sodium metasilicate and trisilicate to be most probable while the glass composed of sodium disilicate is being devitrified. The most probable products of the crystallization of glass composed of sodium trisilicate are crystals of the same composition.

For determining the most stable crystallization products by the relative method the following compositions of two component glasses of the system Na_2O-SiO_2: $Na_2O \cdot SiO_2$, $Na_2O \cdot 2SiO_2$, $Na_2O \cdot 3SiO_2$, $Na_2O \cdot \frac{3}{2}SiO_2$, $Na_2O \cdot 2\frac{1}{2}SiO_2$ and $Na_2O \cdot 5SiO_2$ were chosen.

For calculating the corresponding ΔG of devitrification by the relative method the data for the Gibbs free energies of the crystalline silicate formation from oxides calculated earlier [385] were used within the temperature range of 298 to 1000 K.

The following possible ways of the glass crystallization were examined:

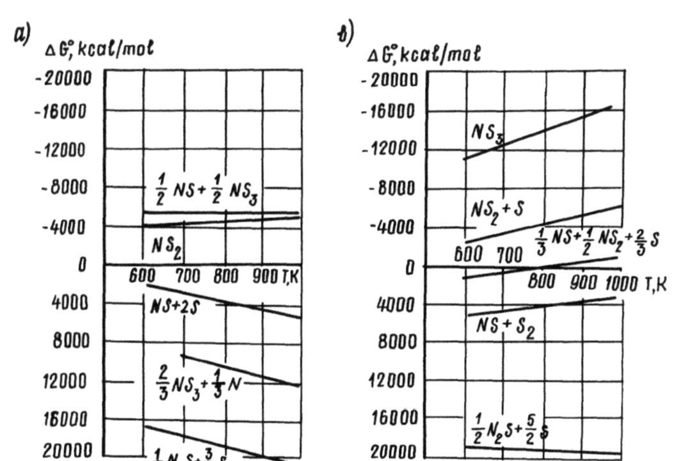

Fig. 67. $\Delta G^0 = f(T)$ of the reaction of the devitrification of glasses composed of: **a** $Na_2O \cdot 2SiO_2$, **b** $Na_2O \cdot 3SiO_2$

I. Glass composed of $Na_2O \cdot SiO_2$

1. $Na_2O \cdot SiO_2\ gl \rightarrow Na_2O\ cr + SiO_2\ cr$
2. $Na_2O \cdot SiO_2\ gl \rightarrow Na_2O \cdot SiO_2\ cr$
3. $Na_2O \cdot SiO_2\ gl \rightarrow \frac{1}{2}(Na_2O \cdot 2SiO_2\ cr) + \frac{1}{2}Na_2O\ cr$
4. $Na_2O \cdot SiO_2\ gl \rightarrow \frac{1}{3}(Na_2O \cdot 3SiO_2\ cr) + \frac{2}{3}Na_2O\ cr$
5. $Na_2O \cdot SiO_2\ gl \rightarrow \frac{1}{4}(Na_2O \cdot 2SiO_2\ cr) + \frac{1}{6}(Na_2O \cdot 3SiO_2\ cr)$
6. $Na_2O \cdot SiO_2\ gl \rightarrow \frac{1}{2}(2Na_2O \cdot SiO_2\ cr) + \frac{1}{2}SiO_2\ cr$

II. Glass composed of $Na_2O \cdot 2SiO_2$

1. $Na_2O \cdot 2SiO_2\ gl \rightarrow Na_2O\ cr + 2SiO_2\ cr$
2. $Na_2O \cdot 2SiO_2\ gl \rightarrow Na_2O \cdot SiO_2\ cr + SiO_2\ cr$
3. $Na_2O \cdot 2SiO_2\ gl \rightarrow Na_2O \cdot 2SiO_2\ cr$
4. $Na_2O \cdot 2SiO_2\ gl \rightarrow \frac{2}{3}(Na_2O \cdot 3SiO_2\ cr) + \frac{1}{3}Na_2O\ cr$
5. $Na_2O \cdot 2SiO_2\ gl \rightarrow \frac{1}{2}(Na_2O \cdot SiO_2\ cr) + \frac{1}{2}(Na_2O \cdot 3SiO_2\ cr)$
6. $Na_2O \cdot 2SiO_2\ gl \rightarrow \frac{1}{2}(2Na_2O \cdot SiO_2\ cr) + \frac{3}{2}SiO_2\ cr$

III. Glass composed of $Na_2O \cdot 3SiO_2$

1. $Na_2O \cdot 3SiO_2\ gl \rightarrow Na_2O\ cr + 3SiO_2\ cr$
2. $Na_2O \cdot 3SiO_2\ gl \rightarrow Na_2O \cdot SiO_2\ cr + 2SiO_2\ cr$
3. $Na_2O \cdot 3SiO_2\ gl \rightarrow Na_2O \cdot 2SiO_2\ cr + SiO_2\ cr$
4. $Na_2O \cdot 3SiO_2\ gl \rightarrow Na_2O \cdot 3SiO_2\ cr$
5. $Na_2O \cdot 3SiO_2\ gl \rightarrow \frac{1}{2}(Na_2O \cdot SiO_2\ cr) + \frac{1}{2}(Na_2O \cdot 2SiO_2\ cr) + \frac{3}{2}SiO_2\ cr$
6. $Na_2O \cdot 3SiO_2\ gl \rightarrow \frac{1}{2}(2Na_2O \cdot SiO_2\ cr) + \frac{5}{2}SiO_2\ cr$

IV. Glass composed of $Na_2O \cdot \frac{3}{2}SiO_2$

1. $Na_2O \cdot \frac{3}{2}SiO_2\ gl \rightarrow Na_2O\ cr + \frac{3}{2}SiO_2\ cr$
2. $Na_2O \cdot \frac{3}{2}SiO_2\ gl \rightarrow Na_2O \cdot SiO_2\ cr + \frac{1}{2}SiO_2\ cr$
3. $Na_2O \cdot \frac{3}{2}SiO_2\ gl \rightarrow \frac{3}{4}(Na_2O \cdot 2SiO_2\ cr) + \frac{1}{4}Na_2O\ cr$
4. $Na_2O \cdot \frac{3}{2}SiO_2\ gl \rightarrow \frac{1}{2}(Na_2O \cdot 3SiO_2\ cr) + \frac{1}{2}Na_2O\ cr$
5. $Na_2O \cdot \frac{3}{2}SiO_2\ gl \rightarrow \frac{1}{2}(Na_2O \cdot SiO_2\ cr) + \frac{1}{2}(Na_2O \cdot 2SiO_2\ cr)$
6. $Na_2O \cdot \frac{3}{2}SiO_2\ gl \rightarrow \frac{1}{2}(2Na_2O \cdot SiO_2\ cr) + SiO_2\ cr$

V. Glass composed of $Na_2O \cdot 2\frac{1}{2}SiO_2$

1. $Na_2O \cdot 2\frac{1}{2}SiO_2\ gl \rightarrow Na_2O\ cr + 2\frac{1}{2}SiO_2\ cr$
2. $Na_2O \cdot 2\frac{1}{2}SiO_2\ gl \rightarrow Na_2O \cdot SiO_2\ cr + 1\frac{1}{2}SiO_2\ cr$
3. $Na_2O \cdot 2\frac{1}{2}SiO_2\ gl \rightarrow Na_2O \cdot 2SiO_2\ cr + \frac{1}{2}SiO_2\ cr$
4. $Na_2O \cdot 2\frac{1}{2}SiO_2\ gl \rightarrow \frac{5}{6}(Na_2O \cdot 3SiO_2\ cr) + \frac{1}{6}Na_2O\ cr$
5. $Na_2O \cdot 2\frac{1}{2}SiO_2\ gl \rightarrow \frac{1}{2}(2Na_2O \cdot SiO_2\ cr) + 2SiO_2\ cr$
6. $Na_2O \cdot 2\frac{1}{2}SiO_2\ gl \rightarrow \frac{1}{2}(Na_2O \cdot SiO_2\ cr) + \frac{1}{2}(Na_2O \cdot 2SiO_2\ cr) + SiO_2\ cr$
7. $Na_2O \cdot 2\frac{1}{2}SiO_2\ gl \rightarrow \frac{1}{2}(Na_2O \cdot SiO_2\ cr) + \frac{1}{2}(Na_2O \cdot 3SiO_2\ cr) + \frac{1}{2}SiO_2\ cr$
8. $Na_2O \cdot 2\frac{1}{2}SiO_2\ gl \rightarrow \frac{1}{2}(Na_2O \cdot 2SiO_2\ cr) + \frac{1}{2}(Na_2O \cdot 3SiO_2\ cr)$
9. $Na_2O \cdot 2\frac{1}{2}SiO_2\ gl \rightarrow \frac{1}{3}(Na_2O \cdot SiO_2\ cr) + \frac{1}{3}(Na_2O \cdot 2SiO_2\ cr) + \frac{1}{3}(Na_2O \cdot 3SiO_2\ cr) + \frac{1}{2}SiO_2\ cr$

VI. Glass composed of $Na_2O \cdot 5SiO_2$

1. $Na_2O \cdot 5SiO_2\,gl \rightarrow Na_2O\,cr + 5SiO_2\,cr$
2. $Na_2O \cdot 5SiO_2\,gl \rightarrow Na_2O \cdot SiO_2\,cr + 4SiO_2\,cr$
3. $Na_2O \cdot 5SiO_2\,gl \rightarrow Na_2O \cdot 2SiO_2\,cr + 3SiO_2\,cr$
4. $Na_2O \cdot 5SiO_2\,gl \rightarrow Na_2O \cdot 3SiO_2\,cr + 2SiO_2\,cr$
5. $Na_2O \cdot 5SiO_2\,gl \rightarrow \frac{1}{2}(2Na_2O \cdot SiO_2\,cr) + 4\frac{1}{2}SiO_2\,cr$
6. $Na_2O \cdot 5SiO_2\,gl \rightarrow \frac{1}{2}(Na_2O \cdot SiO_2\,cr) + \frac{1}{2}(Na_2O \cdot 2SiO_2\,cr) + 3\frac{1}{2}SiO_2\,cr$
7. $Na_2O \cdot 5SiO_2\,gl \rightarrow \frac{1}{2}(Na_2O \cdot 2SiO_2\,cr) + \frac{1}{2}(Na_2O \cdot 3SiO_2\,cr) + 2\frac{1}{2}SiO_2\,cr$
8. $Na_2O \cdot 5SiO_2\,gl \rightarrow \frac{1}{2}(Na_2O \cdot SiO_2\,cr) + \frac{1}{2}(Na_2O \cdot 3SiO_2\,cr) + 3SiO_2\,cr$.

The ΔG values obtained for the devitrification at different temperatures are presented in Table 60.

An analysis of the data from Table 60 for the devitrification of glasses of the system Na_2O-SiO_2 shows that:

1. In the crystallization of glasses corresponding in their stoichiometric composition to the crystalline compounds known in the system Na_2O-SiO_2 the most stable crystallization products will be these crystalline silicates.

2. With intermediate compositions the most stable crystallization products will be the crystalline mixtures of the corresponding stoichiometric composition.

3. In the crystallization of glasses with the silica modulus higher than 3 the most stable devitrification product will be the mixture of the crystalline sodium trisilicate and the excess SiO_2.

The data obtained earlier (see Section 2.11) permit calculating the reactions of the glass devitrification in the system $CaO-Al_2O_3-SiO_2$ [271, 375, 531].

The calculation was carried out in a way similar to that of the system Na_2O-SiO_2 for 16 glasses the composition of which is presented in Fig. 68.

The ΔG_T^0 values of the reactions of the devitrification of these glasses for a temperature of 1.300 K are presented in Table 61.

A comparison of the most stable crystallization products obtained by calculating the final crystallization products obtained by the phase diagram of the system $CaO-Al_2O_3-SiO_2$ demonstrate their full coincidence.

The above examples show that the suggested method of calculating the devitrification products can be used for determining the final products of the devitrification of glasses for which no phase diagrams, in particular multicomponent ones, are available.

4.5 The Calculation of the Heat Consumption of Glass Melting

In view of the great practical importance of the calculation of the theoretically necessary heat consumption of glass melting we found it also necessary to briefly elucidate this essential problem.

Table 60. The Gibbs free energies of the devitrification of glasses of the system of different compositions

Nos. of reac- tions	Glass compo- sition	The ΔG value of devitrification [kcal/mol] at temps. [K]					
		298	600	700	800	900	1000
1		0	0	0	0	0	0
2		−60.00	−60.88	−61.13	−61.40	−61.80	−62.02
3		−33.00	−34.23	−34.62	−35.01	−35.48	−35.95
4	$Na_2O \cdot SiO_2$	−24.33	−25.80	−26.22	−26.64	−27.26	−27.50
5		−28.67	−30.01	−30.42	−30.83	−31.38	−31.72
6		−47.00	−46.90	−46.79	−46.65	−46.49	−46.40
1		0	0	0	0	0	0
2		−60.00	−60.88	−61.13	−61.40	−61.80	−62.02
3		−66.00	−68.45	−69.24	−70.03	−70.95	−71.90
4	$Na_2O \cdot 2SiO_2$	−48.67	−51.60	−52.45	−53.28	−54.59	−55.00
5		−66.50	−69.14	−69.90	−70.66	−71.84	−72.25
6		−47.00	−46.90	−46.79	−46.65	−46.49	−46.40
1		0	0	0	0	0	0
2		−60.00	−60.88	−61.13	−61.40	−61.80	−62.02
3		−66.00	−68.45	−69.24	−70.03	−70.55	−71.90
4	$Na_2O \cdot 3SiO_2$	−73.00	−77.40	−78.67	−79.92	−81.89	−82.49
5		−63.00	−64.67	−65.18	−65.71	−66.38	−66.96
6		−47.00	−46.90	−46.79	−46.65	−46.50	−46.40
1		0	0	0	0	0	0
2		−60.00	−60.88	−61.13	−61.40	−61.80	−62.02
3		−49.50	−51.34	−51.93	−52.52	−53.22	−53.92
4	$Na_2O \cdot \frac{3}{2}SiO_2$	−36.50	−38.70	−39.33	−39.96	−40.94	−41.24
5		−63.00	−64.67	−65.19	−65.96	−66.38	−66.96
6		−47.00	−46.90	−46.79	−46.65	−46.49	−46.40
1		0	0	0	0	0	0
2		−60.00	−60.88	−61.13	−61.40	−61.80	−62.02
3		−66.00	−68.45	−69.24	−70.03	−70.95	−71.90
4		−60.83	−64.49	−65.55	−66.60	−68.24	−68.84
5	$Na_2O \cdot 2\frac{1}{2}SiO_2$	−47.00	−46.90	−46.79	−46.65	−46.49	−46.40
6		−63.00	−64.67	−65.19	−65.71	−66.38	−66.96
7		−66.50	−69.14	−69.90	−70.66	−71.84	−72.25
8		−69.50	−72.92	−73.95	−74.98	−76.42	−77.19
9		−66.33	−68.88	−69.68	−40.45	−71.55	−72.13
1		0	0	0	0	0	0
2		−60.00	−60.88	−61.13	−61.40	−61.80	−62.02
3		−66.00	−68.45	−69.24	−70.03	−70.95	−71.90
4		−73.00	−77.40	−78.67	−79.92	−81.89	−82.49
5	$Na_2O \cdot 5SiO_2$	−47.00	−46.90	−46.79	−46.65	−46.49	−46.40
6		−63.00	−64.67	−65.19	−65.71	−66.37	−66.96
7		−69.00	−72.92	−73.96	−74.98	−76.42	−77.19
8		−66.50	−69.14	−69.90	−70.66	−71.84	−72.25
9		−66.33	−68.88	−69.68	−70.45	−71.55	−72.13

Table 61. The ΔG_T^0 values of devitrification of some glasses of the system $CaO-Al_2O_3-SiO_2$ at a temperature of 1300 K

Presumable crystallization products[a]	$-\Delta G_T^0$ [kcal/mol] for glasses															
	11	12	13	14	21	22	23	24	31	32	33	34	41	42	43	44
A_3S_2+CS+S	46.0	41.8	37.99	34.1	56.4	51.8	47.9	44.1	67.0	62.5	58.1	53.8	77.3	72.9	68.3	64.0
$CS+CA+S(A)$	36.3	30.0	25.0	19.8	46.8	40.5	34.8	29.4	57.4	51.0	44.9	39.0	67.6	61.2	55.2	9.1
$C_2AS+CS+S$	35.5	30.0	24.8	19.7	45.9	40.0	34.3	29.2	56.5	49.9	44.3	38.6	66.0	60.1	54.2	48.2
CAS_2+CS+S	50.2	46.9	39.5	31.4	60.6	57.0	63.7	46.5	71.4	67.6	63.9	60.4	81.7	77.8	74.1	70.5
$CAS_2+A_3S_2$			42.4	44.6	–	–	–	49.4	–	–	–	–	–	–	–	–
The most stable phases according to the phase diagram	CAS_2 S CS	CAS_2 S CS	CAS_2 A_3S_2 S	A_3S_2 CAS_2 S	CAS_2 S CS	CAS_2 S CS	CAS_2 S CS	CAS_2 A_3S_2 S	CAS_2 CS S	CAS_2 CS S	CAS_2 CS S	CAS_2 S CS	CAS_2 CS S	CAS_2 CS S	CAS_2 CS S	CAS_2 CS S

[a] Note: The most stable compounds are enframed

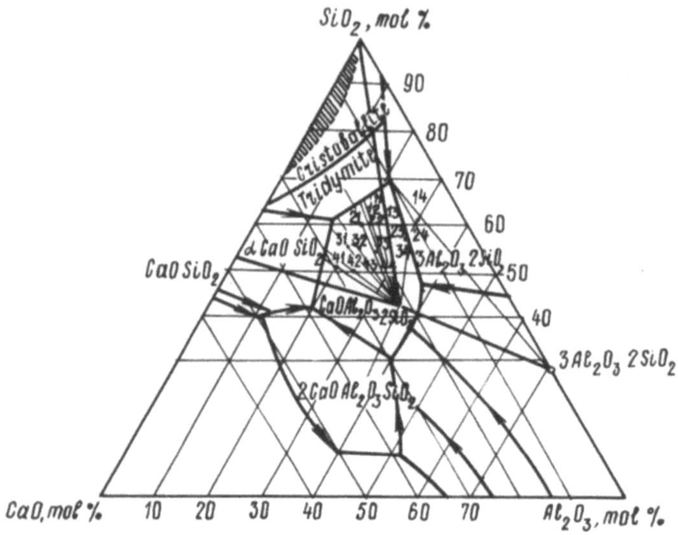

Fig. 68. The state of the system $CaO-Al_2O_3-SiO_2$ and the composition of the glasses

These data are indispensable in evaluating the efficiency of glass melting furnaces and fuel consumption as well as in determining and comparing the heat consumption while melting various glasses.

It should be noted that the method of calculating useful heat consumption has still not been exactly established. The way in which particular heat components constituting the useful heat are viewed differs often. Furthermore, the initial data underlying the calculations, e.g., in case of Zimin's Table [33], have to a large extent become obsolete.

In literature three ways of calculating heat requirements for glass melting are set forth and their essence can be understood from an enthalpy diagram (Fig. 69).

According to the first method [306] the following values are added:

1. the heat required for heating the initial substances up to the temperature of the reaction; 2. the heat of the reactions in the batch at temperatures of each reaction; 3. the heat required for heating the reaction products up to the final temperature of melt. On the diagram (see Fig. 69) this method corresponds to the way a-b-f-k and is expressed as:

$$\Delta H_{\Sigma} = \Sigma N[(H_{T_r} - H_{298})_{\text{init. subst.}} + H_{T_r}] + \bar{C}_p|_{T_r}^{T}(T - T_r) + (i_T - i_{T_r}) , \qquad (161)$$

where H, sum of heat consumption; N, molar fraction of a given component in the glass composition; H_{T_r}, heat content of a component at the temperature of the beginning of the reaction; H_{298}, heat content of a component at the initial temperature of 298 K; H_{T_r}, heat of reaction at the temperature of the reaction; $\bar{C}_p|_{T_r}^{T}$, mean heat capacity of glass within the temperature range under considera-

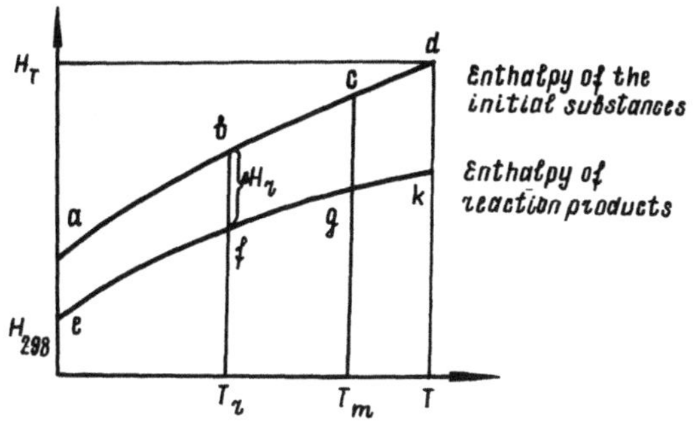

Fig. 69. The enthalpy diagram of the glass-melting process; *a*, initial thermodynamic state of a component of the reaction mixture; *k*, final thermodynamic state of the reaction products; ΔH_r reaction heat; T_r reaction temperature; T_m melting temperature; T_f, final temperature of the process

tion; T_r, temperature of the beginning of the reaction; T, final temperature of glass melting; i_{T_r}, heat content of gases evolved at the temperature of the beginning of the reaction; i_T, heat content of gases evolved at the final temperature of glass melting.

According to the second method employed in the calculations of [457] the increase of the heat content of the components of the batch is calculated beginning with room temperature up to the temperature of melting. To the resulting amount of heat the heat of the reactions in the batch at the temperature of each reaction is added plus the increase of the heat content of the substances in the final melted state. This method corresponds to the way a-c-g-k (see Fig. 69) and is expressed as:

$$H_\Sigma = \Sigma N[(H_{T_m} - H_{298})_{\text{init. subst.}} + H_{T_r}] + \bar{C}_p|_{T_m}^{T}(T - T_m) + (i_T - i_{T_m}), \tag{162}$$

where H_{T_m}, heat content of a component at the temperature of melting; T_m, temperature of melting; i_{T_m}, heat content of the gas evolved at the temperature of melting of a component.

Kröger [306] employed a third method according to which is assumed that the reaction in the batch took place at room temperature. Added to the heat effect obtained is the increment of the amount of heat required to increase the temperature of products resulting from the reaction up to the final temperature of melted glass. On the diagram (see Fig. 69) this method corresponds to the method a-e-k which is expressed as:

$$H_\Sigma = \Sigma N \Delta H_{298}^0 + \bar{C}_p|_{298}^{T}(T - 298) + (i_T - i_{298}), \tag{163}$$

where ΔH_{298}^0, standard heat effect value of the reactions of formation of silicates from oxides at 298 K.

While using this calculation method the melting and transformation heats, the values of which are related to the corresponding transformation temperature are to be recalculated for room temperature. Such a recalculation is possible only in rare cases when their temperature dependence both for crystalline and vitreous phases is known.

Available data required for calculating useful heat expenditure by these methods are rarely reliable enough. Moreover, the methods used require knowing the intermediate mechanism of reactions after which ready glass melt is obtained. In fact, glass melting proceeds not in conditions of a slow rise of the batch temperature, as was accepted in numerous studies, but in conditions of its abruptly heating, because in a glassmaking furnace the batch enters a high temperature chamber.

According to the method we [368] suggested, the heat required to heat the initial substances up to final temperatures is first calculated with regard to all transformations.[12]

To the result obtained, the heat of the reactions at the final temperature of the process is added. On the enthalpy diagram (see Fig. 69) this calculation scheme corresponds to the method a-d-k. The necessary temperature dependences of the enthalpy are preliminarily determined.

This calculation method is expressed as:

$$\Delta H_{\Sigma} = \Sigma N[\Sigma(H_T - H_{298})_{\mathrm{init.\,subst.}} + \Delta H_T]\,, \qquad (164)$$

where ΔH_T, the value of the heat effect of the reactions of formation of silicates from oxides at a given temperature.

For calculating the reaction heat effects we set up the corresponding temperature dependences for the main vitreous silicates.

The advantages of the method are that, firstly, one does not need to know the temperatures at the beginning and at the end of the reactions; secondly, the temperature dependences of enthalpies obtained can be used in calculations for various glass compositions and at various final temperatures of the process. Moreover, this method enables us to substantially simplify all the calculations. Heat consumption for melting 1 kg of glass at a given final temperature is expressed as:

$$\Delta H_{\Sigma} = \Sigma P_i \cdot \Delta H_i\,, \qquad (165)$$

where P_i, content of a given oxide in glass in weight percent; ΔH_i, heat consumption for melting the silicate corresponding to the given oxide.

Using the data obtained in the previous chapters equations were set up of the temperature dependence of enthalpies $\Delta H^0 = f(T)$ for all the individual substances, components of the reactions of formation of silicates listed below:

---

[12] The reasons that underlie the method are that the glass melt may be considered as a "perfect solution" of binary silicates most stable at the temperature of melting

1. $3Al_2O_3 + 2SiO_2 = 3Al_2O_3 \cdot 2SiO_2$
2. $BaCO_3 + 2SiO_2 = BaO \cdot 2SiO_2 + CO_2$
3. $CaO + SiO_2 = CaO \cdot SiO_2$
4. $CaCO_3 + SiO_2 = CaO \cdot SiO_2 + CO_2$
5. $K_2CO_3 + 4SiO_2 = K_2O \cdot 4SiO_2 + CO_2$
6. $2KNO_3 + 4SiO_2 = K_2O \cdot 4SiO_2 + 2NO + 1.5\,O_2$
7. $MgCO_3 + SiO_2 = MgO \cdot SiO_2 + CO_2$
8. $MgO + SiO_2 = MgO \cdot SiO_2$
9. $2MgCO_3 + SiO_2 = 2MgO \cdot SiO_2 + 2CO_2$
10. $2MgO + SiO_2 = 2MgO \cdot SiO_2$
11. $Na_2CO_3 + 3SiO_2 = Na_2O \cdot 3SiO_2 + CO_2$
12. $PbO + SiO_2 = PbO \cdot SiO_2$.

On the basis of the equations obtained equations were set up of $\Delta H_T^0 = f(T)$ of the reactions of silicate formation by which the enthalpy values within the temperature range under consideration were then calculated. The values of the reaction enthalpies are the differences of the values of ΔH_T^0 of the reaction products and those of initial substances according to Hess' law. Equations of temperature dependence of the heat consumption of producing silicates (Table 62) were set up starting from the calculations conducted. With the equations presented in Table 62 the heat consumption of melting the silicates under consideration at temperatures of 1173 to 1800 K can be calculated. The values of heat consumption are expressed in kilocalories per 1 kg of dissociation oxide. The calculation results are presented in Table 63. The alteration of specific heat consumption is shown in Fig. 70.

The data of Tables 62 and 63 permit calculating the useful heat consumption of melting various silicate glasses within a wide range of temperatures and compositions.

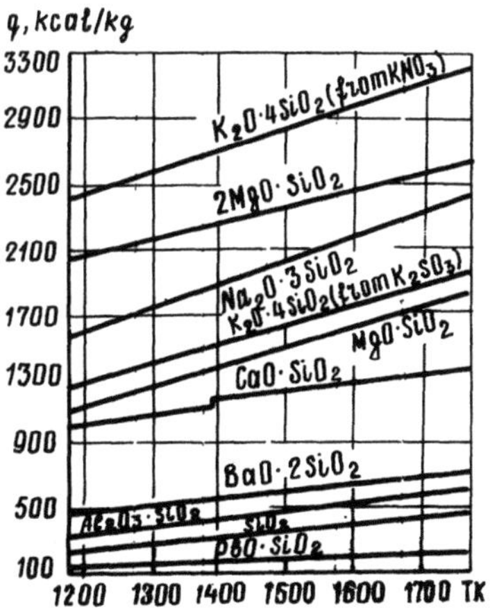

Fig. 70. The temperature dependence of specific consumption of heat by glass production from particular silicates

Table 62. Equations of temperature dependence of specific heat consumption of melting silicates within the temperature range of 1173 to 1800 K (kcal per 1 kg of dissociation oxide)

Compounds	Equations	Dissociation oxides
$Al_2O_3 \cdot SiO_2$	$q = 481.3 \cdot 10^{-3}T - 215.40$	Al_2O_3
$BaO \cdot 2SiO_2$	$q = 387.0 \cdot 10^{-3}T + 2.25$	BaO
$CaO \cdot SiO_2$ (1173 – 1400 K)	$q = 554.7 \cdot 10^{-3}T + 349.67$	CaO
$CaO \cdot SiO_2$ (1400 – 1800 K)	$q = 570.6 \cdot 10^{-3}T + 342.03$	CaO
$K_2O \cdot 4SiO_2$ (from K_2CO_3)	$q = 1190.0 \cdot 10^{-3}T - 62.07$	K_2O
$K_2O \cdot 4SiO_2$ (from KNO_3)	$q = 1359.2 \cdot 10^{-3}T + 823.85$	K_2O
$MgO \cdot SiO_2$	$q = 774.5 \cdot 10^{-3}T + 255.36$	MgO
$2MgO \cdot SiO_2$	$q = 1036.7 \cdot 10^{-3}T + 813.18$	MgO
$Na_2O \cdot 3SiO_2$	$q = 1438.57 \cdot 10^{-3}T - 147.33$	Na_2O
$PbO \cdot SiO_2$	$q = 139.33 \cdot 10^{-3}T - 31.54$	PbO
SiO_2 (heating)	$q = 287.6 \cdot 10^{-3}T - 54.41$	SiO_2

Table 63. Specific heat consumption of melting silicates at various temperatures (kcal per 1 kg of dissociation oxide)

Compounds	Temp. [K]						
	1173	1273	1373	1473	1573	1673	1800
$Al_2O_3 \cdot SiO_2$	349.2	397.3	445.4	493.6	541.7	589.8	638.0
$BaO \cdot 2SiO_2$	456.2	494.9	533.6	572.3	611.0	649.7	688.4
$CaO \cdot SiO_2$	1000.3	1055.8	1111.3	1182.5	1239.6	1296.6	1353.7
$K_2O \cdot 4SiO_2$ (from K_2CO_3)	1233.8	1352.8	1471.8	1590.8	1709.8	1828.8	1947.8
$K_2O \cdot 4SiO_2$ (from KNO_3)	2418.2	2554.1	2690.1	2826.0	2961.9	3097.8	3233.7
$MgO \cdot SiO_2$	1163.8	1241.3	1318.8	1396.2	1473.7	1551.1	1628.6
$2MgO \cdot SiO_2$	2029.2	2132.9	2236.6	2340.2	2443.9	2547.6	2651.3
$Na_2O \cdot 3SiO_2$	1540.0	1683.9	1827.7	1971.6	2115.4	2259.4	2403.1
$PbO \cdot SiO_2$	118.9	132.8	146.8	160.7	174.6	188.6	202.5
SiO_2 (heating)	282.7	311.5	340.2	369.0	598.3	426.5	455.3

Below an instance is presented of different ways of calculating heat consumption of melting glass of the same composition.

With the help of a calorimeter Haggerty, Cooper and Heasley measured and calculated the enthalpy of Lillie glass at different temperatures [181]. The molecular composition of Lillie glass (in molar fractions) is: SiO_2, 0.699; CaO, 0.098; Na_2O, 0.204. The weight composition [%] is: SiO_2, 69.84; CaO, 9.14; Na_2O, 21.02.

In accordance with the results of previous investigations it was assumed that while melting Lillie glass the following silicates are formed: $CaO \cdot SiO_2$, 0.098; $Na_2O \cdot 3SiO_2$, 0.193; $Na_2O \cdot 2SiO_2$, 0.011 (in molar fractions).

Table 64. Distribution of heat consumption of Lillie glass melting

Heat consumption	Calculation method							
	According to Haggerty et al.		According to us		According to Zimin		According to Kröger	
	[kcal/kg]	[%]	[kcal/kg]	[%]	[kcal/kg]	[%]	[kcal/kg]	[%]
Glass melting	414.4	100	413.6	100	383.9	100	423.9	100
Reactions (at 298 K)	102.0	24.6	102.0	24.7	81.3	21.2	121.3	28.6
Component heating	312.3	75.4	311.6	75.3	302.6	78.8	302.6	71.4

The calculations of the heat consumption of melting Lillie glass were carried out at a temperature of 900 °C.

The results of Haggerty, Cooper and Heasley's test are expressed by the equation:

$$H_T - H_{298} = 158.81 + 0.3570\,T \quad \text{(from 850 to 1200 K)} \tag{166}$$

which, the temperature being 1173 K, is 259.95 kcal/kg glass. Heat expenditure for the reactions and for carbon dioxide heating must be added to this amount thus the total heat consumption will be 414.4 kcal/kg. The calculation according to Kröger [306] yields the total heat expenditure value of 423.9 kcal/kg. The calculations according to Zimin [33] produced the value of 383.9 kcal/kg. Our data gives the total heat consumption of 413.6 kcal/kg. All the resulting values of heat consumption of Lillie glass melting as well as that of particular processes are presented in Table 64.

It follows from Table 64 that the calculation results of our data differ 0.02% from those of Haggerty et al., 2.4% from those of Kröger. The greatest deviation, 7.7%, from the calculation according to Zimin is accounted for, mainly, by the errors in the initial data, as well as by the lack of heat effect values of the reactions of formation of compounds more basic or more acid than metasilicates.

Contrary to the statement of the authors of article [334], the data of Table 64 show from 2.5 to 3 times more heat to be consumed by the heating of components of the batch and of gases evolved than by reactions.

We checked our method of calculating heat consumption by comparing our data with those of Kröger for three industrial glass composition (flat, feeder and lead crystal). While comparing, the same assumptions were taken for the two calculation methods.

The results of calculating heat consumption by melting different glasses according to Kröger and to our data are compared in Table 65.

Table 65. Heat consumption of glass melting (each process taken separately) in kcal

Heat consumption	Glass composition		
	Flat	Feeder	Crystal
According to Kröger			
1. Dissociation and transformations in batch	11.02	0.96	12.64
2. Silicate formation	121.27	92.69	51.27
3. Silicate melting	5.49	5.02	21.90
4. Glass heating	436.0	445.30	332.40
5. Heating gases evolved	92.53	71.10	39.05
6. Total heat consumption calculated per 1 kg of glass	670.7	607.8	456.9
According to our data			
1. Dissociations and transformations in batch	11.09	0.95	8.50
2. Heating and dissociation of components, formation and melting of silicates	582.06	501.57	370.30
3. Heating and melting of free silica and feldspar	76.52	39.80	68.39
4. Total heat consumption calculated per 1 kg of glass	669.7	592.3	447.2

Table 66. Calculation methods according to Kröger and to our data

Glass	Calculation method		Deviation	
	According to Kröger	Our data	[kcal/kg]	[%]
Flat	670.7	669.7	1.0	0.01
Feeder	607.8	592.3	15.5	2.62
Crystal	456.9	447.2	9.7	2.17

In Table 66 deviations are presented of the values of heat consumption for melting the glasses under consideration calculated by the two methods. As seen from Table 66, the values calculated by the two methods differ only slightly from one another, the maximum deviation being 2.62%.

The above calculation instances indicate that the $\Delta H^0 = f(T)$ values obtained can be well used for determining the heat consumption of technological processes of the silicate industry. The necessary heat consumption required for making cement, clinker, ceramics and other silicate materials can be similarily theoretically determined.

5 Reactions Proceeding During Firing Binders

According to the above regularities, the reactions of making binders such as lime, caustic magnesite, gypsum hemihydrate and cement are covered by reactions in solid phases and those of dissociation with gaseous phases and melt taking part, therefore all the foregoing theoretical principles and methods of calculations are applicable to them.

As the reactions of $CaCO_3$ and $MgCO_3$ dissociation underlying the production of lime and caustic magnesite were already considered in this chapter, we shall now present the calculation results only for the reactions of making gypsum hemihydrate and anhydrite as well as for those of clinker formation.

We studied the reactions of dehydration of gypsum dihydrate within a temperature range of 298 to 900 K [401] and found that the dehydration of $CaSO_4 \cdot 2H_2O$, i.e., producing gypsum hemihydrate and gaseous water, becomes thermodynamically possible beginning with 380 K (Fig. 71), and the dissociation of β-$CaSO_4 \cdot 0.5\,H_2O$ to soluble anhydrite and water, beginning with 480 K.

The transformation of the soluble anhydrite to the insoluble one is irreversible within the entire temperature range. It is noteworthy that kinetically the correctness of the calculation results obtained is well confirmed by a thermographic investigation of gypsum [401].

The other transformations of $CaSO_4 \cdot 2H_2O$ and of the products of its dissociation were thoroughly studied by Kelley, Southard and Anderson [260].

In the process of portland-cement-clinker firing the raw mixture undergoes gradual transformations, the main ones being: the removal of physically and chemically bound water from argillaceous minerals, carbonate thermal dissociation, solid-phase reactions and those with a melt, all this forming clinker minerals [35, 88, 171, 252].

In the general form the reactions of clinker formation are covered by reactions in a complex multicomponent system:

$$CaCO_3 - Al_2O_3 \cdot 2SiO_2 \cdot 2H_2O - CaO - Al_2O_3 - SiO_2 - Fe_2O_3 - Na_2O - K_2O$$
$$- MgO - P_2O_5 - SO_3(\text{gas}) - CO_2(\text{gas}) - H_2O(\text{gas})\,.$$

Owing to the great difficulties arising while studying this complicated system, we simplified the studies by excluding from it all the oxides entering the raw mixture as impurities (Na_2O, K_2O, MgO, P_2O_5 etc.).

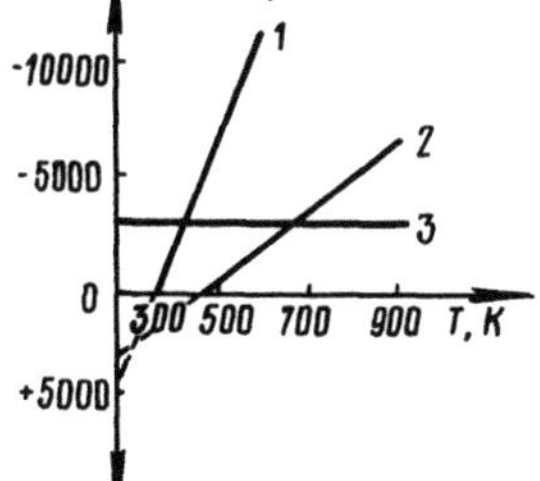

Fig. 71. The temperature dependence of ΔG^0 of the reactions of decomposition of various gypsum modifications with a temperature range of 298 to 900 K. *1* $CaSO_4 \cdot 2H_2O$ (cr) = β-$CaSO_4 \cdot 0.5\,H_2O$ (cr) + $1.5\,H_2O$ (gas); *2* β-$CaSO_4 \cdot 0.5\,H_2O$ (cr) = $CaSO_4$ (cr) + $0.5\,H_2O$ (gas); *3* $CaSO_4$ (soluble) = $CaSO_4$ (insoluble)

Bearing in mind the above transformation stages, we examined the reactions in the mixtures of raw materials and of the products of their thermic dissociation as well as in those of oxides and compared the preference of the formation or reaction products with regard to the amount of initial substances.

The types of reactions examined are presented in Table 67 and the results of the $\Delta G^0 = f(T)$ calculations, in Table 68.

Table 67. Types of primary and secondary reactions in cement batch consisting of calcium oxides, calcite, kaolinite and metakaolinite

Nos. of reactions	Reactions in cement batch
1.	$2CaO + SiO_2 = \beta\text{-}2CaO \cdot SiO_2$
2.	$3CaO + SiO_2 = 3CaO \cdot SiO_2$
3.	$3CaO + Al_2O_3 = 3CaO \cdot Al_2O_3$
4.	$CaO + Al_2O_3 = CaO \cdot Al_2O_3$
5.	$12CaO + 7Al_2O_3 = 12CaO \cdot 7Al_2O_3$
6.	$2CaO + Fe_2O_3 = 2CaO \cdot Fe_2O_3$
7.	$4CaO + Al_2O_3 + Fe_2O_3 = 4CaO \cdot Al_2O_3 \cdot Fe_2O_3$
8.	$CaCO_3 + Al_2O_3 \cdot 2SiO_2 \cdot 2H_2O = CaO + Al_2O_3 \cdot 2SiO_2 + CO_2 + 2H_2O$
9.	$5CaCO_3 + Al_2O_3 \cdot 2SiO_2 \cdot 2H_2O = 2(\beta\text{-}2CaO \cdot SiO_2) + CaO \cdot Al_2O_3 + 2H_2O + 5CO_2$
10.	$5CaO + Al_2O_3 \cdot 2SiO_2 = 2(\beta\text{-}2CaO \cdot SiO_2) + CaO \cdot Al_2O_3$
11.	$5CaCO_3 + Al_2O_3 \cdot 2SiO_2 = 2(\beta\text{-}2CaO \cdot SiO_2) + CaO \cdot Al_2O_3 + 5CO_2$
12.	$2CaCO_3 + Fe_2O_3 = 2CaO \cdot Fe_2O_3 + 2CO_2$
13.	$2CaCO_3 + SiO_2 = \beta\text{-}2CaO \cdot SiO_2 + 2CO_2$
14.	$2CaCO_3 + CaO \cdot Al_2O_3 = 3CaO \cdot Al_2O_3 + 2CO_2$
15.	$2CaO + CaO \cdot Al_2O_3 = 3CaO \cdot Al_2O_3$
16.	$CaO \cdot Al_2O_3 + CaO + 2CaO \cdot Fe_2O_3 = 4CaO \cdot Al_2O_3 \cdot Fe_2O_3$
17.	$CaO \cdot Al_2O_3 + CaCO_3 + 2CaO \cdot Fe_2O_3 = 4CaO \cdot Al_2O_3 \cdot Fe_2O_3 + CO_2$
18.	$40CaCO_3 + 7(Al_2O_3 \cdot 2SiO_2) = 12CaO \cdot 7Al_2O_3 + 14(\beta\text{-}2CaO \cdot SiO_2) + 40CO_2$
19.	$7CaCO_3 + Al_2O_3 \cdot 2SiO_2 = 2(3CaO \cdot SiO_2) + CaO \cdot Al_2O_3 + 7CO_2$
20.	$\beta\text{-}2CaO \cdot SiO_2 + CaCO_3 = 3CaO \cdot SiO_2 + CO_2$
21.	$\beta\text{-}2CaO \cdot SiO_2 + CaO = 3CaO \cdot SiO_2$
22.	$7(CaO \cdot Al_2O_3) + 5CaCO_3 = 12CaO \cdot 7Al_2O_3 + 5CO_2$

Comparing the calculation results with the known generalized experimental data one can state that:

The reactions of clinker mineral formation from oxides (reactions 1 – 7) are thermodynamically possible, but bearing in mind that the initial components of these reactions in cement furnaces, at temperatures below 1000 K, are generally insignificant in amount, these reactions, as a rule, do not play an important role in the production of portland-cement.

Table 68. Results of the $\Delta G^0 = f(T)$ calculations for the reactions in Table 67

Nos. of reactions	Equations $\Delta G^0 = f(T)$	ΔG^0 value [kcal/mol] at various temps.				
		298 K	1000 K	1200 K	1400 K	1500 K
1.	$\Delta G^0 = -31293 - 3.51\,T\ln T - 0.62\cdot10^{-3}T^2 - 1.56\cdot10^5T^{-1} + 25.51\,T$	−30.228	−30.809	−31.571	−32.506	−33.032
2.	$\Delta G^0 = -29096 - 0.43\,T\ln T - 1.72\cdot10^{-3}T^2 + 2.74\cdot10^5T^{-1} + 5.16\,T$	−27.522	−28.353	−28.812	−29.408	−29.760
3.	$\Delta G^0 = -1082 + 0.16\,T\ln T + 0.86\cdot10^{-3}T^2 - 0.52\cdot10^5T^{-1} - 10.6\,T$	−4.067	−9.769	−11.245	−12.651	−13.327
4.	$\Delta G^0 = -2161 + 3.09\,T\ln T - 0.91\cdot10^{-3}T^2 - 1.03\cdot10^5T^{-1} - 24.9\,T$	−4.762	−6.726	−7.144	−7.538	−7.730
5.	$\Delta G^0 = -3770 + 30.09\,T\ln T - 15.56\cdot10^{-3}T^2 - 11.35\cdot10^5T^{-1} - 255.5\,T$	−34.015	−68.075	−77.688	−87.587	−92.697
6.	$\Delta G^0 = -3135 - 1241\,T\ln T + 4.54\cdot10^{-3}T^2 - 3.82\cdot10^5T^{-1} + 71.3\,T$	−8.835	−18.415	−21.952	−25.558	−27.364
7.	$\Delta G^0 = -985 + 8.11\,T\ln T + 4.29\cdot10^{-3}T^2 - 8.69\cdot10^5T^{-1} - 74.0\,T$	−11.804	−15.534	−15.324	−14.542	−13.944
8.	$\Delta G^0 = 118015 - 4.52\,T\ln T + 11.74\cdot10^{-3}T^2 - 5.68\cdot10^5T^{-1} - 95.75\,T$	+80.944	+2.220	−18.913	−39.276	−49.169
9.	$\Delta G^0 = 185740 + 11.21\,T\ln T + 9.02\cdot10^{-3}T^2 - 5.38\cdot10^5T^{-1} - 53.0\,T$	+187.700	+218.400	+229.790	+242.270	+248.880
10.	$\Delta G^0 = -109722 + 4.69\,T\ln T - 6.72\cdot10^{-3}T^2 + 5.58\cdot10^5T^{-1} - 27.76\,T$	−108.760	−111.240	−112.330	−113.790	−114.660
11.	$\Delta G^0 = 111970 + 18.49\,T\ln T - 1.72\cdot10^{-3}T^2 + 0.92\cdot10^5T^{-1} - 322.1\,T$	+46.941	−84.102	−119.650	−154.740	−172.120
12.	$\Delta G^0 = 80111 - 6.89\,T\ln T + 6.54\cdot10^{-3}T^2 - 5.93\cdot10^5T^{-1} - 45.54\,T$	+53.433	−7.084	−24.244	−41.134	−49.465
13.	$\Delta G^0 = 57460 + 3.83\,T\ln T + 2.76\cdot10^{-3}T^2 - 1.89\cdot10^5T^{-1} - 107.3\,T$	+31.597	−20.809	−34.894	−48.645	−55.396
14.	$\Delta G^0 = 96291 + 2.59\,T\ln T + 5.78\cdot10^{-3}T^2 - 2.1\cdot10^5T^{-1} - 101.0\,T$	+70.399	+18.751	+5.276	−7.660	−13.940
15.	$\Delta G^0 = 7740 - 2.93\,T\ln T + 3.78\cdot10^{-3}T^2 + 0.5\cdot10^5T^{-1} + 16.06\,T$	+8.055	+7.387	+7.566	+7.951	+8.226
16.	$\Delta G^0 = 8534 + 17.43\,T\ln T + 2.67\cdot10^{-3}T^2 - 4.32\cdot10^5T^{-1} - 121.39\,T$	+0.739	+9.802	+14.654	+20.293	+23.381
17.	$\Delta G^0 = 42700 + 20.22\,T\ln T + 3.67\cdot10^{-3}T^2 - 5.63\cdot10^5T^{-1} - 181.64\,T$	+21.337	+3.867	+1.599	+0.284	−0.050
18.	$\Delta G^0 = 1112909 + 453.65\,T\ln T - 2.15\cdot10^{-3}T^2 - 17.75\cdot10^5T^{-1} - 4652\,T$	−490.64	−408.880	−614.080	−804.270	−894.580
19.	$\Delta G^0 = 115400 + 20.15\,T\ln T + 1.57\cdot10^{-3}T^2 - 2.17\cdot10^5T^{-1} - 352.4\,T$	+4.002	−96.437	−133.950	−170.670	−188.750
20.	$\Delta G^0 = 37220 + 0.33\,T\ln T + 1.65\cdot10^{-3}T^2 - 0.44\cdot10^5T^{-1} - 11.8\,T$	+34.262	+29.306	+28.207	+27.207	+26.823
21.	$\Delta G^0 = 4590 - 1.93\,T\ln T - 0.65\cdot10^{-3}T^2 + 0.87\cdot10^5T^{-1} - 15.65\,T$	−3.116	−24.957	−31.475	−38.107	−41.462
22.	$\Delta G^0 = 239301 + 22.16\,T\ln T + 9.88\cdot10^{-3}T^2 - 10.61\cdot10^5T^{-1} - 387.8\,T$	+158.670	+13.419	−24.154	−60.258	−77.778

Kaolinite most often found in the cement batch can interact with calcite forming lime and metakaolinite (reaction 8) beginning with 1050 K.

The reaction of β-C_2S and CA formation from kaolinite and calcite (reaction 9) is thermodynamically impossible.

With the occurrence of metakaolinite the formation of β-C_2S and CA (reactions 10, 11), and $C_{12}A_7$ (reaction 18) in particular, becomes possible.

The interaction of calcite with β-C_2S forming C_3S in solid state (reaction 20) is impossible, whereas the interaction of β-C_2S with calcium oxide (reaction 21) is thermodynamically possible within the entire temperature range. Beginning with 1000 K the reaction of the C_3S and CA formation on account of the interaction of metakaolinite with calcite (reaction 19) is also quite probable. It should be particularly emphasized that the formation of $C_{12}A_7$ instead of CA (compare reactions 18 and 19) is thermodynamically more probable.

In addition, at temperatures above 1100 K the formation of $C_{12}A_7$ from CA and calcite becomes possible (reaction 22).

The formation of C_3A from CA and CaO, or $CaCO_3$ (reactions 14 and 15) is impossible which permits supposing that C_3A will probably be formed from melt [21, 105, 355].

The formation of C_2F from ferric oxide and calcite or calcium oxide (reactions 6, 12) is possible within the entire temperature range. The transformation of C_2F, however, into C_4AF from CA and calcite or calcium oxide (reactions 16, 17) is impossible within the entire temperature range, which proves that for these reactions a melt is necessary.

In view of the practical importance of determining the possibility of replacing the argillaceous component by other initial materials, in collaboration with Shchyotkina and Sapozhnikova we examined the reactions of C_2S and C_3S formation in combination with C_3A from mixtures of $CaCO_3$ and CaO not only with kaolinite ($Al_2O_3 \cdot 2SiO_2 \cdot 2H_2O$) and montmorillonite ($Al_2O_3 \cdot 3.7SiO_2 \cdot H_2O$), but also with pseudowollastonite (α-$CaO \cdot SiO_2$), gehlenite ($2CaO \cdot Al_2O_3 \cdot SiO_2$), anorthite ($CaO \cdot Al_2O_3 \cdot 2SiO_2$) – pseudowollastonite, gehlenite and anorthite being slag minerals – as well as with combinations of montmorillonite with gehlenite and of kaolinite with pseudowollastonite [425].

The reactions examined are presented in Table 69.

Presented in Table 70 are the results of the calculations of the temperature dependence (from 573 to 1573 K within each 100 K) of the Gibbs free energies of the reactions of β-C_2S and C_3A as well as C_3S and C_3A formation from mixtures of $CaCO_3$ and CaO with the said argillaceous and slag minerals and with their combinations as well.

As it follows from thermodynamic calculations, the reactions of clinker mineral formation in a mixture of calcium carbonate with argillaceous minerals become possible at temperatures much lower than that of the start of the $CaCO_3$ dissociation. Thus, reaction 1 becomes possible beginning with 873 K, i.e., 300 K before the $CaCO_3$ dissociation proper.

Table 69. The $\Delta G^0 = f(T)$ dependence for the reactions of formation of di- and tricalcium silicates with tricalcium aluminate from mixtures of $CaCO_3$ and CaO with kaolinite ($Al_2O_3 \cdot 2SiO_2 \cdot 2H_2O$), montmorillonite ($Al_2O_3 \cdot 3.7SiO_2 \cdot H_2O$), pseudowollastonite ($\alpha$-$CaO \cdot SiO_2$), gehlenite ($2CaO \cdot Al_2O_3 \cdot SiO_2$), anorthite ($CaO \cdot Al_2O_3 \cdot 2SiO_2$) and with their combinations

Reactions
1. $7CaCO_3 + Al_2O_3 \cdot 2SiO_2 \cdot 2H_2O = 2(\beta\text{-}2CaO \cdot SiO_2) + 3CaO \cdot Al_2O_3 + 7CO_2 + 2H_2O$
2. $7CaO + Al_2O_3 \cdot 2SiO2 \cdot 2H_2O = 2(\beta\text{-}2CaO \cdot SiO_2) + 3CaO \cdot Al_2O_3 + 2H_2O$
3. $7CaCO_3 + Al_2O_3 \cdot 2SiO_2 = 2(\beta\text{-}2CaO \cdot SiO_2) + 3CaO \cdot Al_2O_3 + 7CO_2$
4. $7CaO + Al_2O_3 \cdot 2SiO_2 = 2(\beta\text{-}2CaO \cdot SiO_2) + 3CaO \cdot Al_2O_3$
5. $9CaCO_3 + Al_3O_3 \cdot 2SiO_2 \cdot 2H_2O = 2(3CaO \cdot SiO_2) + 3CaO \cdot Al_2O_3 + 2H_2O + 9CO_2$
6. $9CaO + Al_2O_3 \cdot 2SiO_2 \cdot 2H_2O = 2(3CaO \cdot SiO_2) + 3CaO \cdot Al_2O_3 + 2H_2O$
7. $14.1CaCO_3 + Al_2O_3 \cdot 3.7SiO_2 \cdot H_2O = 3.7(3CaO \cdot SiO_2) + 3CaO \cdot Al_2O_3 + 14.1CO_2 + H_2O$
8. $14.1CaO + Al_2O_3 \cdot 3.7SiO_2 \cdot H_2O = 3.7(3CaO \cdot SiO_2) + 3CaO \cdot Al_2O_3 + H_2O$
9. $10.4CaCO_3 + Al_2O_3 \cdot 3.7SiO_2 \cdot H_2O = 3.7(\beta\text{-}2CaO \cdot SiO_2) + 3CaO \cdot Al_2O_3 + H_2O + 10.4CO_2$
10. $10.4CaO + Al_2O_3 \cdot 3.7SiO_2 \cdot H_2O = 3.7(\beta\text{-}2CaO \cdot SiO_2) + 3CaO \cdot Al_2O_3 + H_2O$
11. $CaO + \alpha\text{-}CaO \cdot SiO_2 = \beta\text{-}2CaO \cdot SiO_2$
12. $CaCO_3 + \alpha\text{-}CaO \cdot SiO_2 = \beta\text{-}2CaO \cdot SiO_2 + CO_2$
13. $2CaO + \alpha\text{-}CaO \cdot SiO_2 = 3CaO \cdot SiO_2$
14. $2CaCO_3 + \alpha\text{-}CaO \cdot SiO_2 = 3CaO \cdot SiO_2 + 2CO_2$
15. $3CaO + 2CaO \cdot Al_2O_3 \cdot SiO_2 = \beta\text{-}2CaO \cdot SiO_2 + 3CaO \cdot Al_2O_3 + 3CO_2$
16. $3CaO + 2CaO \cdot Al_2O_3 \cdot SiO_2 = \beta\text{-}2CaO \cdot SiO_2 + 3CaO \cdot Al_2O_3$
17. $4CaO + 2CaO \cdot Al_2O_3 \cdot SiO_2 = 3CaO \cdot SiO_2 + 3CaO \cdot Al_2O_3$
18. $4CaCO_3 + 2CaO \cdot Al_2O_3 \cdot SiO_2 = 3CaO \cdot SiO_2 + 3CaO \cdot Al_2O_3 + 4CO_2$
19. $6CaO + CaO \cdot Al_2O_3 \cdot 2SiO_2 = 2(\beta\text{-}2CaO \cdot SiO_2) + 3CaO \cdot Al_2O_3$
20. $6CaCO_3 + CaO \cdot Al_2O_3 \cdot 2SiO_2 = 2(\beta\text{-}2CaO \cdot SiO_2) + 3CaO \cdot Al_2O_3 + 6CO_2$
21. $8CaO + CaO \cdot Al_2O_3 \cdot 2SiO_2 = 2(3CaO \cdot SiO_2) + 3CaO \cdot Al_2O_3$
22. $8CaCO_3 + CaO \cdot Al_2O_3 \cdot 2SiO_2 = 2(3CaO \cdot SiO_2) + 3CaO \cdot Al_2O_3 + 8CO_2$
23. $13.4CaCO_3 + Al_2O_3 \cdot 3.7SiO_2 \cdot H_2O + 2CaO \cdot Al_2O_3 \cdot SiO_2 = 4.7(\beta\text{-}2CaO \cdot SiO_2) + 2(3CaO \cdot Al_2O_3) + 13.4CO_2 + H_2O$
24. $13.4CaO + Al_2O_3 \cdot 3.7SiO_2 \cdot H_2O + 2CaO \cdot Al_2O_3 \cdot SiO_2 = 4.7(\beta\text{-}2CaO \cdot SiO_2) + 2(3CaO \cdot Al_2O_3) + H_2O$
25. $18.1CaCO_3 + Al_2O_3 \cdot 3.7SiO_2 \cdot H_2O + 2CaO \cdot Al_2O_3 \cdot SiO_2 = 4.7(3CaO \cdot SiO_2) + 2(3CaO \cdot Al_2O_3) + 18.1CO_2 + H_2O$
26. $18.1CaO + Al_2O_3 \cdot 3.7SiO_2 \cdot H_2O + 2CaO \cdot Al_2O_3 \cdot SiO_2 = 4.7(3CaO \cdot SiO_2) + 2(3CaO \cdot Al_2O_3) + H_2O$
27. $8CaCO_3 + Al_2O_3 \cdot 2SiO_2 \cdot 2H_2O + \alpha\text{-}CaO \cdot SiO_2 = 3(\beta\text{-}2CaO \cdot SiO_2) + 3CaO \cdot Al_2O_3 + 2H_2O + 8CO_2$
28. $8CaO + Al_2O_3 \cdot 2SiO_2 \cdot 2H_2O + \alpha\text{-}CaO \cdot SiO_2 = 3(\beta\text{-}2CaO \cdot SiO_2) + 3CaO \cdot Al_2O_3 + 2H_2O$
29. $11CaCO_3 + Al_2O_3 \cdot 2SiO_2 \cdot 2H_2O + \alpha\text{-}CaO \cdot SiO_2 = 3(3CaO \cdot SiO_2) + 3CaO \cdot Al_2O_3 + 11CO_2 + 2H_2O$
30. $11CaO + Al_2O_3 \cdot 2SiO_2 \cdot 2H_2O + \alpha\text{-}CaO \cdot SiO_2 = 3(3CaO \cdot SiO_2) + 3CaO \cdot Al_2O_3 + 2H_2O$

Table 70. The ΔH^0_{298}, ΔG^0_{298} and ΔG^0 values of the reactions given in Table 69 at temps. of 573 to 1573 K

No. of reactions	ΔH^0_{298} [kcal/mol]	ΔG^0_{298} [kcal/mol]	ΔG^0 [kcal/mol] at temps. [K]										
			573	673	773	873	973	1073	1173	1273	1373	1473	1573
1	278.8	173.4	70	38.59	10.83	−22.79	−54.2	−78.79	−119.59	−148.69	−179.9	−209.69	−241.12
2	−21.1	−45.76	−69.19	−76.3	−83.69	−91.38	−99.02	−107.09	−113.39	−121.29	−128.09	−135.6	−141.98
3	200.8	−1369.4	43.11	17.15	−8.86	−34.01	−60.0	−86.13	−112.3	−138.41	−164.1	−190.1	−216.3
4	−97.9	−100.29	−102.39	−103.17	−103.9	−104.8	−105.6	−106.5	−107.43	−108.39	−109.4	−110.47	−111.59
5	368.3	−242.7	129.89	84.29	47.6	10.0	−29.8	−67.09	−108.6	−143.71	−188.59	−231.69	−252.78
6	−15.7	−110.09	−60.0	−67.24	−73.73	−79.81	−87.32	−93.31	−100.26	−107.26	−115.75	−121.19	−128.4
7	610.5	434.3	217.94	162.11	107.05	52.17	−0.92	−53.87	−106.18	−157.84	−208.88	−259.31	−309.15
8	8.83	−5.96	−23.87	−28.52	−33.12	−37.67	−42.17	−46.61	−50.99	−55.31	−59.57	−63.76	−67.88
9	442.7	305.2	136.59	93.26	50.56	8.46	−33.07	−74.07	−114.54	−154.52	−194.02	−233.06	−271.65
10	−1.03	−19.5	−41.77	−47.35	−52.83	−58.21	−63.51	−68.72	−73.86	−78.92	−83.92	−88.85	−93.72
11	−11.49	−11.5	−11.55	−11.56	−11.59	−11.62	−11.67	−11.75	−11.87	−11.97	−12.13	−12.31	12.52
12	31.2	19.7	9.47	5.84	2.26	−1.29	−4.81	−8.29	−11.78	−15.23	−18.67	−22.09	−25.53
13	−8.82	−7.8	−6.49	−6.41	−6.19	−6.85	5.12	−5.08	−5.02	−4.96	−4.84	−4.76	−4.65
14	76.5	54.7	33.94	26.29	19.52	12.41	5.36	−1.63	−8.56	−15.44	−22.27	−29.06	−35.79
15	−580.4	96.7	66.01	55.01	45.01	34.16	23.85	13.51	3.06	−7.72	−17.61	−28.18	−39.02
16	5.9	5.52	−0.01	−0.079	−0.16	−0.25	−0.31	−0.42	−0.54	−0.66	−0.77	−0.89	−0.91
17	8.46	9.2	−26.02	−26.11	−26.23	−26.31	−26.31	−26.39	−26.42	−26.48	−26.51	−26.59	−26.72
18	473.8	134.1	120.13	86.46	67.48	45.11	23.11	−1.51	−26.51	−50.11	−75.19	−100.08	−136.57
19	−41.02	−34.3	−33.09	−33.96	−34.42	−35.02	−35.54	−35.98	−36.96	−36.39	−36.82	−37.06	−37.08
20	215.0	−323.09	74.59	55.29	36.28	17.79	−1.21	−20.03	−41.13	−61.19	−69.68	−101.96	−119.59
21	−35.7	−26.9	−25.74	−25.41	−25.13	−24.89	−24.69	−24.52	−24.38	−24.28	−24.19	−24.14	−24.11
22	305.7	222.8	140.02	111.07	82.32	53.92	25.88	−1.84	−29.23	−56.31	−83.09	−109.58	−88.04
23	576.5	404.4	235.0	185.0	135.0	85.0	35.0	−15.59	−65.30	−116.0	−167.0	−217.0	−268.0
24	835.2	−14.09	−31.56	−37.38	−42.44	−47.12	−53.61	−59.11	−64.81	−69.80	−74.95	−80.31	−86.01
25	789.6	586.5	402.13	334.13	226.04	157.16	88.27	20.03	−45.13	111.94	−179.96	−241.25	−308.13
26	17.3	3.25	−8.52	−14.11	−20.13	−26.90	−33.12	−38.84	−44.03	−51.13	−55.06	−51.13	−66.22
27	308.78	192.51	84.06	49.67	12.35	−24.92	−62.09	−94.13	−129.41	−165.75	−200.2	−235.91	−271.02
28	−32.57	−57.27	−81.35	−90.84	−95.14	−102.35	−110.01	−117.54	−125.35	−131.94	−140.81	−147.36	−152.75
29	−3714.7	297.3	183.34	114.12	65.29	12.52	−32.15	−82.38	−118.9	−164.42	−202	−252.63	−306.95
30	−24.6	−46.22	−65.32	−72.3	−80.15	−87.32	−87.32	−94.15	−101.34	−115.94	−121.92	−126.14	−135.74

For the study of the rapid firing processes, interesting are comparative thermodynamic calculations of the reactions of the clinker mineral formation in a mixture of calcium carbonate and oxide with kaolinite and metakaolinite.

As one can see from Table 70, beginning with 1073 K reaction 2 becomes more preferable than reaction 4, probably, due to the increased activity of kaolinite resulting from structural changes in it while changing into metakaolinite.

Comparing temperature dependence of the Gibbs free energies of reactions 3 and 4, it is worth noting that up to a temperature of 1073 K the most probable is reaction 4; at the temperature of $CaCO_3$ dissociation, when calcium oxide being formed in statu nascendi is most reactive, the probabilities of both reactions equalize and at higher temperatures reaction 3 becomes the most probable.

A diagrammatic representation of the calculation results is given in Figs. 72–74.

Comparing the probability of reactions in mixtures with montmorillonite and kaolinite, it should be mentioned that the mixtures with kaolinite must be more reactive. Thus, as one can see from Fig. 72, a reaction in the mixture of calcium carbonate with montmorillonite intended to produce β-C_2S and C_3A is possible beginning with 900 K (reaction 9); the formation of the same products from a mixture of calcium carbonate with kaolinite is realized at about 800 K (reaction 1). Reactions meant to form the same products from slag minerals in mixtures with $CaCO_3$ and CaO feature a lesser probability as compared with similar reactions in mixtures with argillaceous minerals. Thus, the formation of dicalcium silicate and tricalcium aluminate from mixtures of calcium carbonate with anorthite and gehlenite becomes possible only at about 1000 and 1200 K respectively.

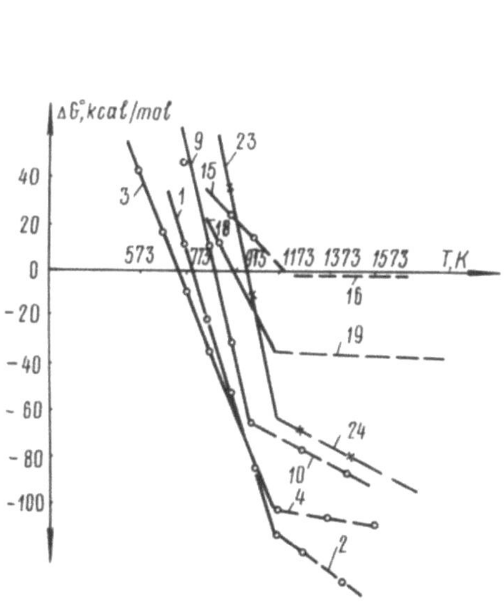

Fig. 72. The temperature dependence of the Gibbs free energy of the reactions meant for producing β-C_2S and C_3A from various initial substances. Line numbering corresponds to the reactions from Table 70; ——— from a mixture with $CaCO_3$; – – – from a mixture with CaO

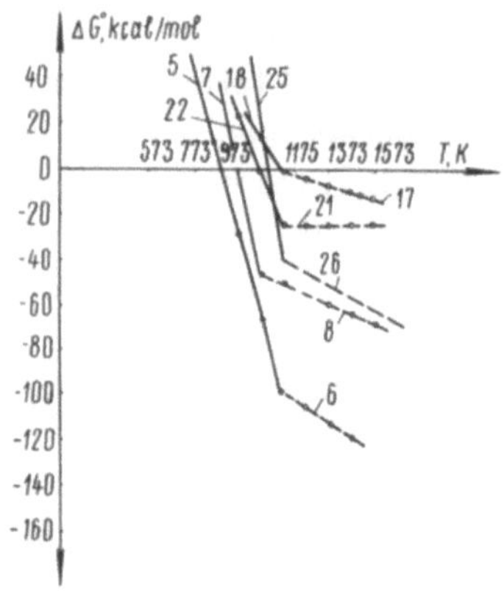

Fig. 73. The dependence $\Delta G^0 = f(T)$ for the reactions of C_3S and C_3A production

Reactions calculated which are to form high-basic calcium silicate and aluminate ($C_3S + C_3A$) is possible beginning with temperatures about 100 K higher, than while forming β-C_2S and C_3A (see Fig. 73).

As one can see from Fig. 72 and 73, in mixtures meant to produce C_2S and C_3A the probability of reactions decreases in succession: kaolinite – montmorillonite – anorthite – gehlenite.

While we first carried out thermodynamic calculations of the clinker formation reactions with montmorillonite, anorthite and gehlenite, similar calculations for the mixtures with pseudowollastonite were carried out independently by other authors [541]. It should be noted that the same authors conducted an experimental work confirming a close agreement of the test and calculation data concerning the processes of producing monomineral binders on the basis of pseudowollastonite [470].

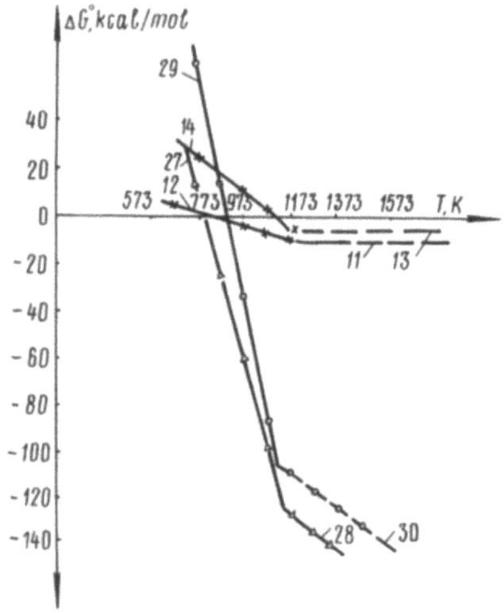

Fig. 74. The temperature dependence of the Gibbs free energy of the clinker mineral formation from mixtures of $CaCO_3$ and CaO with pseudowollastonite only (lines *11–14*) and in combination with kaolinite (lines *27–30*)

As follows from Fig. 74, the formation of dicalcium silicate (the calculated $\Delta G^0 = f(T)$ for β- and γ-modifications of C_2S are very close) from a mixture of calcium carbonate with pseudowollastonite becomes possible beginning with 800 K (reaction 11) and that of tricalcium, with 1100 K (reaction 13).

While adding kaolinite to the reaction mixture (see reactions from 27 to 30) the temperature of the beginning of the β-C_2S synthesis does not essentially change ($\sim$800 K), with the elevation of temperature, however, the reaction substantially intensifies (a steep shape of the $\Delta G^0 = f(T)$ function diagram).

The addition of kaolinite sharply effects the synthesis of C_3S: the reaction of its formation becomes possible 200 K earlier than in mixtures with pseudowollastonite only. The necessity of using combined aluminosilicate constituents is confirmed by an analysis of the results of thermodynamic calculations of the reactions of calcium silicate and aluminate formation in mixtures with combinations of montmorillonite and gehlenite (reactions from 23 to 26). Thus, while the formation of C_3A and C_2S from gehlenite is actually improbable within the entire temperature range, upon adding equal amounts of montmorillonite to the mixture the reaction becomes possible at about 1000 K and activates sharply with the growth of temperature (diagram 23, 24, Fig. 72).

Montmorillonite similarly effects the acceleration of the reactions of the C_3S and C_3A synthesis (curves 25, 26, Fig. 73).

Thus, the investigations of temperature dependences of the Gibbs free energies of the reactions of calcium silicate and aluminate synthesis in mixtures with different initial components and their combinations, revealed the necessity of combining slag and argillaceous minerals in the composition of raw mix, whose combination increases the probability of the reactions of the clinker mineral synthesis.

6 Instances of Using Thermodynamic Calculations for Solving Other Problems of the Study of Materials

In addition to their being advantageous in solving basic problems of the thermodynamics of silicate formation, thermodynamic calculations are of great use in elucidating some actual scientific, technical and practical problems in other fields of the study of materials, based on the similar processes. Such problems are the thermodynamic aspects of choosing solid phases for the dispersion-strengthening of composite materials based on metals, the thermodynamic investigation of the interaction of ceramics with metals and the thermodynamic principles of choosing oxides of enamel binding.

6.1 Composite Materials

The weakening of dispersion-hardening alloys at high-working temperatures is conditioned mainly by the dissolution of the disperse phase. The insolubility of the disperse phase in the metallic matrix, its high chemical stability and hardness are sine qua non for strength increase.

The production of high-temperature dispersion-hardening metallic materials with strengthening phases stable up to the temperature of matrix melting is practicable either by the methods of powder metallurgy and of solid-phase welding or while forming such phases in the crystallization process.

For producing composite materials with high mechanical properties it is necessary that during agglomeration or welding in the solid phase between the atoms of the matrix and of the strengthening phase an interaction should occur resulting in the establishment of chemical bonds strong enough. However, this interaction must not result in the dissolution of particles of the strengthening phase.

As no universal theory of dispersion-hardening is applicable, in particular, to the materials produced in solid phase by the methods of powder metallurgy and welding, the determination of the character of the above chemical interaction and of the conditions of its occurrence [255, 663, 691] is of great interest.

Thermodynamics allows the determination of the energy possibility and the direction of a reaction, i.e., of a chemical interaction between the metallic plastic matrix and disperse inclusions, as well as the preference of reactions and the stability of compounds formed. Therefore we derived certain regularities for producing composite materials on the basis of a number of metals and disperse additives – oxides, carbides, nitrides, silicates – by methods of powder metallurgy and solid phase welding.

Let us first examine the oxide-metal interaction [318].

The character of the chemical interaction and the possibility of a reaction can be found from a general equation [20, 697]:

$$\mathrm{Me'} + \mathrm{Me''O} = \mathrm{Me'O} + \mathrm{Me''}\,, \tag{167}$$

where Me′ and Me″ are different metals (and corresponding oxides).

Thermodynamically this may be expressed as:

$$\Delta G^0 = \Delta G^0_{\mathrm{Me'O}} + \Delta G^0_{\mathrm{Me''}} - \Delta G^0_{\mathrm{Me''O}} - \Delta G^0_{\mathrm{Me'}}\,, \tag{168}$$

where ΔG^0, alteration of the Gibbs free energy of the reaction; $\Delta G^0_{\mathrm{Me'}}$, $\Delta G^0_{\mathrm{Me'O}}$, $\Delta G^0_{\mathrm{Me''}}$, $\Delta G^0_{\mathrm{Me''O}}$, the Gibbs free energy of the reaction components.

While conducting calculations the following assumptions were made:

1. the said reactions were considered as proceeding in condensed phases;
2. there was assumed to be no mutual solubility of components;
3. deviations from stoichiometry were disregarded.

A calculation may be carried out using the thermodynamic properties of individual substances at high temperatures. For this purpose the knowledge of $H_T - H_{298} = f(T)$ and $S_T - S_{298} = f'(T)$ of the components of the reaction as well as of the standard values of ΔH^0_{298} and ΔS^0_{298} is required. Unfortunately, only a limited number of substances offer data of this kind for comparison.

For facilitating and, on the whole, intensifying the efficiency of such calculations the diagrams plotted by Tripp and King [642] were used showing the function of the Gibbs free energy of the oxide formation from elements within a temperature range of 273 to 2673 K.

We calculated the reactions conducted for 1273 and 2073 K.

As shown by Kulikov [317], in order to perform calculations accurately it is necessary to know the dissociation mechanism of the interacting components and to bear it in mind while determining the intermediate interaction reactions. The examination showed that at 1273 K the results of an approximate and a more precise calculation fairly well agree.

The limitation of the quantity of the possible couples forces us to consider separately the constituents of composite materials (for the matrix and for the strengthening phase). As a matrix metals (Al, Ti, Zr, Hf, V, Nb, Ta, Cr, Mo, W, Ni, Co, Fe) were chosen having a wide application range, of which nickel, cobalt, iron and aluminium are more often used than others in making dispersion-strengthened alloys.

Bearing in mind the susceptibility of metals to oxidation, as far as the oxidation stability is concerned, the most advantageous is the maximum ΔG value (the minimum in absolute value). Therefore in Table 71 ΔG of metal oxides, the metals chosen being those used as a matrix are compared. We shall refer to them as to the oxides of the 1st group.

An examination of the data of Table 71 permits formulating a series of metals used as a matrix according to the increase of their susceptibility to oxidation: Ni, Co, Fe, W, Mo, Cr, Nb, Ta, V, Ti, Al, Zn, Hf.

While choosing a strengthening phase in the form of oxides, from the viewpoint of the thermodynamic stability, the most advantageous is the minimum ΔG value (the maximum in absolute value). In Table 72 the most stable oxides that we call the 2nd group are summarized.

The sequence of the thermodynamic stability of oxides used as the strengthening phase can be expressed in the form of a series:

ThO_2, CaO, BeO, MgO, HfO_2, ZrO_2, VO_2, Al_2O_3, TiO_2, SiO_2, Cr_2O_3.

The value of energy capacity, as was shown in [387], is closely interconnected with the strength of ceramics from pure oxides. This fact can also be expressed as a sequence of oxides with decreasing strength:

BeO, MgO, UO_2, HfO_2, ThO_2, Al_2O_3, Cr_2O_3, ZrO_2, CaO, SiO_2.

Table 71. The 1st group metal oxides chosen as a matrix. The ΔG^0 values of the 1st group oxides [kcal/mol]

Formulae of oxides	Al_2O_3	TiO_2	ZrO_2	HfO_2	V_2O_3	NbO	Ta_2O_5	Cr_2O_3	MoO_2	WO_3	FeO	CoO	NiO
Number according to the increase of the metal stability towards oxidation	3	4	2	1	5	7	6	8	9	10	11	12	13
ΔG^0 values at 1273 K	−200	−190	−205	−208	−145	−138	−140	−125	−85	−85	−85	−70	−60
ΔG^0 values at 2073 K	−160	−150	−170	−171	−110	−100	−100	−90	−60	−70	−70	−40	−25

Table 72. The 2nd group oxides used as a disperse phase. The ΔG^0 values of the 2nd group oxides [kcal/mol]

Formulae of oxides	BeO	MgO	CaO	Al_2O_3	SiO_2	TiO_2	ZrO_2	HfO_2	Cr_2O_3	ThO_2	UO_2
Number according to the decrease of the oxide thermodynamic stability	3	4	2	8	10	9	6	5	11	1	7
ΔG^0 values at 1273 K	−230	−220	−240	−200	−153	−190	−205	−208	−125	−240	−205
ΔG^0 values at 2073 K	−200	−150	−190	−160	−115	−150	−170	−171	−900	−210	−150
The value of the energy capacity	2732	2085	1176	1527	535	–	1306	1632	1523	1627	2064

Starting from the above series one can conclude, e.g., that it is of no use to employ, as a disperse strengthening phase, such oxides as (crystalline) SiO_2 and CaO.

While evaluating the general rules of producing composite materials based on metals and on thermodynamically stable oxides, two cases may be considered theoretically.

Case A. In order to avoid the interaction of the matrix with the strengthening oxide it is necessary to take couples consisting of a combination of metals, the oxides of which have the maximum ΔG^0 value – from the 1st group – and of the disperse oxide phase with the minimum ΔG^0 value – from the 2nd group –.

Metallic matrix: Ni, Co, Fe, W, Mo, Cr, Nb, Ta, V, Ti, Al, Zr, Hf.

Disperse oxide phase: ThO_2, BeO, MgO, HfO_2, ZrO_2, UO_2, Al_2O_3, TiO_2, Cr_2O_3.

The latter is confirmed by experimental data. Thus, a composite material from nickel matrix with disperse inclusions of thorium oxide-TD nickel [548] is 3 to 4 times as strong as pure nickel within a range of 1143 K to 1588 K and stronger than standard high-temperature alloys at a temperature above 1368 K.

Case B. Bearing in mind the interaction of the matrix with the disperse oxide phase it is necessary to take couples consisting of a combination of metals, the oxides of which have the minimum ΔG^0 in absolute value – the 1st group oxides –, and of a disperse oxide phase with the maximum ΔG^0 value – the 2nd group oxides –.

Metallic matrix: Hf, Zr, Al, Ti, V, Ta, Nb, Cr, Mo, W, Fe, Co, Ni.

Disperse oxide phase: Cr_2O_3, SiO_2, TiO_2, Al_2O_3, UO_2, ZrO_2, HfO_2, MgO, BeO, CaO, ThO_2.

Thus, by choosing required components while producing composite materials, the properties of the latter can be controlled.

The said regularities which allow estimating whether or not there is any interaction of the metallic matrix and the disperse oxide phase hold only at standard conditions. The examination results are summarized in Table 73.

The data of Table 73 show the increase of the interaction activity and the decrease of the temperature of the beginning of the interaction to take place from left to right and from top to bottom. Thus, in the left upper corner the most stable – non-interacting – couples will concentrate (the metallic matrix oxide), and in the right lower corner, the least stable – interacting –.

At the same time, one should not disregard that the metals used as a matrix (Ni, Co, Fe, W, Mo, Cr, Nb, Ta, V, Ti, Al, Zr, Hf) can to a certain extent oxidize in the oxygen atmosphere, which makes us consider the interaction of the matrix metal and the disperse oxide phase in some cases as a reaction between oxides that results in the formation of binary chemical compounds.

Table 73. ΔG^0 of the interaction of some oxides with metals in kcal/mol (at temperatures of 1000 and 1800 °C)

Oxides \ Metals	Ni	Co	Fe	W	Mo	Cr	Nb	Ta	V	Ti	Al	Zr
ThO_2	180.185	170.170	155.140	155.155	155.150	115.120	102.110	100.110	95.100	50.60	40.50	35.40
CaO	180.165	170.150	155.120[c]	155.135[c]	155.130[c]	115.110[c]	102.90	100.90	95.80	50.40[c]	40.30[c]	38.20[c]
BeO	170.175	160.160	145.130[c]	145.145[c]	145.140[c]	105.110[c]	92.100	90.100	85.90	40.50[c]	30.40[c]	25.30[c]
MgO	160.120[c]	150.110[c]	135.80[c]	135.95[c]	135.90[c]	95.60[c]	82.58[c]	80.50[c]	75.40[c]	30.0[c]	20. − 10[a]	15. − 20[a]
ZrO_2	145.145	135.130	120.100	120.115[c]	120.110	80.80	67.70[c]	65.70	60.60	15.20	5.10	0.0
UO_2	145.125	135.110	120.80	120.95	120.90	80.60	67.50	65.50	60.40	15.0[a]	5. − 10[a]	0. − 20[a]
Al_2O_3	140.135[c]	130.120[c]	115.80[c]	115.105[b]	130.100[b]	75.70	62.60[b]	60.60[b]	55.50	10.10	0. − 0	− 5. − 10[a]
TiO_2	130.125[c]	120.110[c]	105.80[c]	105.95	105.00	65.60	52.50	50.50	45.40	0.0	− 10. − 10[a]	− 15. − 20[a]
SiO_2	93.90[c]	83.75[c]	68.45[c]	68.60[b]	68.55[b]	28.25	15.15[b]	13.15[b]	8.5	− 37. − 55[a]	− 47. − 45[a]	− 52. − 55[a]
Cr_2O_3	65.65[c]	55.50[c]	45.20[c]	40.35	40.30	0.0	− 13. − 10[a]	− 15. − 10[a]	− 30. − 10[a]	− 40. − 60[a]	− 90. − 70[a]	− 100. − 80[a]

Vertical axis (arrow upward): Increase of the interaction temperature / Decrease of the activity of the interaction

Horizontal axis (arrow rightward): Decrease of interaction temperature / Increase of the activity of the interaction

[a] An interaction of the type Me′ + Me″O = Me′O + Me is possible
[b] The same in a high vacuum
[c] A reaction between oxides is possible, binary compounds being formed

Table 74

Oxides	NiO	CoO	FeO
Al_2O_3	$NiAl_2O_4$	$CoAl_2O_4$	$FeAl_2O_4$
TiO_2	$NiTiO_3$	Co_2TiO_4, $CoTiO_3$	Fe_2TiO_4, $FeTiO_3$
SiO_2	Ni_2SiO_4	Co_2SiO_4	$FeSiO_3$, Fe_2SiO_4
Cr_2O_3	$NiCr_2O_3$	$CoCr_2O_4$	$FeCr_2O_4$

This important point must be borne in mind while choosing the metallic matrix and the disperse oxide phase. Thus, e.g., Ni, Co and Fe, oxidizing to NiO, CoO, FeO can interact with Cr_2O_3, SiO_2, TiO_2 and Al_2O_3 forming corresponding compounds given in Table 74 [20, 112, 619, 624, 716].

An examination of ΔG^0 of the formation of these compounds shows that at temperatures above 1273 K they are thermodynamically stable (Table 75).

Similar considerations may be set forward in case of such couples as:

MgO – Co, Fe, W, Mo, Cr, V and Ti;
BeO – Fe, W, Mo, Cr, Ti, Al and Zr;
CaO – Fe, W, Mo, Cr, Ti, Al, Zr, etc.

The high vacuum lowers the temperature of the beginning of the interaction of the metallic matrix W, Mo, Nb, Ta with the disperse oxide phase Al_2O_3 and SiO_2 (crystalline) [348, 714].

Thus, thermodynamic calculations of the interaction of various metallic matrices and disperse oxide phases carried out for 1273 K and 2073 K permitted stating the following:

Table 75. ΔG^0 of the interaction between oxides [kcal/mol]

Oxides	NiO	Literary source	CoO	Literary source	FeO	Literary source
Al_2O_3	−6.1 1)	[332]	−7.7 2)	[332]	−5.6	[332]
TiO_2	−2.7	[619]	−4.4 3)	[619]	−5.5 3)	[619]
	−2.4		−3.7		−4.2	
SiO_2	1.6 3)	[619]	−3.2 4)	[11]	−5.1 3)	[331]
	−1.0		−2.8		−4.6	
Cr_2O_3	−10.1 5)	[327]	−11.5 5)	[327]	−12.6	[455]
	−9.4		−10.5		−11.2	

1) $T = 1273$ K
2) $T = 1273$ K
3) $T = 1273$ and 1373 K
4) $T = 1273$ and 1573 K
5) $T = 1273$ and 1623 K
6) $T = 1273$ and 1773 K

1. in the neutral atmosphere

a) In case there is no interaction of the metallic matrix with the disperse strengthening oxide phase, it is necessary to take couples consisting of metals oxides which have the maximum ΔG^0 value (Ni, Co, Fe, W, Mo, Cr, Nb, Ta, V, Ti, Al, Zr, Hf) and of the disperse oxide phase with the minimum ΔG^0 value (ThO_2, BeO, HfO_2, ZrO_2, UO_2, Al_2O_3, TiO_2, SiO_2 and Cr_2O_3);

b) In case of the interaction of the metallic matrix with the disperse oxide phase it is necessary to take couples consisting of a combination of metals oxides which have the minimum ΔG^0 value (Hf, Zr, Al, Ti, V, Ta, Nb, Cr, Mo, W, Fe, Co, Ni) and of a disperse oxide phase with the maximum ΔG^0 value (Cr_2O_3, SiO_2, TiO_2, Al_2O_3, UO_2, ZrO_2, HfO_2, MgO, BeO, ThO_2).

2. in the oxidizing atmosphere

a) In case there is no interaction of the metallic matrix with the disperse oxide phase, from the above series (1a): Ni, Co, Fe with Al_2O_3, TiO_2, Cr_2O_3; Co, Fe, W, Mo, Cr, V, Ta with MgO; Fe, W, Mo, Cr, Ti, Al, Zr with BeO; Nb, W with ZrO_2 will be excluded.

b) In case of an interaction of the metallic matrix with the disperse oxide phase the series stated for the neutral atmosphere (1b) remains unchanged.

In case the above calculations indicate the impossibility of the redox reactions between the couples under consideration, the interaction may take place owing to the possibility of the solid solution formation [312, 444].

Such an interaction may be presented as a reaction:

$$Me'O = [Me']_{Me''} + [O]_{Me''} \tag{169}$$

according to which the elements of an oxide dissolve in a metal forming on its base a solid solution of the metal of oxide $[Me']_{Me''}$ and a solid solution of oxygen $[O]_{Me''}$. One of the main causes of such an interaction is a substantial alteration of partial entropy values while forming solid solutions. An examination of the calculations carried out by Mozzhukhin [444] shows the sequence of the interaction activity to be in accordance with that of Table 73. The interaction range, however, broadens essentially.

An important problem is in the study of the interaction between silicates of different composition and metals in condensed phases [373].

The interaction of different metals with meta- and orthosilicates may be expressed as:

$$Me' + Me''SiO_3 = Me'SiO_3 + Me'', \tag{170}$$

where Me′, metal interacting with a silicate; Me″, metal entering a silicate.

Thermodynamically that will be expressed as:

$$\Delta G^0 = \Delta G^0_{Me'SiO_3} - \Delta G^0_{Me''SiO_3} + \Delta G^0_{Me''} - \Delta G^0_{Me'}, \tag{171}$$

where ΔG^0, alteration of the Gibbs free energy of the reaction; $\Delta G^0_{Me'SiO_3}$, $\Delta G^0_{Me''SiO_3}$, $\Delta G^0_{Me''}$, $\Delta G^0_{Me'}$, the Gibbs free energy of the components of the reaction.

For calculating ΔG^0 it is necessary to know the values of the Gibbs free energy of the formation of individual silicates from simple substances at high temperatures. For most silicates the temperature dependence of ΔG^0 has still not been calculated. Nonetheless, ΔG^0 can be determined by the data of the formation of silicates from the constituent metal oxides entering the equation of reaction (170). In this case ΔG^0 can be determined by Eq. (172):

$$\Delta G^0 = \Delta G^0(\mathrm{ox})_{Me'SiO_3} - \Delta G^0(\mathrm{ox})_{Me''SiO_3} + \Delta G^0_{Me'O} - \Delta G^0_{Me''O} + \Delta G^0_{Me''} - \Delta G^0_{Me'}\,, \quad (172)$$

$$\Delta G^0(\mathrm{ox})_{Me'SiO_3} - \Delta G^0(\mathrm{ox})_{Me''SiO_3} = \Delta G^0_1\,, \quad (173)$$

$$\Delta G^0_{Me'O} - \Delta G^0_{Me''O} + \Delta G^0_{Me''} - \Delta G^0_{Me'} = \Delta G^0_2\,, \quad (174)$$

$$\Delta G^0 = \Delta G^0_1 + \Delta G^0_2\,. \quad (175)$$

Using a separate calculation of ΔG^0_1 and ΔG^0_2 the alteration of the Gibbs free energy of the reaction of the interaction of metals with silicates can be found.

The Calculation of ΔG^0_1

First the regularities found earlier were taken into account (see Chapter II.1–4).

1. Of the silicates of metals of the first group of the periodic system thermodynamically the least stable are lithium silicates, the most stable, rubidium and caesium silicates.

2. Of the silicates of metals of the second group of the periodic system thermodynamically the least stable are beryllium and magnesium silicates, the most stable, barium silicates.

Therefore later we will be dealing only with the least stable silicates of the metals of the first and the second groups of the periodic system as well as with all other silicates (the ΔG^0 values of which are known) which do not belong to these groups. The calculation was carried out, as a rule, for two temperatures: 1000 and 1600 K. Where it was necessary to use data for other temperatures, it was specified in the text. The ΔG^0 values necessary for the calculation were, as a rule, taken from the preceding paragraphs. Only those of Ni_2SiO_4 were borrowed from [619] and those of Co_2SiO_4, from [11]. The initial data, namely ΔG^0 of the formation of meta- and orthosilicates from oxides, are summarized in Table 76, and the calculation results, in Table 77.

An examination of the data obtained shows the immediate interaction between silicates and oxides of the metals chosen to feature a series: NiO, CoO, BeO, ZrO_2, Al_2O_3, PbO, FeO, MnO etc., dependent on the ΔG value of the formation of the corresponding silicates from oxides.

Table 76. ΔG^0 of formation of different silicates from oxides [kcal/mol] at temperatures of 1000 and 1600 K

Silicates	Temp. [K]		Silicates	Temp. [K]	
	1000	1600		1000	1600
Li_2SiO_4	−49.40	−45.73	Li_2SiO_3	−30.13	−31.09
Mg_2SiO_4	−14.50	−14.39	$MgSiO_3$	−7.53	−6.57
Mn_2SiO_4	−9.90	−10.15[a]	$MnSiO_3$	−6.06	−6.50[a]
Fe_2SiO_4	−8.50	−6.58[a]	$PbSiO_3$	−5.14	−4.17
Pb_2SiO_4	−7.47	−6.38	$FeSiO_3$	−2.40	−1.10[a]
			$BeSiO_3$	−1.06	+5.56
Al_2SiO_5	−3.90	−4.20			
$ZrSiO_4$	−3.30	−1.40			
Be_2SiO_4	−3.26	9.13			
Co_2SiO_4	−3.20	−2.80			
Ni_2SiO_4	−1.60[b]	−1.00[c]			

[a] at 1473 K [b] at 1073 K [c] at 1373 K

The Calculation of ΔG_2^0

Equation (174) characterizes the possibility of reactions according to general Eq. (167).

The calculation may be carried out in the same way as described above. As with ΔG_1^0, the calculations were performed for temperatures of 1000 and 1600 K.

An examination of the data obtained shows an immediate interaction to be possible between the following metals and oxides: Ni with PbO; Co with PbO and NiO; Fe with PbO, NiO and CoO; Mn with PbO, NiO, CoO and FeO; Al with PbO, NiO, CoO, Li_2O and MnO. Be will interact with all the given oxides and Zr, with all but BeO; Pb will interact with none of the given oxides.

Table 77. The character of the interaction of oxides of some metals with silicates

Oxides	Silicates with which an oxide interacts at a temperature [K]	
	1000	1600
NiO		Be
CaO	Ni	Be, Ni, Zr
BeO	Ni, Co	–
ZrO_2	Ni, Co, Be	Be, Ni
Al_2O_3	Ni, Co, Be, Zr	Be, Ni, Zr, Co
PbO	Ni, Co, Be, Zr, Al	Be, Ni, Zr, Co, Al
FeO	Ni, Co, Be, Zr, Al, Pb	Be, Ni, Zr, Co, Al, Pb
MnO	Ni, Co, Be, Zr, Al, Pb, Fe	Be, Ni, Zr, Co, Al, Pb, Fe

A comparison of the ΔG_1^0 and ΔG_2^0 values obtained for the interacting metal-silicate couples shows that in the vast majority of cases the contribution of ΔG_2^0 significantly surpasses in absolute value that of ΔG_1^0. Only in the case of the interaction of the following couples are these values commensurable: Al with Zr and Mg silicates; Ni with Co and Li silicates; Co with Ni and Li silicates and Fe with Li silicates.

Thus, as a first approximation, the interaction between a metal and a silicate will be determined by the ΔG_2^0 value. This fact permits evaluating the possibility of the interaction between silicates and metals which do not form stable silicates, such as Cu, W, Mo, Cr, Nb, Ta and V. The calculation in this case is conducted in a way similar to that of ΔG_2^0.

An examination of all the data obtained reveals that the immediate interaction is possible between the following metals and silicates:

Ni with Pb silicates;
Co with Pb and Ni silicates;
Fe, Mo, W with Pb, Ni, Co silicates;
Cr with Pb, Ni, Co, Fe silicates;
Nb, Ta, V with Pb, Ni, Co, Fe, Mn silicates;
Al, Ti with Pb, Ni, Co, Fe, Mn, Li silicates;
Zr with Pb, Ni, Co, Fe, Mn, Li, Al silicates;
Be with all the silicates in question;

Cu and Pb will not interact with any silicates in question.

It is particularly interesting to consider the possibility of the interaction of the above metals with glasses and materials based on them, e.g., with glass ceramics. The reasons for this attempt were as follows. All industrial glasses are produced starting from a homogeneous melt at temperatures above 1800 K. At these conditions, according to the previous thermodynamic calculations, as far as the glass composition is concerned, binary and ternary compounds being the most stable are formed. Thus, as a first approximation, a silicate melt may be viewed as a "perfect solution" of the most stable, binary and ternary silicates at the melting temperature. The amount of silica in glass being greater than can be bound by the most stable silicates, silica will enter the composition of the melt in the form of the free SiO_2. Oxides, the glass composition of which does not form silicates on entering, such as TiO_2, B_2O_3, SnO_2, CeO_2 etc., will also enter the melt composition in the form of free oxides.

From the above discussion it is clear that an immediate interaction of the metals touched upon in the work may take place both with the silicates making up glass or glass ceramics and with the free SiO_2 and other oxides, if present in the glass.

In Table 78 the qualitative composition of some silicate materials is presented and the evaluation of the immediate interaction of the metals in question is given with them.

Table 78. Qualitative composition of some silicate materials and the possibility of their interaction with metals

Chemical (qualitative) composition of silicate materials	Metals and the possibility of the immediate interaction (marked with +)														
	Pb	Cu	Ni	Co	Fe	W	Mo	Cr	Nb	Ta	V	Ti	Al	Zr	Be
1. SiO_2, Al_2O_3, B_2O_3, BaO	–	–	–	–	–	–	–	–	–	–	–	–	–	+	+
2. SiO_2, Al_2O_3, B_2O_3, PbO	–	–	+	+	+	+	+	+	+	+	+	+	+	+	+
3. SiO_2, Al_2O_3, MgO, TiO_2, CeO_2	–	–	–	–	–	–	–	–	–	–	–	–	–	+	+
4. SiO_2, Al_2O_3, B_2O_3, CaO, Li_2O, PbO, ZrO_2	–	–	+	+	+	+	+	+	+	+	+	+	+	+	+
5. SiO_2, Al_2O_3, CaO, MgO, Li_2O, Na_2O, TiO_2	–	–	–	–	–	–	–	–	–	–	–	+	+	+	+
6. SiO_2, Al_2O_3, Li_2O, K_2O, CeO_2	–	–	–	–	–	–	–	–	–	–	–	+	+	+	+
7. SiO_2, Al_2O_3, CaO, MgO, Na_2O, K_2O, MnO, FeO, TiO_2	–	–	–	–	–	–	–	+	+	+	+	+	+	+	+

An examination of Table 78 enables us to draw the following conclusions. With silicate materials containing PbO practically all metals except Cu and Pb can interact. This fact demonstrates that the interaction of the metals and silicate materials considered in the table is most essentially effected by the presence of lead silicates and oxides in them, which accounts for the composition of numerous electrotechnical and solder glasses of which PbO is a permanent component. With composition 7 (see Table 78), containing FeO, Cr, Nb, Ta, V, Ti and Al can interact. The most active metals are Zr and Be which can interact with almost all the silicate materials in question.

The calculations conducted were used for determining the optimum conditions of welding of ceramics based on aluminium oxide with niobium in solid state [240]; for investigating the problems of the interaction of quartz and sulfide ceramics with structural materials in diffusion welding [239, 240, 241]; for studying the processes of the interaction and of the strength of the bond of substrates from glass and glass ceramics with metals in vacuum sputtering; for investigating the process of the chemical interaction of glass and glass ceramic phases with matrices based on metals and their alloys, as well as for developing the compositions and processes of the production of composite material starting from metals and glass ceramics [372, 374].

6.2 The Interaction of Ceramics with Metals

Recently an increasing amount of attention has been paid by scientists and engineers to the problems of the interaction of ceramics with metals and alloys.

Investigations of the processes of binding various materials [239, 318, 373, 432, 538] show the physico-chemical interaction of the materials in contact to be accompanied by the formation of an intermediate layer which may be a new chemical compound or a solid solution of a variable composition.

These processes may occur only where it is energically favorable, i.e., in case ΔG of the system lessens, which may be judged by conducting a thermodynamic analysis of the possibility of a chemical reaction or of the formation of solid solutions between the materials combined according to the technique developed in [239, 312, 318, 373, 444].

Below results are given concerning the thermodynamic evaluation of the possible interaction of some types of ceramics of pure oxides with a number of metals used in the solid phase binding by method of thermocompression – diffusion – welding.

The thermodynamic data for calculations were taken from [20, 52, 259, 303, 311, 671]. All the calculations were carried out for temperatures of 1273 K and 1650 K – the upper limit of the working temperatures –.

For calculating $\Delta G^0 = f(T)$ Gibbs-Helmholtz equation was used in the form:

$$\Delta G_T^0 = \Delta H_T^0 - T\Delta S_T^0. \tag{176}$$

For most reactions in question first the ΔH_T^0 and ΔS_T^0 values of the individual reactants were calculated throughout the temperature range required:

$$\Delta H_T^0 = \Delta H_{298}^0 + (H_T - H_{298}) , \quad (177)$$

$$\Delta S_T^0 = \Delta S_{298}^0 + (S_T - S_{298}) \quad (178)$$

subsequently finding ΔH_T^0 and ΔG_T^0 of the reactions being studied.

The processes under consideration feature mainly three schemes:

1. A direct interaction according to the Eq. (167)

$$Me' + Me''O = Me'' + Me'O ,$$

where Me′, structural metal and Me″, metal in the ceramics oxide.

2. A direct interaction of oxides and other compounds – spinels, silicates etc. – in ceramics with metals and their oxides, e.g., in the reactions:

$$Me' + Me''SiO_3 = Me'SiO_3 + Me'' \quad \text{[from Eq. (170)]},$$

$$Me'O + Me_2''O_3 = Me'Me_2''O_4 , \quad (179)$$

$$Me'O + Me''O \cdot SiO_2 = Me''O + Me'O \cdot SiO_2 . \quad (180)$$

3. An interaction on account of the formation of solid solutions of ceramic components with some metals according to the scheme

$$Me''O = [Me'']_{Me'} + [O]_{Me'} \quad \text{[from Eq. (169)]} .$$

Below we will examine the interaction of two types of vacuum-tight ceramics Cr22S and A-995 (for composition see Table 79) with some metals, such as Cu, Ni, Fe, Co, Mo, Cr, Mn, Nb, Ta, V, Ti, Al, Zr.

Cr22S Ceramics

An immediate interaction according to scheme (167) is possible:

a) Al_2O_3 only with Zr.
b) SiO_2 with Ti, Al, Zr.
c) MnO with Nb, Ta, V, Ti, Al, Zr.
d) Cr_2O_3 with Mn, Nb, Ta, V, Ti, Al, Zr.
e) Fe_2O_3 with Mo, Cr, Mn, Nb, Ta, V, Ti, Al, Zr.

Predominantly thermodynamically active metals, such as Ti and Zr are capable of an interaction of this kind, the interaction being significantly intensified by impurities MnO, Cr_2O_3, Fe_2O_3 in ceramics. Zirconium, for one interacting very actively with the Cr22S and A-995 ceramics so vigorously oxidizes, that even in a vacuum of $10^{-4}-10^{-5}$ mmHg it turns into a fragile (spongy) easily destructible oxide. Metals of low thermodynamic activity, Cu, Ni, Co, Fe cannot interact with ceramic components according to this scheme.

The formation of compounds between ceramic components and the metals in question according to scheme (170) is possible:

a) spinels

$MnAl_2O_4$	only with Zr.
$MnCr_2O_4$	with Zr, Al, Ti, V, Ta, Nb.
$MnFe_2O_4$	with Zr, Al, Ti, V, Ta, Nb, Cr.
$FeAl_2O_4$	with Zr, Al, Ti, V, Ta, Nb.

b) silicates

$2MnO \cdot SiO_2$	with Zr, Al, Ti, V, Ta, Nb.
$3Al_2O_3 \cdot 2SiO_2$	with Zr.
$2FeO \cdot SiO_2$	with Zr, Al, Ti, V, Ta, Nb, Cr.

Spinel formation in ceramics does not negatively effect the interaction with the metals in question as compared with the reaction of an immediate interaction of metals with oxides.

The mullite formation lowers the ceramics capability of interacting, a fact which can be used for practical purposes in order to control the ceramics activity. Spinel and silicate formation does not essentially effect the interaction with low-activity metals.

The interaction of ceramic oxides with the oxides of the metals under consideration, binary compounds being formed, according to Eq. (179) is possible:

a) Al_2O_3	with NiO, CoO, FeO, Cr_2O_3.
b) SiO_2	with NiO, CoO, FeO, Al_2O_3, MnO, ZrO_2.
c) MnO	with CuO, NiO, CoO, Fe_2O_3, MoO_2, Cr_2O_3.
d) Cr_2O_3	with CuO, NiO, CoO, FeO, MnO, Cr_2O_3, TiO_2.
e) Fe_2O_3	with CuO, NiO, CoO, MoO_2, MnO, Cr_2O_3, TiO_2.

The possibility of oxidation of the low-activity metals in question increases their interaction capability, as do additions of MnO, Cr_2O_3 and Fe_2O_3 to ceramics.

This may be supported by the welding of the Cr18Ni10Ti stainless steel with Cr22S and A-995 ceramics (see Table 79) when the formation of Cr_2O_3 (preferably in vacuum where the oxide film is thinner) and the interaction of Cr_2O_3 with the ceramics oxides (evidenced by the x-ray diffraction [497]) provides a 20 kg/mm^2 strong compound. On the other hand with Fe and Ni, which do not oxidize at welding, the strength is from 1.5 to 2 times lower.

Table 79. The dielectric composition and the strength of dielectric-metal welded joints

Chemical composition [weight %]		Leucosaphire	A-995	Cr22S	102
Al_2O_3		100	99.8	94.4	72
SiO_2			traces	2.76	16.5
MnO				2.35	
CrO_3				0.49	
MgO			0.2		2
Fe_2O_3			traces	traces	
BaO					6.5
CaO					2
Structural metal	**Atmosphere**	**Bending strength [kg/mm^2]**			
MOB copper	hydrogen	10–12	10–15	19–24	10–12
Nickel	hydrogen		15–18		
Cr18Ni10Ti steel	hydrogen		10	10	
	from 10^{-4} to 10^{-5} mmHg (vacuum)		18–20	20	
V T-I titanium	hydrogen		4	7	
	from 10^{-4} to 10^{-5} mmHg (vacuum)		13	16	5

The interaction of compounds with oxides according to scheme (180) is possible:

a) spinels with oxides

$MnAl_2O_4$ with ZrO_2.

$MnCr_2O_4$ with NbO, Ta_2O_5, V_2O_3, TiO_2, ZrO_2.

$MnFe_2O_4$, $FeAl_2O_4$ with Cr_2O_3, MoO_2, NbO, Ta_2O_5, V_2O_3, TiO_2, ZrO_2.

b) silicates with oxides

$3Al_2O_3 \cdot 2SiO_2$ with FeO, MnO.

$2MnO \cdot SiO_2$ with none of oxides.

$2FeO \cdot SiO_2$ only with MnO.

In the case of spinel formation in ceramics and the preliminary oxidation of the metals in question the activity of their interaction practically does not change.

The silicate formation lowers the activity of the ceramics interaction with preliminary oxidized metals as well.

The preliminary oxidation of metals with low thermodynamic activity does not result in the increase of activity either.

A-995 Ceramics

The formulation of the A-995 ceramics unlike that of Cr22S, does not include MnO and Cr_2O_3, however, it has a thermodynamically stable oxide MgO. If the main conclusions drawn for Cr22S still hold for A-995, the peculiarity of its interaction with metals is brought about by the processes with MgO.

An immediate interaction of metals with MgO of ceramics according to scheme (167) is possible only with Al and Zr.

The $MgAl_2O_4$ formation in ceramics does not negatively effect the interaction. The forsterite formation lowers the ceramics activity, whereas the $2FeO \cdot SiO_2$ formation increases it.

Of oxides of the metals in question, MgO of ceramics can interact with FeO, MoO_2, Cr_2O_3, TiO_2 and ZrO_2 to form binary compounds.

A possible oxidation of the low-activity metals under consideration increases the interaction activity, additions of MnO, Cr_2O_3, Fe_2O_3 to ceramics may have the same effect.

The application of manganese oxides by depositing on ceramics or by sputtering on metals increases the seal strength of several metals and alloys [431].

An interaction of compounds making up ceramics with metal oxides according to scheme (180) is possible:

a) spinels with oxides
   $MgAl_2O_4$ with TiO_2 and ZrO_2.
   $MgFe_2O_4$ with MoO_2, Cr_2O_3, TiO_2 and ZrO_2.

b) silicates with oxides
   $2MgO \cdot SiO_2$ with no metals.

Thus, the presence of such a thermodynamically stable oxide as MgO in the A-995 ceramics decreases the ceramics interaction with metals as well as with their oxides. Two groups of the materials in questions may be distinguished in the way they interact with Cr22S and A-995 ceramics: "active" metals, Zr, Ti etc., which actively interact with the ceramics and its impurities and "low-activity" ones, such as Cu, Ni, Co, Fe etc. which do not immediately interact with the ceramics and its impurities. Where the above calculations emphasize the impossibility of the reactions of immediate interaction, the latter may occur due to the formation of solid solutions in the Eq. (169):

$$Me''O = [Me'']_{Me'} + [O]_{Me'}$$

according to which the oxide elements dissolve in a metal forming an oxygen solid solution $[O]_{Me'}$ and $[Me'']_{Me'}$ based on it. One of the main causes of such an interaction is the significant alteration of the entropy partial values while solid solutions are formed.

An examination of Mozzhukhin's [444] calculations shows that this mechanism is peculiar to the interaction of the such ceramics with low-activity metals, such as Cu, Ni etc. (the activity increasing from left to right). While adding oxides with the Gibbs free energy higher than that of Al_2O_3 to the ceramics, the interaction activity increases in the following order:

$$Al_2O_3 \rightarrow SiO_2 \rightarrow MnO \rightarrow Cr_2O_3 \rightarrow Fe_2O_3 .$$

While adding oxides with the Gibbs free energy lower than that of Al_2O_3 to the ceramics, the interaction activity will decrease:

$$Al_2O_3 \rightarrow ZrO_2 \rightarrow MgO \rightarrow CaO .$$

This scheme is well confirmed by the copper interaction with various ceramic materials (see Table). The copper weldability also noticeably improves with "activating" A-995, 102, sapphire by manganese and its oxides, chromium and other activators at sputtering, rubbing on or firing [431].

It is also interesting to consider the effect of atmosphere on the above processes, which will be illustrated by that of hydrogen and vacuum.

A study of a reaction of the type:

$$\underset{\text{solid}}{MeO} + \underset{\text{gas}}{H_2} = \underset{\text{solid}}{Me} + \underset{\text{gas}}{H_2O} \tag{181}$$

shows Cu, Co, Ni and Fe to be the only metals of the set in question to be reduced by hydrogen from oxides, i.e., only these metals are preserved by hydrogen from an oxide film being formed on their surface.

Active metals, like zirconium and titanium, will not be reduced by hydrogen from oxides, i.e., the hydrogen atmosphere will not effect the oxide film of these metals. Thus, with the reaction

$$\underset{\text{solid}}{CuO} + \underset{\text{gas}}{H_2} = \underset{\text{solid}}{Cu} + \underset{\text{gas}}{H_2O}$$

at the hydrogen pressure of 1 atm and at a temperature of 1250 °C the equilibrium partial pressure of water vapour will be $0.4p_{H_2}$, so the metal oxidation will take place if the calculated or a greater amount of water is present. Under real welding conditions in gas – hydrogen – atmosphere the partial pressure of water vapour is always below the value required for the oxidation reaction, and therefore copper oxidation in humid hydrogen is improbable. Contrarily, hydrogen invariably impedes the formation of copper oxides which allows for a reaction with ceramics. The same is true for Co, Ni, Fe.

With zirconium, hydrogen pressure being 1 atm, the equilibrium partial pressure of water vapour is $2.35 \cdot 10^{-8}$ mmHg at 1400 K, i.e., the heating of the metallic zirconium in the hydrogen atmosphere where the partial pressure of water vapour is higher than $2.35 \cdot 10^{-8}$ mmHg will result in its oxidation. The same is true for titanium.

Vacuum effects the oxide stability according to the formula

$$G_T = G_T^0 + 4.576\, T \lg p \,. \tag{182}$$

At a temperature of 1000 K and vacuum of $10 \cdot 10^{-5}$ mmHg correction for vacuum will be $4.576 \cdot 1000 \cdot (-5) = -23$ kcal, i.e., a great value; consequently, vacuum will particularly effect easily oxidizable metals, such as Zr and Ti, which is well confirmed by the "active" brazing practice.

6.3 Binding Oxides for Enamels

The substantiation of the choice of binding oxides for enamels is of particular practical interest as well.

How one can apply the above calculations to the enameling process, will be illustrated by the choice of oxides for binding of enamel with steel. Many years of experience has shown reliable binding of enamel with a metal to be reached by introducing into the enamel a small amount of such oxides as CoO, NiO and some others which known as "binding oxides".

It is of interest to be able to determine the conditions of the appearance of the binding properties of various oxides and, thus, to predict new oxides producing binding effects. We undertook such an investigation starting from the thermodynamic analysis of binding reactions. Several theories have been put forward to explain the bond action of the binding oxides [481]. The most true to the facts, we believe, is the electrochemical concept of the binding mechanism worked out by Dietzel [117, 481, 546]. According to his point of view, while covering a steel substrate with an enamel melt containing nickel monoxide the reaction

$$\mathrm{Fe} + \mathrm{NiO} = \mathrm{Ni} + \mathrm{FeO} \tag{183}$$

takes place.

Nickel released on iron forms local short-circuit galvanic cells with iron in the melt. Iron, as the anode of such cells, dissolves passing over into the melt in the form of corresponding ions, on the steel base, however, micro-checks and cracks are formed and after these the melted enamel. Similar corrosion may be observed while covering aluminium plates with the enamel melt containing FeO or while covering a nickel base with the enamel containing CuO. This enamel binds well with a nickel or cobalt base on which enamels containing CoO and NiO do not hold.

The well-known fact that enamels fired in vacuum do not bind well is due to the oxygen action. Oxygen, in this case, plays the role of a depolarizer, and in its absence the process of the electrochemical corrosion of the steel base does not begin.

Thus, the binding action of any oxide with a steel base will be expressed as:

$$Fe + MeO = Me + FeO\,, \tag{184}$$

where Me, any metal, and MeO, its oxide.

Whether this reaction is possible may be evaluated by a thermodynamic calculation.

Starting from the data of Tripp and King [642] we carried out calculations for various oxides of metals of the periodic system. The examination showed that reaction (184) is possible with the oxides of the following elements:

Groups of the periodic system	I–III	IV	V	VI–VIII
Elements	Ag, Cu, Cd	Pb, Ge, Sn	Sb, As, Bi	Pt, Co, Ni, Mo, W

While applying enamels, however, concurrently with reaction (184) the following reaction takes place

$$2FeO + MeO = Fe_2O_3 + Me\,. \tag{185}$$

In case the equilibrium of this reaction is shifted to the right, in enamel a direct reduction of the metal will occur; no formation of the galvanic cell in the binding of the enamel with the metal being possible is the result.

Thus, along with $\Delta G^0 \leqslant 0$ being indispensable for reaction (184), the following premise is indispensable for reaction (185):

$$\Delta G^0 \geqslant 0 \quad \text{or} \quad \Delta G^0_{Fe_2O_3} - 2\Delta G^0_{FeO} \geqslant \Delta G^0_{MeO}\,.$$

The substitution of corresponding data for 1273 K yields

$$\Delta G^0_{MeO} \leqslant -32\ \text{kcal/mol}\,.$$

In accordance with our calculations the two conditions are met by the oxides of the following elements:

Groups of periodic system	I–III	IV	V	VI–VIII
Elements	Cd	Ge, Sn	As	Co, Ni, Mo, W

Thus, in addition to the wide-spread CoO and NiO one must take Cd, Ge, Sn, As, Mo and W oxides as binding oxides into account as well.

6.4 Water Resistance of Silicates [18, 359, 360]

The destruction of silicate materials in water is connected with complicated physico-chemical processes of the dissolution of the silicates that constitute them, the nature of these processes having not yet been fully revealed. Nonetheless, one cannot deny that in the leaching products of the given materials there are hydrated ions of the corresponding metals and silico-oxygen anions.

For the thermodynamic interpretation of the silicate destruction in water is is necessary to know the Gibbs free energy of the silicates dissolved and of the corresponding ions viewed as the final products of dissolution. The data on thermodynamics of reactions being available, we can rightly evaluate the factors determining the stability of silicate materials in water and in its vapour.

The transition into solution will be exemplified by 14 metasilicates and 12 orthosilicates of certain metals as well as by sillimanite and mullite. The transition of silicates into solution is presumed to proceed according to the following scheme:

$$MeSiO_3 \rightarrow Me^{2+} + SiO_3^{2-} \quad \text{(metasilicates)},$$

$$Me_2SiO_4 \rightarrow 2Me^{2+} + SiO_4^{4-} \quad \text{(orthosilicates)},$$

where Me: metal and Me^{2+}: metal ion (with possibly variable valence).

The following schemes are presumed to reflect the transition of sillimanite and mullite into solution:

$$Al(AlSiO_5) \rightarrow Al^{3+} + AlO_2^{1-} + SiO_3^{2-}$$

$$Al_3(Al_3Si_2O_{13}) \rightarrow 3Al^{3+} + 3AlO_2^{1-} + Si_2O_7^{6-}.$$

The suggested schemes per se allow for the destruction of the lattice of the substance dissolved and for its interaction with the solvent – hydration –, as while calculating the thermochemical properties of hydrated ions were used. The Gibbs free energies of aqueous cations were taken from [20, 242], and those of silicates, from the previous calculations.

The Gibbs free energy calculations at dissolving were carried out by the following equations:

metasilicates

$$\Delta G_{dissol} = \Delta G_{Me^{2+}} + \Delta G_{SiO_3^{2-}} - \Delta G_{MeSiO_3}$$

orthosilicates

$$\Delta G_{dissol} = 2\Delta G_{Me^{2+}} + \Delta G_{SiO_4^{4-}} - \Delta G_{Me_2SiO_4}$$

sillimanite

$$\Delta G_{dissol} = \Delta G_{Al^{3+}} + \Delta G_{AlO_2^{1-}} + \Delta G_{SiO_3^{2-}} - \Delta G_{Al_2SiO_5}$$

mullite

$$\Delta G_{dissol} = 3\Delta G_{Al^{3+}} + 3\Delta G_{AlO_2^{1-}} + \Delta G_{Si_2O_7^{6-}} - \Delta G_{Al_6Si_2O_{13}},$$

where ΔG_{dissol}, alteration of the Gibbs free energy in the transition of a given silicate – aluminosilicate – into solution at 25 °C (298 K); $\Delta G_{Me^{2+}}$, the Gibbs free energy of the corresponding aqueous cation at 298 K; $\Delta G_{Al^{3+}}$, the Gibbs free energy of an aluminium aqueous cation at 298 K; $\Delta G_{SiO_3^{2-}}$, $\Delta G_{SiO_4^{4-}}$, $\Delta G_{Si_2O_7^{6-}}$, the Gibbs free energies of the corresponding aqueous silicate ions at 298 K; $\Delta G_{AlO_2^{1-}}$, the Gibbs free energy of an aqueous aluminate ion.

The initial data and calculation results are summarized in Table 80.

Table 80. The Gibbs free energies of silicates, aluminosilicates, hydrous ions and reactions of the transition into solution

Substance dissolved	$-\Delta G^0_{298}$ [kcal/mol]	Ion	$-\Delta G^0_{298}$ [kcal/mol]	$\Delta G^0_{298\,(dissol)}$ [kcal/mol]
–	–	SiO_3^{2-}	226.9	–
$BaSiO_3$	369.08	Ba^{2+}	134.0	8.02
$BeSiO_3$	347.15	Be^{2+}	78.76	41.49
β-$CaSiO_3$	370.55	Ca^{2+}	132.18	11.47
$FeSiO_3$	269.35	Fe^{2+}	20.30	22.15
K_2SiO_3	350.15	K^+	67.47	–11.69
Li_2SiO_3	368.70	Li^+	70.22	1.36
$MgSiO_3$	349.48	Mg^{2+}	108.99	13.59
$MnSiO_3$	296.55	Mn^{2+}	54.4	16.35
Na_2SiO_3	351.26	Na^+	62.59	–0.82
$PbSiO_3$	255.19	Pb^{2+}	5.81	22.48
$SrSiO_3$	374.64	Sr^{2+}	133.2	14.54
$ZnSiO_3$	287.15	Zn^{2+}	35.18	25.07
Rb_2SiO_3	345.25	Rb^+	67.45	–16.55
Cs_2SiO_3	343.70	Cs^+	67.41	–18.02
	–	SiO_4^{4-}	260.39	–
Ba_2SiO_4	521.83	Ba^{2+}	134.0	–6.56
$BeSiO_4$	488.4	Be^{2+}	78.76	70.49
γ-Ca_2SiO_4	525.19	Ca^{2+}	132.18	0.34
Fe_2SiO_4	333.65	Fe^{2+}	20.30	32.66
Li_4SiO_4	522.8	Li^+	70.22	–18.47
Mg_2SiO_4	492.15	Mg^{2+}	108.99	13.78
Mn_2SiO_4	387.95	Mn^{2+}	54.4	18.76
Na_4SiO_4	476.05	Na^+	62.59	–34.70
Pb_2SiO_4	302.92	Pb^{2+}	5.81	30.91
Sr_2SiO_4	532.39	Sr^{2+}	133.2	5.60
Zn_2SiO_4	363.98	Zn^{2+}	35.18	33.23
$ZrSiO_4$	456.65	Zr^{4+}	141.0	55.26
	–	AlO_2^{1-}	204.7	–
	–	$Si_2O_7^{6-}$	487.29	–
Al_2SiO_5	583.35	Al^{3+}	115.0	36.75
$Al_6Si_2O_{13}$	1543.93	Al^{3+}	115.0	97.54

An examination of the data obtained enables us to draw the following conclusions.

Of all the compounds under consideration only alkali metal silicates (except lithium metasilicate) prove to be capable of transforming into solution at standard conditions (a temperature of 298 K).

The stability of particular silicates can be presented in the form of the following series (chemical stability decreasing from left to right) as a function of the Gibbs free energy value:

metasilicates of
Be, Zn, Pb, Fe, Mn, Sr, Mg, Ca, Ba, Li, Na, K, Rb, Cs

orthosilicates of
Be, Zr, Zn, Fe, Pb, Mn, Sr, Ca, Ba, Li, Na.

Chapter III. Hydration Reactions

1 General

The hydration reactions characterizing the transition of substances from an anhydrous state to an hydrated one are of great importance in silicate technology comprising all the processes of binder hardening under normal and hydrothermal conditions, as well as in geochemistry including processes involved by the effect of water and vapour on the rocks: surface and inside of the earth crust.

Water can be combined with numerous salts and oxides, making up a vast class of compounds – crystal hydrates –. From the viewpoint of the combination mode and bonding force of water crystal hydrates may be subdivided into compounds:

1. in which water is in the form of OH ions (crystal water);
2. in which water is bound with the constituents of the lattice in the form of molecules (coordination bound water);
3. in which molecules of water are not bound with the lattice, but only fill in vacancies in it (zeolitic water).

Crystal hydrates in which water is in more than one form at a time may also occur, e.g., in the form of crystal and zeolitic water etc.

Great importance has been attached recently to research into the structure, chemistry and properties of crystal hydrates [59, 61, 369]. Primarily it concerns calcium silicate hydrates, aluminate hydrates, ferrite hydrates and aluminate sulphate hydrates, being hydration and hardening products of portland cement, which is the principal representative of hydraulic binders [35, 127, 252, 314, 325, 419, 599, 635, 668]. Of great importance in natural hydrates are magnesium hydrosilicates (chrysotile-asbestos, serpentinite, etc.), aluminium silicate hydrates (kaolinite and other argillaceous minerals) and other hydrous silicates and aluminosilicates [533, 582, 670].

A thermodynamic study of the reactions of the synthesis of some hydrate compounds and of technically relevant hydrosilicates will permit working out and putting into practice the most efficient modes of their industrial manufacturing as well as rationally controlling the making of products based on them with set properties [506, 507].

The main hydration reactions of great importance in silicate production and geochemistry are those in the following binary, ternary, quaternary and more complex aqueous systems: SiO_2-H_2O, $CaO-H_2O$, $CaO-SiO_2-H_2O$, $CaO-Al_2O_3-H_2O$, $CaO-Fe_2O_3-H_2O$, $CaO-Al_2O_3-Fe_2O_3-CaSO_4-H_2O$, $MgO-H_2O$, $MgO-SiO_2-H_2O$ etc.

As a result of recent research work the binder hydration process has been found to cover not only their mere chemical interaction with water or with electrolytic solutions, hydrated forms and pore electrolyte of the corresponding composition being formed, but also physical, physico-chemical and colloid-chemical phenomena, such as dissolving, hydrolysis, crystallization, polymerization and polycondensation providing, under proper conditions, the binder setting and hardening, and the formation of spatial capillary-porous structure of certain strength and stability.

This involves not only the consideration of hydration processes as an exclusively chemical aspect meaning a calculation of equilibrium states and thermochemical values, but a calculation of colloid-chemical equilibria at the phase boundary and of the thermodynamic conditions of structure formation with respect to the moisture content of the systems binder-water as well [226].

As several comprehensive reviews dealing at length with the main principles of the theory of binder hardening [87, 95, 153, 166, 299, 314, 335, 396, 398, 419, 422, 484, 505, 534, 574, 593, 601, 612, 613, 665, 676, 679] have been recently published, these problems will not be handled here; we shall only present data concerning the thermodynamics of hydration processes and general conclusions clarifying our ideas on the subject.

It is a matter of common knowledge that hydration processes in general and ionic hydration phenomena in the systems binder-water in particular are extremely involved. Therefore when one views these processes basically from the standpoint of classical thermodynamics one finds definite regularity for them. Leaving aside the methods of classical thermodynamics, inevitably leads to more or less arbitrarily assumptions and simplifications, which may cause errors [437].

The theoretical principles of the classical thermodynamics of binder hydration processes comprise a thermodynamic consideration of the reactions of pure solids and water, as well as that of actual solutions and thermodynamics of mixed crystalline phases coexistance [333, 564, 659].

The application of the classical thermodynamic methods to the reactions of pure solids with water permits the solution of the problem of the preference of some reactions to others, as in the case of pyrosilicate reactions, which necessitates using the method of the ΔG^0 calculation put forward by us with regard to the given ratio of initial substances in the system (see Chapter I.11).

As for the reactions in which only pure solids and liquids take part, the equilibrium constant has no physical sense, their concentrations being constant values, then $\Delta G = \Delta G^0_{298}$; thus the conclusion may be drawn that simply by the sign of ΔG^0_{298} one can judge the possibility or impossibility of a process and by the value of it, the inequilibrium degree of one or another reaction under actual conditions.

The application of thermodynamics to solutions of binders in water is based on empirical data, on the "activity" method in particular. The Gibbs free energy alteration of the binder solution and hydrolysis reaction may be calculated by Eq. (68). In this connection the data on the activity of substances in solutions permit calculating ΔG, on the other hand, the ΔG values being available, one can calculate the active concentrations of ions in a solution. The equilibrium concentrations of ions in a solution can be calculated by Eq. (65) if the ΔG^0 value is known.

The problems of nucleation, structure forming and destruction in the systems binder-water are not considered here [66, 648, 654, 529].

The technique of the thermodynamic calculation of ionic equilibria of the dissolution of solids in water is given in [146] and needs no comment. As to the reactions of anhydrous binder dissolution and hydrolysis, they dissolve irreversibly and the term "solubility" does not physically apply to them [325], thus we have only a hypothetical equilibrium state, which must be borne in mind while carrying out a thermodynamic analysis of these systems.

The technique of rough calculations of ionic equilibria of the binder hydration and hydrolysis reactions will be reported when handling practically all actual systems binder-water, and more precise calculations with regard to complexing will be exemplified by gypsum hemihydrate and calcium oxide hydration.

Considering, that in all the hydration reactions water is involved as an important component of all the transformations in these complex heterogeneous systems, it is necessary to describe the peculiarities of its structure and thermodynamic properties in a separate chapter. A description of the thermodynamic properties of each monomineral binder will be likewise given in corresponding sections.

2 The Basic Properties and Thermodynamic Stability of Water

Water is known to be a chemical substance of paramount importance. The consideration of water from the viewpoint of the structure of its molecule and of its structure as a liquid body revealed several of its anomalies which are caused in the first place by intense mutual attraction of water molecules based on the formation of hydrogen bonds. Although the energy of hydrogen bonds amounts to only 4.5 kcal (valent bonds being normally from 50 to 150 kcal), their role in reactions with water taking part is immense, and they determine to a great extent the structure and properties of hydrated forms arising.

As each molecule of water comprises two hydrogen atoms and two unshared electron pairs, four hydrogen bonds can be formed. Tetrahedral arrangement of electron pairs results in the tetrahedral direction of hydrogen bonds in space.

In works of Samoylov [544], Pauling [486], Coulson [106] and Pople [487] the electrostatic model of the molecule of water was established, which revealed

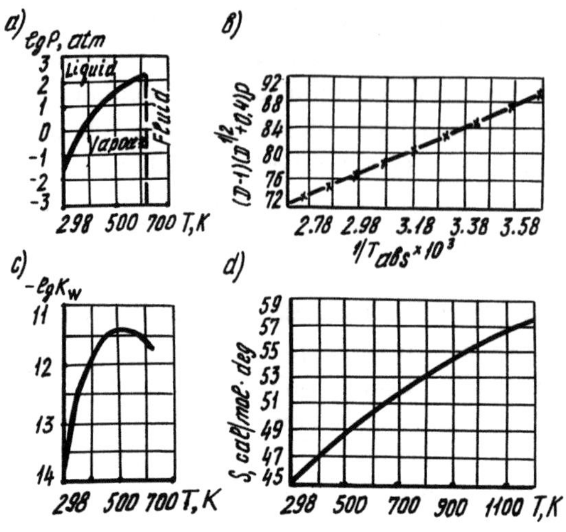

Fig. 75. The properties of water as a function of temperature: **a** phase diagram; **b** dielectric permittivity; **c** ionization constant; **d** entropy (with ideal gas)

its most important properties: great constant dipole moment, great solvation capacity, great dielectric constant and other properties conditioned by the additional energy of hydrogen bonds.

Some physical properties of water and their temperature and pressure dependence in the form of function curves taken from the work by Fife, Turner and Verhoogen [131] are illustrated in Fig. 75. As one can see from Fig. 75a the alterations of the phase state of water are continuous, and at a temperature and pressure above critical ones, only one phase is in the stable state; this phase is called fluid (distinguishable neither from liquid nor vapour). In Fig. 75b the dielectric permittivity of water at temperatures of 273 to 373 K is expressed corresponding to the following ratio:

$$\frac{(D-1)\cdot T}{(D+2)\cdot \varrho} = a + bT + cT^2 . \tag{186}$$

The product of water ionization (as a function of temperature) according to the formula

$$\lg k_{\mathrm{w}} = -\frac{4470.99}{T} + 6.0875 - 0.01706\,T$$

is given in Fig. 75c.

In Fig. 75d the temperature dependence of the water vapour entropy is presented. As found by Holster [207], within the pressure range of 1 to 2000 bar and temperature range of 473 to 1273 K the water vapour markedly differs from an ideal gas, the fugacity coefficient of it being insignificant. Thus, e.g., at 2000 bar and 473 K the fugacity coefficient f amounts to only 0.0172. At high temperatures the deviations from the state of an ideal gas lessen. Thus, at 1273 K and 2000 bar $f = 0.962$.

While dissolving diverse compounds in water ions are introduced into the structure of water, bringing about its distortion, the latter being the more perceptible the smaller the size and the greater the charge of an ion [436] is. The work of the separation of ions of solids is performed chiefly based on the ion hydration process, as the latter is accompanied by a decrease of the Gibbs free energy. The great amount of energy released by the ion hydration considerably contributes to the process, endothermic in itself, of the crystalline lattice destruction and of the electrolyte dissociation, and at the same time stabilizes ions hindering the reverse process of their mutual bonding.

By ion hydration heat the amount of energy is meant which would be released while converting ions from the gaseous state to aqueous solution, it being infinitely diluted. For calculating the ion hydration heats the following thermochemical equation was put forward:

$$L_i = -\Delta H_i(\mathrm{aq}) + \Delta H_i(\mathrm{gas}) - 102.5 \cdot n\,, \tag{187}$$

where H_i(aq), heat of ion formation in aqueous solution infinitely diluted (on the condition that ΔH^0_{298} of a proton equals 0 kcal); ΔH_i(gas), heat of ion formation gaseous state; n, ion charge (with regard to the sign).

E.g., for the ion Ca^{2+} the hydration heat will be equal to:

$$L_i = -(-129.70) + (+461.30) - 102.5 \cdot 2 = 386.0\ \mathrm{kcal/g \cdot ion}$$

An intense electric field of ions orients dipoles of water around them. The degree of hydration (number of hydration), i.e., the number of dipoles of water, forming a hydrate shell around the ion, depends on the features of the latter. This number, as stated above increases with the increase of the valence and decreases with the increase of its radius. The hydration heat, thermochemical radii and hydration numbers of some ions are given in Table 81.

In addition to ions, molecules of substances dissolved can be hydrated as well. Hydration may be intensified by chemical forces of the interaction between water and constituents of a solution as well. Several silicates and related compounds form hydrates not only in solutions, but also in solid state. Almost all salts tend to such complexing [215, 289, 301].

To study hydration reactions one must know the thermodynamic conditions at which water is a stable compound. These relations are easily arrived at, starting from the following equation of the reaction of water decomposition:

$$2H_2O_{(\mathrm{liq})} = O_{2(\mathrm{gas})} + 2H_{2(\mathrm{gas})}; \tag{188}$$

$$\Delta G^0_r = 0 + 0 + 2 \cdot 56.69 = +113.38\ \mathrm{kcal/mol}\,.$$

Table 81. Hydration heat, thermochemical radii and hydration numbers of some ions

Ion	Hydration heat [kcal/g-ion]	Thermo-chemical radius [Å]	Hydration numbers according to authors				
			Wash-burn	Rosen-fel'd	Brint-zinger [63]	Robinson and Stokes	Bernal and Fowler [49]
H^+	264	0	1	0	–	8	2
Li^+	125	0.78	14	158	5 and 10	7	6
Na^+	100	0.98	8.4	71	3 and 5	3.5	6
K^+	80	1.33	5.4	22	4	1.9	–
Mg^{2+}	464	0.78	–	–	6 and 15	13.2	6
Ba^{2+}	313	1.43	–	–	14	7.7	6
Ca^{2+}	391	1.06	–	–	6 and 16	12	6
Al^{3+}	1103	0.57	–	–	18 and 39	–	6
CO_3^{2-}	317	1.85	–	–	–	–	–
SO_4^{2-}	241	2.3	–	–	–	–	–
SiO_3^{2-}	294	2.02	–	–	–	–	–
SiF_6^{2-}	221	2.45	–	–	–	–	–
OH^-	116	1.4	–	–	–	–	–

The equilibrium constant is

$$K = \frac{p_{O_2} \cdot p_{H_2}^2}{[H_2O]^2}$$

or, with regard to $[H_2O] = 1$ for pure water, $p_{H_2}^2 \cdot p_{O_2} = K$, and $113.38 = -1.364 \lg p_{O_2} \cdot p_{H_2}^2$, in other terms

$$p_{O_2} \cdot p_{H_2}^2 = 10^{-83.1}\,. \qquad (189)$$

Equation (189) shows the degree of the water dissociation to oxygen and hydrogen under standard conditions ($p = 1$ atm and $T = 298$ K) to be negligible. Using this equation it is possible to plot a diagram of the stability fields of liquid water as a function of the hydrogen and oxygen partial pressure in the earth atmosphere.

If $p_{O_2} = 1$, $p_{H_2} = 10^{-41.5}$, conversely, when $p_{H_2} = 1$, $p_{O_2} = 10^{-83.1}$. Thus, if actually in the earth atmosphere the hydrogen or oxygen partial pressure is higher than 1 atm, water begins decomposing, an extra amount of oxygen or hydrogen being released, so that the system should retain its equilibrium state.

The limits of the stability of water may be presented also in a diagram with coordinates Eh-pH. To this end it is necessary to put the reactions between water and oxygen or between water and hydrogen in terms of hydrogen ions and electrons. A thermodynamic study of the equilibrium of these reactions was conducted by several scientists. In this context we shall not derive the equations in question (one can easily find them in [145, 146], but put only the initial reactions and the relations arrived at; the emphasis being placed on the results obtained.

As for the upper limit of the water stability, i.e., at $p_{O_2} = 1$ atm, the reaction of the water decomposition is expressed as:

$$2H_2O_{(liq)} = O_{2(gas)} + 4H^+_{(aq)} + 4e^- . \tag{190}$$

For this reaction it was shown in [146] that

$$Eh = 1.23 + \frac{0.059}{4} \lg p_{O_2} - 0.059\,\mathrm{pH} . \tag{191}$$

A similar equation for the reaction of hydrogen decomposition

$$H_2 = 2H^+_{(aq)} + 2e^- \tag{192}$$

is expressed as:

$$Eh = -\frac{0.059}{2} \lg p_{H_2} - 0.059\,\mathrm{pH} . \tag{193}$$

A diagrammatic representation of Eqs. (191) and (193) is given in Fig. 76. One can see from the figure that the uppermost line corresponds to the conditions at which $p_{O_2} = 1$ atm and $p_{H_2} = 10^{-41.7}$ and refers to the upper limit of the water stability. Above this line water is unstable. The lowermost straight line corresponds to the conditions when $p_{H_2} = 1$ atm and $p_{O_2} = 10^{-83.4}$. Below this line water is also unstable.

If similar to the expression $-\lg[H^+] = \mathrm{pH}$ we assume that $-\lg p_{H_2} = \mathrm{rH}$, then starting from the data of Fig. 76 we can conclude that at rH = 27.7 water is neutral, at rH = 0 water has reduction properties, and at rH = 41.7 has oxidation properties.

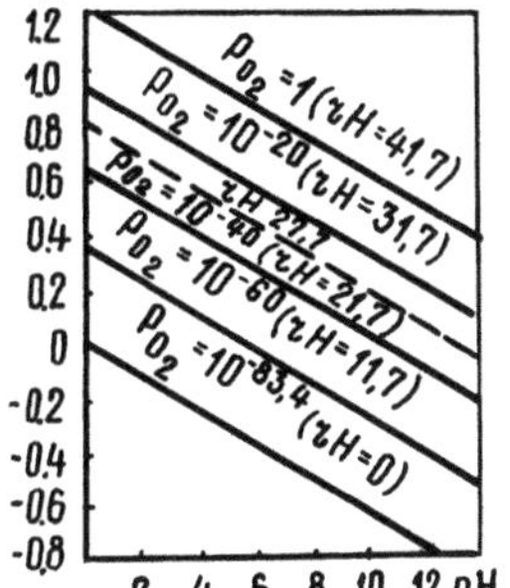

Fig. 76. The limits of the water stability in the form of a function E_h and pH at 25 °C at 1 atm of general pressure

In order to consider the elementary processes of binder hydration we must have thermodynamic data at our disposal concerning the hypothetical ion O^{2-} in water solution. Due to insufficient data on this ion in literature, we carried out estimations to determine $\Delta H^0_{298\,(gas)}$, $\Delta H^0_{298\,(aq)}$, $\Delta G^0_{298\,(aq)}$, $S^0_{298\,(gas)}$, $S^0_{298\,(aq)}$, as well as ΔH^h, ΔG^h and ΔS^h.

Starting from [132, 235] on the energy of the crystalline lattice for CaO: $U_{cr} = 828.8$ kcal/mol and $\Delta H^0_{gas(Ca^{2+})} = 460.29$ kcal/g-ion [528], by Yatsimirskiy's formula [700, 701] we determine $\Delta H^0_{298(O^{2-}\,gas)}$:

$$U_{cr} = -[-\Delta H^0_{CaO(cr)}] + \Delta H^0_{Ca^{2+}\,(gas)} + \Delta H^0_{O^{2-}\,(gas)} \tag{194}$$

as

$$\Delta H^0_{298\,(CaO\ cr)} = -151.9 \text{ kcal/mol}$$

then

$$\Delta H^0_{298\,(O^{2-}\,gas)} = 828.8 - 151.9 - 460.29 = 216.61 \text{ kcal/g-ion}.$$

The chemical heat of the ion O^{2-} hydration (L^h) will be estimated by Yatsimirskiy's formula:

$$L^h = \frac{165.5z^2}{r+l}, \tag{195}$$

where: z, ion charge ($z = 2$); r, thermochemical radius ($r = 1.32$ according to [700]), l, constant (for ions like O^{2-}, S^{2-}, Se^{2-} and Te^{2-} equal to 0.2).

$$L^h = \frac{165.5\ 2^2}{1.32+0.2} = 436.0 \text{ kcal/g-ion}$$

$$H^h = -L^h = -436.0 \text{ kcal/g-ion}.$$

The heat of the formation of the ion O^{2-} in an aqueous solution $\Delta H^0_{298\,(aq)}$ can be determined starting from the equation of Fajans (using the data on CaO [700]):

$$U_{cr} = -L^h_{CaO} + L^h_{Ca^{2+}} + L^h_{O^{2-}}, \tag{196}$$

where L^h_{CaO}, chemical heat of CaO dissolving $[-L^h_{CaO} = \Delta H^h_{CaO} = \Delta H^0_{(aq)\,298}(Ca^{2+}) + \Delta H^0_{(aq)\,298}(O^{2-}) - \Delta H^0_{298\,CaO\,(cr)}]$.

$L^h_{Ca^{2+}}$, chemical heat of the ion Ca^{2+} hydration ($L^h_{Ca^{2+}} = -\Delta H^h_{Ca^{2+}} =$ 386 kcal/g-ion [437, p 71].

$$L^h_{O^{2-}} = +436 \text{ kcal/g-ion}.$$

U_{cr}, energy of the CaO crystalline lattice (828.8).
$L^h_{CaO} = -828.8+386+436 = -6.8;$
$L^h_{CaO} = -\Delta H^h_{CaO}.$

As $\Delta H^h_{CaO} = \Delta H^0_{(aq)\,(O^{2-})} + \Delta H^0_{(aq)\,(Ca^{2+})} - \Delta H^0_{CaO\,(cr)}$, then $+6.8 = \Delta H^0_{(aq)\,(O^{2-})} - 129.7 - (-151.9)$;

$$\Delta H^0_{(aq)\,(O^{2-})} = -15.4\ \text{kcal/g-ion}\,.$$

The entropy of the ion O^{2-} in an aqueous solution may be estimated starting from the data given in [448] for weak acids (water may rank among them). The entropy alteration at the 1st stage of dissociation equals -19 ± 3 cal/mol deg, and at the 2nd, -30 ± 3 cal/mol deg.

For water

$$1.\ H_2O_{(liq)} = H^+_{(aq)} + OH^-_{(aq)}; \tag{197}$$

$$\Delta S_1 = 0+(-2.57)-(+16.71) = -19.28$$

(the comparison of the values -19 ± 3 and -19.28 proves that for water this rule is true).

$$2.\ OH^-_{(aq)} = O^{2-}_{(aq)} + H^+_{(aq)}, \tag{198}$$

$-30 = S^0_{(aq)\,(O^{2-})} + O - (-2.57)$; thus we determine $S^0_{(aq)\,(O^{2-})} = -32.57$ cal/mol deg.

The entropy of the gaseous ion O^{2-} (gas) is determined by the formula from [700]:

$$S^i_{(gas)} = \tfrac{3}{2} R \ln A + 25.7\,, \tag{199}$$

where R, gas constant (1.99 cal/mol); A, atomic mass.

$$S_{O^{2-}\,(gas)} = \tfrac{3}{2}1.99 \cdot 2.302\,\lg 16 + 25.7 = 34.0\ \text{cal/mol deg}\,.$$

The chemical entropy of the hydration of ion O^{2-} can be calculated by Mishchenko's formula [437].

$$\Delta S^h = S^0_{(aq)} - S^0_{(gas)} + 6.35$$

$$\Delta S^h = -32.57 - 34.0 + 6.35 = -60.22\ \text{cal/mol deg}\,.$$

A probable alteration of the Gibbs free energy, when the ion O^{2-} is hydrated, can be found by the formula

$$\Delta G^h = \Delta H^h - T \cdot \Delta S^h$$

$$\Delta G^h = -436.0 - 0.298(-60.22) = -418 \text{ kcal/g-ion}.$$

The calculation of ΔH^h of the ion O^{2-} can be checked with Eq. (187):

$$L^h = -H^h = -H^0_{(aq)} + H^0_{(gas)} - 102.5\,n.$$

$$-\Delta H^h = -(-15.4) + 216.61 - 102.5\,(-2)$$

$$= -15.4 + 216.61 + 205.0 = 437.01,$$

whereas by Eq. (195) we obtained 436.0 which is quite near to this value.

ΔG^0_{298} of the ion O^{2-}_{aq} will be found from the equation

$$\Delta G^0_{298} = \Delta H^0_{298} - T \cdot \Delta S \text{ for the reaction } \tfrac{1}{2}O_{2(gas)} + 2e^- = O^{2-}_{(aq)};$$

$$S = -32.57 - \tfrac{1}{2}49.003 - 2(15.604) = -88.27 \text{ cal/mol deg}.$$

$$\Delta G^0_{298\,O^{2-}(aq)} = -15.4 - 0.298(-88.27) = +10.9 \text{ kcal/mol}.$$

The data obtained for the ion O^{2-} are summarized in Table 82.

On the basis of the data obtained we calculate the 2nd constant of the water dissociation and the thermodynamic possibility of the ion O^{2-} formation in an aqueous solution.

$$1.\ OH^-_{(aq)} = O^{2-}_{(aq)} + H^+_{(aq)} \tag{200}$$

$$\Delta G^0_{298} = +10.9 - (-37.59) = 48.49 \text{ kcal/mol}$$

$$\lg K_1 = -\frac{\Delta G^0_{298}}{1.364} = -\frac{48.49}{1.364} = -35.55.$$

$K_1 = 10^{-35.5}$, i.e., the probability of the water dissociation at the 2nd stage is too low.

2. For the summary dissociation reaction

Table 82. The probable thermochemical data for the hypothetical ion O^{2-}

Ion	State	ΔH^0_{298} [kcal/ g-ion]	ΔG^0_{298} [kcal/ g-ion]	ΔS^0_{298} [cal/ g-ion · deg]	ΔH^h [kcal/ g-ion]	ΔG^h [kcal/ g-ion]	ΔS^h [cal/ g-ion^0]	r [Å]
O^{2-}	gas	+216.61	–	34	–	–	–	1.32
O^{2-}	aq	−15.4	+10.9	−32.6	−436	−418	−60	–

$$H_2O_{(liq)} \rightleftarrows 2H^+_{(aq)} + O^{2-}_{(aq)} \tag{201}$$

$$\Delta G^0_{298} = +10.9 - (-56.69) = 67.59$$

$$\lg K_2 = -\frac{67.59}{1.364} = -49.55 .$$

$K_2 = 10^{-49.55}$, which also emphasizes the low probability of Eq. (201).

Other important characteristics of water are: the constant of the rate of the water electrolytic dissociation at the 1st stage

$$H_2O = H^+ + OH^- \text{ equal to } 2.6 \cdot 10^{-5}\,s^{-1};$$

the value of the hydrogen bond energy equal to 6.0 kcal/mol and the energy of the O – H bond equal to 109.7 kcal/mol. The values of ΔH^h and ΔG^h for the ion O^{2-}_{aq} are used in the following sections while analyzing the hydration processes of monomineral binders.

3 Gypsum Hemihydrate Hydration and Hydrolysis Reactions

Reactions in the system gypsum hemihydrate-water underlie the manufacturing of gypsum products and are therefore of particular theoretical as well as practical importance.

In spite of the seeming simplicity of the processes of gypsum hemihydrate hydration and numerous investigations in this field, certain problems have not been definitely solved [233, 260, 485, 501 – 504, 560, 561, 679]. Thus, the mechanism of hydration processes is still not clear enough. It remains unknown whether the dihydrate formation goes topochemically or if it is preceded by the hemihydrate dissolution to ions, the hydrolysis of ions Ca^{2+}, a rise of supersaturation in the solution after which its crystallization becomes possible.

It should be noted that earlier Le Chatelier found that the solubility of gypsum hemihydrate is from 4 to 5 times as high as that of gypsum dihydrate and amounts to ca. 8 g/l $CaSO_4$ at 20 °C. These solutions are metastable with respect to dihydrate and therefore at normal temperature will precipite. Studying the processes of gypsum hemihydrate solution and hydrolysis, Sakharov and Salomatin found that at the first stage of hardening hemihydrate water solutions have an acid reaction (pH being within the limits of 5.15 to 5.7), whereas this value of dihydrate water solutions is 7.05 [542]. In their opinion gypsum dihydrate owes its anomalous properties to complexing in these solutions.

It is necessary to begin the thermodynamic analysis of the reactions in this system with the general reaction of gypsum hemihydrate hydration.

$$CaSO_4 \cdot 0.5\,H_2O_{(cr)} + 1.5H_2O_{(liq)} = CaSO_4 \cdot 2H_2O_{(cr)} \quad (202)$$

$$\Delta G^0_{298} = -429.54 + 348.18 + 1.5 \cdot 56.69 = -1.33 \text{ kcal}.$$

Since $\Delta G = \Delta G^0_{298} + RT\ln K_p$, and

$$K_p = \frac{[CaSO_4 \cdot 2H_2O]}{[CaSO_4 \cdot 0.5H_2O] \cdot [H_2O]^{1.5}}$$

when $[CaSO_4 \cdot 2H_2O] = 1$; $[CaSO_4 \cdot 0.5H_2O] = 1$ and $[H_2O] = 1$, $K_p = 1$ and $\lg K_p = 0$, then $\Delta G = \Delta G^0_{298}$ and the value, and the sign of the Gibbs free energy at standard conditions characterize the real conditions of the gypsum hemihydrate hydration process. The value and sign of ΔG^0_{298} of Eq. (202) characterize the thermodynamic possibility of its occurrence and proves the irreversibility of the process at normal conditions.

Equation $\Delta G^0 = f(T)$ of Eq. (202) is expressed as:

$$\Delta G^0 = -50.3 + 1.66\,T\ln T + 4.8 \cdot 10^{-3}\,T^2 + 2 \cdot 10^5\,T^{-1} + 20.9\,T. \quad (203)$$

A diagrammatic representation of Eq. (203) is given in Fig. 77, where one sees that the equilibrium state of Eq. (202) is reached at 405 K, which corresponds to experimental data [474].

In connection with the irreversibility of the gypsum hemihydrate hydration reaction at a normal temperature, like hydration reactions in general, as shown below, one should especially concentrate on the problem of the applicability of the law of mass action and of the methods of classical thermodynamics to calculating equilibria in these systems.

F. Rossoti and H. Rossoti [530] were the first to suggest a possible approach to irreversible systems as to equilibrium ones. They presumed this possibility to be connected with the rate of ionic reactions. With low-rate reactions the consideration of ionic equilibria is quite admissible, even if they run irreversibly in the final analysis. This approach was substantiated in works [543, 546] as applied to the systems binder-water. Since the binder setting and hardening process rate

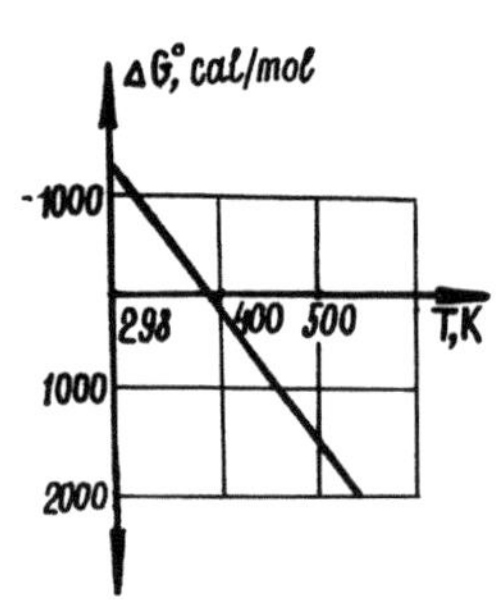

Fig. 77. The temperature dependence of ΔG^0 of the reaction of the gypsum hemihydrate hydration

comes to minutes, and that of clinker minerals, to hours, one certainly may use the methods of classical thermodynamics for calculating ionic equilibria in these systems.

The first calculations of ionic equilibria in the system gypsum hemihydrate-water were done by Kelley, Southard and Anderson [474]. The pH equilibrium value for α-hemihydrate amounted with them to 7.29, which differs greatly from the experimental value (5.5). Their results for dihydrate gypsum agrees with experimental data (pH = 7.05). The most comprehensive calculation of ionic equilibria in the system gypsum hemihydrate-water was carried out in [542].

Starting from Izmaylov's theory [215] on the formation of associates, cationic acids and anionic bases as applied to gypsum hemihydrate solutions, the authors found that while calculating ionic equilibria in these solutions it is necessary to consider the existence of the following complex molecules and ions: $CaOH^+$, $CaSO_4$, HSO_4^-, H^+, OH^-, H_2O [338].

The instability stoichiometric constants in the system gypsum hemihydrate-water determined by them proved to be:

1. $CaSO_4 \rightleftarrows Ca^{2+} + SO_4^{2-}$ $\quad K_1 = 5.2 \cdot 10^{-5}$
2. $CaOH^+ \rightleftarrows Ca^{2+} + OH^-$ $\quad K_2 = 4.6 \cdot 10^{-5}$
3. $HSO_4^- \rightleftarrows H^+ + SO_4^{2-}$ $\quad K_3 = 6.4 \cdot 10^{-2}$.

The anomalous pH value in gypsum hemihydrate water solutions is due to the fact that the hydrolysis by cation

$$Ca^{2+} + HOH \rightleftarrows CaOH^+ + H^+ \tag{204}$$

connected with the release of hydrogen ions predominates over the hydrolysis by anion

$$SO_4^{2-} + HOH \rightleftarrows HSO_4^- + OH^-, \tag{205}$$

which accounts for the self-acidifying.

The experimental determination of pH of gypsum hemihydrate solutions was conducted by these authors in suspensions and in a filtrate. It was demonstrated that the pH value of the filtered solutions decreases with the increase of the gypsum hemihydrate concentration up to about five in the solution. They failed to obtain the pH values lower than 5 in filtered solutions, although the time of the hemihydrate mixing with water and of the suspension filtration was reduced to the minimum.

Measured in suspensions, its pH value was as low as 3.9. The authors ascribe the difference in the pH values in filtered solutions and in concentrated suspensions to hydrolytic adsorption. Comparing these values of stoichiometric constants with those of thermodynamic ones in the system $CaSO_4 \cdot 2H_2O - H_2O$

($K_1 = 4.9 \cdot 10^{-3}$, $K_2 = 5.0 \cdot 10^{-2}$ and $K_3 = 7.6 \cdot 10^{-2}$, the gypsum hemihydrate calcium is in "active" state, and therefore the instability constant of complex molecules and ions with it (i.e., $CaSO_4$ and $CaOH^+$) differs greatly from those in common gypsum dihydrate.

We carried out calculations of ionic equilibria in the system gypsum hemihydrate-water according to standard methods [38], and conducted experimental research to determine the initial and equilibrium concentrations in this system.[13]

On the basis of data given in [185] the equilibrium constant of ionic reactions occurring in the system gypsum hemihydrate-water may be expressed as:

1. $H_2O_{(liq)} \rightleftarrows H^+ + OH^-$ $K_1 = 1 \cdot 10^{-14}$
2. $CaSO_{4(cr)} \rightleftarrows Ca^{2+} + SO_4^{2-}$ $K_2 = 2.4 \cdot 10^{-5}$
3. $Ca^{2+} + SO_4^{2-} \rightleftarrows CaSO_{4(aq)}$ $K_3 = 204$
4. $Ca^{2+} + OH^- \rightleftarrows CaOH^+$ $K_4 = 25$
5. $SO_4^{2-} + H^+ \rightleftarrows HSO_4^-$ $K_5 = 98$.

With the addition of the equation of material balance

$$L = [Ca^{2+}] + [CaOH^+] + [CaSO_4] , \tag{206}$$

(where L, solubility) and the equation of electroneutrality.

$$2[Ca^{2+}] + [H^+] + [CaOH^+] = 2[SO_4^{2-}] + [HSO_4^-] + [OH^-] \tag{207}$$

we obtain seven equations with seven unknowns. Solving them together we obtain the following values of equilibrium concentrations:

$[Ca^{2+}] = 4.86 \cdot 10^{-3}$
$[CaOH^+] = 1.4 \cdot 10^{-8}$
$[CaSO_4] = 4.84 \cdot 10^{-3}$
$[SO_4^{2-}] = 4.86 \cdot 10^{-3}$
$[HSO_4^-] = 1.4 \cdot 10^{-8}$
$[H^+] = 8.72 \cdot 10^{-8}$
$[OH^-] = 1.15 \cdot 10^{-7}$.

The pH equilibrium value is $-\lg 8.72 \cdot 10^{-8} = 7.06$.
Measurements gave the following results:

1. pH initial = 6.05
   pH final = 7.20
   (equilibrium)

[13] The calculations were performed with the assistance of Koshmay

2. $C_{Ca^{2+}\ \mathrm{initial}} = 5.4 \cdot 10^{-2}$ g-ion/l
   $C_{Ca^{2+}\ \mathrm{final}} = 5.1 \cdot 10^{-3}$ g-ion/l
   (equilibrium)

3. $C_{SO_4^{2-}\ \mathrm{initial}} = 4.98 \cdot 10^{-2}$ g-ion/l
   $C_{SO_4^{2-}\ \mathrm{final}} = 5.14 \cdot 10^{-3}$ g-ion/l,
   (equilibrium)

which is in agreement with the above calculated data.

4 Reactions in the System $CaO - H_2O$

The reaction of calcium oxide hydration is of great importance in the processes of the manufacturing of lime binders and products as well as materials based on them [465, 665]. At low temperatures calcium oxide is a compound quite unstable with respect to water, therefore with water it cannot be free. Let us consider the following reaction:

$$CaO_{(sol)} + H_2O_{(liq)} = Ca(OH)_{2\ (sol)} \,. \tag{208}$$

The alteration of the Gibbs free energy of the reaction at standard conditions is

$$\Delta G^0_{298} = -214.39 + 144.4 + 56.69 = -13.30 \text{ kcal}\,.$$

Since the activities of all the reactants are equal to one, the equilibrium constant of the reaction (if equilibrium were reached) is equal to one as well, i.e.,

$$K = \frac{[Ca(OH)_2]}{[CaO] \cdot [H_2O]} = 1\,.$$

With the help of the well-known Eq. (65) we obtain:

$$\Delta G = \Delta G^0 + 1.364 \lg K = \Delta G^0 + 0 = -13.0 \text{ kcal/mol}\,.$$

Thus, at 298 K in the reaction of calcium oxide hydration with liquid water no equilibrium state can be achieved, because all the reactants are solids having activities equal to one. Equilibrium state in such reactions may be achieved only by altering temperature, which is easily found by conducting special calculations. For reaction (207) equation $\Delta G^0 = f(T)$ is

$$\Delta G^0 = -16649 - 0.19\,T \ln T + 3.79 \cdot 10^{-3}\,T^2 + 2.02 \cdot 10^5\,T^{-1} + 9.13\,T\,.$$

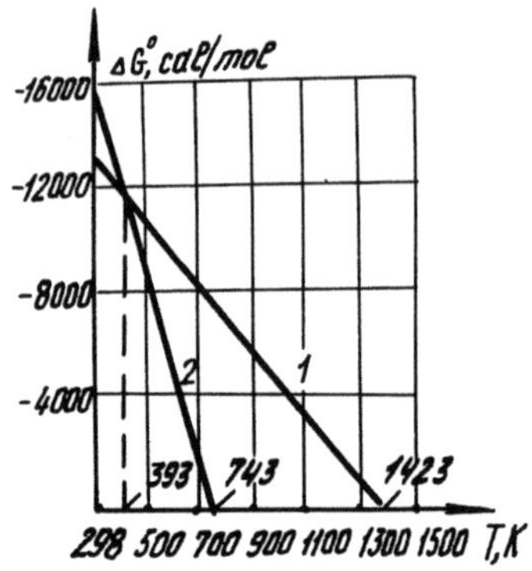

Fig. 78. The temperature dependence of the ΔG^0 of the reactions of calcium oxide hydration: *1* CaO (cr) + H_2O (liq) = $Ca(OH)_2$ (cr); *2* CaO (cr) + H_2O (gas) = $Ca(OH)_2$ (cr)

As at equilibrium ΔG^0 equals zero, solving this equation for temperature, we find its value for equilibrium state. $T_{\text{equilibrium}} = 1423$ K.

Whether it is possible to hydrate lime with vaporous water it also of interest:

$$CaO_{(sol)} + H_2O_{(gas)} = Ca(OH)_{2(sol)}; \tag{209}$$

$$\Delta G^0_{298} = -214.39 + 144.4 + 54.634 = -15.356 \text{ kcal/mol}$$

$K = \dfrac{[Ca(OH)_2]}{[CaO] \cdot [H_2O]}$. Since the activities of $[Ca(OH)_2]$ and [CaO] are equal to one, and $[H_2O]$ may be replaced by P_{H_2O}, then

$$K = \frac{1}{P_{H_2O}} \quad \text{and} \quad K = -\lg P_{H_2O};$$

$$\lg K = \frac{15.356}{1.364} = 11.26;$$

$$11.26 = -\lg P_{H_2O}; P_{H_2O} = 10^{-11.26}.$$

The equilibrium partial pressure of water vapour at 298 K amounts to an insignificant value of $10^{-11.26}$, which reveals the hydration reaction as thermodynamically possible at partial pressure of water vapour $>10^{-11.26}$ atm.

Equation $\Delta G^0 = f(T)$ with reaction (209) is expressed as:

$$\Delta G^0 = -27174 - 0.95\,T\ln T - 3.4 \cdot 10^{-3}T^2 + 0.73 \cdot 10^5 T^{-1} + 45.4\,T. \tag{210}$$

A diagrammatic interpretation of Eqs. (208) and (210) is given in Fig. 78, where one can see that lime hydration is possible within a wide temperature range. With liquid water equilibrium is achieved only at a temperature of 1423 K, with gaseous water – at a water vapour pressure of 1 atm at 743 K. These calculation results suggest that lime hydration with vapour of a temperature of up to 393 K is advantageous from the energy viewpoint as compared to that with liquid water, a fact which specialists should consider in the search of optimum conditions for lime slaking. Let us now continue the thermodynamic analysis of possible stages of the process of the calcium oxide hydration with liquid water at normal conditions.

We will first consider what may occur if calcium oxide is put into liquid water at $T = 298$ K. What kind of active concentrations of ions Ca^{2+} and OH^- will result?

Let us examine the reaction

$$CaO_{(sol)} + H_2O_{(liq)} = Ca^{2+}_{(aq)} + 2OH^-_{(aq)} \tag{211}$$

and discern which conditions allow for the equilibrium state of this reaction.

$$\Delta G^0_{298\ \text{reaction}} = -132.10 - 2 \cdot 37.59 + 144.4 + 56.69 = -6.19\ \text{kcal/mol}$$

$$\Delta G_{\text{reaction}} = \Delta G^0_{298\ \text{reaction}} + 1.364\ \lg \frac{[Ca^{2+}] \cdot [OH^-]^2}{[CaO] \cdot [H_2O]},$$

bearing in mind that the activities of the solid CaO and liquid H_2O are equal to one, $\Delta G^0 = -6.19$, and substituting $10^{-14}/[H^+]$ for $[OH^-]$ we obtain:

$$\Delta G_{\text{reaction}} = -6.19 + 1.364\ \lg\ [Ca^{2+}] \cdot \frac{10^{-28}}{[H^+]^2}$$

or

$$\Delta G_{\text{reaction}} = -6.19 + 1.364\ \lg\ [Ca^{2+}] - 2 \cdot 1.364\ \lg\ [H^+] - 38.18\ .$$

Substituting pH for $-\lg\ [H^+]$ we derive at the end

$$\Delta G_{\text{reaction}} = -44.37 + 1.364\ \lg\ [Ca^{2+}] + 2.728\ \text{pH}\ . \tag{212}$$

In case of equilibrium the equation is expressed as:

$$\lg\ [Ca^{2+}] = 32.53 - 2\ \text{pH}\ . \tag{213}$$

From the diagrammatic representation of this equation given in Fig. 79 one can easily see that at 298 K in the entire range of the pH values calcium oxide with liquid water practically cannot be stable, as the equilibrium activities of calcium ions surpass the value of $10^{+4.6}$. In all cases this guarantees the solubility product of lime hydrate and its precipitation.

With actual activities of calcium ions in the solution (we shall restrict them to 10^{-1} to 10^{-6} g-ions Ca^{2+}/l) the pH values are higher than 14 (from 16.8 to 19.3), which again allow us to conclude that in reaction (211) equilibrium is not practically achieved, but before it, the crystallization of a hydrated form of calcium oxide $Ca(OH)_2$ more stable under these conditions begins.

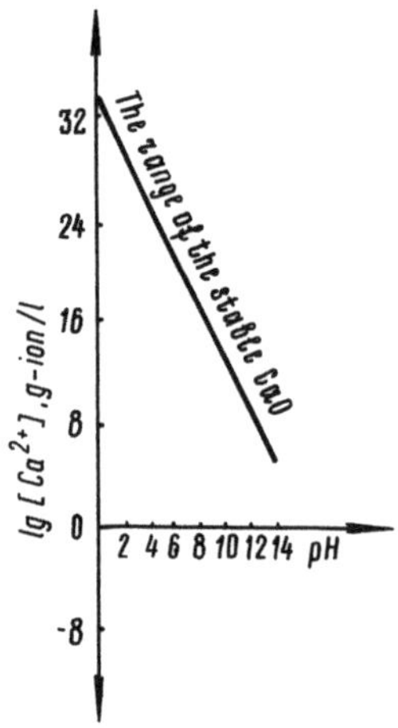

Fig. 79. The ranges of the stability of calcium oxide in the coordinates of pH-lg $[Ca^{2+}]$ at 25 °C and 1 atm of general pressure

As the foregoing statement is true, let us examine the reaction of the calcium hydroxide crystallization in more detail.

$$Ca^{2+}_{(aq)} + 2OH^{-}_{(aq)} = Ca(OH)_{2(cr)} . \tag{214}$$

Having carried out calculations as above, we obtain the following formula:

$$\Delta G^0 = 31.23 - 1.364 \lg [Ca^{2+}] - 2.728 \text{ pH} , \tag{215}$$

and for equilibrium conditions

$$\lg [Ca^{2+}] = 22.9 - 2 \text{ pH} . \tag{216}$$

A diagrammatic representation of Eq. (216) is given in Fig. 80. As one can see, for achieving stable state at pH above 12 absolutely insignificant activities of calcium ions are required. Thus, for pH = 12 the equilibrium active concentration equals $10^{-1.1}$ and for pH = 12.6 even less, namely $10^{-2.3}$. Therefore upon the appearance of $10^{-2.3}$ calcium ions in the solution at pH = 12.6, the precipitation of calcium hydroxide becomes possible. At pH = 13 the precipitation of $Ca(OH)_2$ from the solution may start, the activity of calcium ions being just $10^{-3.9}$ etc.

Comparing Figs. 79 and 80, one can certify that the entire stability range of $Ca(OH)_2$ lies within the instability range of CaO, which testifies to the irreversible character of the lime hydration processes. A stable equilibrium in the system $CaO - H_2O$ is established only after all CaO has been hydrated.

According to Eq. (213) the solubility product of calcium hydroxide is $1 \cdot 10^{-5.1}$. Considering that in the process of $Ca(OH)_2$ crystallization, as well as of its dissolving, the reaction stoichiometry is maintained, on achieving the equi-

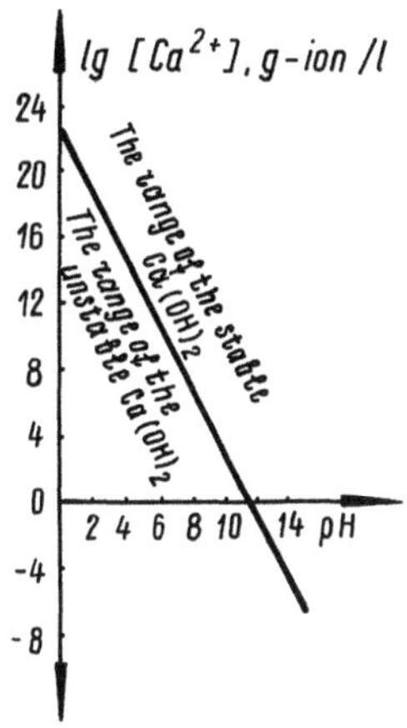

Fig. 80. The ranges of the stability of calcium hydroxide in the coordinates of pH-lg $[Ca^{2+}]$ at 25 °C and 1 atm of general pressure

librium state, 1 g-ion of Ca^{2+} must correspond to $\frac{1}{2}$ g-ion of ions OH. This enables us to calculate the equilibrium value of pH at the crystallization (dissolution) of calcium hydroxide:

$$Ca(OH)_{2(sol)} \rightleftarrows Ca^{2+}_{(aq)} + 2OH^{-}_{(aq)}\,. \tag{217}$$

The solubility product of calcium hydroxide is

$$Sp_{Ca(OH)_2} = [Ca^{2+}] \cdot [OH^-]^2 = 1 \cdot 10^{-5.1}\,.$$

Bearing in mind that $[Ca^{2+}]$ is $\frac{1}{2}[OH^-]$ (see above),

$$Sp = \tfrac{1}{2}[OH^-] \cdot [OH^-]^2 = \tfrac{1}{2}[OH^-]^3\,,$$

$10^{-5.1} = \frac{1}{2}[OH^-]^3$, whence $[OH^-] = \sqrt[3]{2 \cdot 10^{-5.1}} = 10^{-1.6}$ or $[H^+] = 10^{-12.4}$ or $pH = 12.4$. The calculation of pH starting from thermodynamic data ($\Delta G^0_{\text{reaction (206)}} = -132.1 - 2 \cdot 37.59 + 214.39 = +7.11$) also yields 12.37.

For analyzing possible patterns of the interaction of CaO and water we calculated the alterations of ΔH^0 and ΔG^0 of the reactions under standard condi-

Table 83. Thermodynamic data for ions formed at the calcium oxide dissolution and hydration with infinite dilution

Ions	State	ΔH^h [kcal/g-ion]	ΔG^h [kcal/g-ion]	ΔS^h [cal/g-ion deg]
Ca^{2+}	aq	−386.0	−373.0	−43.9
OH^-	aq	−122.0	−111.0	−37.6
H^+	aq	−265.0	–	–
O^{2-}	aq	−436.0	−418.0	−43.9

tions and determined the so-called chemical heats of hydration and the hydration Gibbs free energy at an infinite dilution of ions (ΔH^h and ΔG^h).

Initial data for compounds and ions in the standard state are taken from Appendices 2 and 3, those for infinite dilution, from the work of Mishchenko and Poltoratskiy [437], ΔH^h, ΔG^h and ΔS^h for the hypothetical ion O^{2-} are taken from Table 82. Data for ions at infinite dilution are given in Table 83.

The results of the calculations of the reactions of calcium oxide dissociation, dissolution and hydration are given in Table 84.

Moreover, ΔH^h, ΔG^h, ΔH^0_{298} and ΔG^0_{298} of the reactions of the ion O^{2-} transition to OH^- and those of the reactions of $Ca(OH)_2$ crystallization from solution were calculated. The calculation results are given in Tables 85 and 86.

From Table 84 one can see that although the reaction of the CaO dissolving to ions $Ca^{2+}_{(aq)}$ and $O^{2-}_{(aq)}$ is thermodynamically possible with an infinite dilution (reaction 2, Table 84, $\Delta G^h = -646.0$ kcal/mol), thermodynamically the reaction of the ion O^{2-} transformation to $OH^-_{(aq)}$ is impossible (reaction 3, Table 85, $\Delta G^h = 252.69$ kcal/mol), therefore the summary reaction $CaO + H_2O$ may not include any intermediate stages, and ions Ca^{2+} and $2OH^-$ must be formed at once. Considering that the mechanism of the calcium oxide hydration process is of principal importance, we conducted special research using the theory of ionic equilibria and the methods of electrochemical analysis.[14]

Irrespective of the hydration mechanism the final product in the system $CaO - H_2O$ forms the following ionic equilibria:

$$Ca(OH)_2 \rightleftarrows Ca^{2+} + 2OH^- \qquad P, \tag{218}$$

$$Ca^{2+} + OH^- \rightleftarrows Ca(OH)^+ \qquad K^{-1}. \tag{219}$$

For calculating ionic equilibria we shall introduce the following symbols:

$$C^i_{Ca(OH)_2} = C^i_{Ca^{2+}} = C^i; \quad [Ca^{2+}] = x; \quad [CaOH^+] = y;$$

$[OH^-] = z$, where C^i denotes initial concentrations and square brackets, equilibrium ones. In this way we obtain the following system of equations:

$$xz^2 = P, \tag{220}$$

$$xz = Ky, \tag{221}$$

$$z = 2x + y, \tag{222}$$

the latter equation being arrived at starting from the principle of the solution electroneutrality.

[14] The calculations were conducted with the assistance of Koshmay

Table 84. The results of the calculations of ΔH^h and ΔG^h, ΔH^0_{298} and ΔG^0_{298} of CaO dissociation, dissolution and hydration

Nos.	Dissociation, dissolution and hydration reactions	ΔH^h [kcal/mol]	ΔG^h [kcal/mol]	ΔH^0_{298} [kcal/mol]	ΔG^0_{298} [kcal/mol]
1.	$CaO_{(cr)} = Ca^{2+}_{(gas)} + O^{2-}_{(gas)}$	–	–	+842.19	+831.67
2.	$CaO_{(cr)} \overset{aq}{=} Ca^{2+}_{(aq)} + O^{2-}_{(aq)}$	−670.1	−646.0	+21.5	+37.9
3.	$CaO_{(cr)} + H_2O_{(liq)} = Ca^{2+}_{(aq)} + 2OH^-_{(aq)}$	−409.78	−393.91	−19.42	−6.19
4.	$CaO_{(cr)} \overset{aq}{=} Ca^{2+}_{(gas)} + O^{2-}_{2(aq)}$	+176.19	+177.46	+619.49	+621.06
5.	$CaO_{(cr)} \overset{aq}{=} Ca^{2+}_{(aq)} + O^{2-}_{(gas)}$	+4.1	+7.87	+252.2	+248.77
6.	$CaO_{(cr)} + H_2O_{(liq)} = Ca^{2+}_{(gas)} + 2OH^-_{(aq)}$	+436.51	+430.15	+570.57	+576.97
7.	$CaO_{(cr)} + H^+_{(aq)} = Ca^{2+}_{(gas)} + OH^-_{(aq)}$	−558.51	+541.15	+625.54	+614.56
8.	$CaO_{(cr)} + H^+_{(aq)} = Ca^{2+}_{(aq)} + OH^-_{(aq)}$	−356.1	−339.6	−32.77	−25.29
9.	$CaO_{(cr)} + H_2O_{(liq)} = Ca^{2+}_{(gas)} + 2O^{2-}_{(gas)} + 2H^+_{(gas)}$	+1874.83	+1859.38	+1874.83	+1859.58
10.	$CaO_{(cr)} + H_2O_{(liq)} = Ca^{2+}_{(aq)} + 2O^{2-}_{(aq)} + 2H^+_{(aq)}$	−1037.38	−1003.91	+89.12	+120.19
11.	$CaO_{(cr)} + 2H^+_{(aq)} = Ca^{2+}_{(aq)} + H_2O_{(liq)}$	−302.42	−285.29	−46.17	−44.39
12.	$CaO_{(cr)} + 2OH^+_{(aq)} = Ca^{2+}_{(aq)} + 2O^{2-}_{(aq)} + H_2O_{(liq)}$	−930.01	−859.29	+62.42	+52.91

Table 85. The results of the calculations of the reactions of the ion O^{2-} transition to OH^-

Nos.	Reactions of the ion O^{2-} transformation	ΔH^h [kcal/mol]	ΔG^h [kcal/mol]	ΔH^0_{298} [kcal/mol]	ΔG^0_{298} [kcal/mol]
1.	$O^{2-}_{(gas)} + H_2O_{(liq)} = 2 \cdot OH^-_{(aq)}$	−405.68	−401.78	−271.62	−254.96
2.	$O^{2-}_{(gas)} + H^+_{(aq)} = OH^-_{(aq)}$	−87.0	–	−284.97	−274.06
3.	$O^{2-}_{(aq)} + H_2O_{(liq)} = 2 \cdot OH^-_{(aq)}$	+260.32	+252.69	−40.92	−44.09
4.	$O^{2-}_{(aq)} + H^+_{(aq)} = OH^-_{(aq)}$	+579.0	–	−54.07	−63.19

Table 86. The results of the calculations of the reactions of $Ca(OH)_2$ crystallization from solution

Nos.	Crystallization reactions	ΔH^0_{298} [kcal/mol]	ΔG^0_{298} [kcal/mol]
1.	$Ca^{2+}_{(gas)} + O^{2-}_{(gas)} + H_2O_{(liq)} = Ca(OH)_{2(cr)}$	−857.3	−845.23
2.	$Ca^{2+}_{(aq)} + O^{2-}_{(aq)} + H_2O_{(liq)} = Ca(OH)_{2(cr)}$	−36.61	−51.7
3.	$Ca^{2+}_{(aq)} + 2OH^-_{(aq)} = Ca(OH)_{2(cr)}$	+4.31	−7.11
4.	$Ca^{2+}_{(gas)} + 2O^{2-}_{(gas)} + 2H^+_{(gas)} \overset{aq}{=} Ca(OH)_{2(cr)}$	−1522.7	−1505.55
5.	$Ca^{2+}_{(aq)} + 2O^{2-}_{(aq)} + 2H^+_{(aq)} = Ca(OH)_{2(cr)}$	−104.23	−133.49

Assuming that $K=4\cdot10^{-2}$ and $P=5.6\cdot10^{-6}$ [185], let us calculate equilibrium concentrations.

$$x=\frac{P}{z^2};\quad y=\frac{xz}{K}=\frac{P}{Kz}$$

$$z=\frac{2P}{z^2}+\frac{P}{Kz};\quad z^3-\frac{P}{K}\cdot z-2P=0$$

$$z^3-1.4\cdot10^{-4}\cdot z-11.2\cdot10^{-6}=0\,.$$

Solving the latter equation we have $z=2.44\cdot10^{-2}$, from which $y=5.7\cdot10^{-3}$, $x=9.34\cdot10^{-3}$. The $Ca(OH)_2$ solubility L is then found:

$$L=C^i=x+y=1.504\cdot10^{-2}\,.$$

pH of the solution in the system of ionic equilibria according to scheme is found from conditions

$$[H^+]\cdot[OH^-]=10^{-14}$$

$$[H^+]=\frac{10^{-14}}{Z}=\frac{10^{-14}}{2.44\cdot10^{-2}}=0.41\cdot10^{-12}$$

$$pH=-\lg[H^+]=-\lg 0.41\cdot10^{-12}=12.39\,.$$

Thus, $Ca(OH)_2$ dissociating in water gives rise to pH = 12.39.

If the $Ca(OH)_2$ formation proceeds topochemically, i.e., in solid phase with dissociating, the pH value cannot exceed 12.39. However, if $Ca(OH)_2$ is formed when the CaO solution products interact, the given values may be different on account of $C^i_{OH^-}\neq0$. In this case the $Ca(OH)_2$ formation must proceed by scheme 2.

$$CaO+H_2O\rightarrow CaOH^++OH^-$$

$$CaOH^+\rightleftarrows Ca^{2+}+OH^-\qquad K$$

$$Ca^{2+}+2OH^-\rightleftarrows Ca(OH)_2\qquad P^{-1}\,.$$

In this scheme the first process is nonequilibrium one determining the initial concentrations of ions $CaOH^+$ and OH^-. The other two are equilibria, effected by $C^i_{CaOH^+}$ and $C^i_{OH^-}$. Let us calculate equilibria in accordance with the equations of scheme 2. Using symbols $C^i_{CaOH^+}=S$, $C^i_{OH^-}=S$ for the first equation; $[Ca^{2+}]=a$, $[OH^-]=a$ and $[CaOH^+]=S-a$ for the second one; $C^i_{Ca^{2+}}=a$, $C^i_{OH^-}=2a$, $[Ca^{2+}]=x$, $[OH^-]=b$ for the third one we arrive at

$$\frac{a^2}{S-a}=K$$

$$xb^2=P$$

$$b=2x+S-a\,.$$

As the system of three equations contains four unknowns, in order to solve it we shall make some allowances. In the equation $b=2x+S-a$ component $2x$ corresponds to the $[OH^-]$ equilibrium concentration of the process

$$Ca(OH)_2 \rightleftarrows Ca^{2+}+2OH^-\,,$$

therefore $x(2x)^2=4x^3=P$, from which $x=\sqrt[3]{P/4}=\sqrt[3]{1.4\cdot 10^{-6}}=1.12\cdot 10^{-2}$ and $2x=2.24\cdot 10^{-2}$. In the same equation the component $S-a$ corresponds to $[OH^-]$ of processes in scheme 1. Indeed, ionic equilibria by scheme 2 being established, for both equations we have: $[OH^-]=S+a-2a=S-a$; $[CaOH^+]=y$; $[Ca^{2+}]=x$. In this case

$$x(S-a)^2=P$$

$$x(S-a)=Ky$$

$$S-a=2x+y\,.$$

The system of equations derived corresponds to equilibrium processes by scheme 1. Consequently $S-a=2.44\cdot 10^{-2}$ and $b=2x+S-a=2.24\cdot 10^{-2}+2.44\cdot 10^{-2}=4.68\cdot 10^{-2}$. We arrive at $x=P/b^2=2.56\cdot 10^{-3}$ and $a=\sqrt{(S-a)C_2}=3.12\cdot 10^{-2}$ for the overall process.

pH of the solution, in case of the processes taking place by scheme 2, will increase:

$$pH=-\lg[H^+]=-\lg\frac{10^{-14}}{4.68\cdot 10^{-2}}=12.67\,.$$

Scheme 2 represents the process of $Ca(OH)_2$ formation by way of crystallization from solution. This is obviously possible solely at $S+a-2a>0$ or $S-a>0$ as $b=2x+S-a$ this requires $b-2x>0$. Calculations showed that, if $b=2x$, $b=2.24\cdot 10^{-2}$, i.e., pH = 12.35.[15] Thus, the process of the CaO hydration goes topochemically if in an equilibrium system establishes pH < 12.35.

Studying experimentally the mechanism of the $Ca(OH)_2$ hydration we conducted a simultaneous potentiometric measurement of the pH and pCa values. In

---

[15] Comparing this value with that calculated by us before (12.37) we can see that applying tentative calculations on the basis of thermodynamic data is quite reasonable

so doing one may consider the errors of a glass electrode in the high-alkali range and evaluate the suspension effect [142].

The pCa measurement was carried out by means of an ionoselective electrode of EM-SA-OI type with the EV-74 pH meter. Previously the possibility of such measurements in water-cement suspensions [122] was shown.

The $Ca(OH)_2$ suspension was provided for the test by slaking CaO with water, the water to solid ratio being 1/1. The pH and pCa measurements were carried out directly in the suspension without isolating the liquid phase. This helped to eliminate the difference between the active concentrations of the same solution in suspension and in the free state.

The test demonstrated that in the $Ca(OH)_2$ suspension pH = 12.64 and $[Ca^{2+}] = 2.9 \cdot 10^{-3}$. The decrease of the ions Ca^{2+} concentration as compared to the calculated one follows as the true pH of the suspension is higher than that calculated by the equilibrium. It results from the initial C_{OH^-} concentration in the CaO interaction with water and is conditioned by the kinetics and not thermodynamics of the process.

Thus, the process of $Ca(OH)_2$ formation in the system $CaO + H_2O$ takes place through dissolving, but not topochemically.

The most probable scheme of the calcium oxide hydration process, as in the case of gypsum hemihydrate, is the "through solution" scheme. When using the term "through solution mechanism" of the calcium oxide hydration it should be borne in mind that the first stage of the hydration process is not just the solution of CaO in water to constituent ions, but is a chemical reaction of the CaO and water ionization with Ca^{2+} and $2OH^-$ being formed.

As slaked lime is a constituent of several binders, it is worth detailing its thermodynamic properties and behaviour in solution. Calcium hydroxide dissociation in water takes place in two stages:

$$Ca(OH)_2 \rightleftarrows CaOH^+_{(aq)} + OH^-_{(aq)}; \quad K_1 = \frac{[CaOH^+] \cdot [OH^-]}{[Ca(OH)_2]} = 1.4 \cdot 10^{-4} \quad (A)$$

$$CaOH^+_{(aq)} \rightleftarrows Ca^{2+}_{(aq)} + OH^-_{(aq)}; \quad K_2 = \frac{[Ca^{2+}] \cdot [OH^-]}{[CaOH^+]} = 4.0 \cdot 10^{-2}. \quad (B)$$

The constants of the hydrated lime dissociation available permit calculation of the equilibrium proportions of various forms of lime in the solution. Denoting the overall contents of all types of lime as 100%, we arrive at the following equations for determining the percentage of particular forms of it as a function of the concentration of ions OH^- (or pH) and of the constants of the dissociation of the first and the second stages:

$$\%Ca(OH)_2 = 100 \cdot \frac{[OH^-]^2}{[OH^-]^2 + K_1[OH^-] + K_1 \cdot K_2} \quad (C)$$

Table 87. The contents of various forms of lime as a function of the pH value

Forms of lime	Percentage at various pH							
	6	7	8	10	11	12	13	14
$Ca(OH)_2$	0	0	0	0.18	14.84	93.45	99.86	99.986
$Ca(OH)^+$	0	0	0.025	0.25	2.07	1.310	0.14	0.014
Ca^{2+}	100	100	99.975	99.57	83.09	5.24	0	0

$$\%\mathrm{CaOH}^+ = 100 \cdot \frac{[\mathrm{OH}^-] \cdot K_1}{[\mathrm{OH}^-]^2 + K_1[\mathrm{OH}^-] + K_1 \cdot K_2} \tag{D}$$

$$\%\mathrm{Ca}^{2+} = \frac{K_1 \cdot K_2}{[\mathrm{OH}^-]^2 + K_1[\mathrm{OH}^-] + K_1 \cdot K_2} . \tag{E}$$

The results of the calculations by Eqs. (C), (D) and (E) are given in Table 87. From these data one can see that at pH = 12 the main form of $Ca(OH)_2$ is undissociated $Ca(OH)_2$, because 93.45% of all the lime in the solution is in this form. It is remarkable that at pH = 13 about 100% of lime is in undissociated state, i.e., in the form of $Ca(OH)_2$. pH decreasing to 11, there is about 15% (14.84%) of undissociated $Ca(OH)_2$, and at pH = 9 – is as good as none.

Thus, at elevated pH values (>12) lime hydrate cannot be viewed as a strong electrolyte, because under these conditions it fails to fully dissociate into ions.

The temperature dependence of ΔG^0, ΔH^0 and ΔS of the calcium hydroxide dissolving in water was studied by Greenberg and Copeland [168]. They found that in the process

$$\mathrm{Ca(OH)_{2(sol)}} + n\mathrm{H_2O_{(liq)}} \rightleftarrows \mathrm{Ca^{2+}_{(aq)}} + 2\mathrm{OH^-_{(aq)}}$$

$$\Delta G^0 = -17.67\,RT\ln T + 0.078\,RT^2 + 89\,RT;$$

$$\Delta H^0 = -17.67\,RT - 0.078\,RT^2;$$

$$\Delta S = 17.67\,RT\ln T - 0.156\,RT - 71.3\,R\,;$$

$$\ln K_0 = 17.67\ln T - 0.078\,T - 89\,.$$

The degree of the $Ca(OH)_2$ molecules dissociation at 298 K in a semisaturated solution is 90% according to their calculations, which is in agreement with the data of Table 87.

The $Ca(OH)_2$ solubility in pure water at diverse temperatures is given in Table 88.

Table 88. The $Ca(OH)_2$ solubility in water at diverse temperatures

Temp. [K]	Concentration [g CaO/1000 ml H_2O]	Temp. [K]	Concentration [g CaO/1000 ml H_2O]
273	1.3	372	0.523
288	1.22	393	0.4
298	1.13	398	0.38
313	1	423	0.247
323	0.917	463	0.084
333.8	0.818	473	0.05
354.7	0.657	523	0.037
363	0.591	–	–

5 Reactions in the System SiO_2-H_2O

Silicon dioxide is the final and most stable (at normal conditions) product of silico-oxygen acid polymerization. The molecule of SiO_2 is polymeric and is formulated by $(SiO_2)_n$ where n is a great number. In nature silica occurs not only in the form of quartz – hexagonal structure – and its modifications, such as cristobalite – cubic structure – and tridymite – hexagonal structure, but in the form of amorphous silica and glass as well.

Several acids, being various degrees of water accumulation by silicic anhydride SiO_2, correspond to it, well up to the higher orthosilicic acid H_4SiO_4. As it is not possible to separate in the pure form and isolate silicic acids, one can only determine their formula by starting from the composition and structure of their salts, natural and artificial silicates. From the structure of silicates it is possible to deduce the formulae of silicic acids of any complexity and water accumulation degree.

All transformations of silicic acids having a common formula of $m\,SiO_2 \cdot n\,H_2O$ are at ordinary conditions directed irreversibly towards water detachment and finalize in quartz, the form with the least water content. However, in the presence of metal cations the transformation of some forms of silicic acid to others will be determined by the character of the products derived therewith, by the degree of their solubility and polymerization which are a function of pH of the medium.

We must consider the great difficulties connected with the study of the system silica-water resulting from the colloidal character of most silicic acids and from the variability of their composition due to polymerization and dehydration. This system requires further studies.

The energic correlation between the three different crystalline forms of silica (data taken from Iler [211]) is given in Fig. 81.

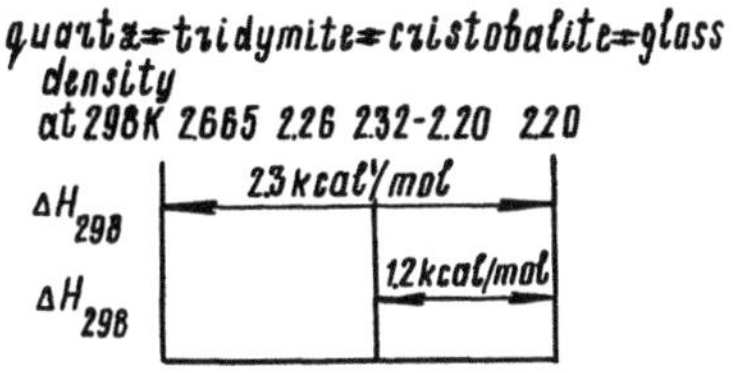

Fig. 81. The correlation between various crystalline forms of silica

The difference of the heats of the formation of low-temperature forms of quartz and cristobalite is about 1 kcal/mol. The difference between the heats of the formation of fine amorphous silica and of cristobalite is about 1.2 kcal/mol. By their total the heat of the crystalline quartz formation from vitreous silica was calculated, whose value is in agreement with the data of Sosman, 2.3 kcal/mol [586].

The heat of β-quartz dissolving in water is 7.34 ± 0.37 kcal/mol and that of amorphous silica, 2.65 ± 0.28 kcal/mol [167]. The solubility of various forms of silica in water was studied by several scientists [170, 261, 262, 445]. On the basis of these data Iler [211] plotted diagrams of the temperature dependence of the crystalline quartz and amorphous silica solubility (Fig. 82). As one can see, the solubility of amorphous silica is higher than that of crystalline quartz. The silica solubility greatly depends on pH of the medium, increasing with the increase of pH.

Different opinions exist as to the nature of silica dissolved in water. If was found [188, 460] that with molybdic acid, photocolorimetric measurements of SiO_2 in solution being conducted, silica reacts in the solution in ionic and not in colloid form. The investigations of Weitz, Frank and Schuchard [686], of Jander and Jahr [217] and of the brothers Brintzinger [63] showed dissolved silica found in the solution in monomeric or dimeric state actually not to break down to ions, but to be in hydrated form.

With alkalies, however, the silica solubility soars at a high pH value (beginning with 10.7), which seems to involve silica depolymerization, silicate ions thus being formed [211]. If, however, the solution is neutralized with acid and pH falls below 10.7, silicate ions are hydrated to silicic acid which then polymerizes to silica. Further investigations conducted by Jander and Heukeshoven [216],

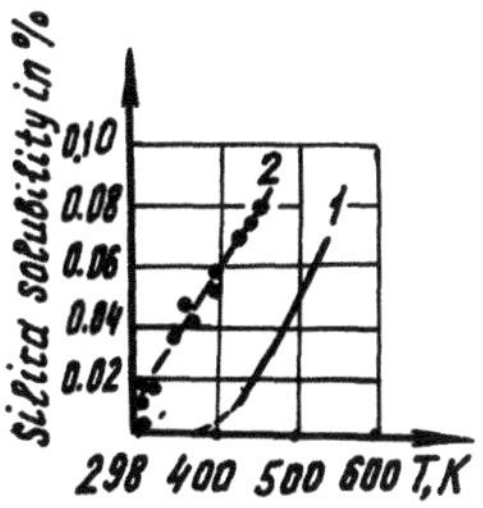

Fig. 82. The solubility in water of crystalline (*1*) and amorphous (*2*) silica at different temperatures

Table 89. The content of various forms of orthosilicic acid as a function of the pH value

Forms of silicic acid	Percentage in the solution at various pH							
	8	9	10	11	12	12.5	13	14
H_4SiO_4	98.33	85.47	36.58	4.60	0.12	0	0	0
$H_3SiO_4^{1-}$	1.67	14.53	62.18	78.23	20.02	3.07	0.45	0.01
$H_2SiO_4^{2-}$	0	0	1.24	15.61	39.93	23.13	9.04	0.98
$HSiO_4^{3-}$	0	0	0	1.56	39.93	73.16	90.38	97.47
SiO_4^{4-}	0	0	0	0	0	0.04	0.13	1.54

based on measuring the diffusion rate, demonstrated that at pH above 13.6 the metasilicate ion corresponding to a monomer is stable. Between pH = 13.6 und pH = 10.9 the disilicate ion is the main anion present in the solution.

The mechanism of the quartz solution in water consists in the depolymerization and hydrolysis of siloxane bonds

$$-\overset{|}{\underset{|}{Si}}-O-\overset{|}{\underset{|}{Si}}-\ ,$$

groups $-\overset{|}{\underset{|}{Si}}-OH$ being formed. Thus, eventually orthosilicic acid $Si(OH)_4$ is formed which will be either in the hydrated or in the ionic form depending on pH.

The dissociation of silicic acids to ions is quite insignificant and depends very little on temperature. In accordance with Greenberg and Price [167], Flint and Wells [137] and with the report by Thilo [625] and coworkers concerning the fourth constant, for calculating the percentage of different forms of orthosilicic acid as a function of pH we accepted the following dissociation constants:

$$K_1 = 1.7 \cdot 10^{-10},\ K_2 = 1.995 \cdot 10^{-12},\ K_3 = 1 \cdot 10^{-12} \text{ and } K_4 = 1.583 \cdot 10^{-16}$$

(as ΔG^0_{298} for ion $SiO_4^{4-} = -245.5$ kcal/g-ion [448]).

The calculation results are given in Table 89.

While carrying out calculations for metasilicic acid the following constants were accepted in accordance with the data of Roller and Ervin [525]: $K_1 = 10^{-9.8}$ and $K_2 = 10^{-12.16}$ near to those of [225, 458].

The calculation results are given in Table 90.

From Tables 89 and 90 one can see that at pH less than 8 the two forms of silicic acid are in undissociated state. Using thermodynamic data on undissociated molecules of $H_4SiO_{4\,(aq)}$ and $H_2SiO_{3\,(aq)}$ we calculated equilibrium concentrations of SiO_2 (β-quartz), $H_4SiO_{4\,(cr)}$, $H_2SiO_{3\,(cr)}$ and $SiO_{2\,(am)}$ in contact with water. The results of these tentative calculations are given in Fig. 83. The β-

Table 90. The content of various forms of metasilicic acid as a function of the pH value

Forms of silicic acid	Percentage in the solution at various pH									
	4	6	8	9	10	11	12	12.5	13	14
H_2SiO_3	100	99.98	97.96	82.02	31.03	3.80	0.18	0.02	0	0
$HSiO_3^-$	0	0.02	2.14	17.95	67.89	83.04	38.62	16.62	5.93	0.6
SiO_3^{2-}	0	0	0	0.03	1.08	13.16	61.20	83.36	94.07	99.4

quartz solubility is $10^{-3.97}$ g-mol/l of H_4SiO_4, which is close to the experimental value of $2.8 \cdot 10^{-4.2}$ g-mol [154]. As one can see from Fig. 83, amorphous silica has the highest solubility, $10^{-2.84}$ g-mol/l H_4SiO_4.

Thermodynamic calculations show that orthosilicic acid is more stable than metasilicic one.

$$H_2SiO_{3\,(cr)} + H_2O_{(liq)} = H_4SiO_{4\,(cr)} \tag{223}$$

$$\Delta G^0_{298} = -318.0-(-261.1)-(-56.69) = -0.21 \text{ kcal/mol}\,.$$

The polycondensation process of silica alkaline solutions is widely considered to be more probable from the monomeric anion of metasilicic acid [213]:

$$\begin{matrix} O^- \\ | \\ Si{=}O\,. \\ | \\ O^- \end{matrix}$$

As the most probable scheme of polycondensation the following is suggested:

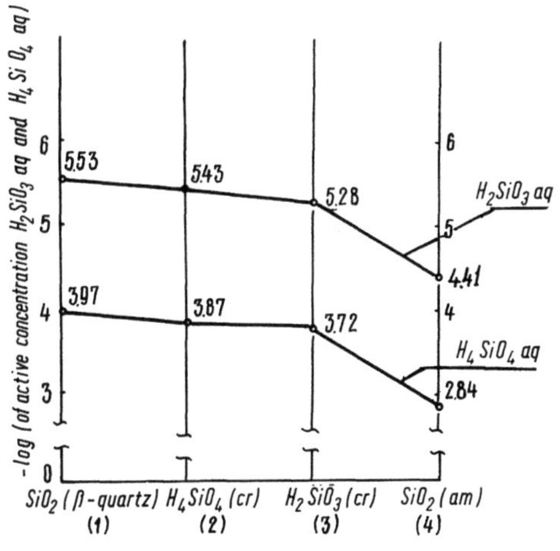

Fig. 83. Equilibrium active concentrations of metasilicic (*upper broken line*) and orthosilicic (*lower broken line*) acids in the solution, formed at the dissolution in water: 1. SiO_2 (β-quartz); 2. H_4SiO_4 (cr); 3. H_2SiO_3 (cr); 4. SiO_2 (am)

1. hydrolysis by the 1st stage

$$SiO_3^{2-} + H_2O = HSiO_3^{1-} + OH^-, \quad (224)$$

2. condensation

$$HSiO_3^{1-} + HSiO_3^{1-} = Si_2O_5^{2-} + H_2O\,, \quad (225)$$

then again hydrolysis of the ion $-O-\overset{\overset{\displaystyle O}{\|}}{Si}-O-\overset{\overset{\displaystyle O}{\|}}{Si}-O-$ by the 1st stage and its subsequent condensation by the active group OH^-.

A thermodynamic calculation also shows a greater probability of hydrolysis proceeding by the 1st stage:

$$1.\quad SiO^{2-}_{3\,(aq)} + H_2O_{(liq)} \rightleftarrows HSiO^-_{3\,(aq)} + OH^-_{(aq)} \quad (226)$$

$$\Delta G^0_{298} = -240.7 - 37.59 + 224.6 + 56.69 = +3.0$$

$$\lg K_1 = -\frac{3.0}{1.364} = -2.2; \quad K_1 = 10^{-2.2}; \quad (pH = 11.8)$$

$$2.\quad HSiO^{1-}_{3(aq)} + H_2O_{(liq)} \rightleftarrows H_2SiO_{3\,(aq)} + OH^-_{(aq)} \quad (227)$$

$$\Delta G^0_{298} = -253.9 - 37.59 + 240.7 + 56.69 = +5.9\ \text{kcal}$$

$$\lg K_2 = -\frac{5.9}{1.364} = -4.33; \quad K_2 = 10^{-4.33};$$

The pH equilibrium value is 9.67.

Therefore $K_1/K_2 = 10^{-2.2}/10^{-4.33} = 10^{+2.13}$, i.e., two orders or more.

A calculation of the processes of polycondensation of the ions $HSiO_3^{1-}$ gives the following results:

$$HSiO^{1-}_{3(aq)} + HSiO^{1-}_{3(aq)} \rightleftarrows Si_2O^{2-}_{5(aq)} + H_2O_{(liq)} \quad (228)$$

$$\Delta G^0_{298} = -428.3 - 56.69 + 2 \cdot 240.7 = -3.59\ \text{kcal}$$

$$\lg K = +\frac{3.59}{1.364} = +2.63; \quad K = 10^{+2.63}.$$

The further hydrolysis of the ion $Si_2O^{2-}_{5(aq)}$ brings about the formation of the ion $HSi_2O^{1-}_{5(aq)}$:

$$Si_2O_5^{2-} + H_2O \rightleftarrows HSi_2O^{1-}_{5(aq)} + OH^-_{(aq)} \quad (229)$$

$$\Delta G^0_{298} = -441.6 - 37.59 + 428.3 + 56.69 = +5.8\ \text{kcal}$$

$$\lg K = -\frac{5.8}{1.364} = -4.25; \quad K = 10^{-4.25},$$

which then again undergoes the process of polycondensation with the formation of a four-dimensional anion $Si_4O_9^{2-}$ having linear structure etc.

The processes of polycondensation of orthosilicic acid ions must result chiefly in the formation of ribbon and net-form ions, whereas metasilicic acid ions provide mainly the formation of chain ions ($Si_2O_5^{2-}$, $Si_3O_7^{2-}$ etc.).

Calculations also show that ions SiO_3^{2-} and $HSiO_3^{1-}$ are unstable towards the transition to ions $H_2SiO_4^{2-}$ ($K = 10^{+1.57}$) and $H_3SiO_4^{1-}$ ($K = 10^{+1.47}$) respectively, which corresponds to the pH limit of the ion $H_2SiO_4^{2-} \sim 12$ and ion $H_3SiO_4^{1-} \sim 11$, whereas ions $HSiO_4^{3-}$ are unstable towards SiO_3^{2-} (equilibrium pH = 10.43) and $HSiO_3^{1-}$ (equilibrium pH = 11.12). As at pH less than 10.43 the amount of the ion $HSiO_4^{3-}$ (see Table 89) is practically zero, and at 11.12 that of the ion $HSiO_4^{3-}$ only about 1.6% (see Table 90), these transitions seem to be of no importance. It concerns more the possibility of SiO_4^{4-} transition to SiO_3^{2-}.

6 Reactions in the Systems $CaO-SiO_2-H_2O$, β-$2CaO \cdot SiO_2-H_2O$ and $3CaO \cdot SiO_2-H_2O$

6.1 General

New formations in the system $CaO-SiO_2-H_2O$ at normal and hydrothermal conditions are dealt with in several works [9, 10, 22, 71, 72, 87, 136, 137, 143, 144, 172, 212, 232, 299, 475, 578, 592, 617, 640]. We shall adduce the results of the latest, most reliable works in this field briefly stating the characteristics of the main physical and thermodynamic constants of calcium hydrosilicates available. Flint and Wells [137] studied the system $Ca(OH)_2-SiO_2-H_2O$ analyzing the filtrate composition after shaking up lime and silica mixtures. They believed hydrosilicates arising in the process to be salts of orthosilicic acid: $Ca(H_3SiO_4)_2$, CaH_2SiO_4, $Ca_3(HSiO_4)_2$ and Ca_2SiO_4 aq. The works by Belov and Mamedov [344, 345], and Bernal [48] demonstrated, however, that the structure of calcium hydrosilicates is based not on common silico-oxygen tetrahedra (afwillite is an exception), but on more complex formations such as $[Si_6O_{17}]^{10-}$, $[Si_{12}O_{31}]^{14-}$ and $[Si_6O_{15}]^{6-}$. From these works we acquired a better understanding of the composition and crystallochemical formulae of calcium hydrosilicates. In Table 91 the formulae of calcium hydrosilicates are given.

All these hydrosilicates are representatives of hydrated wollastonites. In wollastonite, according to Belov and Mamedov a structural unit, block, is sharply cut. The block is built by a column of calcium octahedra and two columns of calcium prisms. Two chains of silico-oxygen tetrahedra $[Si_3O_9]^{6-}$ not linked with each other run along them.

In the xonotlite structure, according to Belov and Mamedov [344, 345], only a third of calcium cations entering the formula are in oxygen octahedra stretch-

Table 91. Crystallochemical and stoichiometric formulae of calcium hydrosilicates according to Belov and Mamedov

Names of calcium hydrosilicates	Crystallochemical formula	Stoichiometric formula
Hillebrandite	$Ca_{12}[Si_6O_{17}]_2(OH)_4 \cdot 12Ca(OH)_2$	$2CaO \cdot SiO_2 \cdot 1.17H_2O$
Afwillite	$Ca_3 2H_2O[SiO_3(OH)]_2$	$3CaO \cdot 2SiO_2 \cdot 3H_2O$
Foshagite	$Ca_8[Si_6O_{17}](OH)_6$	$4CaO \cdot 3SiO_2 \cdot 1.5H_2O$
Xonotlite	$Ca_6[Si_6O_{17}](OH)_2$	$6CaO \cdot 6SiO_2 \cdot H_2O$
Riversideite	$2\{Ca_{10}[Si_{12}O_{31}](OH)_6 \cdot 3H_2O\}$	$5CaO \cdot 6SiO_2 \cdot 3H_2O$
Tobermorite	$Ca_{10}[Si_{12}O_{31}](OH)_6 \cdot 8H_2O$	$5CaO \cdot 6SiO_2 \cdot 5.5H_2O$
Plombierite	$Ca_{10}[Si_{12}O_{31}](OH)_6 \cdot 18H_2O$	$5CaO \cdot 6SiO_2 \cdot 10.5H_2O$
Gyrolite	$Ca_4[Si_6O_{15}](OH)_2 \cdot 4H_2O$	$2CaO \cdot 3SiO_2 \cdot 2.5H_2O$
Okenite	$3\{Ca_3[Si_6O_{15}]2H_2O \cdot 4H_2O\}$	$CaO \cdot 2SiO_2 \cdot 2H_2O$
Truscottite	$4\{Ca_3[Si_6O_{15}] \cdot 2H_2O\}$	$CaO \cdot 2SiO_2 \cdot 0.67H_2O$
Nekoite	$Ca_3[Si_6O_{15}] \cdot 2H_2O \cdot 4H_2O$	$CaO \cdot 2SiO_2 \cdot 2H_2O$

ing in columns. The remaining two thirds of calcium cations also feature sixfold coordination, their polyhedra being not octahedra, but trigonal prisms. Prisms in xonotlite rest on their longitudinal edges and are bound by them in vertical columns located on both sides of the column of calcium octahedra. The columns of the two kinds are connected in endless wide ribbons or walls, layers parallel to 001.

Columns of calcium octahedra, interlaid with double chain ribbons of silico-oxygen tetrahedra $(Si_6O_{17})^{10-}$ predetermine the peculiar fibrousness of xonotlite. Two hydroxyl groups entering the xonotlite structure belong both to calcium of octahedra and to calcium of prisms, each being common to three calcium polyhedra ($3\frac{2}{6} = 1$).

The crystallochemical formula of xonotlite is $Ca_6[Si_6O_{17}](OH)_2$. The radical peculiar to xonotlite is $[Si_6O_{17}]^{10-}$, which underlies the so-called xonotlite chains. Xonotlite occurs in nature and is synthesized as well from a stoichiometric mixture of lime and quartz at 473 – 673 K. As reported by Butt and coworkers [85], samples of synthetic xonotlite feature high physicomechanical properties ($R_{compress} = 122$ kg/cm^2, $R_{bend} = 74$ kg/cm^2) and are hardly susceptible to carbonization, although McConnell [393] found xonotlite in which part of SiO_2 was replaced by CO_2.

In hillebrandite $Ca_{12}[Si_6O_{17}]_2(OH)_4 \cdot 12Ca(OH)_2$ a cell is made up of alternating layers of xonotlite and portlandite. Hillebrandite occurs in nature and may be synthesized as well. The artificial fibrous product received the designation $C_2SH(B)$. Humid carbon dioxide gradually destroys hillebrandite, calcite and amorphous SiO_2 being formed. Hillebrandite is a member of a series of calcium dibasic hydrosilicates having different structures. In addition, $C_2SH(A)$ and $C_2SH(C)$ are known in this series, $C_2SH(A)$ being artificial. Unlike $C_2SH(B)$ this mineral crystallizes in the form of prismatic plates poorly intergrowing; therefore the strength of samples of this mineral is low and with those of 1 g/cm^3 volume density amounts to 5 to 20 kg/cm^2 only. This hydrosilicate carbonizes rather slowly, but the strength of samples grows tenfold therewith.

The mineral $C_2SH(C)$ is unknown in nature. It is a product of the β and γ-C_2S hydration crystallized in granules. $C_2SH(C)$ samples of 0.98 g/cm^3 volume density have $R_{compress} = 14\ kg/cm^2$ and $R_{bend} = 24\ kg/cm^2$.

In foshagite $Ca_8[Si_6O_{17}](OH)_6$, as reported by Belov and Mamedov, four columns of calcium octahedra (two extra calcium hydroxides) fall on each xonotlite chain. This mineral occurs in nature, but is very difficult to synthesize.

In tobermorites (riversideite, tobermorite and plombierite) xonotlite ribbons at condensing overthrust forming seam corrugations:

$$2[Si_6O_{17}]_{\infty}^{10-} - 3O^{2-} = [Si_{12}O_{31}]_{\infty}^{14-}.$$

The chemical formula of the constant part of tobermorites is expressed as follows: $Ca_{10}[Si_{12}O_{31}](OH)_6$.

These minerals occur in nature. Minerals of this group produced artificially in hydrothermal conditions are denoted as CSH(B). Natural plombierite corresponds to CSH(B) 14 Å hydrate, crestmoreite is a mixture of 14 and 10.4 Å hydrates, riversidite is identical to 9.3 Å hydrate.

It is universally acknowledged at present that the difference between tobermorites and CSH(B) is that the first are more crystallized. In their structure tobermorites resemble argillaceous minerals, montmorillonite and vermiculite. The C/S ratio in synthetic tobermorites may vary from 0.83 to 1.75 and even more.

Brunauer and Greenberg [67], using the well-known constants of orthosilicic acid H_4SiO_4 dissociation and the Debye and Hückel rule, calculated the Ca^{2+} and $H_3SiO_4^{1-}$ activities for tobermorite with $C/S = 0.8$ and on this basis computed the product of activities: $K = a_{Ca^{2+}} \cdot a^2_{H_3SiO_4^{1-}}$. At 298 K K was $10^{-8.5}$. These authors also demonstrated that at $pH = 10.8$ $H_3SiO_4^{1-}$ maximum concentration is reached.

Studying reactions in the system $Ca(OH)_2 - SiO_2 - H_2O$ Steinour [594] and Taylor [612] plotted a diagram of phase equilibria at room temperature (Fig. 84).

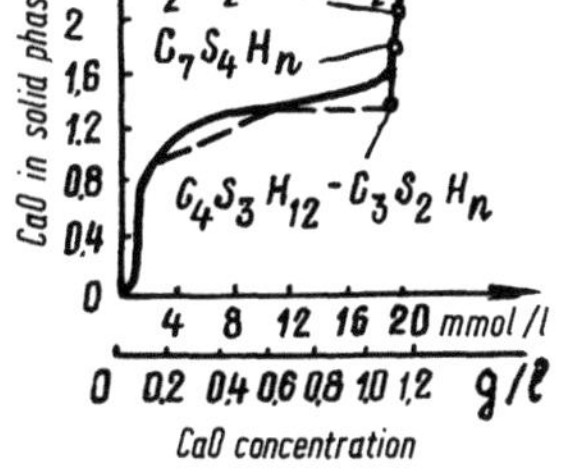

Fig. 84. Phase equilibrium in the system $CaO - SiO_2 - H_2O$ at room temperature by Taylor (*continuous line*) and Kalousek (*dotted line*)

These researchers are of the opinion that at room temperature only one compound, CSH(B), is stable in the system $CaO-SiO_2-H_2O$. With lime concentration making up saturated and supersaturated states hydrosilicate C_2SH_2 is formed there. The lime concentration in the solution decreasing, C_2SH_2 breaks down to CSH(B) and $Ca(OH)_2$.

Taylor's data were specified by Kalousek and Prebus [232] (dotted curve in Fig. 84). They demonstrated that hydrosilicate Ca_8SH_n may coexist with silica gel only in a solution with 0.07 g CaO per litre and about 0.08 g SiO_2 per litre. The stable $C_4S_3H_n$ has an invariant point with the solid $Ca(OH)_2$ at the solution of full saturation with lime.

The reactions in the system $Ca(OH)_2-SiO_2-H_2O$ at 423 K are researched in the work of Flint, McMurdie and Wells [136]. They plotted an equilibrium curve of phase coexistence and found one invariant point with the concentration of SiO_2 0.73 g/l and CaO 0.033 g/l.

As one can see from Fig. 85, while the CaO concentration in the solution increases, the SiO_2 concentration abruptly decreases to quite insignificant values. Peppler [475] tested the products of reactions at 453 K within a wide composition range, from C/S = 0 to C/S = 3. The test results are given in Tables 92 and 93.

For the test in question (Table 92) samples were prepared by adding silica gel to calcium hydroxide solution of different concentrations, 2 g of SiO_2 being added per litre of solution. Steaming was performed with conditioning of 4 to 17 days.

In Table 93 the results of testing quite different samples are presented: a mixture of silica gel and calcium oxide in different proportions were steamed and conditioned for 6 months. Starting from the results of the tests, Peppler stated that at 453 K only xonotlite and hillebrandite are stable in the system $CaO-SiO_2-H_2O$.

Kalousek et al. (Table 94) studied the interaction of quartz with lime at 448, 473 and 498 K [230–232].

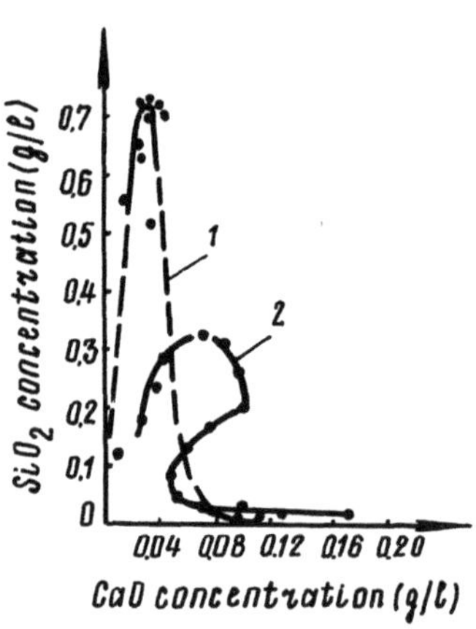

Fig. 85. The silica solubility in the lime solution: *1* at 150 °C; *2* at 30 °C

Table 92. Peppler's results in determining the character of new formations in the system $CaO-SiO_2-H_2O$ at 453 K

C/S in the initial mixture	Conditioning in days	Refractive index of the reaction products	Composition of steamed samples	Composition of steamed samples after heating up to 1373 K
0	6	1.45	Silica gel	α-quartz
0.26	4	1.42 – 1.54	Gyrolite + xonotlite + silica gel	α-CS, β-CS α-quartz
0.26	12	1.46 – 1.54	Idem	Idem
0.63	7	1.54 – 1.59	Idem	Idem
0.8	7	1.57	Gyrolite + xonotlite	Idem
1.05	13	1.57 – 1.59	Xonotlite	α-CS, β-CS
1.5	15	1.61	Hillebrandite	α-CS, β-CS
1.73	7	1.58 – 1.61	Idem	β-C_2S, β-CS
2	11	1.6	Idem	β-C_2S
3	14	1.6	Hillebrandite + calcium hydroxide	β-C_2S + C

Table 93. Peppler's results in determining the character of new formations in the system $CaO-SiO_2-H_2O$ at 453 K

C/S in the initial mixture	Product of the reaction
from 0.1 to 0.66	Xonotlite + silica gel
from 0.8 to 1	Xonotlite
1.5	Xonotlite + hillebrandite
from 1.66 to 2	Hillebrandite
3 to 4	Hillebrandite + calcium hydroxide

Table 94. Kalousek's results in determining the character of new formations in the system $CaO-SiO_2-H_2O$ at different temperatures

C/S in the initial mixture	Temp. of treatment [K]	Time of conditioning	Products of the reactions
0.8	448	2 hours	C_2SH_2
0.8	448	2 – 6 hours	$C_5S_4H_n$
0.8	448	6 – 7 hours	CSH_n
0.8	448	7 – 8 hours	$C_4S_5H_n$
0.8	448	8 days	$C_4S_5H_5$
1.8 – 2.4	448	3 days	$C_2SH(A)$
2	473	3 days	$C_2SH(A)$
1.5	473	24 hours	$C_2SH(C)$
1.75 – 2	498	2 days	$C_2SH(C)$
1.75	498	5 days	$C_2SH(B)$

Butt and coworkers [85, 87] studied the synthesis of separate phases in the system $CaO-SiO_2-H_2O$. They synthesized most known calcium hydrosilicates: gyrolite, CSH(B), xonotlite, tobermorite, $C_2SH(A)$, $C_2SH(B)$ and some others. Afwillite synthesis, however, was not successful. The synthesis was conducted in suspension (Table 95).

The works devoted to the study of compounds formed in lime-sand products are those of Taylor and others [612–616]. He tested a lime-sand light-weight block (C/S = 1/2 by weight) steamed at a temperature of 456 K. All the new-formations were identified by Taylor as hydrosilicate I ($C_{0.8\text{ to }1.33}SH_{1\text{ to }2.5}$). According to chemical analysis the block was 58.3% hydrosilicate, 38.3% quartz, 3.2% calcium carbonate and 0.2% free lime.

Heller and Taylor [197] investigated the reactions of calcium hydrosilicate formation with a composition of C/S = 3/2. At a temperature below 413 K hydrosilicate I was found to appear and at temperatures of 413 to 453 K is a primary product. Curing at 413 to 433 K results in the formation of afwillite

Table 95. Butt and coworkers' results in determining the character of new formations in the system $CaO-SiO_2-H_2O$ at different temperatures

C/S	Siliceous component	Water to solid ratio	Temp. of treatment [K]	Time of conditioning	Products of the reaction	Notes
0.66	Quartz	20	448	20 h	CSH(B) + tobermorite	With stirring
0.66	Quartz	4	573	10 d	Xonotlite	–
0.66	SiO_2-aqueous	5	423	20 d	Gyrolite	–
0.66	SiO_2-aqueous	17	448	20 h	Tobermorite + gyrolite	–
0.66	SiO_2-aqueous	17	448	10 d	Gyrolite	–
0.66	SiO_2-aqueous	5	448	20 d	Gyrolite	–
0.66	SiO_2-aqueous	5	523	10 d	Gyrolite + ?	–
0.8	Quartz	20	448	4 h	Tobermorite + $C_{0.8}SH_n$	With stirring
0.8	Quartz	4	448	6 d	Tobermorite	–
1	SiO_2-aqueous	6	448	4 h	CSH_n	–
1	SiO_2-aqueous	4	473	7 d	Xonotlite	–
1.5	Quartz	6	423	31 d	$C_2SH(B)$ + xonotlite	–
1.5	Quartz	4	433	9 d	$C_2SH(B)$ + xonotlite	–
2	Quartz	6	473	12 d	$C_2SH(B)$	–
2	SiO_2-aqueous	6	448	10 d	$C_2SH(C)$	–
3	Quartz	4	523	10 d	$C_2SH(C)$ + CH	–
3	SiO_2-aqueous	4	523	15 d	$C_2SH(C)$ + CH + C_3SH_2	–

$C_3S_2H_3$ and at 453 to 473 K mainly xonotlite $CSH_{0.18}$ and hillebrandite C_2SH are formed.

According to Newman [450] heats of the formation of hillebrandite $2CaO \cdot SiO_2 \cdot H_2O$, foshagite $5CaO \cdot 3SiO_2 \cdot 3H_2O$ and xonotlite $5CaO \cdot 5SiO_2 \cdot H_2O$ calculated by the heats of their dissolving in a mixture of nitric and hydrofluoric acids proved to be 34.7; 94.6 and 92.6 kcal/mol respectively.

A consideration of all the data available brings us to the conclusion that for carrying out comprehensive thermodynamic calculations of reactions in the system $Ca(OH)_2 - SiO_2 - H_2O$ at normal and hydrothermal conditions it is necessary to determine the main thermic constants: $C_p = f(T)$, ΔH^0_{298}, ΔG^0_{298}, S^0_{298} of nearly all calcium hydrosilicates as well as of silicate and polysilicate ions in water solution.

As experiments on directly determining the thermic constants of calcium hydrosilicates are long and laborious and are impeded by the synthesis of monocrystals pure enough, to obtain initial thermodynamic data on calcium hydrosilicates and silicic and polysilicic acid ions we [403] used approximate calculation methods. Although we do not claim to have achieved high accuracy, in a short space of time we were able to evaluate qualitatively the processes of the $Ca(OH)_2$, SiO_2 and H_2O interaction, as well as to investigate the thermodynamics of the reactions of the C_3S and β-C_2S hydration at normal and hydrothermal conditions.

6.2 The Calculation of Thermochemical Values of Calcium Hydrosilicates

For calculating the heats of formation of inorganic compounds, as mentioned earlier, several methods of differing accuracy were suggested (see Chap. 1). Here, in the case of calcium hydrosilicates, we followed the method of calculating by mean bond energies [295], the one most accurately covering all the crystallochemical peculiarities of their structure, and for comparison, a method of calculation by the hydrosilicate solubility in water was applied as well. As the technique of such calculations for silicates has specific features of its own, we thought it necessary to describe it in more detail in this part of the book rather than in Chapter I. It should be noted that our calculations proved to be in agreement with later experiments [618]. As the objects of investigations hillebrandite, foshagite, xonotlite, riversidite, tobermorite, plombierite, gyrolite and okenite, being representatives of hydrated wollastonites were chosen, as according to Belov and Mamedov [344, 345] they are based on polymeric silico-oxygen wollastonite chains $Si_3O_9^{6-}$. As an exception we also exmined afwillite[16] interpreted by Megow [427].

[16] Thermochemical data for afwillite are not reliable because its structure to a great extent differs from that of the rest of hydrosilicates; therefore the method of mean bond energies accepted here for calculations failed to yield data comparable with those for other hydrosilicates

On considering the structures of calcium hydrosilicates we were able to find the following bond types and energy shares in them: Si – O, Ca – O, Ca – O – H and the energy share for adding crystal water. We determined the binding energy of Si – O starting from the well-known value of the heat of the β-quartz formation (– 217.75 kcal/mol), the following reaction being taken as a basis:

$$Si_{(cr)} + 2 \cdot O_{(gas)} = SiO_{2\,(cr)} . \quad (230)$$

Describing this process one can express, as stated by Cottrell,

$$-4\varepsilon_{Si-O} = \Delta H^0_{qu} - 2\Delta U_{oxyg.} - \Delta U_{silic} , \quad (231)$$

from which

$$\varepsilon_{Si-O} = \frac{217.75 + 59.16 \cdot 2 + 88.04}{4} = 106.03 \text{ kcal/bond} .$$

Since the calcium hydrosilicates in question are representatives of hydrated wollastonites, the binding energy of Ca – O was found on the basis of a known value, the heat of formation of β-wollastonite (– 390.95 kcal/mol). We arrived at the mean binding energy of Ca – O 139.22 kcal/bond. The binding energy of Ca – O – H was found starting from the heat of calcium hydroxide formation (– 235.8 kcal/mol) and was 252.16 kcal/bond. The mean binding energy of O – H found from ΔH^0_{298} of liquid water was 110.57 kcal/bond. The share of energy of the addition of crystal water was found as a result of separately processing these values for several various hydrates. Thus, the share of energy of adding a molecule of crystal water for hydrosilicates was taken equal to 75.7 kcal/mol, for those containing from 2 to 8 molecules of water, 72.5 kcal/mol, and for those containing more than 8 molecules, 71.8 kcal/mol. Entropies of calcium hydrosilicates were calculated by the method of additivity. As a basis entropy shares for ions given by Latimer [323] were taken. The shares of entropy terms for various type bonds in silico-oxygen anions $[Si_{12}O_{31}]^{14-}$, $[Si_6O_{17}]^{10-}$ and $[Si_6O_{15}]^{6-}$ were found in the following way. Share S_1 falling on the bond of silicon with oxygen was found from the entropy of β-quartz as a quotient of an entropy unit divided by 4 ($S_1 = 10/4 = 2.5$ cal/mol deg). Share S_2 accounting for the energy of the bond of silicon with oxygen which is in its turn bound with calcium was found from the data on the entropy of β-wollastonite. The wollastonite chain is characterized by the endless radical $[Si_3O_9]^{6-}$. In this radical out of nine oxygen atoms present three are bound only with silicon atoms and six are bound with both silicon and calcium. The entropy of $Ca_3Si_3O_9$ equals $3 \cdot 19.6 = 58.8$ cal/mol deg where 19.6 cal/mol deg is the entropy of $CaSiO_3$.

The above discussion brings us to the following equation:

$$6S_2 = 58.8 - 3S_{Ca^{2+}} - 3S_1 = 58.8 - 3 \cdot 9.3 - 3 \cdot 2.5 = 23.4 \text{ cal/mol deg}$$

whence

$$S_2 = \frac{23.4}{6} = 3.9 \text{ cal/mol deg}.$$

Carrying out entropy calculations we considered that, while summing up the entropy terms of silico-oxygen anions, not the number of bonds Si – O, Ca – O etc. must be added, but the number of oxygen anions bound only with Si ($S_1 = 2.5$ cal/mol deg), and the number of oxygen ions that remained after that in a silico-oxygen anion and bound both with Si and Ca ($S_2 = 3.9$ cal/mol deg).

The value of the Gibbs free energy at standard conditions was calculated by the well-known formula

$$\Delta G^0_{298} = \Delta H^0_{298} - T\Delta S,$$

where ΔH^0_{298}, the heat of formation from elements in kcal/mol; ΔS, alteration of entropy in cal/mol deg; T, standard temperature (298 K); ΔG^0_{298}, alteration of the Gibbs free energy in kcal/mol.

Equations $C_p = f(T)$ were calculated by Landiya's method [320] by entropies. The results of the calculations of ΔH^0, ΔG^0, S^0 and $C_p = f(T)$ are given in Appendix 2.

6.3 Thermodynamic Calculations of the Main and Secondary Reactions in the System $Ca(OH)_2 - SiO_2 - H_2O$ at Normal and Hydrothermal Conditions

In the system $Ca(OH)_2 - SiO_2 - H_2O$ we considered the reactions of formation of nine calcium hydrosilicates: hillebrandite, afwillite, foshagite, xonotlite, riversidelite, tobermorite, plombierite, gyrolite and okenite from $Ca(OH)_2$, β-quartz and water as well as possible secondary reactions of the transformation of hydrosilicates into one another at the stoichiometric ratio of $Ca(OH)_2$ and SiO_2, SiO_2 or $Ca(OH)_2$ being in excess [17, 26, 85, 186, 404, 410, 532].

It must be noted here that while studying secondary reactions in the system $Ca(OH)_2 - SiO_2 - H_2O$ only tobermorite proper ($5CaO \cdot 6SiO_2 \cdot 5.5H_2O$) was examined out of tobermorite group hydrosilicates.

The calculation of $\Delta G^0 = f(T)$ of the reactions was conducted by the common technique by compositions assuming p = const. [17]

[17] According to Gonikberg [159] pressure noticeably effects chemical reactions beginning only with that of 500 atm. This enabled us to employ the technique of the calculations of $\Delta G^0 = f(T)$ assuming that p = const.

The main reactions in the system $Ca(OH)_2-SiO_2-H_2O$ were examined at the following ratios of C/S in the solution: 2/1 (and more: 3/1, 4/1 etc.); 3/2, 4/3, 6/6, 5/3, 1/2 (and less: 1/3, 1/4 etc.) at a constant (for all the reactions of the same composition) amount of moles of water.

In order to understand the character of reactions in this system more clearly a list is given of all the reactions at the ratio of $Ca(OH)_2$ and SiO_2 being 5/6:

1. $5Ca(OH)_2+6SiO_2+5.5H_2O \rightarrow \frac{5}{2}(2CaO\cdot SiO_2\cdot 1.17H_2O) + \frac{7}{2}SiO_2+\frac{25.15}{2}H_2O;$

2. $5Ca(OH)_2+6SiO_2+5.5H_2O \rightarrow \frac{5}{3}(3CaO\cdot 2SiO_2\cdot 3H_2O)+\frac{8}{3}SiO_2+5.5H_2O;$

3. $5Ca(OH)_2+6SiO_2+5.5H_2O \rightarrow \frac{5}{4}(4CaO\cdot 3SiO_2\cdot 1.5H_2O)+\frac{9}{4}SiO_2+\frac{34.5}{4}H_2O;$

4. $5Ca(OH)_2+6SiO_2+5.5H_2O \rightarrow \frac{5}{6}(6CaO\cdot 6SiO_2\cdot H_2O)+SiO_2+\frac{58}{6}H_2O;$

5. $5Ca(OH)_2+6SiO_2+5.5H_2O \rightarrow 5CaO\cdot 6SiO_2\cdot 3H_2O+7.5H_2O;$

6. $5Ca(OH)_2+6SiO_2+5.5H_2O \rightarrow 5CaO\cdot 6SiO_2\cdot 5.5H_2O+5H_2O;$

7. $5Ca(OH)_2+6SiO_2+5.5H_2O$ [18] $\rightarrow 5CaO\cdot 6SiO_2\cdot 10.5H_2O;$

8. $5Ca(OH)_2+6SiO_2+5.5H_2O \rightarrow 2(2CaO\cdot 3SiO_2\cdot 2.5H_2O) + Ca(OH)_2+4.5H_2O;$

9. $5Ca(OH)_2+6SiO_2+5.5H_2O \rightarrow 3(CaO\cdot 2SiO_2\cdot 2H_2O) + 2Ca(OH)_2+2.5H_2O\,.$

In Fig. 86 the calculation results are given in the form of the ΔG of the reactions of formation of calcium hydrosilicates as a function of the composition (of the ratio of CaO/SiO_2 in the solution) at 298 and 448 K. Curves 1 to 9 denote the curves of $\Delta G=f(T)$ of the reactions of hydrothermal synthesis of hillebrandite (1), afwillite (2), foshagite (3), xonotlite (4), riversidite (5), tobermorite (6), plombierite (7), gyrolite (8) and okenite (9) respectively.

An analysis of direct reactions in the system $Ca(OH)_2-SiO_2-H_2O$ permits drawing the following conclusions.

The ratios of C/S being 2/1 and more basic ones (3/1; 4/1; 5/1 etc.), within the entire temperature range hillebrandite is thermodynamically the most stable. The ratios of C/S being 3/2; 4/3 and intermediate ones, foshagite is preferable. The ratio of C/S being 1/1, at low temperatures (up to 363 K) tobermorites (plombierite and tobermorite) are stable and above 363 – 373 K the xonotlite formation becomes thermodynamically the most probable. The ratio of C/S being 5/6, at a low temperature plombierite is the most stable thermodynamically,

[18] The amount of moles of water 5.5 was taken in the seventh reaction as it is the greatest one for all the reactions of this composition

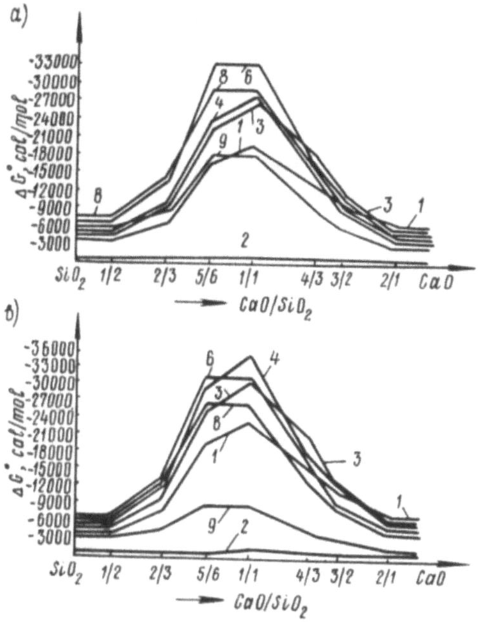

Fig. 86. ΔG^0 of reaction in the system $Ca(OH)_2-SiO_2-H_2O$ as a function of the ratio CaO/SiO_2 in a solution at a temperature of 25 °C **(a)** and 175 °C **(b)**

from 323 to 493 K, tobermorite proper and above 220 °C, xonotlite. The ratio of C/S being 2/3; 1/2 and more acid (1/3; 1/4; 1/5 etc.) at temperatures up to 503 – 513 K the gyrolite formation is the most preferable, and above 503 – 513 K that of xonotlite.

An analysis of side reactions in the system $Ca(OH)_2-SiO_2-H_2O$ allows to draw the following conclusions.

1. Neither SiO_2 nor $Ca(OH)_2$ being in excess, stable hydrosilicates are hillebrandite, foshagite and xonotlite, but only at temperatures above 373 K: at low temperatures with water they are able to transform into tobermorite. Tobermorites and gyrolite are thermodynamically stable at low temperatures, but tend to transform into xonotlite at temperatures above 473 – 493 K. Afwillite and okenite are thermodynamically unstable compounds within the entire temperature range, afwillite tending to transform into foshagite and okenite, into gyrolite and tobermorite.

2. $Ca(OH)_2$ being in excess, all hydrosilicates tend to transform into hillebrandite, and SiO_2 being in excess, into gyrolite. Afwillite, as an exception, with SiO_2 tends to convert to gyrolite only at temperatures up to 448 K, however, at higher temperatures its transition into foshagite is preferable.

The calculation results are in agreement with available experimental data obtained by several scientists: Peppler [475], Kalousek et al. [231], Taylor [612], Budnikov and Petrovykh [80], Butt, Rashkovich and Berkovich et al. [45, 87].

6.4 The Energetics of the Processes of the C_2S and C_3S Hydration [319, 411, 424]

First the values of C_3S and C_2S hydration heat were arrived at by Woods, Steinour and Starke [695] by an indirect method. They found the heats of hydration of portland cement of a certain chemical composition by the difference of heats of dissolving in acid the initial and hydrated cement during a specific interval. Having processed their data with the statistical method of least squares, they found a relation between the cement heat release and mineralogical composition. Thus the hydration heat of 136 cal/g for C_3S and 62 cal/g for C_2S was first obtained. These are the values of a year's term of hydration at 308 K.

A direct determination of the C_3S and C_2S hydration heat was carried out by Lerch and Bogue [329]. They hydrated separately pure clinker minerals C_3S, C_2S, C_3A and C_4AF until complete hydration and obtained 120 cal/g for C_3S, 62 for C_2S, 207 for C_3A and 100 for C_4AF.

Using Woods' method, Verbeck and Foster [669] after a 6.5 year hydration, obtained 120 cal/g for C_3S_2 and 53 cal/g for C_3S.

Powers and Brownyard [490] found the heat released at the portland cement hydration to be a sum of the hydration heat as such and of the heat of adsorption of water of the hydration products. They also found that the heat of the water adsorption may reach as much as a quarter of the total hydration heat.

Brunauer, Hayes and Hoss [68] determined the chemical heat of hydration using hydrated substances dried with the so-called α-drying. As a result they obtained 96.5 cal/g for C_3S and 24.5 for C_2S. For the heat of hydration into tobermorite they suggested the following formula:

$$H_0 = H_C - S + H_{H_2O} + H_{Ca(OH)_2} \quad (232)$$

where H_0, total heat of hydration; H_C, chemical heat of hydration; S, surface energy of tobermorite; H_{H_2O}, heat of water adsorption on tobermorite particles; $H_{Ca(OH)_2}$, heat of the $Ca(OH)_2$ adsorption on tobermorite.

As the $Ca(OH)_2$ adsorption heat is insignificant, $H_{Ca(OH)_2}$ may be neglected. If the specific surface of tobermorite particles is small, we may neglect the latter three terms in Eq. (232). If the specific surface of tobermorite is large, S must be taken into consideration.

The surface energy of solids with a specific surface of 4 m^2/g is 1 cal/g. Thus, the specific surface energy of the initial C_3S and C_2S may be neglected. The surface energy of $Ca(OH)_2$ formed as a result of the C_3S hydration may be neglected as well. However, the specific surface energy of tobermorite may not be neglected, as it is 386 ± 20 erg/cm^2 at 296.5 K, as stated by Brunauer, Kantro and Weise [69, 70].

The surface energy of tobermorite with a specific surface of approximately 300 m^2/g is about 27.7 cal/g. The heat of the hydration of dihydrate tobermorite into trihydrate one will be

$$Ca_3Si_2O_7 \cdot 2H_2O_{(sol)} + H_2O_{(liq)} = Ca_3Si_2O_7 \cdot 3H_2O_{(sol)}\,, \tag{233}$$

$\Delta H_{(hydr)} = -7600$ cal/mol at 296.5 K.

The calculation of the hydration heats of pure C_3S and C_2S was conducted by the above authors with equations derived from the study of reactions at full hydration of pure C_3S and C_2S in the form of paste (Water/solid = 0.7) and suspension (Water/solid = 9) shaken up in polyethylene bottles and ground in a mill at 297 ± 1 K. It was found that the main product of C_3S in paste and suspension shaken in polyethylene bottles is tobermorite and that of C_2S in suspension ground in a mill, afwillite. With C_2S in both cases it was tobermorite, but when shaken up in bottles had different compositions (C/S from 1.54 to 1.73) depending on the time of treatment. The amount of $Ca(OH)_2$ formed was determined by the method elaborated by Copeland and Breaker. In nearly all the cases the amount of $Ca(OH)_2$ was less than that by the stoichiometric reaction:

in case of C_3S:

$$2(Ca_3SiO_5) + 5H_2O = Ca_3Si_2O_7 \cdot 2H_2O + 3Ca(OH_2); \tag{234}$$

in case of C_2S:

$$2(Ca_2SiO_4) + 3H_2O = Ca_3Si_2O_7 \cdot 2H_2O + Ca(OH)_2\,. \tag{235}$$

All tobermorites contained more water[19] than required by Eqs. (234) and (235), therefore corrections were made (starting from 420 cal/g). Tobermorites also had different specific surface, but for uniformity were reduced to that of 300 m^2/g. As a result the mean hydration heat of C_3S was found to be 77.2 cal/g of C_3S and that of C_2S, 7.9 cal/g of C_2S.

The hydration heat of the reaction

$$2Ca_3SiO_5 + 6H_2O = Ca_3Si_2O_7 \cdot 3H_2O + 3Ca(OH)_2 \tag{236}$$

may be obtained by combining Eqs. (233) and (234):

$$\Delta H_{296.5\,K} = -21.386 \text{ kcal/mol } C_3S\,.$$

That of

$$2(Ca_2SiO_4) + 4H_2O \rightarrow Ca_3Si_2O_7 \cdot 3H_2O + Ca(OH)_2\,, \tag{237}$$

[19] The water in these hydrosilicates is in two forms: in the form of hydration water and in that of adsorption one. A full separation of these forms of water by physical means still seems impossible. At heating the water content diminishes gradually and the free energy of water binding varies greatly. Although the mean value of the free energy of water binding in the structure is greater than that of adsorption, the removal of the adsorption water was noticed to take place often later than that of the hydration.

combining Eqs. (235) and (236):

$$\Delta H_{296.5\,\mathrm{K}} = -5.16\ \mathrm{kcal/mol}\ C_2S\,.$$

The chemical heats of C_3S and C_2S hydration were determined for the case when the specific surface of tobermorite was 300 m^2/g: if different, a correction of 386 erg/cm^2 must be taken into account.

The same authors made approximate calculations of the entropy and free energy alterations at the C_3S and C_2S hydration. First the difference of the free energies of the C_3S and C_2S hydration was calculated by subtracting equation

$$2\Delta G^0_{C_2S\ \text{(hydration reaction)}} = \Delta G^0_{\text{(tob)}} + \Delta G^0_{Ca(OH)_2} - 2\Delta G^0_{C_2S} - 4\Delta G^0_{H_2O} \qquad (238)$$

from equation

$$2\Delta G^0_{C_3S\ \text{(hydration reaction)}} = \Delta G^0_{\text{(tob)}} + 3\Delta G^0_{Ca(OH)_2} - 2\Delta G^0_{C_3S} - 6\Delta G^0_{H_2O}\,. \qquad (239)$$

Thus, we arrive at

$$\begin{aligned}&\Delta G^0_{C_3S\ \text{(hydration reaction)}} - \Delta G^0_{C_2S\ \text{(hydration reaction)}}\\ &\quad = \Delta G^0_{Ca(OH)_2} + \Delta G^0_{C_2S} - \Delta G^0_{C_3S} - \Delta G^0_{H_2O}\\ &\quad = -214.33 - 524.19 + 665.47 + 56.69 = -16.36\ \mathrm{kcal/mol}\,.\end{aligned}$$

The ΔG values for C_3S and C_2S, being -665.47 and -524.19 respectively, were calculated by the data of Barany, King and Todd [32], those for the other compounds, however, were taken from the handbook of Rossini et al. [528] taking into account the new ΔH^0_{298} and ΔG^0_{298} values for β-quartz.

The entropy alteration at hydration could be calculated if the entropy of tobermorite were known. The authors calculated its approximate value by the additivity method taking the following entropy shares: 9.8 for CaO, 13.8 for SiO_2, 8.7 for OH and 10.4 for interplanar water. Considering the stoichiometric coefficients, the tobermorite entropy found was 84.8 cal/mol grad. To this value 3.3 cal/mol grad was added for the surface entropy of tobermorite with a specific surface of 300 m^2/g and rounding off they arrived at the entropy of tobermorite of 90 cal/mol grad, accurate within ± 5 cal/mol grad.

Having taken the $Ca(OH)_2$ and H_2O entropies at 298 K from the handbook of Rossini [528] (18.2 and 16.72 cal/mol grad respectively), and the C_3S and C_2S from Todd [629] (40.3 and 30.5 cal/mol grad respectively) they obtained $\Delta S = -36.3$ for reaction (236) and $\Delta S = -19.7$ for reaction (234) falling on 2 moles of the initial C_3S and C_2S. Using ΔH of $2C_3S$ being -42.85 kcal and ΔH of $2C_2S$ being 10.32 kcal they calculated the alteration of the Gibbs free enery of $2C_3S$: $\Delta G_{23.5\,^\circ C} = -4.4$ kcal. The difference between ΔG_{C_3S} and ΔG_{C_2S} is

$(32-4.4)/2 = 13.8$ kcal. Comparing this value with more exact data (15.3 kcal), the authors found a deviation of 1.5 kcal.

The heats of hydration of clinker minerals by the heat of dissolution in a mixture of HF(4 n) and HNO_3 (11 n) at a ratio of 1/1 were determined also by Schwiete and Tan Tik-Jen [557]. They showed that introducing aluminium and magnesium oxides into the lattice of C_3S involves an increase of the dissolution heat of 716.4 cal/g for pure C_3S to 802.4 cal/g for alite.

They prepared calcium hydrosilicates for the test by two methods: by the interaction of $Ca(OH)_2$ and SiO_2 or of $Ca(OH)_2$ and silicic acid salts (alkaline silicates) and by hydration of C_3S and C_2S at hydrothermal conditions. Tobermorite was obtained in nitrogen atmosphere at 100 atm. When the temperature is raised to 583 K, tobermorite changes into xonotlite. It was noticed that the transition of hydrosilicates from an amorphous state to a crystalline one was connected with a gradual alteration of the dissolution heat.

The heats of the water adsorption are endothermic and amount to 4.31 kcal/mol of H_2O. After the first dehydration and first wetting the adsorption heat falls from 4.31 to 1.39 kcal/mol. The adsorption water was found to be removed entirely at 430 K. C/S in the solution being 1.6, the hydration heat obtained was 120 cal/g. The heat of the hydrosilicate decomposition into $Ca(OH)_2$ and SiO_2 arrived at was 62 cal/g.

The dissolution (hydration) heat of β-C_2S is 52.7 cal/g (9.06 kcal/mol). It was found that C_3S cannot convert into a silicate hydrate at the ratio $C/S < 1.6$ and β-C_2S, at $C/S < 1.7$.

The heat of the C_3S hydration at the set ratio $C/S = 0.81$ amounts to 69.9 cal/g of C_3S with a hydrothermally synthesized hydrate and 115 cal/g with normally hydrated C_3S. The hydration heat of C_2S hydrated at normal conditions is 22.7 cal/g C_2S. The most probable values of C_3S and C_2S hydration heats agree with those found by Brunauer et al. [68] (96.5 cal/g of C_3S and 24.5 cal/g of C_2S). A very low value of the C_3S hydration heat at hydrothermal conditions (69.6 cal/g) as compared with the hydration heat at normal conditions is possibly accounted for by calcium hydrosilicates differing from tobermorite appearing at these conditions [133].

At temperatures of 353 to 393 K C_3S is hydrated with $C_2SH(B)$ being formed, within 393 to 473 K, with the formation of $C_2SH(A)$, and within 473 to 623 K the C_3S hydration brings about first the formation of $C_2SH(C)$ and $Ca(OH)_2$ which then change into C_3SH_2; at above 573 K calcium chondrodite $Ca_5(SiO_4)_2 \cdot (OH)_2$ is formed. At elevated temperatures up to 393 K C_2S is hydrated into C_2SH_2 and CSH(B), tobermorite, and at temperatures of 433 to 523, into $C_2SH(C)$.

We have considered the following reactions:

1. $\beta\text{-}2CaO \cdot SiO_2 + 1.17H_2O = 2CaO \cdot SiO_2 \cdot 1.17H_2O$;
2. $\beta\text{-}2CaO \cdot SiO_2 + 2H_2O = \frac{1}{2}(3CaO \cdot 2SiO_2 \cdot 3H_2O) + \frac{1}{2}Ca(OH)_2$;

3. $\beta\text{-}2CaO \cdot SiO_2 + \frac{7}{6}H_2O = \frac{1}{3}(4CaO \cdot 3SiO_2 \cdot 1.5H_2O) + \frac{2}{3}Ca(OH)_2$;
4. $\beta\text{-}2CaO \cdot SiO_2 + \frac{7}{6}H_2O = \frac{1}{6}(6CaO \cdot 6SiO_2 \cdot H_2O) + Ca(OH)_2$;
5. $\beta\text{-}2CaO \cdot SiO_2 + \frac{12.5}{6}H_2O = \frac{1}{6}(5CaO \cdot 6SiO_2 \cdot 5.5H_2O) + \frac{7}{6}Ca(OH)_2$;
6. $\beta\text{-}2CaO \cdot SiO_2 + \frac{6.5}{3}H_2O = \frac{1}{3}(2CaO \cdot 3SiO_2 \cdot 2.5H_2O) + \frac{4}{3}Ca(OH)_2$;
7. $\beta\text{-}2CaO \cdot SiO_2 + 2.5H_2O = \frac{1}{2}(CaO \cdot 2SiO_2 \cdot 2H_2O) + \frac{3}{2}Ca(OH)_2$.

Equations $\Delta G^0 = f(T)$ for these reactions and the results of calculations with them are given in Table 96. As one can see, the most stable of all the calcium hydrosilicates under consideration at the $\beta\text{-}C_2S$ hydration within the temperature range of 298 to 473 K is calcium dibasic hydrosilicate, hillebrandite. At normal temperature, $\beta\text{-}C_2S$ being hydrated, aside from hillebrandite, the formation of foshagite ($\Delta G^0_{298} = -1.71$ kcal/mol), xonotlite ($\Delta G^0_{298} = -0.42$ kcal/mol), tobermorite ($\Delta G^0_{298} = -1.38$ kcal/mol) and gyrolite ($\Delta G^0_{298} = -0.49$ kcal/mol) is thermodynamically possible.

Thus, it is insignificant which of the hydrosilicates is formed first – it is determined mainly by the kinetic conditions and structural peculiarities –, eventually each of them will change into the form most stable in the given system, hillebrandite. Therefore one can easily understand the phenomenon invariably observed in tests where the composition of tobermoritelike hydrosilicates hydrated at 298 K tend towards an increase of the ratio of CaO/SiO_2, its limit being two. As for the $\beta\text{-}C_2S$ hydration at elevated temperatures, in this case the experiment results fully agree with those of thermodynamic calculations. The formation of

Table 96. Equations $\Delta G^0 = f(T)$ and the ΔG^0 values of the reactions in the system $2CaO \cdot SiO_2 - H_2O$ at different temperatures

Nos. of reactions	Equations $\Delta G^0 = f(T)$	ΔH^0_{298} [kcal/mol]	ΔG^0_{298}	ΔG^0_{373}	ΔG^0_{473}
1	$\Delta G^0 = -7450 - 3.13\,T\ln T + 3.59 \cdot 10^{-3}T^2 + 2.13 \cdot 10^5 T^{-1} + 31.3\,T$	−5.4	−2.41	−1.6	−0.54
2	$\Delta G^0 = -8986 - 0.07\,T\ln T + 7.94 \cdot 10^{-3}T^2 + 3.9 \cdot 10^5 T^{-1} + 37.1\,T$	−7.12	+3.94	+6.82	+11.13
3	$\Delta G^0 = -6017 + 1.66\,T\ln T + 3.81 \cdot 10^{-3}T^2 + 1.68 \cdot 10^5 T^{-1} + 1.95\,T$	−5.72	−1.71	−0.646	+0.96
4	$\Delta G^0 = -3740 + 2.29\,T\ln T + 4.1 \cdot 10^{-3}T^2 + 1.42 \cdot 10^5 T^{-1} - 4.75\,T$	−3.83	−0.42	+0.49	+1.92
5	$\Delta G^0 = -4710 + 9.94\,T\ln T + 0.67 \cdot 10^{-3}T^2 + 1.37 \cdot 10^5 T^{-1} - 47.2\,T$	−6.81	−1.38	+0.18	+2.19
6	$\Delta G^0 = -9545 - 0.78\,T\ln T + 10.25 \cdot 10^{-3}T^2 + 4.65 \cdot 10^5 T^{-1} + 22.5\,T$	−7.1	−0.49	+1.27	+2.07
7	$\Delta G^0 = -7705 + 2.57\,T\ln T + 13.55 \cdot 10^{-3}T^2 + 5 \cdot 10^5 T^{-1} + 6.14\,T$	−6.32	+1.38	+3.5	+6.78

calcium dibasic hydrosilicate is provided immediately, any kind of by-processes being ruled out.

Let us consider in more detail possible stages of the process of hydration of β-dicalcium silicate.

As demonstrated above, in water solutions orthosilicic acid is highly unstable, water is soon separated, and orthosilicic acid is transformed into metasilicic one. We shall take the main forms of its existence as a basis of all the transformations of calcium silicates.

Obtained from experiments, the pH value of hardening pastes of β-C_2S is about 11.9 units [25]. Using the same methods as in the case of lime hydration, it is possible to clarify the mechanism of the β-C_2S hydration processes, whether they undergo the dissolution stage or proceed in solid phase.

If the scheme of the β-C_2S hydration in solid phase is taken as real, the experimentally established pH value may turn out to be a result of the process of the dissolution of the newly formed (on account of a solid-phase reaction) calcium hydrosilicate. In order to check the correctness of the solid-phase scheme of hardening, using the methods of tentative calculation set forth above, let us determine the pH equilibrium values at the dissolution of hillebrandite ($2CaO \cdot SiO_2 \cdot 1.17\,H_2O$) and tobermorite ($5CaO \cdot 6SiO_2 \cdot 5.5H_2O$) in water.[20]

$$5CaO \cdot 6SiO_2 \cdot 5.5H_2O_{(cr)} + 11.5H_2O_{(liq)} \rightleftarrows 5Ca^{2+}_{(aq)} + 10 \cdot OH^-_{(aq)} + 6H_4SiO_{4(aq)} \tag{240}$$

$$\Delta G^0_{\text{reaction}} = -5 \cdot 132.10 - 10 \cdot 37.59 - 6 \cdot 312.72 + 11.5 \cdot 56.69 + 2361.45$$

$$= +100.62\ \text{kcal/mol}\,.$$

$$\lg K = -\frac{100.62}{1.364} = -73.76\,.$$

Taking into consideration that the activities of all solids amount to one and that at equilibrium conditions $[Ca^{2+}] = \frac{5}{10}[OH^-]$ and $[H_4SiO_4] = \frac{6}{10}[OH^-]$ and substituting for $\lg K$ its value equal to -73.76 we arrive at pH = 10.63.

Such a calculation of a reaction of the hillebrandite dissolution by the scheme

$$2CaO \cdot SiO_2 \cdot 1.17\,H_2O_{(cr)} + 2.83 \cdot H_2O_{(liq)} = 2Ca^{2+}_{(aq)} + 4OH^-_{(aq)} + H_4SiO_{4(aq)}\,. \tag{241}$$

gives the pH equilibrium value of -11.44.

Thus, in both cases the pH value, being experimentally determined as 11.9, with tobermorite is 1.27 pH units less than needed and with hillebrandite 0.46 less, which reveals the solid-phase scheme to be unreal.

[20] In both reactions instead of $SiO^{4-}_{4(aq)} H_4SiO_{4(aq)}$ was taken as a final form of the silicic acid formed at the hydrolysis of ions in a solution

If, however, we admit, as the first stage of the process of the dicalcium silicate hydration, its chemical interaction with water, ions Ca^{2+}_{aq}, $SiO^{2-}_{3\,aq}$ and OH^-_{aq} being formed, the calculations show the following:

$$\beta\text{-}2CaO\cdot SiO_{2(sol)}+4H_2O \rightleftarrows 2Ca^{2+}_{(aq)}+H_4SiO_{4(aq)}+4OH^-_{(aq)} \qquad (242)$$

$$\Delta G^0 = -2\cdot 132.1-312.72-4\cdot 37.59+524.19+4\cdot 56.69 = +23.67\ \text{kcal};$$

$$\lg K = -\frac{23.67}{1.364} = -17.35\,.$$

Considering that the activities of β-C_2S and H_2O are equal to one, $[Ca^{2+}] = \frac{1}{2}[OH^-]$ and $[H_4SiO_4] = \frac{1}{4}[OH^-]$ we obtain the value of pH = 11.69[21], which is nearer to the experimentally determined pH value of 11.9. The results obtained convince that the through-solution mechanism of the process of the β-C_2S hydration is more probable.

Thus, at the β-C_2S slaking with water first comes its chemical interaction with H_2O with the formation of ions Ca^{2+}, OH^- and SiO_4^{4-} in the solution which, being hydrolized, form H_4SiO_4. In this case the solution pH increases, and, the value of 11.9 being reached, a supersaturation is already reached great enough for the crystallization of calcium hydrosilicates in question, the equilibrium pH values of the formation of which are much lower than 11.9.

Thus the cause of the primary formation of tobermorite at ordinary temperatures is clear, with tobermorite the pH equilibrium value is 10.63, with hillebrandite 11.44. It is obvious that in the case of tobermorite the required supersaturation value is achieved earlier than in the case of hillebrandite, which results in the primary formation and precipitation of tobermorite, although under these conditions thermodynamically the most stable form is hillebrandite.

We also carried out similar tentative calculations for the reactions of C_3S hydration. As thermodynamic data for C_3SH were not available and the data for afwillite are unreliable, the study of all possible reactions in the system C_3S-H_2O is impeded. Bearing this in mind we considered only the following reactions:

1. $3CaO\cdot SiO_2+2.17H_2O = 2CaO\cdot SiO_2\cdot 1.17H_2O+Ca(OH)_2$
2. $3CaO\cdot SiO_2+3H_2O = \frac{1}{2}(3CaO\cdot 2SiO_2\cdot 3H_2O)+\frac{3}{2}Ca(OH)_2$
3. $3CaO\cdot SiO_2+\frac{13}{6}H_2O = \frac{1}{3}(4CaO\cdot 3SiO_2\cdot 1.5H_2O)+\frac{5}{3}Ca(OH)_2$
4. $3CaO\cdot SiO_2+\frac{13}{6}H_2O = \frac{1}{6}(6CaO\cdot 6SiO_2\cdot H_2O)+2Ca(OH)_2$
5. $3CaO\cdot SiO_2+\frac{18.5}{6}H_2O = \frac{1}{6}(5CaO\cdot 6SiO_2\cdot 5.5H_2O)+\frac{13}{6}Ca(OH)_2$
6. $3CaO\cdot SiO_2+\frac{9.5}{3}H_2O = \frac{1}{3}(2CaO\cdot 3SiO_2\cdot 2.5H_2O)+\frac{7}{3}Ca(OH)_2$
7. $3CaO\cdot SiO_2+3.5H_2O = \frac{1}{2}(CaO\cdot 2SiO_2\cdot 2H_2O)+\frac{5}{2}Ca(OH)_2$.

[21] β-quartz (β-$2CaO\cdot SiO_2+4H_2O = 2Ca^{2+}+\beta\text{-}SiO_2+4OH^-$), i.e., up to the most stable form of silicic acid, has the pH value of 11.89, which practically coincides with the experimentally found (11.9)

Table 97. Equations $\Delta G^0 = f(T)$ and the ΔG^0 values for reactions in the system $3CaO \cdot SiO_2 - H_2O$ at different temperatures

Nos.	Eqs. $\Delta G^0 = f(T)$	ΔH^0_{298}	ΔG^0_{298}	ΔG^0_{373}	ΔG^0_{473}
		[kcal/mol]			
1	$\Delta G^0 = -32.395 - 10.47 T \ln T + 7.48 \cdot 10^{-3} T^2 + 8.07 \cdot 10^5 T^{-1} + 94.9 T$	−24.2	−18.76	−16.87	−14.52
2	$\Delta G^0 = -29720 - 14.46 T \ln T + 11.84 \cdot 10^{-3} T^2 + 9.87 \cdot 10^5 T^{-1} + 126.5 T$	−19.86	−12.15	−10.02	−7.2
3	$\Delta G^0 = -32210 - 12.96 T \ln T + 8.65 \cdot 10^{-3} T^2 + 7.6 \cdot 10^5 T^{-1} + 111 T$	−24.02	−17.86	−16.16	−14.08
4	$\Delta G^0 = -30936 - 12.32 T \ln T + 3.5 \cdot 10^{-3} T^2 + 7.36 \cdot 10^5 T^{-1} + 109.6 T$	−22.63	−16.58	−15.07	−13
5	$\Delta G^0 = -32051 - 5.71 T \ln T + 4.61 \cdot 10^{-3} T^2 + 7.31 \cdot 10^5 T^{-1} + 72.4 T$	−25.85	−17.68	−15.04	−12
6	$\Delta G^0 = -36870 - 15.44 T \ln T + 14.12 \cdot 10^{-3} T^2 + 10.12 \cdot 10^5 T^{-1} + 136.3 T$	−25.4	−17.74	−15.35	−12.09
7	$\Delta G^0 = -34460 - 11.94 T \ln T + 17.47 \cdot 10^{-3} T^2 + 10.93 \cdot 10^5 T^{-1} + 116.3 T$	−25.1	−14.85	−11.8	−7.95

Equations $\Delta G^0 = f(T)$ of these reactions and the results of calculations by them are given in Table 97. As one can see, the $3CaO \cdot SiO_2$ hydration appears with a marked negative alteration of the Gibbs free energy (~18 kcal/mol). Thermodynamically the formation of all the above hydrosilicates is possible.

Here, as in the previous case, ΔG^0 of all the reactions under consideration are quite close to each other, which characterizes to a certain extent their equal energy probability. If we approach this problem formally and believe, as in the former case, the difference of 0.5 kcal to be enough for a certain reaction to be preferred, then, as one can see from the table, hillebrandite is again the most stable compound in the system C_3S-H. All other hydrosilicates must convert into it if they have been formed earlier for any reason.

The lack of data on C_3SH prevented us from checking the limits of stability of this compound. A comparison of ΔG^0 of the reactions of the C_2S and C_3S hydration elucidates the greater reactivity of C_3S as compared to that of C_2S. The calculation results quite agree with the data of Brunauer et al. [67] obtained experimentally (Table 98).

Thus, ΔG^0 of the reaction of the C_3S decomposition to C_2S and CaO according to our data amounts to 16.3 kcal/mol, which differs only by 1 kcal from the exact value of 15.3 kcal arrived at by Brunauer in his experiments. The same value determined by Brunauer starting from the data on the C_2S and C_3S hydration is 13.8 kcal, which differs by 1.5 kcal from the exact one.

Considering possible schemes of the tricalcium silicate hydrolysis at normal temperature and pressure, one must bear in mind that, as with β-C_2S, undis-

Table 98. A comparison of the results of our calculations of the reactions of C_2S and C_3S hydration with the experimental data of Brunauer

Reactions	ΔH_{298} [kcal/mol]		Deviation [kcal]	ΔG_{298} [kcal/mol]			Deviation [kcal]
	Our data	Brunauer's data		Our data	By Brunauer from the data on hydration	By Brunauer from a direct experiment	
β-C_2S hydration with the formation of tobermorite	−6.81	−5.16	+1.65	−1.38	−2.2	–	−0.82
C_3S hydration with the formation of tobermorite	−25.85	−21.4	+4.45	−17.68	−16	–	+1.68
C_3S dissociation into C_2S and CaO	–	–	–	−16.3	−13.8	−15.3	1.5 (by Brunauer) and 1 (our data)

sociated orthosilicic acid is to be taken for the starting form of silicic acid in the solution. Accordingly, as in the former case, we shall consider only the following reaction as the main one:

$$3CaO \cdot SiO_{2(sol)} + 5H_2O_{(liq)} = 3Ca^{2+}_{(aq)} + H_4SiO_{4(aq)} + 6OH^-_{(aq)} \qquad (243)$$

$$\Delta G^0_{298\,(reaction)} = -3 \cdot 132.1 - 312.72 - 6 \cdot 37.59 + 665.47 + 5 \cdot 56.69 = +14.36$$

$$\lg K = -\frac{14.36}{1.364} = -10.53$$

$$K = \frac{[Ca^{2+}]^3[H_4SiO_4][OH^-]^6}{[3CaO \cdot SiO_2][H_2O]^5}.$$

As the activities of C_3S and H_2O are equal to one, $[Ca^{2+}] = \frac{1}{2}[OH^-]$ and $[H_4SiO_4] = \frac{1}{6}[OH^-]$, by use of proper substitutions and transformations we derive the pH equilibrium value of 13.12 units for reaction (243).[22]

[22] A similar calculation for the reaction of the C_3S hydrolysis to β-SiO_2 ($3CaO \cdot SiO_2 + 3H_2O = \beta\text{-}SiO_2 + 3Ca^{2+} + 6OH^-$) yields the pH value of 12.37

The experimentally found pH value of the reaction of the pure C_3S hydration is 12.8 units [25]. A comparison of the calculated and experimental values shows that, C_3S being hydrated in reaction (243), no equilibrium state is reached. Much earlier the product of the solubilities of the formation of the corresponding calcium hydrosilicates is reached. In order to discern which of the calcium hydrosilicates is most apt to crystallize, in addition to the previous calculations for determining the pH equilibrium values at the tobermorite (pH = 10.63) and hillebrandite (pH = 11.44) dissolution in water, the calculations carried out with afwillite $3CaO \cdot SiO_2 \cdot 3H_2O \rightleftarrows 3Ca^{2+}_{(aq)} + 2H_4SiO_{4(aq)} + 6OH^-_{(aq)}$ show the pH value to be 11.59.

Thus, in the case of tricalcium silicate hydration at normal conditions, first conditions are created for the required supersaturation with tobermorite as well (pH = 10.63).

7 Reactions of Hydration of Calcium Aluminates and of Iron-Containing Phases of Portland Cement Clinker

7.1 General

In the system $CaO - Al_2O_3$, as mentioned above, the formation of five compounds: C_3A, $C_{12}A_7$, CA, CA_2 and CA_6, has been found, the first and the third being of paramount technical importance; the first, for the portland cement production and the third, for that of alumina cement.

Reactions of the interaction of calcium aluminates with water both in the case of individual minerals and as a part of cement were studied by several scientists [78, 122, 224, 518, 519, 535, 567, 592, 606].

Their results, however, sometimes disagree. As to the iron-containing phases, in portland cement clinker two compounds have been found: $2CaO \cdot Fe_2O_3$ and $4CaO \cdot Al_2O_3 \cdot Fe_2O_3$ [88].

At present there is no generally accepted opinion on the problem of what products are formed in their interaction with water.

As no trustworthy thermodynamic data for most calcium hydroaluminates and hydrosulphoaluminates exists, we carried out special calculations to coordinate the initial data in these systems [19].

For calculating the thermodynamic data of calcium hydroferrites and ferrite sulphate hydrates, and to coordinate the available data on calcium aluminate hydrates and aluminate sulphate hydrates we started from the enthalpies of the cubic C_3AH_6 and hexagonal C_4AH_{13} calcium aluminate hydrates experimentally found by Houtepen and Stein [208]. [23] An analysis of their data proved the heat of

[23] To take into consideration activity coefficients one has to know the laborously definable values of the supersaturations of ions Ca^{2+}, $Ca(OH)^+$ and OH^- with respect to $Ca(OH)_2$. This results in greater errors than the calculation with common concentration values for the analysis of ionic equilibria of C_3A, C_2S, C_3S et al. [591]

the cubic aluminate hydrate dissolution $\Delta H_{dissol} = 139.2$ kcal/mol to agree with the value of $\Delta H_{dissol} = 139.163$ kcal/mol found by Berman and Newman [46], and the calculated value of the C_3AH_6 formation enthalpy to be 3.4 kcal/mol less. This is, probably, accounted for by the fact that the authors of work [46] used the wrong value of the enthalpy of the aluminium oxide formation (in work [47] by the same authors the value of the Al_2O_3 enthalpy $\Delta H^0_{298} = 399.09$ kcal/mol is given) and did not consider the formation of a complex in the solution after the dissolving of the aluminate in hydrochloric acid [681]. There are no experimental thermodynamic data of C_4AH_{13} apart from those given in the work under consideration [208]. Thus, we must assume that the calculation is based on the most reliable experimental data concerning the enthalpies of the C_3AH_6 and C_4AH_{13} formation.

The authors of [208] found the heats of the dissolution of compounds determined by the formula $2[Ca_2Al(OH)_6 \cdot 2H_2O]^+ \cdot 2(X^- + 2H_2O)$, where $X^- = Cl^-, Br^-, J^-, NO_3^-, ClO_3^-, ClO_4^-$, to depend only slightly on X^-. The difference of the dissolution heats results from the difference of the energies of the interaction between anions and positively charged main layers $Ca_2Al(OH)_6^+$ which depends on the nature and dimensions of the anion. Similarly, some link must exist between the heats of the dissolution of the corresponding calcium ferrite hydrates and aluminate hydrates, as these compounds are of the same structure [617] and differ only in the kind of an ion in the hypothetical layer $Ca_2M(OH)_6^+$ where M = Al, Fe.

It follows from the diagram of the determination of the enthalpies of hydrates by the heats of their dissolution in hydrochloric acid [208] that the difference of the heats of the dissolution of a mole of calcium aluminate hydrate and of a mole of the corresponding calcium ferrite hydrate is determined, all other things being equal, by two values: first, by the difference of the enthalpies of the aluminate hydrate and ferrite hydrate formation and, second, by the difference of the enthalpies of aluminium chloride and ferric chloride formation in the mixture after the dissolution of calcium aluminate hydrate and ferrite hydrate respectively. Owing to the increase of complexes the enthalpy of $AlCl_3$ and $FeCl_3$ formation in the solution is a function of the amount of water and of the hydrochloric acid concentration, the effect of these factors being much greater for ΔH^0_{298} [$FeCl_3$, $a H_2O$, bHCl] [681]. In order to rule out the effect of the solution composition, ΔH_{dissol} of ferric and aluminium hydrates are to be considered in the extreme case, when a is very small and b very great, i.e., in the case close to the standard one (MCl_{aq}[std. state, $m = 1$]). The equation for calculating the heat of ferrite hydrate formation may be expressed as:

$$\Delta H^0_f[Fe] = \Delta H^0_f[Al] - 2(\Delta H^0_f[AlCl_{3(aq)}] - \Delta H^0_f[FeCl_{3(aq)}]) + C, \qquad (245)$$

where ΔH^0_f[Fe], enthalpy of the calcium ferrite hydrate formation; ΔH^0_f[Al], enthalpy of the calcium aluminate hydrate formation; $\Delta H^0_f[AlCl_{3(aq)}]$, the $AlCl_3$

enthalpy in the standard state ($m = 1$); $\Delta H_f^0[FeCl_{3(aq)}]$, the $FeCl_3$ enthalpy in the standard state ($m = 1$); C, constant.

The constant value is the difference between the heats of the dissolution of calcium aluminate hydrate and of the corresponding ferrite hydrate. This difference is caused by the difference of the energy of the ions OH^- interaction with the ions Al^{3+} and Fe^{3+} on account of the difference of the dimensions of the radii of the latter. As a result of processing this value was for several hydrates accepted as being 10.48 kcal/mol.

Note that while using a unitary system of ΔH_f^0 and other functions the substitution of a new, though more exact, value for a certain one, may result in greater errors if such a substitution is not applied to the values of the other compounds of the system [277]. In the present work the system of coordinated thermodynamic constants of ferric compounds given by Mel'nik [428, 429] was used. For coordinating the enthalpy of the $FeCl_3$ formation the data given in works [428, 448, 537, 681] were analyzed. The value $\Delta H_f^0[FeCl_3(aq)]$ obtained is -134.16 kcal/mol in the standard state.

The enthalpy of the formation of the cubic and hexagonal calcium ferrite hydrates was found by Eq. (245):

$$\Delta H^0_{C_3FH_6} = -1110.8 \text{ kcal/mol},$$

$$\Delta H^0_{C_4FH_{13}} = -1772.8 \text{ kcal/mol}.$$

The enthalpy of the formation of aluminate hydrates C_4AH_{19} and C_2AH_8, aluminate sulphate hydrates $C_3ACs_3H_{32}$ and C_3ACsH_{12}, ferrite sulphate hydrates $C_3FCs_3H_{32}$ and C_3FCsH_{12} was calculated on the structural analogy principle [414] in accordance with the data on the structure of these hydrates [464, 617].

The calculation of the entropy of the aluminium and ferric hydrates was carried out by the additivity method. For cations the entropy shares given by Latimer [323] were taken; the specified anion shares and the thermodynamic data relevant to the calculation were taken from handbook [448]. The equations of the temperature dependence of heat capacity were determined by Landiya's method [320] with a computer by the entropy of compounds.

The Gibbs free energy was calculated by the well-known Eq. (70).

The calculation results are given in Table 99.

As we were the first to obtain the thermodynamic characteristics of calcium ferrite hydrates, we checked the values of the Gibbs free energy by the data on the composition of solutions given by Berezky [42] and H. Hoffman [204]. For this purpose at the beginning the ionic composition of the calcium ferrite hydrate solutions was considered. From the diagram Eh-pH of the system $Fe-H_2O$ (plotted using the system of coordinated thermodynamic data [428, 429]) it follows that at $pH > 11.51$ and $Eh > -0.6$ iron in the solution is mainly in the form of a hydrox complex $Fe(OH)_4^-$ and at $pH < 11.51$, in the form of undissociated

Table 99. Standard enthalpies, Gibbs free energies, entropies and equations of the heat capacities of aluminium and ferric hydrates

Compounds	$-\Delta H^0_{298}$ [kcal/mol]	$-\Delta G^0_{298}$ [kcal/mol]	S^0_{298} [cal/mol deg]	$C_p = f(T)$ [cal/mol deg]		
				a	$b \cdot 10^3$	c
C_3AH_6	1326	1198.4	96.7	68.91	124.17	–
C_4AH_{13}	1988	1756.17	171.6	109.6	242.19	–
C_4AH_{19}	2411	2096.33	228.0	122.36	394.0	–
C_2AH_8	1299.2	1151.54	106.4	68.38	153.45	–
C_3ACsH_{12} [a]	2096.44	1859.14	178.6	113.71	246.84	–
$C_3ACs_3H_{32}$	4201.32	3634.26	417.6	208.0	740.93	–
C_3FH_6	1110.8	984.79	101.5	78.09	116.26	–
C_4FH_{13}	1772.8	1542.57	176.4	118.15	234.67	–
C_3FCsH_{12}	1881.24	1645.53	183.4	122.83	239.14	–
$C_3FCs_3H_{32}$	3986.12	3420.55	422.4	217.93	730.92	–

[a] $CaSO_4$ will be denoted by $C\bar{S}$ or by Cs

molecules $Fe(OH)^0_3$. Molecular forms cannot be viewed as very stable because of their susceptibility to aggregation and condensation [429], therefore the iron ions in the solution of calcium ferrite hydrates were considered in the form of the hydrox complex $Fe(OH)_4^-$.

A detailed analysis performed in [535] showed that the ionic composition of a solution of calcium aluminate hydrates includes mainly ions of hydroxyl OH^-, aluminium $Al(OH)_4^-$ and calcium cations Ca^{2+}. On this basis we considered the composition of the liquid phase of the calcium ferrite hydrate solutions to be constituted by ions Ca^{2+}, OH^-, $Fe(OH)_4^-$.

Given one of the characteristics of a metastable or stable solution of ferrite hydrates, aluminate hydrates and some other substances, one can determine the Gibbs free energy of the crystalline compound. The Gibbs free energy calculation was conducted in the following manner. According to pH value or to the concentration of one of the ions on the basis of the material balance of the ions in the solution and/or the equation of electroneutrality, the numerical value of the concentration of the rest of ions is determined. Then using known relations [535] the solubility product of the compound is calculated by which its Gibbs free energy is found. Based on this technique of determining ΔH^0_f of a crystalline compound for the pH values of 12.16 and 12.36 in one case and for the lime concentration of 1.06 g/l in the other, the Gibbs free energies of tetracalcium ferrite hydrate were calculated. The data for the calculation were taken from [338, 428, 448, 535].

The values ΔG^0_{298} C_4FH_{13} obtained as a result are:

-1541.5 kcal/mol calculated by pH = 12.16;

-1539.15 kcal/mol calculated by pH $= 12.36$;
-1536.2 kcal/mol calculated by $C_{CaO} = 1.06$ g/l.

The extreme lime concentration in a solution of the cubic hydroferrite is 0.25 g/l. A calculation similar to that of the determination of the Gibbs free energy of calcium hexagonal ferrite hydrate yielded the $\Delta G^0_{C_3FH_6}$ value of -987.5 kcal/mol.

Difficulties while examining the iron-containing systems were great, nevertheless the values of the Gibbs free energy of the cubic and hexagonal hydroferrites found by formula:

$$\Delta G^0_{298} = \Delta H^0_{298} - T\Delta S_{298}$$

quite agree with those calculated by experimental data on the composition of solutions.

As stated in the works of Schwiete, H. F. W. Taylor and others, aluminium hydrates form solid solutions among themselves and with ferrite analogues. This fact should be borne in mind as one of major importance while carrying out a thermodynamic analysis of aluminoferrite phase hydration from the standpoint of the quantitative energy theory of isomorphism [346, 659]. The basis of the quantitative theory of isomorphism is the principle that the behaviour of an isomorphic mixture is governed by the thermodynamic functions excessive in respect to the corresponding functions of a mechanical mixture of pure components of the same composition:

$$\Delta G_{mix}(x_1, x_2, T) = \Delta G_{ss}(x_1, x_2) - x_1 \Delta G_1 - x_2 \Delta G_2 = \Delta H_{mix} - T\Delta S_{mix}, \quad (246)$$

where x_1, molar fraction of the first component; x_2, molar fraction of the second component; $\Delta G_{mix}(x_1, x_2, T)$, the Gibbs free energy of mixing; ΔH_{mix}, enthalpy of mixing; ΔS_{mix}, entropy of mixing; T, absolute temperature; $\Delta G_{ss(x_1, x_2)}$, the Gibbs free energy of the solid solution; ΔG_1, the Gibbs free energy of the first component; ΔG_2, the Gibbs free energy of the second component.

The entropy of mixing of a regular solution of two components is definitely determined by the composition of the solution [659]:

$$\Delta S_{mix} = -R(x_1 \ln x_1 + x_2 \ln x_2) \quad (247)$$

where R, the universal gas constant. An analysis of Eq. (247) shows that the ΔS_{mix} value is positive with all the values of x_1 and corresponding ones of x_2, reaching its maximum at $x_1 = x_2 = 0.5$. Then $-T\Delta S_{mix}$ in Eq. (246) is a negative value.

A quantitative evaluation of the heat of mixing ΔH_{mix} by the data of the lattice energy of of the atomization energy or by the heat of the crystalline compound formation from elements [341] is a complicated problem. Starting from the general rule that the greater the positive value of ΔH_{mix}, the narrower are the limits of miscibility at a given temperature [341], and that at $\Delta H_{mix} < 0$ solid solutions are stable down to very low temperatures [659], one can give a qualitative characteristic of ΔH_{mix} of solid solutions of the hydrates under examination: the heat of mixing will be negative or near to zero. Thus, the Gibbs free energy of mixing turns out to be a negative value and the absolute value of the Gibbs free energy of a solid solution $|\Delta G_{ss(x_1, x_2)}|$ proves to be greater than the sum of the absolute values of the Gibbs free energy of a mechanical mixture of components of similar composition ($|x_1 \Delta G_1| + |x_2 \Delta G_2|$).

It follows then that a thermodynamic analysis of the reactions of aluminoferrite phase hydration may be conducted for a mechanical mixture of aluminium and ferric hydrates. If, however, the Gibbs free energies of the reactions being compared turn out to be equal, the mixture whose products are capable of forming solid solutions will be thermodynamically preferable.

7.2 Reactions of Calcium Aluminates with Water

On the basis of the initial thermodynamic data found we examined the reactions of the C_4AH_{19}, C_3AH_6, C_2AH_8 and CAH_{10} formation at the hydration of tricalcium and monocalcium aluminates as well as the formation of the hydrates in the system $CaO-Al_2O_3-H_2O$, the ratio CaO/Al_2O_3 being 3/1 and 1/1.

The thermodynamic analysis of the processes of the calcium aluminate hydration at a temperature of 298 K was conducted by the reactions given in Table 100.

Formerly the hydration products were believed to be crystalline aluminate hydrates, portlandite and amorphous aluminium hydroxide.

The negative value of the Gibbs free energy of the reactions under consideration (Table 101) demonstrates that all of them are thermodynamically possible. The preference of the hydrate formation was analyzed by compositions, according to the principle of the ΔG^0 calculation. As water is both the medium and a component of the reactions, its amount was not kept constant.

The calculations carried out demonstrate that when C_3A is hydrated the formation of hexagonal hydrates C_4AH_x is preferable. The emergence of the hydrates C_2AH_8 and CAH_{10} in particular, is thermodynamically less probable as a result of a higher ΔG^0 value of the reaction. When CA is hydrated the formation of the hydrate C_2AH_8 is most preferable.

The results obtained are in agreement with the known experimental data. Thus, Larionova and coauthors [322] recorded the hydration of the powder of C_3A to result in the formation of mostly C_4AH_4 crystals. Talaber [606], Kondo and Daimon [290] report that the main product of the CA hydration at 295 to 303 K is C_2AH_8.

As Ludwig [337] states, C_3A interacting with water, C_4AH_x and C_2AH_8 are formed, and in the presence of lime, only C_4AH_x. This is corroborated by the analysis of Eqs. 1, 2 and 4 (Table 101). In a lime solution reactions 1 and 2 do not

Table 100. Reactions of the aluminate hydration and of the calcium aluminate hydrate transformation

Nos. of reactions	Equations of the reactions	ΔG^0 of the reactions [cal/mol]
A. System C_3A-H_2O		
1.	$C_3A+15H_2O=\frac{3}{4}C_4AH_{19}+\frac{1}{2}Al(OH)_3$	−43229
2.	$C_3A+10.5H_2O=\frac{3}{4}C_4AH_{13}+\frac{1}{2}Al(OH)_3$	−48190
3.	$C_3A+6H_2O=C_3AH_6$	−42968
4.	$C_3A+9H_2O=C_2AH_8+Ca(OH)_2$	−40437
5.	$C_3A+12H_2O=CAH_{10}+2Ca(OH)_2$	−36920
B. System $CA-H_2O$		
6.	$CA+7H_2O=\frac{1}{4}C_4AH_{19}+\frac{3}{2}Al(OH)_3$	−9161
7.	$CA+5.5H_2O=\frac{1}{4}C_4AH_{13}+\frac{3}{2}Al(OH)_3$	−9148
8.	$CA+4H_2O=\frac{1}{3}C_3AH_6+\frac{4}{3}Al(OH)_3$	−9073
9.	$CA+5.5H_2O=\frac{1}{2}C_2AH_8+Al(OH)_3$	−9287
10.	$CA+10H_2O=CAH_{16}$	−8933
C. Aluminate hydrate transformation		
11.	$C_4AH_{19}=C_4AH_{13}+6H_2O$	−0.038 [a]
12.	$2OH^-+C_3AH_6+Ca^{2+}+6H_2O=C_4AH_{13}$	−10.36
13.	$C_4AH_{13}+\frac{1}{2}Ca^{2+}+Al(OH)_4^-=\frac{3}{2}C_3AH_6+6H_2O$	−3.5
14.	$C_2AH_8+2Ca^{2+}+4OH^-+3H_2O=C_4AH_{13}$	−20.0
15.	$C_4AH_{13}+2Ca^{2+}+4Al(OH)_4^-+3H_2O=3C_2AH_8$	−16.19

[a] For water in the liquid state ($H_2O_{(liq)}$)

Table 101. The temperature dependence of ΔG^0 of the reactions in the systems C_3A-H_2O and $Ca-H_2O$

Nos. of reaction	$-\Delta G^0_{298}$ of the reactions [cal/mol] at temps. [K]					
	298	323	348	373	396	423
A 1	44889	41808	38585	35226	31743	28140
2	44850	42608	40265	37821	35284	32653
3	42968	41623	40218	38755	37231	35656
4	40437	38565	36611	34574	32455	30262
5	36920	34526	32024	29413	26702	23894
B 6	14142	12553	10893	9165	7374	5524
7	14124	12813	11445	10022	8545	7020
8	13496	12484	11426	10829	9191	8016
9	12599	11338	10020	8645	7217	5742
10	8933	6925	4821	2617	323	−2058

take place, because ions $Al(OH)_4^-$ are bound not into hydroxide, but into the corresponding hydroaluminate, which intensifies the preference of these reactions. The presence of $Ca(OH)_2$, in accordance with the Le Chatelier principle, lessens the probability of the C_2AH_8 formation by reaction 4. Consequently, C_3A being hydrated in lime solution, only tetracalcium aluminate hydrates are formed.

The stability of calcium aluminate hydrates is a product of a number of factors, temperature and the composition of the solution being the principal ones among them. Out of quite a variety of possible reactions of transformation at normal temperature (298 K) some of them, related to C_4AH_{13}, are considered. The data obtained demonstrate that the transformation of C_2AH_8 and C_3AH_6 into C_4AH_{13} as well as reverse reactions are thermodynamically possible.

A transformation of C_3AH_6 into C_4AH_{13} and the reverse reaction were established by Stein in his experiments [592]. With the transformation of the hydrate C_4AH_{19} into C_4AH_{13} the Gibbs free energy of the reaction (Eq. 11, Table 100) is as good as zero, i.e., the stability of these hydrates is a function of the relative humidity, and the data adduced in the summary works by Turriciani [649] and Ludwig [337] corroborate it.

Stable hydrate phases in the system $CaO-Al_2O_3-H_2O$, at a high temperature (up to 523 K) and pressure of the saturated water vapour, are calcium hydroxide, gibbsite, up to the temperature of 423 K, and cubic aluminate hydrate, at a temperature below 498 K [592]. Therefore for a study of the thermodynamic stability of calcium aluminate hydrates at high temperatures in Eqs. (1 – 5)A – (1 – 5)B (see Table 100) aluminium hydroxide was considered in the form of gibbsite. The results are given in Table 101 and Figs. 87 and 88. The analysis showed that in the systems C_3A-H_2O and $CA-H_2O$ within a temperature range of 298 to 323 K, thermodynamically the most stable are hexagonal hydrates, and at temperatures above 323 to 348 K, cubic calcium aluminate hydrate, which quite agrees with the known experimental data.

The results of the study of the system $CaO-Al_2O_3-H_2O$, the ratio CaO/Al_2O_3 being 3/1 and 1/1, differ little from those of the study of hydration of C_3A and CA.

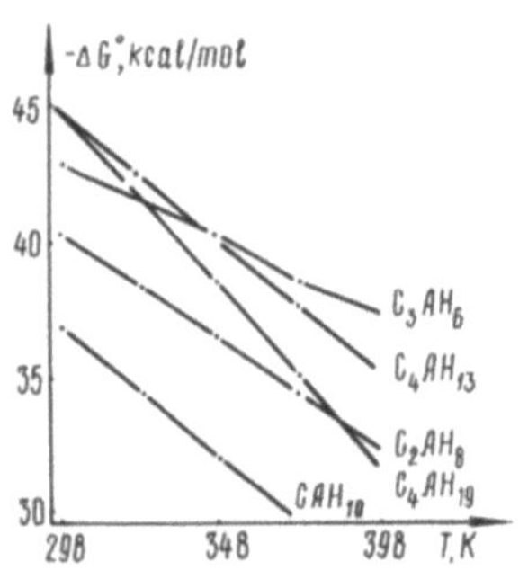

Fig. 87. The temperature dependence of ΔG^0 of the reactions in the system C_3A-H_2O

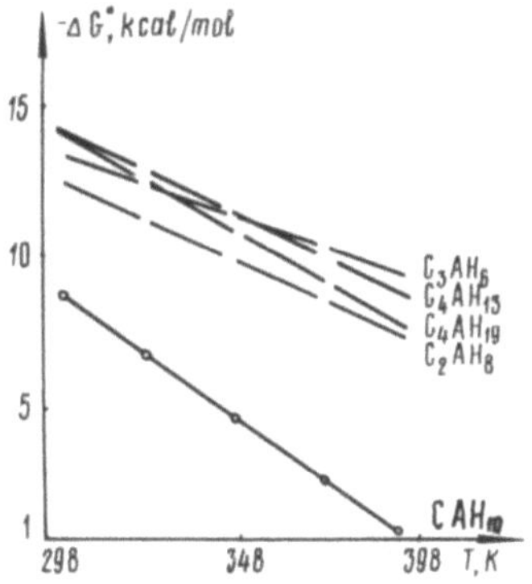

Fig. 88. The temperature dependence of ΔG^0 of the reactions in the system $CA-H_2O$

7.3 Hydration of Calcium Aluminates in the Presence of Gypsum

Calculations were carried out for the reactions of the formation of $C_3ACs_3H_{32}$ and C_3ACsH_{12}, the ratio $C_3A/CaSO_4 \cdot 2H_2O$ being 3/1, for the reactions of the ettringite formation and of its transformation into calcium monosulphoaluminate (Table 102).

The data obtained show that at a temperature of 298 K the formation of both forms of calcium aluminate sulphate hydrate is thermodynamically possible, the primary product being ettringite independent of the ratio $C_3A/CaSO_4 \cdot 2H_2O$. If the amount of gypsum is sufficient for binding C_3A present into ettringite it will be the only stable aluminate sulphate hydrate form. In case of a shortage of gypsum, after its binding into $C_3ACs_3H_{32}$, the latter will interact with the liquid phase forming calcium monosulphoaluminate (reaction 5). No transformation of a high-sulphate form of the aluminate sulphate hydrate into a low-sulphate one takes place and gypsum is the result of a positive value ($\Delta G_{298} = 9.048$ kcal/mol) of the Gibbs free energy (reaction 6).

Table 102. Hydration reactions in the system $C_3A-CaSO_4 \cdot 2H_2O-H_2O$

Nos. of reaction	Equations of the reactions	ΔG^0_{298} of the reactions [kcal/mol]
1.	$C_3A + CaSO_4 \cdot 2H_2O + \frac{56}{3}H_2O = \frac{1}{3}C_2ACs_3H_{32} + \frac{1}{2}C_4AH_{19} + \frac{1}{3}Al(OH)_3$	−48750
2.	$C_3A + CaSO_4 \cdot 2H_2O + \frac{38}{3}H_2O = \frac{1}{3}C_3ACs_3H_{32} + \frac{2}{3}C_3AH_6$	−47470
3.	$C_3A + CaSO_4 \cdot 2H_2O + 10H_2O = C_3ACsH_{12}$	−47420
4.	$C_3A + + 3(CaSO_4 \cdot 2H_2O) + 26H_2O = C_3ACs_3H_{32}$	−56468
5.	$C_3ACs_3H_{32} + \frac{1}{3}C_4AH_{19} + \frac{2}{3}Ca^{2+} + \frac{4}{3}Al(OH)_4^- = 3C_3ACsH_{12} + 16H_2O$	−4570
6.	$C_3ACs_3H_{32} = C_3ACsH_{12} + 2(CaSO_4 \cdot 2H_2O) + 16H_2O$	+9048

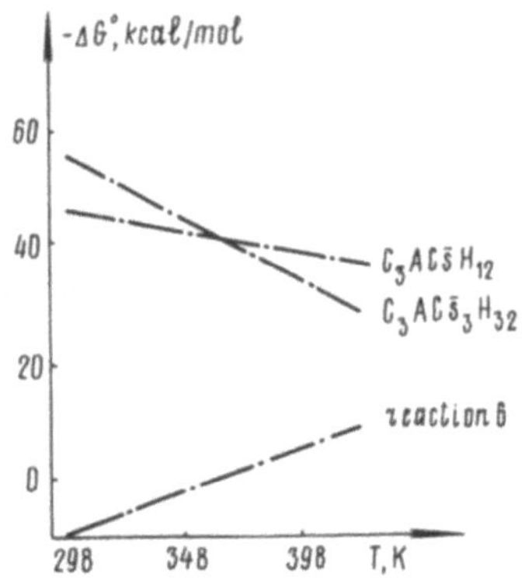

Fig. 89. The temperature dependence of ΔG^0 of the reactions in the system $C_3A-CaSO_4\cdot 2H_2O-H_2O$

For studying the thermodynamic stability of calcium aluminate sulphate hydrates at high temperatures reactions 3 and 4 (Table 102) were considered. As in the reactions in question only a constant amount of the reacting aluminate was present, in addition the temperature dependence of the Gibbs free energy of reaction 6 was determined. From the results given in Table 103 and Fig. 89 it follows that up to temperatures of 343 to 363 K ettringite is thermodynamically stable and above that temperature, calcium monosulphoaluminate [86].

The calculations conducted agree with the experimental data reviewed by Larionova and coauthors [322].

7.4 Dicalcium Ferrite Hydration

We considered the reactions of the formation of cubic (C_3FH_6) and hexagonal (C_4FH_{13}) calcium hydroferrites, of amorphous ferric hydroxide [$Fe(OH)_3$] and portlandite [$Ca(OH)_2$] from C_2F as well as the reactions of the transformation of C_3FH_6 into C_4FH_{13} and of C_4FH_{13} into C_3FH_6. The forms of the reactions in question and calculation results are given in Table 104.

Table 103. The temperature dependence of ΔG^0 of the reactions in the system $C_3A-CaSO_4\cdot 2H_2O-H_2O$

Nos. of reaction	$-\Delta G^0$ of the reactions [cal/mol] at temps. [K]					
	298	323	348	373	398	423
3.	47420	45270	43020	40673	38281	35700
4.	56468	51219	45732	40013	31081	27945
6.	−9048	−5945	−2717	858	4145	7747

Note: see Table 101.

Table 104. The values of the Gibbs free energy at 298 K of the reactions of calcium hydroferrite hydration and transformation

Nos. of reaction	Equations of the reactions	$-\Delta G^0_{298}$ of the reactions [cal/mol]
1.	$2CaO \cdot Fe_2O_3 + 8H_2O = \frac{1}{2}4CaO \cdot Fe_2O_3 \cdot 13H_2O + Fe(OH)_3$	12279
2.	$2CaO \cdot Fe_2O_3 + 7H_2O = \frac{1}{2}4CaO \cdot Fe_2O_3 \cdot 13H_2O + FeOOH$	16820
3.	$2CaO \cdot Fe_2O_3 + \frac{13}{2}H_2O = \frac{1}{2}4CaOFe_2O_3 \cdot 13H_2O + \frac{1}{2}Fe_2O_3$	16510
4.	$2CaO \cdot Fe_2O_3 + 5H_2O = \frac{2}{3}CaO \cdot Fe_2O_3 \cdot 6H_2O + \frac{2}{3}Fe(OH)_3$	11184
5.	$2CaO \cdot Fe_2O_3 + \frac{13}{3}H_2O = \frac{2}{3}3CaO \cdot Fe_2O_3 \cdot 6H_2O + \frac{2}{3}FeOOH$	14211
6.	$2CaO \cdot Fe_2O_3 + 4H_2O = \frac{2}{3}3CaOFe_2O_3 \cdot 6H_2O + \frac{1}{3}Fe_2O_3$	14002
7.	$2CaO \cdot Fe_2O_3 + 5H_2O = 2Ca(OH)_2 + 2Fe(OH)_3$	9025
8.	$3CaO \cdot Fe_2O_3 \cdot 6H_2O + Ca^{2+} + 2OH^- + 6H_2O$ $= 4CaO \cdot Fe_2O_3 \cdot 13H_2O$	10370
9.	$4CaO \cdot Fe_2O_3 \cdot 13H_2O + \frac{1}{2}Ca^{2+} + Fe(OH)_4^-$ $= \frac{3}{2}\, 3CaO \cdot Fe_2O_3 \cdot 6H_2O + 6H_2O$	7487

As can be seen from Table 104, when the C_2F hydration takes place in water at normal temperature, the most preferable is the formation of an association of the hexagonal hydrate C_4FH_{13} and goethite (reaction 2). Depending on conditions (pH, calcium ion concentration), C_4FH_{13} may be transformed into C_3FH_6. Ions $Fe(OH)_4^-$ being present in the solution, a reverse transformation is possible.

From Table 104 one can see that C_2F can be hydrated with the formation of C_4FH_{13} and amorphous ferric hydroxide (reaction 1). The amorphous hydroxide, however, with time and with the elevation of temperature, in particular, converts into its most stable and low water varieties. The final product of the $Fe(OH)_3$ aging in alkaline medium is crystalline goethite.

At elevated temperatures only reactions 2, 3, 5 and 6 were examined. The calculation results are given in Table 105.

As can be seen from Table 105, at temperatures up to 373 K thermodynamically most stable is an association of C_4FH_{13} with goethite, up to 423 K, that with hematite, and at higher temperatures, an association of cubic C_3FH_6 with hematite.

Table 105. The temperature dependence of ΔG^0 of the hydration reactions

Nos. of reactions	$-\Delta G^0$ of the reactions [cal/mol] at temps. [K]						
	298	323	348	373	398	423	448
2.	16820	15297	13676	11958	10151	8259	6286
3.	16510	15094	13585	11988	10312	8560	6733
5.	14211	13179	12075	10899	9661	8356	6994
6.	14002	13043	12015	10920	9768	8556	7296

While examining the processes of the dicalcium ferrite hydrolysis, by analogy with aluminate hydrolysis, the problem of what maximum concentration of the ions $Fe(OH)_4^-$ and $Al(OH)_4^-$ may be achieved in the solutions of calcium ferrite hydrates and aluminate hydrates was first considered. In both cases this value may be next to the concentration great enough for the precipitation of an amorphous hydroxide, which is expressed as:

$$Fe(OH)_3 + OH^- = Fe(OH)_4^- \quad \text{and} \tag{248}$$

$$Al(OH)_3 + OH^- = Al(OH)_4^- \, . \tag{249}$$

A calculation showed that at $pH = 12$ the equilibrium concentration of aluminate ions is by a factor of about 10^5 higher than that of ferrite ones (0.1 and 0.000001 g-ions/l respectively) and is sufficient for the formation of metastable solutions of calcium aluminate hydrates [110, 126, 223]. A low concentration of ferrite ions testifies to the fact that iron-containing hydrates dissolve incongruently and their metastable solubility is not realized. While carrying out thermodynamic calculations of the reactions of the C_2F hydration with $Ca(OH)_2$, the molar ratio of CaO/Fe_2O_3 being 3/1 and 4/1, no qualitative differences as compared with the reactions in the system $CaO-Fe_2O_3-H_2O$ were found.

7.5 Reactions of the Tetracalcium Aluminoferrite Hydration

In a thermodynamic analysis of the reactions of the C_4AF hydration the reactions accompanied by the formation of both hexagonal hydrates: C_4AH_{13}, C_2AH_8, C_4FH_{13}, and cubic ones: C_3AH_6 and C_3FH_6, were considered as well as of amorphous aluminium and ferric hydroxides $Al(OH)_3$ and $Fe(OH)_3$.

The C_4AF hydration reactions and the calculation results are given in Table 106.

As one can see from Table 106, in the interaction of C_4AF with water at normal temperature, thermodynamically, the most preferable is the formation of hexagonal calcium aluminate hydrate C_4AH_{13} and goethite (reaction 3). Although ferric hydroxide may be the primary product, it will nevertheless be transformed into goethite. A calculation of the reactions at elevated temperatures (Table 107) showed that above 423 K in reaction 7 as compared with reaction 4, at thermodynamically more stable product is a mechanical mixture of cubic calcium hydroaluminates and ferrite hydrates C_3AH_6 and C_3FH_6, and hematite.

The possibility of the formation of a solid solution of C_3AH_6 and C_3FH_6 results in the products of reactions 5, 6 and 7 (Table 106) being not a mechanical mixture of these hydrates, but a solid solution of $C_3A_{0.75}-F_{0.25}H_6$. The results of thermodynamic calculations quite agree with the experimental data concerning the primary formation of hexagonal hydrates and their instability at elevated temperatures [89]. Thus, according to Shikiryanskiy [570], in the process of the

Table 106. The values of the Gibbs free energy of the reactions of the C_4AF hydration at 298 K

Nos.	Equations of reactions	$-\Delta G^0_{298}$ [cal/mol]
1.	$4CaO \cdot Al_2O_3 \cdot Fe_2O_3 + 10H_2O = 4Ca(OH)_2 + 2Al(OH)_3 + 2Fe(OH)_3$	30370
2.	$4CaO \cdot Al_2O_3 \cdot Fe_2O_3 + 16H_2O = 4CaO \cdot Al_2O_3 \cdot 13H_2O + 2Fe(OH)_3$	42860
3.	$4CaO \cdot Al_2O_3 \cdot Fe_2O_3 + 14H_2O = 4CaO \cdot Al_2O_3 \cdot 13H_2O + 2FeOOH$	51930
4.	$4CaO \cdot Al_2O_3 \cdot Fe_2O_3 + 13H_2O = 4CaO \cdot Al_2O_3 \cdot 13H_2O + Fe_2O_3$	51310
5.	$4CaO \cdot Al_2O_3 \cdot Fe_2O_3 + 10H_2O$ $= 3CaO \cdot Al_2O_3 \cdot 6H_2O + \frac{1}{3}\ 3CaO \cdot Fe_2O \cdot 6H_2O + \frac{4}{3}Fe(OH)_3$	40680
6.	$4CaO \cdot Al_2O_3 \cdot Fe_2O_3 + \frac{26}{3}H_2O$ $= 3CaO \cdot Al_2O_3 \cdot 6H_2O + \frac{1}{3}\ 3CaO \cdot Fe_2O_3 \cdot 6H_2O + \frac{4}{3}FeOOH$	46740
7.	$4CaO \cdot Al_2O_3 \cdot Fe_2O_3 + 8H_2O$ $= 3CaO \cdot Al_2O_3 \cdot 6H_2O + \frac{1}{3}CaO \cdot Fe_2O_3 \cdot 6H_2O + \frac{2}{3}Fe_2O_3$	46320
8.	$4CaO \cdot Al_2O_3 \cdot Fe_2O_3 + 15H_2O$ $= 2CaO \cdot Al_2O_3 \cdot 8H_2O + \frac{1}{2}\ 4CaO \cdot Fe_2O_3 \cdot 13H_2O + Fe(OH)_3$	40320
9.	$4CaO \cdot Al_2O_3 \cdot Fe_2O_3 + 16H_2O$ $= 2CaO \cdot Al_2O_3 \cdot 8H_2O + \frac{1}{2}\ 4CaO \cdot Fe_2O_3 \cdot 13H_2O + FeOOH$	44870
10.	$4CaO \cdot Al_2O_3 \cdot Fe_2O_3 + \frac{29}{2}H_2O$ $= 2CaO \cdot Al_2O_3 \cdot 8H_2O + \frac{1}{2}4CaO \cdot Fe_2O_3 \cdot 13H_2O + \frac{1}{2}Fe_2O_3$	44550
11.	$4CaO \cdot Al_2O_3 \cdot Fe_2O_3 + 16H_2O = 4CaO \cdot Fe_2O_3 \cdot 13H_2O + 2Al(OH)_3$ am.	36880
12.	$4CaO \cdot Al_2O_3 \cdot Fe_2O_3 + 16H_2O = 4CaO \cdot Fe_2O_3 \cdot 13H_2O + 2Al(OH)_3$	43880
13.	$4Ca \cdot Al_2O_3 \cdot Fe_2O_3 + 16H_2O$ $= 3CaO \cdot Fe_2O_3 \cdot 6H_2O + \frac{1}{3}\ 3CaO \cdot Al_2O_3 \cdot 6H_2O + \frac{4}{3}Al(OH)_3$ am.	36690
14.	$4CaO \cdot Al_2O_3 \cdot Fe_2O_3 + 10H_2O$ $= 3CaO \cdot Fe_2O_3 \cdot 6H_2O + \frac{1}{3}\ 3CaO \cdot Al_2O_3 \cdot 6H_2O + \frac{4}{3}Al(OH)_3$	41350

Table 107. Temperature dependence of ΔG^0 of the reactions of the C_4AF hydration

Nos. of reactions in Table 106	$-\Delta G^0$ of the reactions [cal/mol] at temps. [K]						
	298	323	348	373	398	423	448
3	51930	49080	46070	42930	39650	36240	32700
4	51310	48670	45890	42990	39970	36830	33600
6	46740	44860	42890	40820	38660	36410	34090
7	46320	44590	42760	40850	38870	36810	34690
9	44870	4790	38560	35180	31650	27990	24210
10	44550	41580	38460	35200	31800	28280	24650
12	43860	40390	36740	32930	28970	24570	20650
14	41350	39060	36650	34150	31540	28850	26070

C_4AF hydration tetracalcium aluminate hydrate is first formed, then, as the hydration progresses, $C_3A_{0.75}F_{0.25}H_6$, the composition of which is similar to that found by thermodynamic calculations. According to T. Carlson [89] the composition of hydrates at the C_4AF hydration approaches, but does not reach the ratio Fe_2O_3/Al_2O_3 in anhydrous materials which is also in agreement with the data obtained. We carried out the thermodynamic analysis of the reactions of the C_4AF hydration with lime at the molar ratio $Ca(OH)_2/C_4AF$ being 2. As the products of the reactions the same compounds were taken as in the system C_4AF-H_2O.

The equations of the reactions and the Gibbs free energy values at normal and elevated temperatures are given in Tables 108 and 109.

As can be seen from Tables 108 and 109, at normal temperature, thermodynamically most preferable, is the formation of a mechanical mixture of 1 mol of C_4AH_{13} and 0.5 mol C_4AFH_{13}, which corresponds to the solid solution of the

Table 108. Equations of hydration reactions in the system $C_4AF-Ca(OH)_2-H_2O$, the ratio $Ca(OH)_2/C_4AF$ being 2/1, and their ΔG^0_{298} values

Nos.	Equations of reactions	$-\Delta G^0_{298}$ [cal/mol]
1.	$4CaO \cdot Al_2O_3 \cdot Fe_2O_3 + 2Ca(OH)_2 + 19H_2O$ $= 4CaO \cdot Al_2O_3 \cdot 13H_2O + \frac{1}{3}\, 4CaO \cdot Al_2O_3 \cdot 13H_2O + Fe(OH)_3$	46120
2.	$4CaO \cdot Al_2O_3 \cdot Fe_2O_3 + 2Ca(OH)_2 + 18H_2O$ $= 4CaO \cdot Al_2O_3 \cdot 13H_2O + \frac{1}{2}\, 4CaO \cdot Al_2O_3 \cdot 13H_2O + FeOOH$	50650
3.	$4CaO \cdot Al_2O_3 \cdot Fe_2O_3 + 2Ca(OH)_2 + \frac{35}{2}H_2O$ $= 4CaO \cdot Al_2O_3 \cdot 13H_2O + \frac{1}{2}\, 4CaO \cdot Al_2O_3 \cdot 13H_2O + Fe_2O_3$	50350
4.	$4CaO \cdot Al_2O_3 \cdot Fe_2O_3 + 2Ca(OH)_2 + 19H_2O$ $= 2CaO \cdot Al_2O_3 \cdot 8H_2O + 4CaO \cdot Fe_2O_3 \cdot 13H_2O$	43580
5.	$4CaO \cdot Al_2O_3 \cdot Fe_2O_3 + 2Ca(OH)_2 + 10H_2O$ $= 3CaO \cdot Al_2O_3 \cdot 6H_2O + 3CaO \cdot Fe_2O_3 \cdot 6H_2O$	42840
6.	$4CaO \cdot Al_2O_3 \cdot Fe_2O_3 + 2Ca(OH)_2 + 19H_2O$ $= \frac{1}{2}\, 4CaO \cdot Al_2O_3 \cdot 13H_2O + 4CaO \cdot Fe_2O_3 \cdot 13H_2O + Al(OH)_3$ am.	43120
7.	$4CaO \cdot Al_2O_3 \cdot Fe_2O_3 + 2Ca(OH)_2 + 19H_2O$ $= \frac{1}{2}\, 4CaO \cdot Al_2O_3 \cdot 13H_2O + 4CaO \cdot Fe_2O_3 \cdot 13H_2O + Al(OH)_3$	46630

Table 109. Temperature dependence of the Gibbs free energy of the reactions in the system $C_4AF-Ca(OH)_2-H_2O$, the ratio $Ca(OH)_2/C_4AF$ being 2/1

Nos. of reactions in Table 108	$-\Delta G^0$ of the reactions [cal/mol] at temps. [K]						
	298	323	348	373	398	423	448
2	50650	46930	43020	38930	34660	30230	25640
3	50350	46740	42950	38970	34830	30540	26105
4	43580	39650	35520	31190	26690	22020	17190
5	42840	40600	38290	35780	33210	30540	27790
7	46630	42600	38370	33950	29350	24580	19650

composition $C_4A_{0.67}F_{0.33}H_{13}$ and 1 mol of FeOOH. At temperatures above 423 K a mixture of cubic hydrates C_3AH_6 and C_3FH_6 is thermodynamically stable, which corresponds to the solid solution $C_3A_{0.5}F_{0.5}H_6$. As to $Fe(OH)_3$ being a co-product of the C_4AF hydration, one can state that its formation is thermodynamically possible similar to that of goethite. However, it must later change into a more stable form, α-goethite. This follows from a consideration of the reactions of their mutual transformation

$$Fe(OH)_3 = FeOOH + H_2O \qquad (250)$$

having $\Delta G^0_{298} = -4530$ cal/mol.

7.6 Hydration of Iron-Containing Phases with Gypsum

Reactions were first examined in the system $CaO - Fe_2O_3 - CaSO_4 \cdot 2H_2O - H_2O$. With the ratio 3/1 the reactions of the formation of calcium trisulphoferrite $3CaO \cdot Fe_2O_3 \cdot 3CaSO_4 \cdot 32H_2O$, of monosulphoferrite $3CaO \cdot Fe_2O_3 \cdot CaSO_4 \cdot 12H_2O$, of hexagonal C_4FH_{13} and cubic C_3FH_6 calcium hydroferrites, of amorphous ferric hydroxide $Fe(OH)_3$, of goethite FeOOH and reactions of the trisulphate transformation into monosulphate were considered.

The calculation results are given in Tables 110 and 111.

Table 110. Equations of the reactions in the system $CaO - Fe_2O_3 - CaSO_4 \cdot 2H_2O - H_2O$, the ratio CaO/Fe_2O_3 being 3/1, and of the reactions of calcium ferite sulphate hydrate transformation

Nos. of reactions	Equations of reactions	$-\Delta G^0_{298}$ of the reactions [cal/mol]
1.	$3CaO + Fe_2O_3 + CaSO_4 \cdot 2H_2O + \frac{47}{3}H_2O = \frac{1}{3}(3CaO \cdot Fe_2O_3 \cdot 3CaSO_4 \cdot 32H_2O) + \frac{1}{2}(4CaO \cdot Fe_2O_3 \cdot 13H_2O) + \frac{1}{3}Fe(OH)_3$	40440
2.	$3CaO + Fe_2O_3 + CaSO_4 \cdot 2H_2O + \frac{46}{3}H_2O = \frac{1}{3}(3CaO \cdot Fe_2O_3 \cdot 3CaSO_4 \cdot 32H_2O) + \frac{1}{2}(4CaO \cdot Fe_2O_3 \cdot 13H_2O) + \frac{1}{3}FeOOH$	41950
3.	$3CaO + Fe_2O_3 + CaSO_4 \cdot 2H_2O + \frac{38}{3}H_2O = \frac{1}{3}(3CaO \cdot Fe_2O_3 \cdot 3CaSO_4 \cdot 32H_2O) + \frac{1}{3}(3CaO \cdot Fe_2O_3 \cdot 6H_2O)$	39340
4.	$3CaO + Fe_2O_3 + CaSO_4 \cdot 2H_2O + 10H_2O = 3CaO \cdot Fe_2O_3 \cdot 3CaSO_4 \cdot 12H_2O$	39330
5.	$3CaO + Fe_2O_3 + 3(CaSO_4 \cdot 2H_2O) + 26H_2O = 3CaO \cdot Fe_2O_3 \cdot 3CaSO_4 \cdot 32H_2O$	48270
6.	$3CaO \cdot Fe_2O_3 \cdot 3CaSO_4 \cdot 32H_2O = 3CaO \cdot Fe_2O_3 \cdot CaSO_4 \cdot 12H_2O + 2(CaSO_4 \cdot 2H_2O) + 16H_2O$	−8950
7.	$3CaO \cdot Fe_2O_3 \cdot 3CaSO_4 \cdot 32H_2O + \frac{4}{3}(4CaO \cdot Fe_2O_3 \cdot 13H_2O) + \frac{2}{3}Ca^{2+} + \frac{4}{3}Fe(OH)_4^- = 3CaO \cdot Fe_2O_3 \cdot CaSO_4 \cdot 12H_2O + 16H_2O$	+9940

Table 111. Temperature dependence of the Gibbs free energy of the reactions in the system $CaO - Fe_2O_3 - CaSO_4 \cdot 2H_2O - H_2O$ (the ratio CaO/Fe_2O_3 being 3/1) and of the reaction of the trisulphate transformation into monosulphate

Nos. of reactions in Table 110	$-\Delta G^0$ of the reactions [cal/mol] at temps. [K]				
	298	323	348	373	398
2	41950	38920	35740	32430	28980
3	39340	36800	34140	31370	28490
4	39330	37290	35150	32910	30580
5	48270	43120	37720	32090	26250
6	-8950	-5840	-2600	730	4280

As can be seen from Table 110, in the system $CaO - Fe_2O_3 - CaSO_4 \cdot 2H_2O - H_2O$, independent of the ratio of the amount of gypsum to the other solid phases, at normal temperatures the most preferable is the formation of a high-sulphate form of calcium ferrite sulphate hydrate. At temperatures above 363 K monosulphate becomes the thermodynamically stable sulphate form. The positive value of the Gibbs free energy of reaction 6, Table 110 testifies to the transformation of a high-sulphate form of ferrite sulphate hydrate into that of low-sulphate gypsum then being impossible. That is, with gypsum the ions SO_4^{2-} are bound only into C_3FCsH_{32}. However, after all the gypsum is bound into C_3FCsH_{32} the latter turns out to be thermodynamically unstable with respect to the ions which are in the liquid phase, and interacting with them it is partly or completely transformed into monosulphoferrite C_3FCsH_{12}. This is schematically shown by reaction 7 in Table 110. Thus, the same regularities of the behaviour of sulphate-containing hydrates are characteristic of aluminate as well as of ferrite systems.

In the system $C_4AF - CaSO_4 \cdot 2H_2O$ reactions of the formation of all possible products and of their combinations with molar ratios $C_4AF/CaSO_4 \cdot 2H_2O$ being 1/4 and 1/1 at normal and elevated temperatures were also analyzed.

Results of the investigation are given in Tables 112 and 113. As one can see from Table 112, in the system $C_4AF - CaSO_4 \cdot 2H_2O - H_2O$ at normal temperature independent of the ratio of C_4AF to gypsum the most preferable is the formation of a trisulphate (reaction 2), hexagonal phases and of goethite. Associations of aluminium hydroxide and of calcium hydrates rich in iron are thermodynamically less preferable (reaction 5).

At elevated temperatures associations of the monosulphate with hexagonal and cubic hydrates in particular (reactions 7 and 10 in Table 113) are thermodynamically stable. An analysis of the reactions of the C_4AF hydration in the presence of gypsum is complicated by the presence of solid solutions among the

Table 112. Equations of the reactions in the system $C_4AF - CaSO_4 \cdot 2H_2O - H_2O$, the ratio $C_4AF/CaSO_4 \cdot 2H_2O$ being 1/1

Nos. of reactions	Equations of reactions	$-\Delta G^0_{298}$ of reactions [cal/mol]
1.	$4CaO \cdot Al_2O_3 \cdot Fe_2O_3 + CaSO_4 \cdot 2H_2O + \frac{127}{6}H_2O = \frac{1}{3}(3CaO \cdot Al_2O_3 \cdot 3CaSO_4 \cdot 32H_2O) + \frac{2}{3}(4CaO \cdot Al_2O_3 \cdot 13H_2O) + \frac{1}{12}(4CaO \cdot Fe_2O_3 \cdot 13H_2O) + \frac{11}{6}Fe(OH)_3$	46810
2.	$4CaO \cdot Al_2O_3 \cdot Fe_2O_3 + CaSO_4 \cdot 2H_2O + \frac{58}{3}H_2O = \frac{1}{3}(3CaO \cdot Al_2O_3 \cdot 3CaSO_4 \cdot 32H_2O) + \frac{2}{3}(4CaO \cdot Al_2O_3 \cdot 13H_2O) + \frac{1}{12}(4CaO \cdot Fe_2O_3 \cdot 13H_2O) + \frac{11}{6}FeOOH$	55140
3.	$4CaO \cdot Al_2O_3 \cdot Fe_2O_3 + CaSO_4 \cdot 2H_2O + \frac{127}{6}H_2O = \frac{3}{12}(3CaO \cdot Al_2O_3 \cdot 3CaSO_4 \cdot 32H_2O) + \frac{1}{12}(3CaO \cdot Fe_2O_3 + 3CaSO_4 \cdot 32H_2O) + \frac{3}{4}(4CaO \cdot Al_2O_3 \cdot 13H_2O) + \frac{11}{6}Fe(OH)_3$	46800
4.	$4CaO \cdot Al_2O_3 \cdot Fe_2O_3 + CaSO_4 \cdot 2H_2O + \frac{58}{3}H_2O = \frac{3}{12}(3CaO \cdot Al_2O_3 \cdot 3CaSO_4 \cdot 32H_2O) + \frac{1}{12}(3CaO \cdot Fe_2O_3 \cdot 3CaSO_4 \cdot 32H_2O) + \frac{3}{4}(4CaO \cdot Al_2O_3 \cdot 13H_2O) + \frac{11}{6}FeOOH$	55120
5.	$4CaO \cdot Al_2O_3 \cdot Fe_2O_3 + CaSO_4 \cdot 2H_2O + \frac{127}{6}H_2O = \frac{1}{3}(3CaO \cdot Fe_2O_3 \cdot 3CaSO_4 \cdot 32H_2O) + \frac{2}{3}(4CaO \cdot Fe_2O_3 \cdot 13H_2O) + \frac{1}{12}(4CaO \cdot Al_2O_3 \cdot 13H_2O) + \frac{11}{6}Al(OH)_3$	41300
6.	$4CaO \cdot Al_2O_3 \cdot Fe_2O_3 + CaSO_4 \cdot 2H_2O + 14H_2O = 3CaO \cdot Al_2O_3 \cdot CaSO_4 \cdot 12H_2O + \frac{1}{3}(3CaO \cdot Fe_2O_3 \cdot 6H_2O) + \frac{4}{3}Fe(OH)_3$	45130
7.	$4CaO \cdot Al_2O_3 \cdot Fe_2O_3 + CaSO_4 \cdot 2H_2O + \frac{38}{3}H_2O = 3CaO \cdot Al_2O_3 \cdot CaSO_4 \cdot 12H_2O + \frac{1}{3}(3CaO \cdot Fe_2O_3 \cdot 6H_2O) + \frac{4}{3}FeOOH$	51180
8.	$4CaO \cdot Al_2O_3 \cdot Fe_2O_3 + CaSO_4 \cdot 2H_2O + 14H_2O = \frac{5}{6}(3CaO \cdot Al_2O_3 \cdot CaSO_4 \cdot 12H_2O) + \frac{1}{6}(3CaO \cdot Fe_2O_3 \cdot CaSO_4 \cdot 12H_2O) + \frac{1}{6}(3CaO \cdot Al_2O_3 \cdot 6H_2O) + \frac{1}{6}(3CaO \cdot Fe_2O_3 \cdot 6H_2O) + \frac{4}{3}Fe(OH)_3$	45130
9.	$4CaO \cdot Al_2O_3 \cdot Fe_2O_3 + CaSO_4 \cdot 2H_2O + \frac{31}{2}H_2O = 3CaO \cdot Al_2O_3 \cdot CaSO_4 \cdot 12H_2O + \frac{1}{4}(4CaO \cdot Fe_2O_3 \cdot 13H_2O) + \frac{3}{2}Fe(OH)_3$	45680
10.	$4CaO \cdot Al_2O_3 \cdot Fe_2O_3 + CaSO_4 \cdot 2H_2O + 14H_2O = 3CaO \cdot Al_2O_3 \cdot CaSO_4 \cdot 12H_2O + \frac{1}{4}(4CaO \cdot Fe_2O_3 \cdot 13H_2O) + \frac{3}{2}FeOOH$	52490
11.	$4CaO \cdot Al_2O_3 \cdot Fe_2O_3 + CaSO_4 \cdot 2H_2O + \frac{31}{2}H_2O = \frac{3}{4}(3CaO \cdot Al_2O_3 \cdot CaSO_4 \cdot 12H_2O) + \frac{1}{4}(3CaO \cdot Fe_2O_3 \cdot CaSO_4 \cdot 12H_2O) + \frac{1}{4}(4CaO \cdot Al_2O_3 \cdot 13H_2O) + \frac{3}{2}Fe(OH)_3$	45680

Table 113. Temperature dependence of the Gibbs free energy of the reactions in the system $4CaO \cdot Al_2O_3 \cdot Fe_2O_3 - CaSO_4 \cdot 2H_2O$, the ratio $C_4AF/CaSO_4 \cdot 2H_2O$ being 1/1

Nos. of reactions in Table 112	$-\Delta G^0$ of the reactions [cal/mol] at temps. [K]					
	298	323	348	373	398	423
2	55140	51220	47110	42810	38330	33680
4	55120	51200	47090	42790	38300	33660
7	51180	48500	45680	42730	39650	36450
10	52490	49570	46490	43270	39910	36410

products of the reactions. At present, however, due to insufficient data on the mixing enthalpy of the hydrates under consideration, conducting a complete thermodynamic analysis of the formation of solid solutions seems impossible.

8 Reactions in the System $MgO - SiO_2 - H_2O$

The reaction of the magnesium oxide interaction with water underlies the processes of the magnesian cement hardening, and is also of particular importance in producing magnesia refractories. The chemical activity of magnesium oxide depends to a great extent on the temperature of its production. The magnesium oxide interaction with water results in the formation of magnesium hydroxide:

$$MgO + H_2O = Mg(OH)_2$$

$$\Delta G^0_{298} = -200.05 + 136.13 + 56.69 = -7.23 \text{ kcal/mol;}$$

$$\Delta H^0_{298} = -221.77 + 143.84 + 68.32 = -9.61 \text{ kcal/mol}.$$

Thus, the magnesium oxide hydration is a process advantageous in the thermodynamic respect ($\Delta G^0_{298} = -7.23$ kcal/mol). The hydration reaction proceeds with a significant heat release ($\Delta H^0_{298} = -9.61$ kcal/mol). The dissolution of magnesium oxide in water must be accompanied by the alkalization of the solution.

$$MgO + H_2O = Mg^{2+} + 2OH^- .$$

$$\Delta G^0_{298} = -108.80 - 2 \cdot 37.59 + 136.13 + 56.69 = +8.84 \text{ kcal/mol;}$$

$$\lg K_p = -\frac{8.84}{1.364} = -6.48 .$$

A tentative calculation by the above technique yields pH = 11.94.

The solubility product of $Mg(OH)_2$ calculated starting from thermodynamic data is $10^{-11.03}$, which is very close to the value of $10^{-11.15 \pm 0.3}$ determined experimentally [84]. The low value of the solubility product implies that a very small amount of magnesium ions and of hydroxyl ions is sufficient to facilitate the precipitation of magnesium hydroxide.

In aqueous solutions magnesium hydroxide does not completely dissociate. The second constant of dissociation at room temperature is $2.3 \cdot 10^{-7}$.

The solubility of magnesium hydroxide decreases with the increase of temperature [87] as follows:

Temperature [K]	303	333	383	473	453
Solubility [mol/l]	0.21	0.14	0.1	0.08	0.05 .

Table 114. Crystallochemical and stoichiometric formulae of magnesium hydrosilicates

Mineral	Formula	
	crystallochemical	stoichiometric
Serpentine	$Mg_6(OH)_8[Si_4O_{10}]$	$3MgO \cdot 2SiO_2 \cdot 2H_2O$
Talc	$Mg_3[Si_4O_{10}](OH)_2$	$3MgO \cdot 4SiO_2 \cdot H_2O$
Attapulgite	$Mg_5(H_2O)_4[Si_4O_{10}]_2 \cdot (OH)_2 \cdot 4H_2O$	$5MgO \cdot 8SiO_2 \cdot 9H_2O$
Sepiolite	$Mg_7[Si_4O_{10}]_2(OH)_2$	$7MgO \cdot 8SiO_2 \cdot H_2O$
Anthophyllite	$Mg_3(H_2O)_3[Si_4O_{11}] \cdot H_2O$	$3MgO \cdot 4SiO_2 \cdot 4H_2O$
Kerolite	$Mg_4(OH)_4[Si_4O_{10}] \cdot 2H_2O$	$MgO \cdot SiO_2 \cdot H_2O$

Reactions in the system $MgO-SiO_2-H_2O$ are of great importance for the production of autoclaved magnesian materials [665] as well as for geochemistry [56]. Magnesium hydrosilicates are quite wide-spread in nature, some of them, such as serpentine group minerals, talc and others, were obtained artificially. The most wide-spread magnesium hydrosilicates are serpentine (chrysotile, antigorite), talc, attapulgite, anthophyllite, sepiolite and kerolite, the crystallochemical and common formulae of them being given in Table 114.

The thermodynamic calculations of the reactions of the magnesium hydrosilicate formation were carried out by Budnikov and Bludov [74] and the following ratios of $Mg(OH)_2$ to SiO_2 were taken: 2/1; 3/2; 1/1; 7/8; 3/4; 1/2. It was found that magnesium hydrosilicates can be formed at as low a temperature as a normal one, the most preferable of the hydrosilicates being sepiolite. Chrysotile having a fibrous structure was found to be a less stable form of serpentine as compared to its other forms, antigorite having a laminar structure. These facts are in complete agreement with the well-known experimental data of Young and Roy. The possibility of the talc transformation into anthophyllite and kerolite also proved to be true, which agrees with the results of the tests of Bowen and Tuttle [57].

9 Thermochemistry and the Energy Principles of the Cement Hydration Processes [250, 411]

An artificial conglomerate made up by C_3S, C_2S, C_3A, $C_{12}A_7$ and C_4AF is known to be formed in the process of firing mixtures consisting of lime, alumina, silica and ferric oxide meant for portland cement.

As cement clinker cools down rapidly, at low temperatures equilibrium is barely reached, therefore minerals are partly in a vitreous state. In addition, a study of phase equilibria in the systems of portland cement clinker showed some amount of secondary components (MgO, Al_2O_3) to be apt to enter the lattice of the main compounds and to form solid solutions with them. Therefore energy characteristics of pure clinker minerals differ to a certain extent from those of the minerals produced at industrial conditions, which is to be taken into account in thermochemical studies.

The works dealing with the determination of the dissolution heats for establishing the heat release at the hydration of pure clinker minerals are those of Thorvaldson et al. [628], Lerch and Bogue [329], Woods, Steinour and Starke [695, 696], as well as of V. A. Kind, Okorokov and Vol'fson [268], Verbeck and Foster [669]. The most comprehensive data were obtained by Verbeck and Foster, as tabulated in Table 115.

According to Lerch and Bogue the total hydration heat of clinker minerals differs from that given in Table 115 and is: with C_3S 1.2, with C_2S 0.62, with C_3A 2.07 and with C_4AF 1 cal per 0.01 g of the compound.

One can see from Table 115 that tricalcium aluminate greatly outdoes other clinker minerals in the amount and rate of heat release.

Data on the dissolution and hydration heat of aluminates and aluminate sulphate hydrates were published by Murakami, Tanaka et al. [446].

The most comprehensive representation of the thermochemical data on the reactions of the clinker mineral production is given in [325, 598].

The heat of alite dissolution (the lattice of the alite in question containing 0.97% MgO and 2.5% Al_2O_3) was found by Schwiete and Tan Tik-Ien [557]. They showed that as a result of the introduction of Al_2O_3 and MgO into the lattice of C_3S, the dissolution heat increases from 716.4 cal/g with pure C_3S to 802 cal/g with alite.

In addition they demonstrated that the introduction of C_3A into C_3S is accompanied by heat release (57.9 cal/g).

A great variety of the mineralogical composition of cements and the uncertainty of the composition of their hydration products do not yet permit deriving their exact thermochemical equations, nevertheless the practical importance of the quantitative evaluation of the hydration heat of cements of various composition reveals the need for more research work in this field.

Table 115. The heat of the mineral hydration by Verbeck and Foster

Hydration time	Heat release in cal per 0.01 g of a compound			
	C_3S	C_2S	C_3A	C_4AF
3 days	0.58	0.12	2.12	0.69
7 days	0.53	0.1	3.72	1.18
28 days	0.9	0.25	3.29	1.18
3 months	1.04	0.42	3.11	0.98
1 year	1.14	0.54	2.79	0.9
6.5 years	1.17	0.53	3.28	1.11

The works dealing with the heat release of cement as a function of temperature, grinding fineness and water-cement ratio are those of Carlson and Forbrich, Davis et al., Lerch and Bogue [90, 111, 329]. They found the increase of the temperature, of the grinding fineness and of water-cement ratio to bring about the growth of the rate of heat release at the early age of hydration; however, the greater the time elapse since the start of hydration, the less is the effect of these factors.

After 6.5 years the hydration heat of cement at $W/C = 0.8$ is 8% higher than at $W/C = 0.4$. This occurs owing to the rate of the water diffusion through the reaction products.

The relative rate of heat release at different temperatures, according to Carlson and Forbrich, proved to be the same. According to Verbeck and Foster [669], however, the relative rate of hydration at different temperatures turned out to be inversely proportional to the time required to reach the equivalent degree of hydration at each temperature. The rates of heat release of cements were first thoroughly investigated by Lerch [328] and were subsequently studied by several scientists [411, 414, 416, 421, 496, 559].

Note that of great theoretical importance is the establishment of a dependence between the alteration of the internal energy of binders at hydration and the thermodynamic properties of water at a set temperature.

Thus, together with Radvinskiy, we examined a model of the hydration process, starting from the conditions of phase equilibria in these systems, the model of which allowed deriving an equation for calculating the hydration heats of binders.

As we demonstrated earlier, binders in their initial state are metastable in respect to water. In their turn, hydrated forms gradually equilibrated with the liquid phase, and the chemical potentials of water in both phases equalize as a result of mass transport. Consequently, the work of mass transport in the hydration process may be expressed as follows:

$$A = (\mu_2 - \mu_1)\Delta m\,, \tag{251}$$

where μ_2 and μ_1, chemical potential of water in solid and liquid phase respectively; Δm, mass of combined water.

We considered as a model of the hydration process, the distribution of water vapour between two vessels of equal volume separated by a membrane penetrable only for vapour. It is supposed that prior to the beginning of the process under consideration water is evaporated and at the end condensed. The sum of the evaporation and condensation work is zero. The first vessel is filled with water vapour under a constant pressure p', the second, with a gas inert in respect to water vapour under a pressure $p'' > p'$. The contents of both vessels obey the laws of an ideal gas. As a result of a quasistatic process Δm grams of water (as part of water in the hydrated phase is dissociated, the alteration not of molar, but of specific chemical potentials of water is considered) in the form of vapour pass over from the first vessel to the second one (just as in the process of hydration m grams of water pass over from the liquid phase into the solid one). The pressure equalization is a result of the vapour expansion in the second vessel. In this case, according to Dalton's law, the vapour pressure will alter from $p_0 = p'$ to

$$p_m = Cp',$$

where $C = \Delta m/m$, portion of water condensed in the second vessel, or in the case of hydration, fraction of the chemically bound water of the total mass m of the mixing water. In addition

$$C = \frac{\Delta m}{m} \cdot \frac{\omega}{W/S},$$

where m, mass of the mixing water; ω, amount of the chemically bound water; W/S, water-solid ratio.

The work of vapour expansion in the second vessel is the work of the phase equilibration, and applying this to the hydration process we can express the value of this work, by means of the alteration of the Gibbs free energy as:

$$\Delta G = \omega \int_{p_0}^{p_m} V dp = \omega \frac{RT}{M} \int_{p_0}^{p_m} \frac{dp}{p} = \omega \frac{RT}{M} \ln\left(\frac{p_m}{p_0}\right)$$
$$= \omega \frac{RT}{M} \ln\left(\frac{\Delta m}{m}\right) = \omega \frac{RT}{M} \ln\left(\frac{\omega}{W/S}\right), \quad (252)$$

where M, molecular mass of water in grams.

Conversely, as stated above, the alteration of the internal energy of binders as a result of the combination of water, the chemical potentials of which are equalized in both phases, depends linearly on the mass of the chemically bound water and on its thermodynamic properties at a given temperature

$$\Delta U = \omega \left(L - \frac{RT}{M}\right), \quad (253)$$

where L, latent heat of the water evaporation; RT, work of the ideal gas expansion.

Equations (252) and (253) enable us to calculate the alteration of enthalpy per gram of the initial binder.

$$\Delta H = \Delta U + \Delta G = \omega \left[\left(L - \frac{RT}{M}\right) + \frac{RT}{M} \ln\left(\frac{\omega}{W/S}\right)\right]. \quad (254)$$

On the basis of Eq. (254) we calculated the heat release value of a polymineral cement which had hardened for 6.5 years. In the calculations experimental data on the amount of bound water borrowed from [594] were used. The calculation results are given in Table 116.

The table demonstrates that the calculated data agree quite well with the experimental ones, at an early age of hardening in particular, which shows the theoretical ideas underlying the accepted model of the hydration processes to be true. The practical significance of thermochemical investigations is widely acknowledged, therefore the study of the heat release of cements of different mineralogical compositions is of great importance. As this subject has been dealt with in detail in several reviews [594, 669], we shall not concentrate on this problem in this work.

Thermographic investigations of cements [161, 400, 401] are of particular interest in the study of the processes of heat release as well. Of no lesser importance for evaluating the manifestation of binding properties in the systems binder-water is a comparison of such thermochemical characteristics as the mean binding energies of Ca – O and Mg – O in the transition of binders from anhydrous state into a hydrated one. The values of the mean binding energies of Ca – O and Mg – O were calculated starting from the heats of formation of anhydrous and hydrated compounds assuming that the binding energies of Si – O, Al – O, S – O and O – H in the transition from its anhydrous state to the hydrated one alter their values so insignificantly that these alterations may be neglected.

The results of these calculations are given in Table 117. As one can see from the table, the greatest gain of energy per bond (13.9 kcal) corresponds to C_3A, followed by CaO (13.05 kcal), Ca (11.2 kcal), $CaSO_4 \cdot 0.5H_2O$ (10.2 kcal), MgO (9.68 kcal), C_3S (7.5 kcal) and β-C_2S (4.84 kcal).

It is noteworthy that ΔG^0_{298} of these reactions have a significant negative value as well, which characterizes great motive forces of these processes providing their spontaneous nature.

Such compounds as γ-C_2S, β-CS and $CaSO_4$ are quite peculiar. As one can see from Table 117, they have such mean values of binding energies of Ca – O that a gain of energy exists as compared to hydrated forms, but ΔG^0_{298} of these processes (Table 117) turns out to be either close to zero (γ-C_2S and $CaSO_4$), or positive (β-CS), thus the energy of activation of the hydration processes seems to be very significant. Therefore these processes cannot occur spontaneously with water at ordinary conditions.

Since binders are substances capable of hardening into a stone-like body, a binder must be capable not only of obtaining its hydrated form stable in water, but of turning into a strong concretion of these hydrates. Therefore the ability to create supersaturations in water solutions while hardening is a necessary condition for the given compounds to be binders.

Sometimes at normal conditions a given compound does not yield supersaturation, at hydrothermal conditions, however, either with additives or fine grind-

Table 116. Experimental and calculated data on the heat release of cement

Nos.	Mineralogical composition [%]				Hydration heat [cal/g] at the time of hardening														
	C_3S	C_2S	C_3A	C_4AF	7 days			28 days			90 days			1 year			6.5 years		
					ω	ΔH calc.	ΔH exp.	ω	ΔH calc.	ΔH exp.	ω	ΔH calc.	ΔH exp.	ω	ΔH calc.	ΔH exp.	ω	ΔH calc.	ΔH exp.
1.	64.5	10.0	12.1	7.5	0.173	90.9	87.5	0.201	106.6	108.2	0.216	115.1	114.3	0.218	116.2	117.8	0.241	129.2	126.8
2.	64.0	10.5	5.7	10.1	0.162	84.8	84.0	0.179	94.3	96.6	–	–	–	0.211	112.2	109.0	0.219	116.8	115.6
3.	60.0	13.0	10.4	7.7	0.178	93.7	94.5	0.193	102.1	100.7	–	–	–	0.221	117.9	112.8	0.234	125.3	122.9
4.	57.0	16.0	10.4	7.3	0.177	93.1	92.2	0.195	103.3	101.5	–	–	–	0.216	115.1	113.5	0.226	120.7	120.5
5.	56.0	17.0	10.8	6.4	0.179	94.3	95.6	0.202	107.2	106.8	–	–	–	0.231	123.6	120.0	0.239	128.1	125.8
6.	51.0	21.0	12.2	7.1	0.165	86.5	88.5	0.194	102.7	100.5	0.202	107.2	106.1	0.21	111.7	111.3	0.232	124.1	123.1
7.	53.5	21.0	7.5	10.7	0.149	77.6	79.3	0.173	90.9	91.3	0.189	99.9	98.0	0.197	104.4	104.2	0.214	113.9	112.1
8.	50.0	22.0	12.1	7.2	0.157	82.1	81.0	0.18	94.8	94.2	0.197	104.2	104.4	0.21	111.7	110.1	0.227	121.3	114.0
9.	52.5	22.0	7.9	10.4	0.149	77.6	73.8	0.166	87.0	88.9	0.182	95.9	95.1	0.196	103.8	102.7	0.21	111.7	109.5
10.	52.0	23.0	10.4	9.3	0.147	76.5	79.2	0.186	98.2	102.0	0.20	106.1	106.4	0.204	108.3	109.4	0.22	117.3	114.5
11.	51.0	24.0	3.7	16.6	0.113	57.8	54.8	0.158	82.6	82.7	–	–	–	0.195	103.3	99.0	0.20	106.1	103.3
12.	50.0	26.0	10.1	6.5	0.119	61.1	64.3	0.154	80.3	83.4	0.18	94.8	96.6	0.193	102.1	103.6	0.21	111.7	115.5
13.	46.0	27.0	12.5	7.1	0.148	77.0	78.4	0.194	102.7	98.4	0.209	111.1	109.4	0.219	116.8	111.6	0.237	127.0	120.9
14.	45.0	28.0	12.6	7.3	0.162	84.8	83.2	0.192	101.6	99.7	0.208	110.6	107.2	0.218	116.2	114.6	0.24	128.7	123.1
15.	44.5	28.0	13.2	6.8	0.162	84.8	81.2	0.192	101.6	99.9	0.209	111.1	107.7	0.217	115.6	110.7	0.226	118.5	118.3
16.	44.0	28.5	13.2	6.7	0.17	89.2	82.2	0.202	107.2	102.3	0.210	111.7	106.9	0.217	115.6	109.8	0.224	119.6	117.6

Note: ΔH calc., calculated data; ΔH exp., experimental data given in [594]

Table 117. Gain of energy in the transition into the hydrated state

Anhydrous compounds			Hydration products			Gain of energy in the transition into the hydrated state [kcal/mol]	ΔG^0_{298} of the hydration reaction [kcal/mol]
Formula of a compound	Anion and its charge	Mean binding energy [kcal/mol]	Formula of a compound	Anion and its charge	Mean binding energy [kcal/mol]		
CaO	O^{2-}	128.55	$Ca(OH)_2$	OH^-	141.6	13.05	−13.21
$3CaO \cdot SiO_2$	SiO_4^{4-}	133.05	$2CaO \cdot SiO_2 \cdot 1.17H_2O$	$Si_6O_{17}^{10-}$	140.6	7.55	−18.76
β-$2CaO \cdot SiO_2$	SiO_4^{4-}	135.76	$2CaO \cdot SiO_2 \cdot 1.17H_2O$	$Si_6O_{17}^{10-}$	140.6	4.84	−2.46
γ-$2CaO \cdot SiO_2$	SiO_4^{4-}	136	$2CaO \cdot SiO_2 \cdot 1.17H_2O$	$Si_6O_{17}^{10-}$	140.6	4.6	−0.72
β-$CaO \cdot SiO_2$	$Si_3O_9^{6-}$	139.21	$2CaO \cdot SiO_2 \cdot 1.17H_2O$	$Si_6O_{17}^{10-}$	140.6	1.39	+4.89
$3CaO \cdot Al_2O_3$	O^{2-}	127.7	$4CaO \cdot Al_2O_3 \cdot 19H_2O$	$Al(OH)_6^-$	141.6	13.9	−48.62
$CaO \cdot Al_2O_3$	O^{2-}	130.4	$4CaO \cdot Al_2O_3 \cdot 19H_2O$	$Al(OH)_6^-$	141.6	11.2	−10.27
$CaSO_4 \cdot 0.5H_2O$	SO_4^{2-}	155.8	$CaSO_4 \cdot 2H_2O$	SO_4^{2-}	166	10.2	−1.32
$CaSO_4$ (anh.)	SO_4^{2-}	153.4	$CaSO_4 \cdot 2H_2O$	SO_4^{2-}	166	12.6	−0.25
MgO	O^{2-}	119.45 [a]	$Mg(OH)_2$	OH^-	129.13 [a]	9.86 [a]	−6.45

[a] With MgO and $Mg(OH)_2$ the values of the mean binding energy refer to Mg−O

ing this can be provided; therefore when defining the concept "binder" it is necessary to bear in mind the conditions of hardening. With most binders the process of hardening is thermodynamically possible under normal conditions [406, 407, 413] ($T = 298$ K and $p = 1$ atm).

Belonging to this group are CaO, β-C_2S, C_3S, C_3A, CA, $CaSO_4 \cdot 0.5H_2O$, MgO. It is somewhat different with substances entering the system $Ca(OH)_2 - SiO_2 - H_2O$. Here, on the contrary, ΔG^0_{298} of the reactions has a rather large negative value, and the alteration of the coordination of the Ca ions at normal conditions is impeded; therefore in this case special conditions in the form of hydrothermal treatment are necessary for the manifestation of binding properties.

The thermodynamic instability and potential capability of being transformed into stable hydrated forms with water do not sufficiently testify to the presence of binding properties, for the actual process of their hardening, i.e., the process of the formation of a strong stone-like body from powdery material in presence of water or electrolytes depends on a great number of factors: the overcoming of internal resistances in the system, the crystallochemical and physical peculiarities of the hydrates formed and the character of their concretion, which is determined by supersaturation etc.

Thus we can conclude that in order to determine whether inorganic compounds possess binding properties we must know the following:

1. the value and sign of the Gibbs free energy of the reaction of the interaction of the binder with water (ΔG^0) determining the motive forces of the process:

2. the capability of a given compound of yielding supersaturations in water, which is to a great extent characterized by the amount of the gain of energy of the mean binding energy of Ca – O (with calcium binders), Mg – O (with magnesium binders), Ba – O (with barium binders) etc., at the transition of anhydrous compounds into hydrated forms.

The efficiency of using binders in industry depends on the rate of the hydration process which in its turn depends on temperature[24], physical state of reactants, catalysts etc. Therefore in all cases along with obtaining the energy characteristics of a process, consideration of its kinetics under the conditions in question and knowledge of the mechanism of the hardening process is necessary.

[24] As the book by Mchedlov-Petrossyan and Chernyavskiy "Strukturoobrazovaniye i tverdeniye tsementnykh past i betonov pri ponizhennykh temperaturakh" ("Structure formation and hardening of cement pastes and concretes at low temperatures", Kiev, Budivel'nik, 1974) has been published, the problems of the energetics of hydration reactions at subzero temperatures, given in the 3rd edition of the present book, are not considered here [416–418]

10 A Calculation of Colloid-Chemical Equilibria in the Systems Binder-Water

A thermodynamic analysis of the reactions of the monomineral binder hydration carried out in the preceding sections showed non-equilibrium to be their characteristic feature. This non-equilibrium results in the breaking of the balancing of the difference of the chemical potentials of the bound and freely diffusing ions and molecules by the interphase difference of electric potentials and by the osmotic pressure. Therefore the hydration processes are accompanied by the volume separation of charges and by the appearance of an electric double layer (EDL) on the boundary of the solid and liquid phases [28, 142]. The same electrochemical phenomena observed in colloids, suspensoids and capillary porous bodies are peculiar to the systems binder-water, but these systems have their own unique features. These features are conditioned by the crystallochemical structure of the initial binders, by their metastability in respect to water and by the exothermic character of the hydration processes.

In equilibrium colloid system the difference of the chemical potentials of the bound and free liquid phases is known to be balanced by osmotic pressure. In this process the difference of the chemical potentials of the bound and freely diffusing ions of the same sign is balanced by the osmotic pressure and interphase difference of electric potentials. Thus, it is the calculation of the osmotic pressure connected with the colloid-chemical properties of the systems binder-water, that is important in these systems.

The magnitude of this pressure is a function of the concentration of the external solution (pore electrolyte) and is connected with the alteration of the electrokinetic properties of hydrated new formations passing through a colloid degree of dispersion in the process of their development and growth [24, 28, 38, 58, 128, 390, 590, 666].

Let us consider these processes in more detail.

At the contact of solid free and bound disperse particles with the electrolyte solution an exchange of ions and molecules is always observed. However, as this exchange is accompanied by the appearance of an electric charge (of some origin) on their surface which gives a potential jump on the phase boundary, achieving complete equilibrium is impeded and there is a so-called partial or membrane equilibrium.

An obvious example of such is the osmotic equilibrium, establishing between two solutions separated by a semipermeable membrane, the membrane allowing the passage of some ions and detaining others or allowing only the solvent to pass through and not ions. As in these cases the concentration both of ions and of the solvent on both sides of the membrane will be different, a complete equilibrium will be established in the systems only with the appearance of some additional factor; this being either the weight of the solution column above the membrane according to the value of the osmotic pressure, or the difference of potentials on the phase boundary.

For colloid systems such an additional factor of equilibrium will also be the potentials jump on the boundary of the solid (the surface of solid particles) and liquid (electrolyte) phases. In this connection it is peculiar to them that in each phase a distribution of ions corresponding to the above equilibrium is established. This distribution in the external layer (in the liquid phase) can be compared with that of gas molecules in the atmosphere (or with the distribution of particles in a solution with a semipermeable membrane) with the difference that in the case of a gas the equilibrium is established under the effect of gravity (or of the column of a liquid with a solution). With ions dissociated by a colloid micelle the equilibrium is established when the electric attraction of the charged ions of the internal layer (on the surface of the solid phase) is equal to the counterion osmotic pressure. It should be remembered that in any aqueous solution ions are in hydrated form. The degree of the hydration of an ion, as stated above, depends on its size and valence; the smaller the size of an ion and the greater its valence, the more dipoles of water it can orient near itself having thereby a greater degree of hydration. The increase of the degree of hydration of ions brings about their lesser autopotential, as a result they become less sensitive to the electrostatic attraction of other charged particles and acquire a greater diffusion capability. Therefore highly hydrated ions, such as Na^+, will form the thickest, although loose, layer around colloid particles. As stated above, the electrostatic attraction is similar to a membrane: detaining potential-determining ions near the surface of particles it does not let them pass from the spots with a higher concentration to those with a lower one. The difference of potentials which determines this phenomenon is determined by the following formula:

$$\Delta E = \pm \frac{RT}{nF} \ln K_{\mathrm{m.e.}},$$

where ΔE, membrane potential;[25] R, gas constant; T, absolute temperature; F, Faraday unit; n, valence of an ion with respect to its charge; $K_{\mathrm{m.e.}}$, constant of the membrane equilibrium.

Substituting the R, T and F values into this formula and a decimal logarithm for a natural one, we have

$$\Delta E = \pm \frac{0.059}{n} \ln K_{\mathrm{m.e.}} \qquad (255)$$

for univalent ions.

[25] Sign (+) or (−) of the membrane potential is determined by the sign of the charge of a colloid particle

For colloid systems to which solutions of some salts are added as well as for solutions separated by a semipermeable membrane, the equations of the Donnan distribution of ions in the diffuse layer and on both sides of the conventional membrane are given in Table 118 and the calculation results, in Table 119 [27]. The formulae were derived for electronegative colloids and membranes, such as cement stone. To electronegative colloids clays, silicic acid gels, products of the C_2S and C_3S hydration also belong (as for the products of the $CaSO_4 \cdot 0.5H_2O$, CaO, C_3A, C_4AF hydration as well as particles of trisulphate forms of aluminate sulphate hydrate they have a positive charge).

Substituting concentrations (x and y) for partial pressures P_2 and P_1 proportional to them we find that $K_{\text{m.e.}} = x/y$ and Eq. (255) for univalent ions acquires its final form:

$$\Delta E = \pm 0.0591 \lg (x/y) \text{ [V]} . \tag{256}$$

Suppose a $CaCl_2$ solution is introduced into a gel saturated by calcium ions. Now we find the formula of the Donnan equilibrium for this instance in Table 118:

$$x^3 = y^2(y+z) . \tag{257}$$

In this instance: y, molar concentration of chlorine ions in the micellar solution; x, molar concentration of chlorine ions in the exterior solution; z, molar concentration of potential-determining ions on the surface of a particle.

Starting from this formula we find the excess of the concentration of ions of the micellar solution over the concentration of those of the exterior solution to be

$$C = \left[y + \left(\frac{y}{2} + \frac{z}{2}\right)\right] - \left(x + \frac{x}{2}\right) = \frac{3}{2}y + \frac{z}{2} - \frac{3}{2}x . \tag{258}$$

Thus, the addition of calcium chloride to a gel will decrease the difference of the osmotic pressure between the micellar and external solution. This also features a lesser ΔE value between the internal part of the micellar atmosphere and the external solution, which is easy to find by the following transformations of Eqs. (257) and (256).

Expressing x/y as y and z in Eq. (257) we obtain

$$\frac{x}{y} = \sqrt[3]{\left(1 + \frac{z}{y}\right)} .$$

Substituting this x/y value into Eq. (256) we have

$$\Delta E = \frac{0.059}{3} \lg \left(1 + \frac{z}{y}\right) \text{ [V]} . \tag{259}$$

Table 118. The formulae for calculating equilibrium concentrations and the Donnan potential of electronegative and electropositive colloids[a]

Electronegative colloids			Electropositive colloids			Formula of the Donnan equilibrium	Formula of determining the difference of potentials (E)	Ratio of the differences of potentials
The charge sign of colloid	Counter-ion	Salt added	The charge sign of colloid	Counter-ion	Salt added			
(−)	K^+	K^+A^-	+	A^-	K^+A^-	$x^2=(z+y)\cdot y$	$E=\pm\frac{0.059}{2}\lg\left(1+\frac{z}{y}\right)$	1.0
(−)	K^+	$K_2^+A^{2-}$	+	A^-	$K^{2+}A_2^{1-}$	$x^3=(z+y)^2\cdot y$	$E=\pm\frac{0.059}{3}\lg\frac{(z+y)^2}{y}$	1.33
(−)	K^+	$K_3^+A^{3-}$	+	A^-	$K^{3+}A_3^{1-}$	$x^4=(z+y)^3\cdot y$	$E=\pm\frac{0.059}{4}\lg\frac{(z+y)^3}{3}$	1.5
(−)	K^{2+}	$K^{2+}A_2^{1-}$	+	A^{2-}	$K_2^{1+}A^{2-}$	$x^3=(z+y)\cdot y^2$	$E=\pm\frac{0.059}{3}\lg\left(1+\frac{z}{y}\right)$	0.667
(−)	K^{2+}	$K^{2+}A^{2-}$	+	A^{2-}	$K^{2+}A^{2-}$	$x^2=(z+y)\cdot y$	$E=\pm\frac{0.059}{2}\lg\left(1+\frac{z}{y}\right)$	1.0
(−)	K^{2+}	$K_3^{2+}A_2^{3-}$	+	A^{2-}	$K_2^{3+}A_3^{2-}$	$x^5=(z+y)^3\cdot y^2$	$E=\pm\frac{0.059}{5}\lg\frac{(z+y)^3}{y^3}$	1.2
(−)	K^{3+}	$K^{3+}A_3^{1-}$	+	A^{3-}	$K_3^+A^{3-}$	$x^4=(z+y)\cdot y^3$	$E=\pm\frac{0.059}{4}\lg\left(1+\frac{z}{y}\right)$	0.5
(−)	K^{3+}	$K_2^{3+}A_3^{2-}$	+	A^{3-}	$K_3^{2+}A_2^{3-}$	$x^5=(z+y)^2y^3$	$E=\pm\frac{0.059}{5}\lg\frac{(z+y)^2}{y^2}$	0.8
(−)	K^{3+}	$K^{3+}A^{3-}$	+	A^{3-}	$K^{3+}A^{3-}$	$x^2=(z+y)y$	$E=\pm\frac{0.059}{2}\lg\left(1+\frac{z}{y}\right)$	1.0
(−)	K^{4+}	$K^{4+}A_4^{1-}$	+	−	−	$x^5=(z+y)y^4$	$E=\pm\frac{0.059}{5}\lg\left(1+\frac{z}{y}\right)$	0.4

[a] Symbols: K^+, K^{2+}, K^{3+}, K^{4+}, cations; A^-, A^{2-}, A^{3-}, anions; x, concentration of the external salt at the moment of equilibrium; z, counterion concentration in the micellar solution; y, salt concentration in the micellar solution at the moment of equilibrium; E, difference of potentials between the micellar and external solution

Table 119. Instances of calculations of equilibria in colloid systems[a]

Electronegative colloid		Electropositive colloid		i at $\alpha = 1$		Set concentration value		Values calculated by the formulae given in Table 118					
Common ion and its concentration	Formula of salt added	Common ion and its concentration	Formula of salt added	in the micellar solution	in the external solution	of counterions (z) [m/l]	of the salt added (c) [m/l]	x [g-ion/l]	y [g-ion/l]	$\pm E$ [mv]	P_0 [atm]	P [atm]	ΔP [atm]
$[K^+] = Z$	K^+A^-	$[A^-] = Z$	K^+A^-	1	2	0.4	0.2	0.15	0.05	28.1	9.76	4.88	4.88
$[K^+] = Z$	K^+A^-	$[A^-] = Z$	K^+A^-	1	2	0.4	1.0	0.583	0.417	8.7	9.76	8.10	1.66
$[K^+] = Z$	K^+A^-	$[A^-] = Z$	K^+A^-	1	2	0.4	2.0	1.09	0.91	4.7	9.76	8.8	0.96
$[K^+] = 2Z$	$K_2^+A^{2-}$	$[A^-] = 2Z$	$K^{2+}A_2^-$	2	3	0.4	0.2	0.172	0.028	46.5	19.52	10.52	9.0
$[K^+] = 2Z$	$K_2^+A^{2-}$	$[A^-] = 2Z$	$K^{2+}A_2^-$	2	3	0.4	1.0	0.62	0.38	12.5	19.52	17.55	1.97
$[K^+] = 2Z$	$K_2^+A^{2-}$	$[A^-] = 2Z$	$K^{2+}A_2^-$	2	3	0.4	2.0	1.127	0.873	6.5	19.52	18.6	0.92

[a] Symbols: α, degree of the salt dissociation; $[\alpha = (i-1)/(\nu-1)]$ i, van't Hoff's coefficient where ν, number of ions formed at the dissociation of a molecule of salt; P_0, counterion osmotic pressure; P, counterpressure caused by the external electrolyte; ΔP, observed osmotic pressure equal to the difference $P_0 - P$

An analysis of Eq. (259) shows that the higher the concentration of the electrolyte added to a colloid is, i.e., the higher the y value, the z value being constant, the greater is the decrease of ΔE. Accordingly the greatest osmotic pressure (corresponding to the greatest ΔE value) and the greatest swelling are observed when the gel surface is exactly saturated by corresponding ions. An excess of electrolyte suppresses the swelling and lowers the osmotic pressure. The measure of swelling is determined by the valence of ions cations, anions) as well as by the concentration of the added electrolyte. Thus, the existence of the diffuse distribution of ions around colloid particles detained by their own electric force provides the formation of a solvate shell.

The ions of the shell create an osmotic pressure in it which is interrelated with their electrostatic attraction to the particle. This phenomenon accounts for a greater potential with highly hydrated univalent ions as well as for a decrease of the diffuse layer with multivalent ions.

Lyophilic colloids to which most colloid systems binder-water belong, unlike lyophobic ones can bind substantial amounts of water and possess a substantial osmotic pressure and swelling effect.

According to Ostwald the osmotic pressure of colloids is a sum of two independent pressures: of pressure P_1, expressed by Eq. (260), i.e., true osmotic pressure and of the swelling pressure P_2 occurring as a result of a further interaction of particles with a pore electrolyte or external solution calculated by the same formula (260); but considering the Donnan distribution of ions of the external electrolyte in the micellar solution.

$$P = iCRT\,, \tag{260}$$

where P, osmotic pressure; C, concentration of the corresponding solution; i, van't Hoff's coefficient [$\alpha = (i-1)/(\nu-1)$, where α, degree of dissociation and ν, number of ions into which a salt dissociates]; R, gas constant; T, temperature [K].

We estimated with the assistance of Zinov the value of the total osmotic pressure in the diffuse shells of the particles of the colloid degree of dispersion with the help of the well-known relation of thermodynamics, conditioning the equilibrium of the particle ensemble by the constancy of the chemical potentials and temperatures of all the parts of the ensemble. The potential energy of the ions of the solution or molecules of water in the field of a colloid particle was calculated by the Debye-Hückel method assuming the spherical symmetry of the field at the point ($r = a$) of the ξ potential measurement. As a result a formula was obtained:

$$P(r) = \frac{\xi a^2 \exp[-\varkappa(r-a)]}{(\varkappa a + r) r^2} \sum_{i=1}^{N} \left(e z_i \cdot n_i + P_\mathrm{w} \cdot n_\mathrm{w} \cdot \varkappa + \frac{2 P_\mathrm{w} n_\mathrm{w}}{r} \right)$$

$$\varkappa = \left(\frac{4\pi e^2}{\varepsilon k T} \sum_{i=1}^{N} Z_i^2 \cdot n_i \right)^{1/2} \left[\left(1 + \frac{\pi n_\mathrm{w}}{\varepsilon k T} P_\mathrm{w}^2 \right)^{1/2} - \left(\frac{\pi}{\varepsilon k T} P_\mathrm{w}^2 n_\mathrm{w} \right)^{1/2} \right], \tag{261}$$

where a, radius of a particle of the colloid degree of dispersion; ε, dielectric permittivity of the medium; k, Boltzmann constant; T, absolute temperature; Z_i, n_i, valence and concentration of ions of i-kind solution; n_w, P_w, concentration and dipole moment of the molecules of water.

Supposing that the colloid particle occupies the entire volume of a pore, starting from the conditions of the compatibility of deformations with the help of the relations of the theory of the strength of materials, it is possible to estimate the deformation values $\delta = (r-a)/a$ of the cement stone at the development of the osmotic pressure of the particles of the colloid degree of dispersion with the formula

$$P(\delta) = 10^9 \delta .$$

With this formula and formula (261) the osmotic pressure and deformations caused by these pressures in the cement stone for calcium hydroxide, trisulphate form of aluminats sulphate hydrate and gypsum were evaluated. All the initial data and calculation results are given in Table 120 and Fig. 90.

Calculations conducted showed that as a result of the osmotic pressures of the particles of the colloid degree of dispersion in the cement stone deformations may develop the values of which surpass those of elastic deformations being $\sim 1 \cdot 10^{-4}$ m. With the substances under consideration this takes place when the radius of particles is 10^{-9} m.

In unhardened concrete the values of elastic deformations are lower and the possibility of the realization of osmotic pressure in the expansion of the sample, there being a corresponding degree of the filling of the pore space, increase. If in a pore particles of colloid degree of dispersion of different sizes appear, due to the different density of charges on the surface of the particles the thicknesses of diffuse shells and pressures will be greater when particles are smaller.

Thus, with 10^{-9} and 10^{-7} m radius particles of calcium hydroxide, the space between the particle and the pore wall being $5 \cdot 10^{-10}$ m, the pressure will be

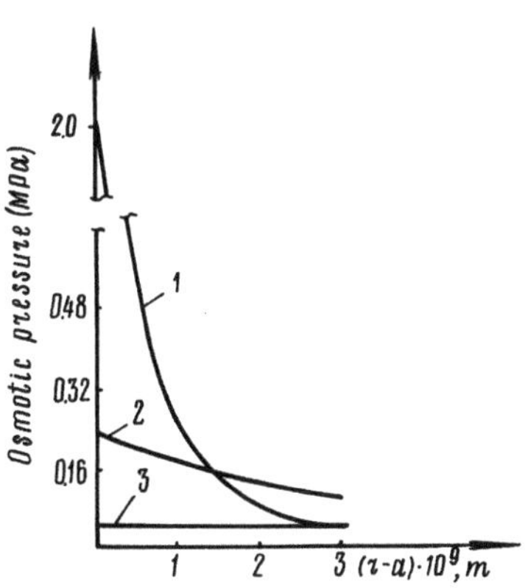

Fig. 90. The osmotic pressure developed by the particles of calcium hydroxide of the colloid degree of dispersion at a radius of particles *1* $a = 10^{-9}$ m; *2* $a = 10^{-8}$ m; *3* $a = 10^{-7}$ m as a function of the diffuse layer thickness $(r-a)$

Table 120. Initial data for calculating the osmotic pressure and the results of calculations of relative deformations resulting from it in the cement stone

Colloid particles	ξ potential [V]	Concentration of counterions [g-ion/l]	Concentration of water molecules [g-ion/l]	Relative deformation with colloid particles of the size of 10^{-9} m	10^{-8} m	10^{-7} m
$Ca(OH)_2$	$+10 \cdot 10^{-3}$	$[OH^-] = 4.32 \cdot 10^{-2}$	55.47	$4.11 \cdot 10^{-4}$	$0.44 \cdot 10^{-4}$	$0.06 \cdot 10^{-4}$
$CaSO_4 \cdot 2H_2O$	$+20 \cdot 10^{-3}$	$[SO_4^{2-}] = 1.5 \cdot 10^{-2}$	55.47	$4.15 \cdot 10^{-4}$	$0.48 \cdot 10^{-4}$	$0.09 \cdot 10^{-4}$
Ettringeite	$+5 \cdot 10^{-3}$	$[SO_4^{2-}] = 1.02 \cdot 10^{-2}$ $[OH^-] = 3.3 \cdot 10^{-2}$ $[AlO_2^-] = 7.3 \cdot 10^{-3}$	55.47	$2.06 \cdot 10^{-4}$	$0.23 \cdot 10^{-4}$	$0.03 \cdot 10^{-4}$

0.6 MPa and 0.03 MPa respectively. In a $2.2 \cdot 10^{-8}$ m radis pore a $2 \cdot 10^{-8}$ m particle of calcium hydroxide will exert a 0.18 MPa pressure on the walls. Ten $2 \cdot 10^{-9}$ m diameter particles of this substance will exert a 1.9 MPa pressure in a like pore.

Therefore the size, intensity of formation, the time of appearance and lifetime of the particles of the colloid degree of dispersion effect the stress in the pore walls of concrete at phase and chemical transformations of substances in the pore space. Colloid chemical phenomena and the calculation of colloid-chemical equilibria in the systems binder-water are of great theoretical and practical importance.

Chapter IV. Corrosion Reactions

Silicate materials, such as cement and its by-products (concrete and reinforced concrete), silicate autoclaved products, building bricks and ceramic pipes as well as glass and glass products are widely used in building at widely differing working conditions in contact with a number of liquid and gaseous media, comprising aggressive ones.

For providing the longevity of these silicate materials and their products it is of importance to know at which working conditions these materials are most stable. This mostly concerns such materials as concrete and reinforced concrete which are of major importance in modern building [16]. While studying corrosion reactions the thermodynamic method permits finding the conditions of stability of the compounds and minerals depending on the active concentration of ions of a solution or gases and also determines the stability of some compounds as compared to others.

We considered the behaviour of the following constituents of cement stone and concrete based on portland and alumina cement in contact with aggressive media[26]:
tobermorite, $5CaO \cdot 6SiO_2 \cdot 5.5H_2O$, calcium aluminate hydrates and ferrite hydrates, $4CaO \cdot Al_2O_3 \cdot 19H_2O$, $4CaO \cdot Fe_2O_3 \cdot 13H_2O$, $3CaO \cdot Al_2O_3 \cdot 6H_2O$, $3CaO \cdot Fe_2O_3 \cdot 6H_2O$, $CaO \cdot Al_2O_3 \cdot 10H_2O$, $2CaO \cdot Al_2O_3 \cdot 8H_2O$, calcium aluminate and ferrite sulphate hydrates, $3CaO \cdot Al_2O_3 \cdot 3CaSO_4 \cdot 32H_2O$, $3CaO \cdot Fe_2O_3 \cdot 3CaSO_4 \cdot 32H_2O$, calcium hydroxide, $Ca(OH)_2$ and calcium carbonate, $CaCO_3$.

As the aggressive media effecting these compounds, gaseous substances SO_2, CO_2 and H_2S as well as hydrogen ions H^+ and sulphate ions SO_4^{2-} were considered.

The reactions considered in this chapter belong to combined ones and comprise those with gaseous phase and those with dissolved components. The peculiarities of the technique of calculating the equilibria of corrosion reactions will be illustrated below by specific examples.

[26] The problems of the thermodynamic stability of reinforcing-bar steel in concrete are examined in the book by Babushkin [16], therefore they are not handled here, as well as other types of corrosion [31, 196, 420, 495, 584]

1 Reactions of the Interaction of the Constituents of Cement Stone and Concrete with Acid Gases

Such reactions are characteristic of the life of concerete in most industrial media, therefore the study of conditions determining their course is an important problem. We shall begin our examination with the reactions of the interaction of constituents of cement stone and concrete with SO_2. Let us first consider the reaction of the interaction of sulphur dioxide with tobermorite.

The reaction of the tobermorite corrosion with sulphur dioxide, as in any other reaction, can be thermodynamically studied at any stage. As the first variant we shall consider the reaction of the solid tobermorite with gaseous SO_2 in the presence of the oxygen of the air and of water.

$$\tfrac{1}{5}(5CaO \cdot 6SiO_2 \cdot 5.5H_2O\,(cr) + SO_2\,(gas) + 0.5 \cdot O_2\,(gas) + 0.9H_2O\,(liq) = CaSO_4 \cdot 2H_2O\,(cr) + \tfrac{6}{5}SiO_2\,(cr)\,. \qquad (262)$$

As the activities of all solid reactants and liquid water are equal to one, the equilibrium constant of reaction (262) will be determined by the relation

$$K = (p_{SO_2} \cdot p_{O_2}^{0.5})^{-1}\,, \qquad (263)$$

$$\lg K = -\lg p_{SO_2} - 0.5 \lg p_{O_2}\,, \qquad (264)$$

$$\Delta G_{298}^0 = -79.79 \text{ kcal/mol}\,.$$

As ΔG of the reaction $= \Delta G_{298}^0 + 1.364 \lg K$, substituting the value of $\lg K$ from (264) we obtain

$$\Delta G \text{ of the reaction} = -79.79 - 1.364 \lg p_{SO_2} - 0.682 \lg p_{O_2}\,. \qquad (265)$$

At equilibrium conditions Eq. (265) is expressed as:

$$\lg p_{SO_2} = -58.51 - 0.5 \lg p_{O_2}\,. \qquad (266)$$

Given the partial pressure of one of the gases, it is easy to calculate the equilibrium partial pressure of the other. A diagrammatic representation of Eq. (266) is given in Fig. 91.

In the conditions of the Earth surface ($p_{O_2} = 0.21$ atm) the equilibrium partial pressure of sulphur dioxide calculated by Eq. (266) corresponds to $10^{-58.44}$. This emphasizes the fact that at atmospheric conditions tobermorite is an unstable compound, and, the amount of sulphur dioxide in the air reaching the value determined by the partial pressure of $10^{-58.44}$, it will undergo a transformation

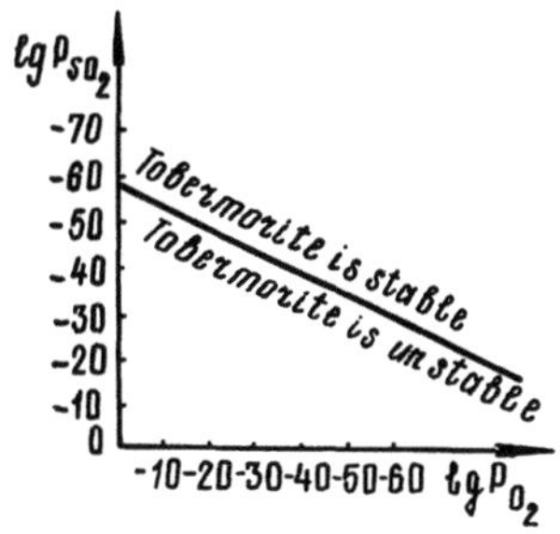

Fig. 91. The fields of the tobermorite stability as a function of the partial pressure of sulphur dioxide and oxygen

into gypsum dihydrate. Thus, if in the air surrounding concrete structures the amount of sulphur dioxide corresponding to a partial pressure of more than $10^{-58.44}$ atm should appear, then tobermoritelike calcium hydrosilicates become unstable.

We carried out similar calculations for other hydration products as well, and, as an exception, for calcium carbonate, as at present carbonate rocks are widely being used as fine and coarse aggregate for concrete. The results of these calculations are summarized in Table 121.

All the above compounds are thermodynamically unstable in respect to sulphur dioxide. Considering that in the products of the hydration of ordinary portland cement there are mainly tobermorites, C_4AH_{19}, aluminate sulphate hydrates and calcium hydroxide, the data of Table 121 permit concluding that for the beginning of the interaction an aluminate sulphate hydrate requires the greatest amount of SO_2 as compared to others $p_{SO_2} = 10^{-57.69}$, tobermorite then follow ($p_{SO_2} = 10^{-58.44}$), C_4AH_{19} ($p_{SO_2} = 10^{-60.94}$), the least stable being calcium hydroxide with which the interaction of cement stone with SO_2 practically begins ($p_{SO_2} = 10^{-63.23}$). The products of the alumina cement (CAH_{10} and C_2AH_8) hydration show the same stability in respect to SO_2 as other aluminate hydrates. Thus, concretes with alumina cement are close to ordinary cements as to their chemical stability towards SO_2.

The presence of aluminate sulphate hydrates in the products of the portland cement hydration must not effect the decrease of the chemical resistance. However a lower content of C_3A in portland cements is desirable, for calcium aluminate hydrates are in fact less stable than tobermorite.

Considering the resistance of calcium carbonate, the data of Table 121 show that even at the concentration of SO_2 corresponding to its partial pressure of more than $10^{-53.67}$ atm it becomes unstable and in the presence of the oxygen of the air ($p = 0.21$ atm) and of the liquid-drop water it will be transformed into gypsum dihydrate. At the same time in the concrete with a carbonate aggregate the constituents of the cement stone will be more active towards SO_2, because for the beginning of the reaction they require several degrees lower concentrations of SO_2 as compared to those of $CaCO_3$. Similar calculations were also carried out for the reactions of the interaction of the concrete constituents with carbon di-

Table 121. The results of the calculations of reactions of corrosion of the cement stone constituents with sulphur dioxide

Nos. of reactions	Equations of corrosion reactions	ΔG^0_{298} of the reactions [kcal/mol]	Equations of the type $\lg p_{SO_2} = f(\lg p_{O_2})$	Equilibrium partial pressure of SO_2 at $p_{O_2} = 0.21$ atm
1.	$\frac{1}{5}(5CaO \cdot 6SiO_2 \cdot 5.5H_2O)(cr) + SO_2(gas) + 0.5O_2(gas) + 0.9H_2O(liq) = CaSO_4 \cdot 2H_2O(cr) + \frac{6}{5}SiO_2(cr)$	−80.18	$\lg p_{SO_2} = -58.78 - 0.5\lg p_{O_2}$	$10^{-58.44}$
2.	$\frac{1}{4}(4CaO \cdot Al_2O_3 \cdot 19H_2O)(cr) + SO_2(gas) + 0.5O_2 = CaSO_4 \cdot 2H_2O(cr) + 0.5Al(OH)_3(am) + 2H_2O(liq)$	−83.59	$\lg p_{SO_2} = -61.28 - 0.5\lg p_{O_2}$	$10^{-60.94}$
3.	$\frac{1}{3}(3CaO \cdot Al_2O_3 \cdot 6H_2O)(cr) + SO_2(gas) + 0.5O_2(gas) + H_2O(liq) = CaSO_4 \cdot 2H_2O(cr) + \frac{2}{3}Al(OH)_3(am)$	−83.63	$\lg p_{SO_2} = -61.31 - 0.5\lg P_{O_2}$	$10^{-60.97}$
4.	$\frac{1}{3}(3CaO \cdot Al_2O_3 \cdot 3CaSO_4 \cdot 32H_2O)(cr) + SO_2(gas) + 0.5O_2(gas) = 2(CaSO_4 \cdot 2H_2O)(cr) + \frac{2}{3}Al(OH)_3(am) + \frac{17}{3}H_2O(liq)$	−79.15	$\lg p_{SO_2} = -58.03 - 0.5\lg p_{O_2}$	$10^{-57.69}$
5.	$\frac{1}{3}(3CaO \cdot Fe_2O_3 \cdot 3CaSO_4 \cdot 32H_2O) + SO_2 + 0.5O_2 = 2(CaSO_4 \cdot 2H_2O)(cr) + \frac{2}{3}Fe(OH)_3(am) + \frac{17}{3}H_2O(liq)$	−82.89	$\lg p_{SO_2} = -60.77 - 0.5\lg p_{O_2}$	$10^{-60.43}$
6.	$CaO \cdot Al_2O_3 \cdot 10H_2O(cr) + SO_2(gas) + 0.5O_2(gas) = CaSO_4 \cdot 2H_2O(cr) + 2Al(OH)_3(am) + 5H_2O(liq)$	−83.37	$\lg p_{SO_2} = -61.12 - 0.5\lg p_{O_2}$	$10^{-60.78}$
7.	$\frac{1}{2}(CaO \cdot Al_2O_3 \cdot 8H_2O)(cr) + SO_2(gas) + 0.5O_2(gas) = CaSO_4 \cdot 2H_2O(cr) + Al(OH)_3(am) + 0.5H_2O(liq)$	−83.37	$\lg p_{SO_2} = -61.12 - 0.5\lg p_{O_2}$	$10^{-60.78}$
8.	$Ca(OH)_2(cr) + SO_2(gas) + 0.5O_2(gas) + H_2O(liq) = CaSO_4 \cdot 2H_2O(cr)$	−86.71	$\lg p_{SO_2} = -63.57 - 0.5\lg p_{O_2}$	$10^{-63.23}$
9.	$CaCO_3(cr) + SO_2(gas) + 0.5O_2(gas) + 2H_2O(liq) = CaSO_4 \cdot 2H_2O(cr) + CO_2(gas)$	−68.89	$\lg p_{SO_2} = -50.51 - 0.5\lg p_{O_2} + \lg p_{CO_2}$	$10^{-53.67}$ (at $p_{CO_2} = 10^{-3.5}$ atm)

oxide and hydrogen sulphide. The results of the calculations of the corrosion reactions with these gases are given in Tables 122 and 123.

As one can see from Table 122, all the constituents of cement stone are unstable towards CO_2 even at the partial pressure of more than 10^{-7} atm. Considering that the air contains CO_2 at $10^{-3.5}$ atm, all the constituents of cement stone will carbonize even at normal conditions.

Table 122 shows that the most vulnerable, as far as its stability towards CO_2 is concerned, is calcium hydroxide the p_{CO_2} equilibrium pressure of which is $10^{-13.1}$.

According to the value of the minimum equilibrium pressure of CO_2 needed for the beginning of the reaction the constituents of cement stone rank as follows:

$$Ca(OH)_2 < C_3AH_6 < C_4AH_{19} < CAH_{10} < CF\bar{S}H < C_5S_6H_{5.5} < CA\bar{S}H$$

where $CF\bar{S}H$, calcium ferrite sulphate hydrates; $CA\bar{S}H$, calcium aluminate sulphate hydrates.

As compared to CO_2 and SO_2 hydrogen sulphide is the most aggressive towards concrete, because the equilibrium values of its minimum partial pressure

Table 122. Calculation results of the carbonization reactions of the constituents of cement stone and concrete

Nos. of reactions	Equations of reactions	ΔG^0_{298} of reactions [kcal/mol]	p_{CO_2} equilibrium values [atm]
1.	$\frac{1}{5}(5CaO \cdot 6SiO_2 \cdot 5.5H_2O)(sol) + CO_2(gas) = CaCO_3(sol) + \frac{6}{5}SiO_2(sol) + 1.1H_2O(liq)$	−11.29	$10^{-8.28}$
2.	$\frac{1}{4}(4CaO \cdot Al_2O_3 \cdot 19H_2O)(sol) + CO_2(gas) = CaCO_3(sol) + \frac{1}{2}Al(OH)_3(am) + 4H_2O(liq)$	−14.70	$10^{-10.78}$
3.	$\frac{1}{3}(3CaO \cdot Al_2O_3 \cdot 6H_2O)(sol) + CO_2(gas) = CaCO_3(sol) + \frac{2}{3}Al(OH)_3(am) + H_2O(liq)$	−14.76	$10^{-10.81}$
4.	$\frac{1}{3}(3CaO \cdot Al_2O_3 \cdot 3CaSO_4 \cdot 32H_2O)(sol) + CO_2(gas) = CaCO_3(sol) + CaSO_4 \cdot 2H_2O(sol) + \frac{2}{3}Al(OH)_3(am) + \frac{23}{3}H_2O(liq)$	−10.26	$10^{-7.52}$
5.	$\frac{1}{3}(3CaO \cdot Fe_2O_3 \cdot 3CaSO_4 \cdot 32H_2O) + CO_2 = CaCO_3 + CaSO_4 \cdot 2H_2O + \frac{2}{3}Fe(OH)_3 + \frac{23}{3}H_2O$	−14.07	$10^{-10.31}$
6.	$CaO \cdot Al_2O_3 \cdot 10H_2O(sol) + CO_2(gas) = CaCO_3(sol) + 2Al(OH)_3(am) + 7H_2O(liq)$	−14.65	$10^{-10.74}$
7.	$\frac{1}{2}(2CaO \cdot Al_2O_3 \cdot 8H_2O)(sol) + CO_2(gas) = CaCO_3(sol) + Al(OH)_3(am) + \frac{5}{2}H_2O(liq)$	−14.65	$10^{-10.74}$
8.	$Ca(OH)_2(sol) + CO_2(gas) = CaCO_3(sol) + H_2O(liq)$	−17.82	$10^{-13.1}$

Table 123. Calculation results of the corrosion reactions of the constituents of cement stone and concrete with hydrogen sulphide

Nos. of reactions	Equations of corrosion reactions	ΔG^0_{298} [kcal/mol]	Equations $\lg p_{H_2S} = f(\lg p_{O_2})$	H_2S equilibrium partial pressure at $p_{O_2} = 0.21$ atm
1.	$\frac{1}{5}(5CaO \cdot 6SiO_2 \cdot 5.5H_2O)(cr) + H_2S(gas) + 2O_2(gas) = CaSO_4 \cdot 2H_2O(cr) + \frac{6}{5}SiO_2(cr) + 0.1\,H_2O(liq)$	−200.60	$\lg p_{H_2S} = -147.07 - 2\lg p_{O_2}$	$10^{-145.70}$
2.	$\frac{1}{4}(4CaO \cdot Al_2O_3 \cdot 19H_2O)(cr) + H_2S(gas) + 2O_2(gas) = CaSO_4 \cdot 2H_2O(cr) + \frac{1}{2}Al(OH)_3(am) + 3H_2O(liq)$	−203.0	$\lg p_{H_2S} = -148.83 - 2\lg p_{O_2}$	$10^{-147.47}$
3.	$\frac{1}{3}(3CaO \cdot Al_2O_3 \cdot 6H_2O)(cr) + H_2S(gas) + 2O_2(gas) = CaSO_4 \cdot 2H_2O(cr) + \frac{2}{3}Al(OH)_3(am)$	−204.05	$\lg p_{H_2S} = -149.6 - 2\lg p_{O_2}$	$10^{-148.24}$
4.	$\frac{1}{3}(3CaO \cdot Al_2O_3 \cdot 3CaSO_4 \cdot 32H_2O)(cr) + H_2S(gas) + 2 \cdot O_2 = 2(CaSO_4 \cdot 2H_2O)(cr) + \frac{2}{3}Al(OH)_3(am) + \frac{20}{3}H_2O$	−199.57	$\lg p_{H_2S} = -146.31 - 2\lg p_{O_2}$	$10^{-144.96}$
5.	$\frac{1}{3}(3CaO \cdot Fe_2O_3 \cdot 3CaSO_4 \cdot 32H_2O)(cr) + H_2S(gas) + 2O_2(gas) = 2(CaSO_4 \cdot 2H_2O)(cr) + \frac{2}{3}Fe(OH)_3(am) + \frac{20}{3}H_2O$	−203.31	$\lg p_{H_2S} = -149.05 - 2\lg p_{O_2}$	$10^{-147.69}$
6.	$CaO \cdot Al_2O_3 \cdot 10H_2O(cr) + H_2S(gas) + 2O_2(gas) = CaSO_4 \cdot 2H_2O(cr) + 2Al(OH)_3(am) + 6H_2O(liq)$	−203.96	$\lg p_{H_2S} = -149.53 - 2\lg p_{O_2}$	$10^{-148.17}$
7.	$\frac{1}{2}(2CaO \cdot Al_2O_3 \cdot 8H_2O)(cr) + H_2S(gas) + 2O_2(gas) = CaSO_4 \cdot 2H_2O(cr) + Al(OH)_3(am) + \frac{3}{2}H_2O(liq)$	−203.79	$\lg p_{H_2S} = -149.41 - 2\lg p_{O_2}$	$10^{-148.05}$
8.	$Ca(OH)_2(cr) + H_2S(gas) + 2O_2(gas) = CaSO_4 \cdot 2H_2O(cr)$	−207.13	$\lg p_{H_2S} = -151.85 - 2\lg p_{O_2}$	$10^{-150.5}$
9.	$CaCO_3(cr) + H_2S(gas) + 2O_2(gas) + H_2O(liq) = CaSO_4 \cdot 2H_2O(cr) + CO_2(gas)$	−188.77	$\lg p_{H_2S} = \lg p_{CO_2} - 2\lg p_{O_2} - 138.39$	$10^{-140.54}$ (at $p_{CO_2} = 10^{-3.5}$ atm)

needed for the reaction to begin are about 10^{65} times as low as those of SO_2 and 10^{135} times as low as those of CO_2. As seen in Table 123, the most vulnerable compound in concrete is again calcium hydroxide for which the p_{H_2S} equilibrium value is $10^{-150.5}$ atm.

According to the value of the minimum equilibrium pressure of H_2S needed for the reaction to begin the constituents of cement stone rank as follows:

$$Ca(OH)_2 < C_3AH_6 < CAH_{10} < C_2AH_8 < CF\bar{S}H < C_4AH_{19} < CA\bar{S}H$$
$$< C_5S_6H_{5.5} < CaCO_3 .$$

2 Reactions of the Interaction of the Constituents of Cement Stone and Concrete with Hydrogen Ions

Reactions of the interaction of the constituents of cement stone and concrete with ions H^+ are most typical of the life of concrete and reinforced concrete constructions effected by sewage. This is the most serious type of concrete corrosion, as it covers as extremely wide scale of conditions, from pH = 0 to pH = 14, i.e., it includes general acid and alkali types of corrosion as well as every corrosion process connected with an alteration of pH (lixiviation, magnesian corrosion etc.) [269, 442, 443].

As a first example let us regard a reaction of tobermorite interacting with H^+(aq):

$$5CaO \cdot 6SiO_2 \cdot 5.5H_2O\,(sol) + 1OH^+(aq)$$
$$= 10.5H_2O\,(liq) + 5Ca^{2+}(aq) + 6SiO_2\,; \qquad (267)$$

$$K = \frac{[Ca^{2+}]^5}{[H^+]^{10}}\,; \quad \lg K = 5\lg[Ca^{2+}] - 10\lg[H^+]\,;$$

$$\Delta G^0_{298} = -10.5 \cdot 56.69 - -5 \cdot 132.18 - 6 \cdot 204.75 + 2361.45 = -123.2\ \text{kcal}$$

$$\lg K = \frac{123.2}{1.364} = +90.3; \quad 90.3 = 5\lg[Ca^{2+}] + 10\,pH\,;$$

$$\lg[Ca^{2+}] = 18.06 - 2\,pH\,. \qquad (268)$$

The results of the calculations by formula (268) are summarized in Table 124.

As one can see from the table tobermorite can be a thermodynamically stable compound only at pH > 10, as only on these conditions can the needed value of the active concentration of calcium ions be created in a solution. At pH = 9, e.g., the active concentration of calcium ions may not be less than $10^{0.06}$, i.e., 0.4 g-

Table 124. The active concentration of calcium ions as a function of pH in a reaction between tobermorite and the ions H^+

pH	lg [Ca^{2+}]	[Ca^{2+}]
1	16.06	$10^{16.06}$
2	14.06	$10^{14.06}$
3	12.06	$10^{12.06}$
4	10.06	$10^{10.06}$
5	8.06	$10^{8.06}$
6	6.06	$10^{6.06}$
7	4.06	$10^{4.06}$
8	2.06	$10^{2.06}$
9	0.06	$10^{0.06}$
10	−1.94	$10^{-1.94}$
11	−3.94	$10^{-3.94}$
12	−5.94	$10^{-5.94}$
13	−7.94	$10^{-7.94}$
14	−9.94	$10^{-9.94}$

ion/l Ca^{2+}, which is not provided in the pore liquid. At pH = 10 [Ca^{2+}] may not be less than $10^{-1.94}$, i.e., 0.01 g-ion/l Ca^{2+} and at pH = 11 [Ca^{2+}] must be ca. 0.0001 g-ion/l Ca^{2+}. While dissolving lime at 25 °C the ion concentration of about 0.02 mol/l Ca^{2+} is reached, which, the activity coefficient being 0.48, will result in the active concentration of $0.02 \cdot 0.48 = 0.0096$, i.e., close to 0.01 g-ion/l Ca^{2+}, thus, according to Table 124 tobermorite will be stable at pH > 10.4.

The calculations of equations of the active concentrations of calcium ions as a function of pH which we did for other hydrate compounds are given in Table 125.

Taking the active concentrations of calcium ions of hydrates in a solution, starting from the data on their solubility (see Table 126), one can determine the minimum pH values at which they will be stable.

The results of these calculations are given in Table 127.

As one can see from Table 127, all the hydrates which are constituents of cement stone become unstable even at pH less than 10.

3 Corrosion Resistance of Aluminium and Iron-Containing Phases of Cement Stone in Sulphate Media

We took sulphate corrosion as an example for conducting an analysis of the corrosion resistance of aluminate and iron-containing phases of cement stone by the

Table 125. The results of the calculations of the corrosion reactions of the constituents of cement stone and concrete with ions H^+

Nos. of reactions	Equations of reactions	ΔG^0_{298} of reactions [kcal/g-ion]	ΔG^0_{298} per 1 g-ion H^+	Equations of the calcium ion activity as a function of pH and the Al^{3+} and Fe^{3+} activities
1.	$\frac{1}{5}(5CaO \cdot 6SiO_2 \cdot 5.5H_2O)(sol) + 2H^+(aq) = 2.1H_2O(liq) + Ca^{2+}(aq) + \frac{6}{5}SiO_2(sol)$	−24.64	−12.32	$lg[Ca^{2+}] = 18.06 - 2pH$
2.	$\frac{1}{4}(4CaO \cdot Al_2O_3 \cdot 19H_2O)(sol) + \frac{14}{4}H^+(aq) = Ca^{2+}(aq) + \frac{1}{2}Al^{3+}(aq) + \frac{26}{4}H_2O(aq)$	−35.30	−10.09	$lg[Ca^{2+}(aq)] = 25.88 - 3.5pH - 0.5\,lg[Al^{3+}]$
3.	$\frac{1}{4}(4CaO \cdot Al_2O_3 \cdot 19H_2O) + 2H^+ = Ca^{2+} + \frac{1}{2}Al(OH)_3 + 5H_2O$	−27.97	−13.99	$lg[Ca^{2+}] = 20.51 - 2pH$
4.	$\frac{1}{3}(3CaO \cdot Fe_2O_3 \cdot 3CaSO_4 \cdot 32H_2O) + 4H^+ = Ca^{2+} + \frac{2}{3}Fe^{3+} + CaSO_4 \cdot 2H_2O + \frac{32}{3}H_2O$	−29.0	−7.25	$lg[Ca^{2+}] = -21.26 - 0.67Fe^{3+} - 4pH$
5.	$\frac{1}{3}(3CaO \cdot Fe_2O_3 \cdot 3CaSO_4 \cdot 32H_2O) + 2H^+ = Ca^{2+} + \frac{2}{3}Fe(OH)_3(am) + CaSO_4 \cdot 2H_2O + \frac{26}{3}H_2O$	−27.27	−13.63	$lg[Ca^{2+}] = +19.96 - 2pH$
6.	$\frac{1}{3}(3CaO \cdot Al_2O_3 \cdot 3CaSO_4 \cdot 32H_2O)(sol) + 4H^+(aq) = Ca^{2+}(aq) + \frac{2}{3}Al^{3+}(aq) + CaSO_4 \cdot 2H_2O(sol) + \frac{32}{3}H_2O$	−33.3	−8.33	$lg[Ca^{2+}] = 24.41 - 4pH - 0.67 \cdot [Al^{3+}]$
7.	$\frac{1}{3}(3CaO \cdot Al_2O_3 \cdot 3CaSO_4 \cdot 32H_2O) + 2H^+ = Ca^{2+} + \frac{2}{3}Al(OH)_3 + CaSO_4 \cdot 2H_2O + \frac{26}{3}H_2O$	−23.53	−11.77	$lg\,Ca^{2+} = 17.25 - 2pH$
8.	$(CaO \cdot Al_2O_3 \cdot 10H_2O)(cr) + 8H^+(aq) = Ca^{2+}(aq) + 2Al^{3+}(aq) + 14H_2O(liq)$	−57.24	−7.155	$lg\,Ca^{2+} = 41.96 - 8pH - 2lg\,Al^{3+}$
9.	$(CaO \cdot Al_2O_3 \cdot 10H_2O)(cr) + 2H^+ = Ca^{2+} + 2Al(OH)_3(am) + 8H_2O$	−27.92	−13.96	$lg\,Ca^{2+} = +20.47 - 2pH$
10.	$Ca(OH)_2(cr) + 2H^+(aq) = Ca^{2+}(aq) + 2H_2O(liq)$	−31.09	−15.55	$lg\,Ca^{2+} = 22.79 - 2pH$
11.	$CaCO_3(cr) + 2H^+(aq) = Ca^{2+}(aq) + CO_2(gas) + H_2O(liq)$	−13.27	−6.64	$lg\,Ca^{2+} = 13.23 - 2pH$ (at $p_{CO_2} = 10^{-3.5}$ atm)

Table 126. The equilibrium concentrations of calcium ions of various hydrates at 298 K

Hydrate compound	Temp. [K]	Equilibrium concentration of ions Ca^{2+} in a solution [g-ion/l]
$C_5S_6H_{5.5}$	298	$1.8 \cdot 10^{-3}$
C_4AH_{19}	298	$4.7 \cdot 10^{-3}$
Aluminate sulphate hydrates	298	$0.8 \cdot 10^{-3}$
$Ca(OH)_2$	298	$21.43 \cdot 10^{-3}$

Table 127. The minimum pH values at which hydrates are stable

Hydrate compounds	pH equilibrium values
$C_5S_6H_{5.5}$	10.4
C_4AH_{19}	11.43
Aluminate sulphate hydrates	10.17
$Ca(OH)_2$	12.23

thermodynamic method. The ionic equilibria were calculated with regard to the following limiting conditions of the initial data:

1. Ion concentrations found from thermodynamic data were taken equal to their active concentrations.

2. The active concentration of calcium ions determining the pore liquid was taken equal to $10^{-1.936}$ g-ion/l, that of $[Al(OH)_4^-]$ at $\mu = 0.06$ was $10^{-4.378}$ g-ion/l and that of $[Fe(OH)_4^-] = 10^{-6}$ g-ion/l.

The thermodynamic possibility of the formation of gypsum dihydrate, of both forms of calcium aluminate sulphate hydrate and calcium ferrite sulphate hydrate was analyzed by the reactions given in Table 128. By the values of the Gibbs free energy of the reactions given in Table 128 with a set limiting value of the initial data, a calculation of the minimum values of the active concentration of the ions SO_4^{2-} was conducted providing the thermodynamic stability of solid phases (Table 129). From the data obtained it follows that the formation of ettringite and its iron-containing analogue is possible at pH greater than 10 and 9.5 respectively and at a low concentration of sulphate ions.

As one can see from Table 129, the formation of a monosulphate is thermodynamically possible at pH above 12.5 and 12.0 respectively for aluminate and ferrite hydrates and at the deficiency of ions SO_4^{2-}; therefore in the number of the products of the sulphate corrosion it is not generally found.

At the concentration of ions SO_4^{2-} above that determined by the solubility product of $CaSO_4 \cdot 2H_2O$ mainly the formation of gypsum dihydrate takes place.

The crystallization in pores and capillaries of ettringite is considered to be one of the chief causes of the destruction of cement stone effected by sulphate ions of

Table 128. Equations of the reactions of the formation of calcium aluminate sulphate hydrates, calcium ferrite sulphate hydrates and gypsum

Nos. of reac-tions	Equations of reactions	$-\Delta G^0_{298}$ of reactions [cal/mol]
1	$6Ca^{2+} + 2Al(OH)_4^- + 3SO_4^{2-} + 4OH^- + 26H_2O = 3CaO \cdot Al_2O_3 \cdot 3CaSO_4 \cdot 32H_2O$	60080
2	$4Ca^{2+} + 2Al(OH)_4^- + SO_4^{2-} + 4OH^- + 6H_2O = 3CaO \cdot Al_2O_3 \cdot CaSO_4 \cdot 12H_2O$	38460
3	$Ca^{2+} + SO_4^{2-} + 2H_2O = CaSO_4 \cdot 2H_2O$	6290
4	$6Ca^{2+} + 2Fe(OH)_4^- + 3SO_4^{2-} + 4 \cdot OH^- + 26H_2O = 3CaO \cdot Fe_2O_3 \cdot 3CaSO_4 \cdot 32H_2O$	67970
5	$4Ca^{2+} + 2Fe(OH)_4^- + SO_4^- + 4OH^- + 6H_2O = 3CaO \cdot Fe_2O_3 \cdot CaSO_4 \cdot 12H_2O$	46450

Table 129. The minimum values of the active concentration of ions providing the thermodynamic stability of aluminate sulphate hydrates, ferrite sulphate hydrates and gypsum

Nos. of reactions from Table 128	lg $[SO_4^{2-}]$ at pH					
	8	9	10	11	12	13
1	0.111	−1.222	−2.555	−3.888	−5.221	−6.554
2	12.303	8.303	4.303	0.303	−3.303	−7.697
3	−2.672	−2.672	−2.672	−2.672	−2.672	−2.672
4	−0.735	−2.064	−3.401	−4.734	−6.067	−7.400
5	9.689	5.689	1.689	−2.311	−6.311	−10.311

aggressive media. Therefore a calculation of the minimum values of the active concentrations of the ions SO_4^{2-} was carried out, at which the corrosion reactions given in Table 130 are possible; it follows from the calculated values of the Gibbs free energy of the reactions that all the hydrates under consideration can interact with sulphate ions forming ettringite or its iron-containing analogue within the range of their thermodynamic stability. And as one can see from the data of Tables 129 and 131, with regard to the thermodynamic instability towards sulphate ions, aluminium and iron monosulphates $C_3AC\bar{S}H_{12}$ and $C_3FC\bar{S}H_{12}$ are practically the same as calcium cubie and hexagonal aluminate hydrates and ferrite hydrates respectively.

Thus, in the stability range of trisulphates, monosulphates are similar to the rest of calcium aluminate and ferrite hydrates, aggressively effected by ions SO_4^{2-}.

Of particular interest is a comparison of the thermodynamic stability of aluminate and ferrite phases towards sulphate corrosion with their solid solutions, for example cubic hydrates (reactions 2, 5, 7, 8, 9 in Table 130). It would seem that with reaction 2 reactions 7 and 8 must be compared, hydrox complexes $Fe(OH)_4^-$ by analogy with $Al(OH)_4^-$ being written in the right part of the equation.

The analysis shows, however, that primarily, the formation of the ion $Al(OH)_4^-$ is thermodynamically more preferable than that of the amorphous aluminium hydroxide, as ΔG_{298}^0 of the reaction

$$Al(OH)_4^- = Al(OH)_3(am) + OH^- \tag{267}$$

is positive (+1.406 kcal/mol). Meanwhile the formation of the ion $Fe(OH)_4^-$ is thermodynamically less preferable as compared to that of the amorphous ferric hydroxide, as ΔG_{298}^0 of the reaction

$$Fe(OH)_4^- = Fe(OH)_3 + OH^- \tag{268}$$

is negative (−5.584 kcal/mol). Secondly, from the diagram Eh-pH of the system iron-water it follows that only at $pH > 11.51$ and $Eh > 0.6$ is iron predominantly in the form of the ion $Fe(OH)_4^-$ and at $pH < 11.51$, in the form of undissociated molecules $Fe(OH)_3^0$. Therefore the data of Table 131 for reactions 7 and 8 at $pH < 11$ are not quite correct. Thus, reaction 2 in Table 130 is to be compared with reactions 5 and 9.

As one can see from Table 131 for reaction 9, in the range of the thermodynamic stability of ettringite a somewhat greater concentration of ions SO_4^{2-} as compared with reactions 2 and 5 is needed. If one could take into consideration the mixing heat at the formation of $C_3A_{0.5} \cdot F_{0.5}H_6$ in the Gibbs free energy of reaction 9, the $\lg[SO_4^{2-}]$ value would additionally increase by the $-\Delta H_{mix}/1.364$, where $\Delta H_{mix} < 0$, i.e., for the reaction of the interaction of $3CaO \cdot 0.5Al_2O_3 \cdot 0.5Fe_2O_3 \cdot 6H_2O$ with the formation of ettringite $C_3ACS_3H_{32}$ a higher active concentration of SO_4^{2-} is needed than for the mixture of C_3AH_6 and C_3FH_6, which reveals a higher sulphate-resistance of the solid solution as compared to individual hydrates.

As calculations showed, aluminoferrite hydrates have a higher thermodynamic corrosion resistance than the corresponding aluminate hydrates.

Some scientists suppose that the concept of sulphate resistance of aluminates with a different molar ratio of Fe_2O_3 to A_2O_3 in their composition is a kinetic concept connected with the slowing down of the reactions of the chemical binding of ions SO_4^{2-} proper. This factor, though important, alone does not explain, e.g., the observed low sulphate resistance of ferrite cement as compared to aluminoferrite one.

We believe that the high corrosion resistance of the aluminate ferrite hydrate phases is conditioned by a kinetic factor connected with the slowing down of chemical reactions with the increase of the molar ratio of Fe_2O_3 to Al_2O_3 in

Table 130. Equations of the reactions of the interaction of calcium aluminate hydrates and calcium ferrite hydrates with sulphate ions

Nos. of reactions	Equations of reactions	$-\Delta G^0_{298}$ of reactions [cal/mol]
1	$\frac{1}{2}3CaO \cdot Al_2O_3 \cdot 13H_2O + SO_4^{2-} + \frac{17}{3}H_2O = \frac{1}{3}3CaO \cdot Al_2O_3 \cdot 3CaSO_4 \cdot 32H_2O + \frac{1}{3}Al(OH)_4^- + \frac{5}{3}OH^-$;	980
2	$\frac{1}{2}3CaO \cdot Al_2O_3 \cdot 6H_2O + SO_4^{2-} + \frac{26}{3}H_2O = \frac{1}{3}3CaO \cdot Al_2O_3 \cdot 3CaSO_4 \cdot 32H_2O + \frac{2}{3}Al(OH)_4^- + \frac{4}{3}OH^-$;	1540
3	$3CaO \cdot Al_2O_3 + SO_4^{2-} + \frac{34}{3}H_2O = \frac{2}{3}3CaO \cdot Al_2O_3 \cdot 3CaSO_4 \cdot 32H_2O + \frac{2}{3}Al(OH)_4^- + \frac{4}{3}OH^-$;	1590
4	$\frac{1}{2}4CaO \cdot Fe_2O_3 \cdot 13H_2O + SO_4^{2-} + \frac{17}{3}H_2O = \frac{1}{3}3CaO \cdot Fe_2O_3 \cdot 3CaSO_4 \cdot 32H_2O + \frac{1}{3}Fe(OH)_3 + 2OH^-$;	1479
5	$\frac{2}{3}3CaO \cdot Fe_2O_3 \cdot 6H_2O + SO_4^{2-} + \frac{26}{3}H_2O = \frac{1}{3}3CaO \cdot Fe_2O_3 \cdot 3CaSO_4 \cdot 32H_2O + \frac{2}{3}Fe(OH)_3 + 2OH^-$;	2570
6	$3CaO \cdot Fe_2O_3 \cdot 6H_2O + SO_4^{2-} + \frac{34}{3}H_2O = \frac{2}{3}3CaO \cdot Fe_2O_3 \cdot 3CaSO_4 \cdot 32H_2O + \frac{2}{3}Fe(OH)_3 + 2OH^-$;	2680
7	$\frac{2}{3}3CaO \cdot Fe_2O_3 \cdot 6H_2O + SO_4^{2-} + \frac{26}{3}H_2O = \frac{2}{3}Fe(OH)_4^- + \frac{4}{3}OH^- + \frac{1}{3}3CaO \cdot FeO_3 \cdot 3CaSO_4 \cdot 3H_2O$;	−1150
8	$\frac{2}{3}3CaO \cdot 0.5Al_2O_3 \cdot 0.5Fe_2O_3 \cdot 6H_2O + SO_4^{2-} + \frac{26}{3}H_2O = \frac{1}{3}3CaO \cdot Fe_2O_3 \cdot 3CaSO_4 \cdot 32H_2O + \frac{2}{3}Fe(OH)_4^- + \frac{4}{3}OH^-$;	−1120
9	$\frac{2}{3}3CaO \cdot 0.5Al_2O_3 \cdot 0.5Fe_2O_3 \cdot 6H_2O + SO_4^{2-} + \frac{26}{3}H_2O = \frac{1}{3}3CaO \cdot Fe_2O_3 \cdot 3CaSO_4 \cdot 32H_2O + \frac{2}{3}Fe(OH)_3 + 2OH^-$.	2604

Table 131. The minimum values of the active concentration of ions providing corrosion reactions

Nos. of reactions in Table 130	$\lg[SO_4^{2-}]$ at pH					
	8	9	10	11	12	13
1	−12.181	−10.514	−8.847	−7.180	−5.513	−3.846
2	−12.050	−10.717	−9.384	−8.051	−6.718	−5.385
3	−12.085	−10.752	−9.419	−8.086	−6.747	−5.420
4	−13.082	−11.082	−9.082	−7.082	−5.082	−3.082
5	−13.884	−11.884	−9.884	−7.884	−5.884	−3.884
6	−13.894	−11.894	−9.894	−7.894	−5.894	−3.894
7	−11.158	−9.825	−8.492	−7.159	−5.826	−4.493
8[a]	−10.983	−9.650	−8.317	−6.984	−5.651	−4.318
9[a]	−13.709	−11.709	−9.709	−7.709	−5.709	−3.709

[a] In these data only the entropy contribution of a solid solution to the Gibbs free energy of mixing is considered

hydrates. It is also conditioned by the fact that solid solutions are less reactive with sulphate ions than a mechanical mixture of aluminate hydrates and aluminoferrite hydrates.

4 Reactions in the Acid-Sulphate Corrosion of Concrete

At plants in the atmosphere of which there are vapours of sulphuric acid concrete and reinforced concrete structures are effected by corrosion which takes place on account of hydrogen ions H^+ and anions SO_4^{2-} simultaneously attacking the constituents of cement stone. The acid-sulphate corrosion of concrete is considered to proceed in two stages [29, 30]:

$$Ca(OH)_2 + H_2SO_4 = CaSO_4 \cdot 2H_2O \,, \tag{269}$$

$$3CaO \cdot Al_2O_3 \cdot 6H_2O + 3(CaSO_4 \cdot 2H_2O) + 20H_2O$$
$$= 3CaO \cdot Al_2O_3 \cdot 3CaSO_4 \cdot 32H_2O \,. \tag{270}$$

While carrying out calculations, as reactants we considered the ions of calcium, aluminium, OH^-, H^+; the reaction medium was the liquid filling in pores in the cement stone of concrete. The thermodynamic activity of the vapours of sulphuric acid is expressed by its partial pressure.

The quantitative (and qualitative) composition of the pore liquid of cement stone depending on a number of technical factors (type of cement, additives etc.) generally follows certain regularities, which allows an accurate characterization.

The content of calcium and aluminium in the pore liquid can be found on the basis of the conditions of their equilibrium with the main products of the cement hydration. In the process of the cement hydration due to the C_3S decomposition a supersaturation of the liquid phase in respect to the crystalline $Ca(OH)_2$ takes place, during a long working time of concrete. The system however approaches an equilibrium, and one can quite safely consider the lime concentration in the pore liquid to be close to saturation as compared to calcium hydroxide.

The equilibrium concentration of aluminate ions in the solutions of the system aluminate-$Ca(OH)_2$-solution is of low value. Thus, for the system with C_3AH_6 Wells and coauthors give a concentration 0.0005 g/l of Al_2O_3 and with C_4AH_{19}, from 0.005 to 0.01 g/l of Al_2O_3. According to d'Ans and Eick with the cubic aluminate the Al_2O_3 concentration amounts to 0.00325 g/l and with the hexagonal one, 0.00332 g/l.

In conditions of corrosion reactions aluminium ions were considered in the form $Al(OH)_4^-$ (according to Rumyantsev and coauthors), calcium ions were considered in two forms: Ca^{2+} and $Ca(OH)^+$.

The thermodynamic calculations were carried out allowing for the following limiting conditions for the initial data:

1. The concentrations of ions found from the thermochemical data were taken equal to their active concentrations.

2. The pore liquid was considered in the form of a solution saturated in respect to $Ca(OH)_2$. The following values of the active concentration of calcium ions were taken:

$\lg[Ca^{2+}(aq)] = -1.53 \qquad \lg[Ca(OH)^{+}(aq)] = -1.84$.

3. The concentration of aluminium ions in the pore liquid was assumed to be 0.00332 g/l, which recalculated to the active concentration of ions $Al(OH)_4^-$ with respect to the ionic strength of the lime saturated solution of $\mu = 0.06$ will be $\lg[Al(OH)_4^-(aq)] = -4.56$.

4. The active concentration of the sulphuric acid vapours is expressed by the partial pressure of the latter.

The thermodynamic possibility of the formation of gypsum dihydrate and of both forms of calcium aluminate sulphate hydrate was analyzed in the following instances (Table 132).

Table 132. Reactions of the formation of gypsum dihydrate, and of the high-sulphate and low-sulphate forms of calcium aluminate sulphate hydrate

Nos. of reactions	Equations of reactions	Gibbs free energy of the reactions ΔG^0_{298} [kcal/mol]
1	$Ca^{2+}(aq) + H_2SO_4(gas) + 2OH^-(aq) \rightleftarrows CaSO_4 \cdot 2H_2O$	−63.706
2	$Ca(OH)^+(aq) + H_2SO_4(gas) + 2OH^-(aq) \rightleftarrows CaSO_4 \cdot 2H_2O$	−62.04
3	$6Ca^{2+}(aq) + 3H_2SO_4(gas) + 2Al(OH)_4^-(aq) + 10OH^-(aq) + 20H_2O \rightleftarrows 3CaO \cdot Al_2O_3 \cdot 3CaSO_4 \cdot 32H_2O$	−228.262
4	$6Ca(OH)^+(aq) + 3H_2SO_4(gas) + 2Al(OH)_4^-(aq) + 4OH^-(aq) + 20H_2O \rightleftarrows = CaO \cdot Al_2O_3 \cdot 3CaSO_4 \cdot 32H_2O$	−218.266
5	$4Ca^{2+}(aq) + H_2SO_4 + 2Al(OH)_4^-(aq) + 6OH^-(aq) + 4H_2O \rightleftarrows 3CaO \cdot Al_2O_3 \cdot CaSO_4 \cdot 12H_2O$	−96.882
6	$4Ca(OH)^+(aq) + H_2SO_4(gas) + 2Al(OH)_4^-(aq) + 2OH^-(aq) + 4H_2O \rightleftarrows 3CaO \cdot Al_2O_3 \cdot CaSO_4 \cdot 12H_2O$	−90.218

The Gibbs free energy values allow in their turn evaluating the partial pressure of the H_2SO_4 vapours as a function of the active concentrations of the other constituents of a solution in each reaction at the conditions of the equilibrium with the solid phase (Table 133).

The results of the calculation of the minimum partial pressures of the sulphuric acid vapours in the air providing the thermodynamic stability of gypsum and calcium aluminate sulphate hydrates in the liquid phase are presented in Fig. 92.

The content of the sulphuric acid vapours in the air of acid plants and some workshops of chemical plants varies from $2.5 \cdot 10^{-7}$ to $1.25 \cdot 10^{-6}$ atm, which considerably surpasses the minimum partial pressure of the H_2SO_4 vapours providing the thermodynamic stability of the products of the corrosion of constituents of cement stone.

Table 133. The relationship between the partial pressures of the H_2SO_4 vapours and the active concentrations of ions providing the thermodynamic stability of the solid phases

Nos. of reactions	$\lg p_{H_2SO_4}$
1	$-\lg[Ca^{2+}]-2pH-18.705$
2	$-\lg[Ca(OH)^+]-pH-31.483$
3	$-2\lg[Ca^{2+}]-0.667\lg[Al(OH)_4^-]-3.333\,pH-9.114$
4	$-2\lg[Ca(OH)^+]-0.667\lg[Al(OH)_4^-]-1.333\,pH-34.672$
5	$-4\lg[Ca^{2+}]-2\lg[Al(OH)_4^-]-6pH-12.973$
6	$-4\lg[Ca(OH)^+]-2\lg[Al(OH)_4^-]-2pH-38.141$

The results of the calculations given in Table 132 prove that both forms of calcium ions fit the crystallization of the hydrates under consideration, the most reactive being the ions $Ca(OH)^+$. The percentage of calcium ions as a function of pH of the solution was found from Hopkins' data for the equilibrium constant $-\lg C = 1.22$ of the reaction

$$Ca(OH)^+ \rightleftarrows Ca^{2+} + OH^- . \tag{271}$$

From the quantitative content of ions given in Table 134 it follows that the amount of the ions $Ca(OH)^+$ becomes considerable beginning with $pH > 12$, while taking into account the undissociated molecules of $Ca(OH)_2 \cdot aq$ (see Table 87) we can consider their role even at $pH > 12$ insignificant.

The formation of gypsum dihydrate (Fig. 92) is more preferable in a solution with $pH < 10.5$. Within the range of $pH = 10.5 + 11.5$ the formation of ettringite is thermodynamically most preferable and at $pH > 11.5$, that of calcium aluminate monosulphate.

A petrographic analysis of concrete samples taken from structures effected by sulphuric acid vapours showed the presence of gypsum dihydrate. Crystals of aluminate sulphate hydrates were not found. Starting from the calculations this may be accounted for by the fact that with the diffusion of the sulphuric acid vapours a considerable decrease of the pH value of the pore liquid occurs and al-

Table 134. The contents of calcium ions in the solution as a function of pH

Ions	Percentage at pH				
	9	10	11	12	13
Ca^{2+}	99.98	99.83	98.36	85.77	37.61
$Ca(OH)^+$	0.02	0.17	1.64	14.23	62.39

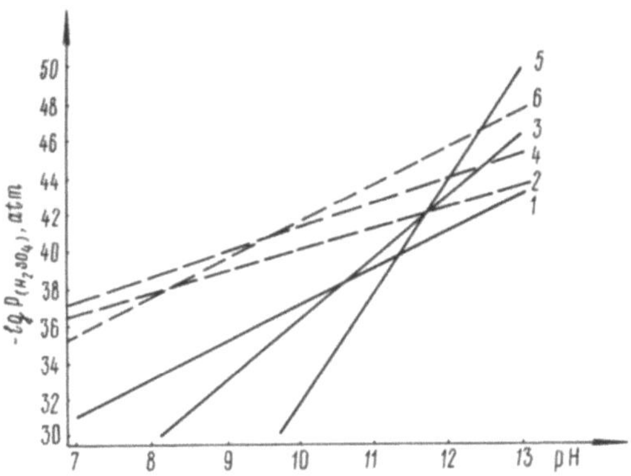

Fig. 92. The partial pressure of the sulphuric acid vapours as a function of pH providing the thermodynamic stability of liquid phases. The numbers of the lines correspond to the numbers of reactions in Tables 132 and 133; ——— calcium ions in the form Ca^{2+}; – – – calcium ions in the form $CaOH^+$

though gypsum dihydrate in contact with the ions $Al(OH)_4^-$ is capable in principle of forming calcium aluminate hydrate, the crystals of the latter cannot appear in the contact zone of concrete because of a low pH value.

If for some reason an increase of pH in the sections of concrete effected by the H_2SO_4 vapours occur, ettringite will be formed in them. Although at pH = −11.5 the formation of aluminate monosulphate is possible, the latter does not appear in the contact zone, a gypsum being in excess, ettringite is the thermodynamically stable aluminate sulphate hydrate, which is confirmed by the negative value of the Gibbs free energy (−3.968 kcal/mol) of the reaction:

$$3CaO \cdot Al_2O_3 \cdot CaSO_4 \cdot 12H_2O + 2(CaSO_4 \cdot 2H_2O) + 16H_2O \rightleftarrows 3CaO \cdot Al_2O_3 \cdot 3CaSO_4 \cdot 32H_2O\,. \quad (272)$$

Experimental data arrived at by Larionova and coauthors also show ettringite to be the main aluminium – containing phase while hydrating mixtures with an addition of gypsum, the molecular ratio $CaSO_4 \cdot 2H_2O/C_3A$ (or C_4AF) being 3.

Thus, the thermodynamic analysis allowed establishing the formation and stability conditions of the products of the acid-sulphate corrosion and we were able to clarify some peculiarities of these processes, among them the considerable decrease of the pH value when the H_2SO_4 vapours effect concrete and the preferred formation of gypsum dihydrate under these conditions.

Chapter V. Thermodynamics of Irreversible Processes[27]

It was shown above that the classical thermodynamics built on two main principles, the law of conservation of energy and the law of increase of entropy results in relationships linking equilibrium (thermodynamic) parameters of a system for any isolated system as a whole. The system of equations obtained is not however, complete. A certain relationship between parameters, state equation, must be added to it. The state equation is either chosen by the empirical way (e.g., Grüneisen's equation for solid bodies) or derived by methods of statistical mechanics based on a certain microscopic model of the system.

While deriving the relationships of thermodynamics the concept of quasistatic reversible processes is used, i.e., of the processes proceeding infinitely slowly in time and at infinitesimal alterations of the potentials. Equations of the classical thermodynamics do not comprise the time factor, thus they are inadequate to solve the problems of the kinetics of the processes, but do reliably determine the initial and final state of quasistatic systems. The strength of the thermodynamic method, however, is that it allows arriving at the main regularities of the processes without knowledge of the details of their molecular mechanism.

As most actual processes are irreversible (e.g., transfer processes and several chemical reactions), it would be of great interest to extend the thermodynamic method to the study of such processes proceeding at finite difference of potentials and a finite rate.

In the past 25 years rapid progress has been made in this field, the theory having received the name – the thermodynamics of irreversible processes – [96, 114, 115, 116, 155, 174, 313, 339, 491]. Until recently however, the methods developed were practically restricted to the range of phenomena close to equilibrium ones; the thermodynamic forces (such as temperature gradient, chemical affinity etc.) and thermodynamic fluxes (such as heat flux, rates of thermodynamic reactions etc.) being connected by linear relationships.

Belonging to the linear non-equilibrium thermodynamics are both Onsager's reciprocity relation [462] and Prigogine's theorem of the minimum entropy production (see, e.g., [492]). This part of the thermodynamics of irreversible processes is at present classical and was most fully set forth in [116]. It has become clear, however, (mainly due to the works of the Brussels school investigators headed by Prigogine) that it is possible to further generalize thermodynamics in

[27] The authors thank P. O. Mchedlov-Petrossyan for helpful discussion of this chapter

studying the entire range of phenomena from equilibrium to non-linear effects [115]. Such a generalization is possible for a whole range of phenomena in which the local entropy may be expressed by the same independent variables as with a system being in equilibrium. This is a so-called presumption of a "local equilibrium" which will be later discussed below (2).

In this chapter we shall give a concise though incomprehensive review of the non-equilibrium thermodynamics (keeping mainly to [155]) and make references to some of its applications to the study of silicate systems. For a more thorough familiarization with the non-equilibrium thermodynamics we recommend monographs [116, 155] as well as Gurov [174] written in a good and clear style (unfortunately, it does not cover problems involving chemical reactions).

1 Local Macrocharacteristics of a System, Conservation Laws and Balance Equations

The equilibrium thermodynamics is a branch of the macroscopic physics dealing with macroscopic characteristics of a system (i.e., mean corresponding characteristics of particles in a system; e.g., temperature T is connected with the mean energy of the thermal motion of particles). The same is true for the phenomenological thermodynamics of irreversible processes, macronotions, however, requiring considerable specification [174].

Let us subdivide the whole volume of the system under consideration into a number of subvolumes each of which, although very small as compared to the whole system, contains a number of particles great enough, thus being macroscopic. One can then attach macroscopic characteristics to each subvolume, i.e., one can speak of the temperature, density etc. of a subvolume, those characteristics being arrived at due to averaging only in the subvolume under consideration. As an individual subvolume is not an isolated system, the mean values of the subvolume may differ from those of the whole volume of the system (fluctuations), the less the size of the subvolume, the greater the differences. However, while averaging over a great interval of time a mean value will coincide with that of the whole. That is, local macrocharacteristics are related to a certain scale in space and a certain interval of time.

Considering some process, we may introduce characteristic time during which an alteration of a property representing the process in question becomes noticeable. Then the fluctuations, the existence time of which is much shorter than the characteristic time of a process, may be ruled out of the theory formalism, averaging the characteristics by such an interval of time on which small-scale fluctuations are barely noticeable.

In a similar way small-scale space fluctuations may be ruled out.

This way "hydrodynamic" space and time variables (x_1, x_2, x_3, τ) are introduced whose each value is a representative of a space part and time interval on which the averaging has been conducted levelling off small-scale fluctuations. The introduction of such a coordinate system and time scale is called coarse-grained roughening of time and space. The above subvolumes are called physical elementary volumes.

If the values of macroscopic characteristics in different physical elementary volumes are different, the system is in a non-equilibrium state in which transfer processes take place.

A transition from some physical elementary volumes to another generally corresponds to an abrupt change of coordinates as well as of thermodynamic variables (the same is true for transitions from some moment of time to the next). If, however, the properties in question do not greatly alter, the coordinates and time may be considered as altering continuously, and instead of finite difference equations differential equations may be used.

Let us consider an extensive value $I(\tau)$ for a system with volume v limited by surface Ω.

Assume that

$$I(\tau) = \int f dV, \tag{273}$$

i.e., f is an intensive value which does not depend on the system as a whole and corresponds to volume density I. Then

$$f \equiv \frac{\delta I}{\delta V}, \tag{274}$$

where δ denotes a variation derivative.

It is wise to discriminate between two mechanisms of the I alteration in time [155]:

$$\frac{\partial I}{\partial \tau} = P[I] + \phi[I]\,. \tag{275}$$

The first term in the right part of this equation describes the production of the value I in a unit time in the volume V; it may be expressed as a volume integral

$$P[I] = \int \sigma[I]\, dV, \tag{276}$$

where $\sigma[I]$ denotes a source in a unit of time per unit of volume.

The second term of the right part is a flux of the value I trough the limiting surface Ω and may be expressed in the form of a surface integral

$$\phi[I] = \int j_n[I]\, d\Omega, \tag{277}$$

where $j_n[I]$ is the projection of the vector of the flux $\boldsymbol{j}[I]$ density on the internal normal to the surface Ω (later we shall omit the word "density" and use simply "flux").

Using formulae from Eqs. (275) to (277) we derive the so-called balance equation corresponding to the extensive variable I.

$$\frac{\partial I}{\partial \tau} = \int \sigma[I]\, dV + \int j_n[I]\, d\Omega\,. \tag{278}$$

Sometimes it is convenient to express Eq. (275) or Eq. (278) by symbols

$$dI = d_i I + d_e I\,, \tag{279}$$

where $d_i I$ corresponds to the source and $d_e I$, to the flux.

As Eq. (278) holds for any volume in the system, by the Green formula we can at once obtain the balance equation in the local form.

$$\frac{\partial f}{\partial \tau} = \sigma[I] - \operatorname{div} \boldsymbol{j}[I]\,. \tag{280}$$

In these symbols, any I value being retained, it follows that the source has disappeared:

$$\sigma[I] = 0\,. \tag{281}$$

For example, if $f = \rho$, where ρ is the mass density,

$$I = M; \quad \sigma[M] = 0\,. \tag{282}$$

If $\boldsymbol{v}$ is the velocity of the substance transfer, then

$$\boldsymbol{j}[M] = \rho \boldsymbol{v}\,. \tag{283}$$

From Eqs. (280), (282) and (283) the classical continuity equation follows:

$$\frac{\partial \rho}{\partial \tau} + \operatorname{div} \rho \boldsymbol{v} = 0\,. \tag{284}$$

Often it is useful to introduce the so-called substantional or hydrodynamic derivative

$$\frac{d}{d\tau} = \frac{\partial}{\partial \tau} + \sum_i v_i \frac{\partial}{\partial x_i} \tag{285}$$

or in the vector form

$$\frac{d}{d\tau} = \frac{\partial}{\partial\tau} + (v \cdot \text{grad}) \,. \tag{286}$$

Then with any intensive variable φ the continuity equation may be expressed (with respect to Eq. (284)) as:

$$\rho\frac{d\varphi}{d\tau} = \rho\frac{\partial\varphi}{\partial\tau} + \sum_i \rho v_i \frac{\partial\varphi}{\partial x_i} = \frac{\partial(\rho\varphi)}{\partial\tau} + \sum_i \frac{\partial}{\partial x_i}(\rho v_i \cdot \varphi) \,. \tag{287}$$

Now let us apply the formalism of the balance equations to the most important values characterizing the system [155], beginning with the conservation of mass.

Allow the system be a continuum consisting of components γ ($\gamma = 1, 2, \dots n$). The macroscopic velocity of the component γ will be denoted by v_γ. The mass concentration N_γ and partial density ρ_γ of component γ will be determined by relations (see Eq. (274))

$$N_\gamma = \frac{\delta m_\gamma}{\delta m} \qquad \sum_\gamma N_j = 1 \,. \tag{288}$$

$$\rho_\gamma = \rho N_\gamma \qquad \sum_\gamma \rho_\gamma = \rho \,. \tag{289}$$

The velocity of the mass centres v (which is also called the barycentrie velocity) and the diffusion flux Δ_γ of component γ are determined by equations

$$\rho v = \sum_\gamma \rho_\gamma v_\gamma \,, \tag{290}$$

$$\Delta_\gamma = v_\gamma - v \,. \tag{291}$$

Having multiplied Eq. (291) by ρ_γ and summarized by all γ we shall obtain

$$\sum_\gamma \rho_\gamma \Delta_\gamma = 0 \,. \tag{292}$$

Flux $\boldsymbol{j}[M] = \rho_\gamma v_\gamma$ in balance Eq. (280) for component γ may thus be broken up into a conductive (here diffusion) $\rho_\gamma \Delta_\gamma$ and convective $\rho_\gamma v$ fluxes.

Now, let component γ take part in a chemical reaction the stoichiometric equation for which is

$$0 = \sum_\gamma \nu_\gamma M_\gamma \,. \tag{293}$$

Stoichiometric coefficient ν_γ corresponding to the component with molar mass M_γ is positive if the component is produced in a chemical reaction, it is negative in the contrary case and zero if γ does not take part in the reaction.

The alteration of density ρ_γ in a unit of time due to the reaction, i.e., the amount of component γ transformed in a unit of time per unit of volume is expressed by the source

$$\sigma[M] = \nu_\gamma M_\gamma w\,, \tag{294}$$

where w is the rate of the chemical reaction in moles in a unit of time per unit of volume. If component γ takes part in r different reactions, then summing up the contributions of individual reactions we obtain

$$\sigma[M_\gamma] = \sum_\rho \nu_{\gamma\rho} M_\gamma w_\rho\,. \tag{295}$$

Balance Eq. (280) will be expressed as:

$$\sum_\rho \nu_{\gamma\rho} M_\gamma w_\rho = \frac{\partial \rho_\gamma}{\partial \tau} + \sum_i \frac{\partial}{\partial x_i}(\rho_\gamma \Delta_{\gamma_i} + \rho_\gamma v_i)\,. \tag{296}$$

Summarizing the equations for all components γ we shall arrive at the law of conservation of the full mass Eq. (282).

Without going into the details of the derivation (see [155]) we shall write down the balance equations for energy too. The full energy of a system U generally includes the kinetic (macroscopic) energy E_{kin}, the potential energy Ω_γ corresponding to external forces $\boldsymbol{F}_\gamma$ and the interal energy E, i.e.,

$$U = E_{\mathrm{kin}} + \sum_\gamma \Omega_\gamma + E\,. \tag{297}$$

The balance equation of the kinetic energy is expressed as:

$$\rho \sum_{i=1}^{3} F_i v_i + \sum_{i,j=1}^{3} P_{ij} \frac{\partial v_i}{\partial x_j} = \frac{\partial}{\partial \tau}\left(\rho \frac{v^2}{z}\right) + \sum_{j=1}^{3} \frac{\partial}{\partial x_j}\left[\sum_i P_{ij} v_i + \rho \frac{v^2}{2} v_j\right]. \tag{298}$$

A comparison with Eq. (280) shows the source to be set by the expression

$$\sigma[E_{\mathrm{kin}}] = \rho \sum_i F_i v_i + \sum_{i,j} P_{ij} \frac{\partial v_i}{\partial x_j}\,. \tag{299}$$

The first term is related to the work of external forces in a unit of time per unit of volume; the second is connected with pressure tensor P_{ij}.

The balance equation for potential energy Ω_γ is

$$-\rho_\gamma \sum_i F_{\gamma_i} v_{\gamma_i} = \frac{\partial}{\partial \tau}(\rho_\gamma \omega_\gamma) + \sum_i \frac{\partial}{\partial x_i}[\rho_\gamma \omega_\gamma v_{\gamma_i}] \tag{300}$$

where ω_γ denotes the potential energy of a unit of mass of component γ. The source of potential energy Ω_γ corresponds to the work in a unit of time per unit of volume performed by force $\boldsymbol{F}_\gamma$ which effects component γ:

$$\sigma[\Omega_\gamma] = -\rho_\gamma \sum_i F_{\gamma_i} v_{\gamma_i}. \tag{301}$$

Finally, in the balance equation for the internal energy the source must be chosen so that the law of conservation of the full energy should be satisfied:

$$\sigma[U] = \sigma[E] + \sigma[E_{\text{kin}}] + \sum_\gamma \sigma[\Omega_\gamma] = 0 \tag{302}$$

i.e.,

$$\sigma[E] = -\sigma[E_{\text{kin}}] - \sum_\gamma \sigma[\Omega_\gamma] = \sum_\gamma \sum_i F_{\gamma_i} \rho_\gamma \Delta_{\gamma_i} - \sum_{i,j} P_{ij} \frac{\partial v_i}{\partial x_j}. \tag{303}$$

This gives the following balance equation:

$$\sum_\gamma \rho_\gamma \sum_i F_{\gamma_i} \Delta_{\gamma_i} - \sum_{i,j} P_{ij} \frac{\partial v_i}{\partial x_j} = \frac{\partial}{\partial \tau}(\rho e) + \sum_j \frac{\partial}{\partial x_j}[w_j + \rho e v_j], \tag{304}$$

where e, density of the internal energy per unit of mass and w_j, j-th component of the thermal flux. It may be shown that in case the external force is the same for all the components and P_{ij} is replaced by the equilibrium pressure P, Eq. (304) is reduced to

$$-P \frac{d}{d\tau}\left(\frac{1}{\rho}\right) = \frac{de}{d\tau} + \frac{1}{\rho} \operatorname{div} \boldsymbol{W}. \tag{305}$$

Supposing that the pressure is homogeneous and integrating the latter equation throughout the system we find

$$dQ = dE + pdV,$$

where dQ, amount of heat received by the system in time $d\tau$. This expression corresponds to the common formulation of the first law of thermodynamics given in Ch. 1.

2 The Second Law of Thermodynamics, Local Equilibrium and the Entropy Balance Equation

As stated earlier in Ch.1, entropy is an extensive value. If a system consists of several parts, the full entropy is the sum of the entropies of these parts. It is also postulated in accordance with Eq. (279) that the alteration of entropy dS of any macroscopic part of a system breaks up into the production of entropy d_iS resulting from alterations within this part of a system and the flux of entropy d_eS resulting from the interaction of this part with all other parts of a system:

$$dS = d_iS + d_eS . \tag{306}$$

In addition the production of entropy d_iS is postulated to never be negative

$$d_iS \geqslant 0 . \tag{307}$$

The notion "entropy" is introduced with the purpose of discriminating between two types of processes: reversible and irreversible. The production of entropy d_iS in any macroscopic part of a system disappears only if reversible processes take place, and is positive in all other cases. In an isolated system as a whole the flux of entropy is zero, and the given "local" formulation of the second law comes to a "global" formulation of the classical thermodynamics, i.e., from Eqs. (306) and (307) (for the entire system) at $d_eS = 0$ we have

$$dS \geqslant 0 \quad \text{(for the entire system)}.$$

In essence, the local formulation of the second law of thermodynamics together with balance equations (V.1) and the assumption of the local equilibrium is the foundation of the non-equilibrium thermodynamics.

In order to clarify local equilibrium let us consider an example given in [174]. Assume that we have two similar size boxes in one of which are black balls and in the other, white. Let us exchange balls from both boxes one by one taking the balls from the upper layer, doing this within equal intervals of time without looking into the boxes. Evidently, in this case no levelling off of the number of balls of both colours in the boxes will occur. To level off, the boxes must be shaken up between every two exchanges, i.e., each box must be brought to an "equilibrium" state corresponding to a random distribution of balls of different colours. As this takes place, it is, obviously, necessary that the time for the equilibrium to establish in each box at shaking up must be much less than the interval between the two exchanges.

The phenomenological non-equilibrium thermodynamics deals only with systems in each small volume element of which is a state of local equilibrium for which local entropy is the same function of local macroscopic variables as for an equilibrium system.

Thus from the above instance with balls, dissipative processes – shaking up – must be substantial enough to avoid considerable deviations from an equilibrium state. If the dissipation mechanism may be described with the help of "collision" processes, then a sufficient number of collisions is necessary to make up for the effect of gradients applied or of chemical affinity. This is, evidently, the case with gases of medium density and in particular with liquids and solids, however, highly rarefied gases do not meet this requirement.

Choosing as thermodynamic variables describing the state of a small element of a system the density of the internal energy per unit of mass e, the specific volume $V_\rho = 1/\rho$ (or pressure p) and mass concentrations N_γ, let us express the assumption of the local equilibrium by equations

$$S = S(e, V_\rho, N_\gamma)\,, \tag{308}$$

$$\delta g = \delta e - TdS + p\delta V - \sum_\gamma \mu_\gamma \delta N_\gamma = 0\,, \tag{309}$$

where g is the Gibbs thermodynamic potential. Using the equations of the equilibrium thermodynamics, equations of the balance of mass and of internal energy (V.1) and introducing the affinity of the ρ-th chemical reaction by the formula

$$A_\rho = -\sum_\gamma \nu_{\gamma\rho} M_\gamma \mu_\gamma \quad (\rho = 1, 2, \dots \tau) \tag{310}$$

we arrive at the equation of the entropy balance [155]:

$$\begin{aligned}
&\sum_j w_j \frac{\partial}{\partial x_j}\left(\frac{1}{T}\right) - \sum_\gamma \sum_j (\rho_\gamma \Delta_{\gamma_j}) \left[\frac{\partial}{\partial x_j}\left(\frac{\mu_\gamma}{T}\right) - \frac{F_{\gamma_j}}{T}\right] \\
&\qquad - \frac{1}{T} \sum_{i,j} P_{ij} \frac{\partial v_i}{\partial x_j} + \sum_\rho w_\rho A_\rho T^{-1} \\
&\qquad = \frac{\partial}{\partial \tau}(\rho S) + \sum_j \frac{\partial}{\partial x_j}\left[\frac{w_j}{T} - \sum_\gamma (\rho_\gamma \Delta_{\gamma_j}) \frac{\mu_\gamma}{T} + \rho_\delta v_j\right].
\end{aligned} \tag{311}$$

This equation has the form (280), while

$$\begin{aligned}
\sigma[S] = {} & \sum_j w_j \frac{\partial}{\partial x_j}\left(\frac{1}{T}\right) - \sum_\gamma \sum_j (\rho_\gamma \Delta_{\gamma_j}) \left[\frac{\partial}{\partial x_j}\left(\frac{\mu_\gamma}{T}\right) - \frac{F_{\gamma_i}}{T}\right] \\
& - \frac{1}{T} \sum_{i,j} P_{ij} \frac{\partial v_i}{\partial x_j} + \frac{1}{T} \sum_\rho w_\rho A_\rho
\end{aligned} \tag{312}$$

and the flux

$$\boldsymbol{L} = \frac{\boldsymbol{w}}{T} - \sum_\gamma (\rho_\gamma \boldsymbol{\Delta}_\gamma) \left(\frac{\mu_\gamma}{T}\right) + \rho S \boldsymbol{v}\,. \tag{313}$$

The division into the flux and source was performed so that the production of entropy should disappear in the thermodynamic equilibrium. Such a division allows for a choice. The choice of a source in Eq. (312) may also be based on the fact that by using empirical relations (e.g., between $\boldsymbol{w}$ and grad T, Fourier's law) it may be demonstrated [174] that $\sigma[S]$ is always non-negative, i.e., corresponds to its physical sense.

Analyzing the structure of Eq. (312) it can be noticed that each term has a bilinear form containing two types of multipliers: flux or rate of an irreversible process and gradient or chemical affinity. The latter values are considered as "generalized" thermodynamic forces. Denoting fluxes or rates as J_α we shall express the local production of entropy (312) as:

$$\sigma[S] = \sum_\alpha J_\alpha X_\alpha \geqslant 0 . \tag{314}$$

In equilibrium both forces and fluxes disappear.

3 The Linear Thermodynamics of Irreversible Processes

In irreversible processes near the equilibrium state the relations between fluxes J_α and generalized forces X_α may be considered as linear. The empirical laws such as Fick's law for diffusion or Fourier's law for heat conductivity are such linear relations (phenomenological relations). Let us express them as:

$$J_\alpha = \sum_\beta L_{\alpha\beta} X_\beta , \quad \alpha, \beta = 1, 2, \dots n . \tag{315}$$

$L_{\alpha\beta}$ are called phenomenological coefficients; coefficients in which $\alpha = \beta$, i.e., $L_{\alpha\alpha}$ will be called self coefficients and the remaining reciprocity coefficients; they describe the interference between irreversible processes α and β. Even on the basis of linear relations the thermodynamic method offers important information about coefficients $L_{\alpha\beta}$ without appealing to any kinetic model whatsoever. We shall exemplify this by an interaction of two irreversible processes for which phenomenological relations (315) are expressed as:

$$\left.\begin{aligned} J_1 &= L_{11} X_1 + L_{12} X_2 \\ J_2 &= L_{21} X_1 + L_{22} X_2 \end{aligned}\right\} . \tag{316}$$

If these processes are, e.g., heat conductivity and diffusion, coefficient L_{21} corresponds to thermal diffusion meaning that the concentration gradient arises in a homogeneous mixture in response to the temperature gradient. Substituting Eq. (316) in the expression for the production of entropy (314) we derive

$$\sigma[S] = L_{11}X_1^2 + (L_{12}+L_{21})X_1X_2 + L_{22}X_2^2 . \tag{317}$$

Such a quadratic form must be positive for all (both positive and negative) values of X_1, X_2, excluding the case of $X_1 = X_2 = 0$ when the production of entropy disappears.

According to elementary algebra this is satisfied if

$$L_{11} > 0 \quad L_{22} > 0 \quad (L_{12}+L_{21})^2 > 4L_{11}L_{22} . \tag{318}$$

Thus, phenomenological self coefficients L_{11}, L_{22} are positive. This is in agreement with experimental data: coefficients such as heat conductivity or electric conductivity are always positive, whereas the sign of the coefficient of thermal diffusion is undetermined.

The most important (from the practical viewpoint) property of phenomenological coefficients is determined by Onsager's reciprocity relations (610)

$$L_{\alpha\beta} = L_{\beta\alpha} (\alpha \neq \beta) \tag{319}$$

which is deduced on the basis of the microscopic reversibility. The latter implies the invariance of all the mechanical equation of motion in respect to the substitution of τ for $-\tau$, i.e., to the change of the direction of the time course.

Onsager's reciprocity relations express the property that, if flux J_α corresponding to irreversible process α is effected by force X_β of irreversible process β, then flux J_β is effected by force X_α by means of the same interference coefficient $L_{\alpha\beta}$.

Using reciprocity relations it is possible to show (see [155]) that within the limits of the applicability of the linear thermodynamics of irreversible processes the production of entropy reaches the minimum value in a stationary state compatible with the set conditions (which are determined by the character of the problem).

It is worth noting that the requirement of linearity (315) is much stronger than the assumption of the local equilibrium (308) and (309). There are systems for which this assumption is valid, the fluxes are, however, expressed by generalized forces non-linearly. If e.g., the rate of a chemical reaction is so small that the equilibrium distributions of each component are not essentially broken, a macroscopic description in terms of mean concentration is allowable and the Gibbs formula (309) may be used; the rate of reactions, however, depends on the affinity non-linearly. The range of linearity is determined by condition (see [492])

$$\frac{|A|}{RT} \ll 1 . \tag{320}$$

It should be noted that sometimes a reaction may be devided into a certain number of elementary reactions for each of which

$$\frac{|A_\rho|}{RT} \ll 1 .$$

We shall then still remain within the range of the applicability of linear laws, even if for the summary reaction inequality (320) does not hold.

As for the non-linear area, the evolution of a system is determined by the alteration of the production of entropy in time. It would be convenient to present the time alteration of the production of entropy as a sum

$$\frac{dP}{d\tau} = \frac{d_x P}{d\tau} + \frac{d_J P}{d\tau} , \tag{321}$$

where the first term corresponds to the alteration of generalized forces and the second one to that of generalized fluxes. Then it is possible to demonstrate that always (see [155])

$$\frac{d_x P}{d\tau} \leqslant 0 . \tag{322}$$

That is, the alteration of generalized forces takes place always so that the value of the production of entropy should decrease.

4 The Application of the Thermodynamics of Irreversible Processes to the Study of Silicate Systems

The application of the non-equilibrium thermodynamics to the study of actual physico-chemical processes proved to be quite fruitful. The methods of the thermodynamics of irreversible processes may be applied to the study of silicates as well. However, as in silicate systems numerous chemical reactions and transfer processes are intricately combined, the limiting conditions for the latter often being unknown (thus, e.g., at heat treatment), this field has to date been insufficiently researched, although partial investigations have been conducted. Thus, starting from the regularities of the linear thermodynamics of irreversible processes, Ralko [498] first carried out an experimental investigation of the heat and mass exchange in kaolinite firing.

The technique of the test was as follows. Hot junctions of three differential thermocouples (see in the bottom right corner of Fig. 93) were put into a cylindric

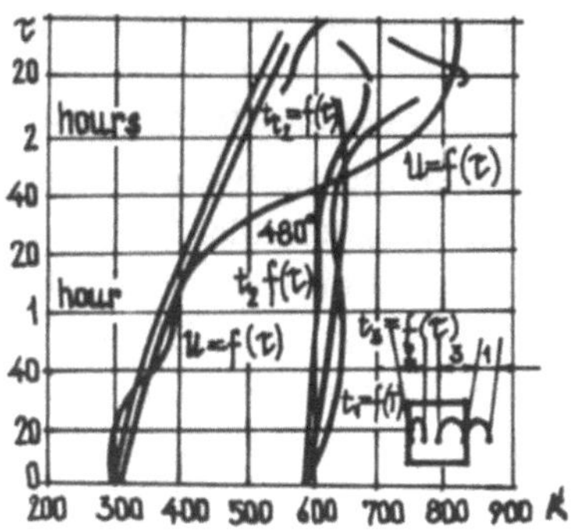

Fig. 93. A thermogram of a kaolin cylinder firing

sample being heated for measuring the difference of temperatures: a: of the environment and the surface of the cylinder 1; b: between the surface of the cylinder and at the bottom being half of its radius 2; c: between the surface of the cylinder and its centre 3. The weight loss was measured in the second kiln.

The location of thermocouples permits fixing the "motion" first of the endothermic and then of the exothermic (breaks of the curves) reaction.

These curves entirely determine the heat and mass exchange characteristics of the body: a, coefficient of temperature conductivity; a^*, efficient coefficient of the substance transfer; Bi and Bi* the Biot criteria of heat and substance transfer respectively; α, external coefficient of heat exchange; λ, coefficient of heat conductivity (Table 135).

Of great interest is the application of the methods of the non-equilibrium thermodynamics to the hardening of binders. The structure formation at hydration is a result of the combined action of transfer processes (in the field of surface forces) as well as of a number of concurrent chemical reactions. The application of the main evolution criterion

$$\sigma[S] \geqslant 0$$

Table 135. Thermophysical characteristics of kaolin

Characteristics	Temperature [K]									
	373	473	573	673	773	873	973	1073	1173	1273
$a \cdot 10^5$ [m²/h]	108	108	101	–	101	108	108	116	–	122
a^*	–	138	–	–	125	110	–	–	125	–
Bi	–	2.7	2.7	2.7	3.75	4.65	5	5.75	5.75	6.84
Bi*	–	1.5	–	–	1.8	–	–	2.5	–	–
α [kcal/m² h deg]	27	40	50	63	90	120	154	170	190	200
$\lambda \cdot 10^3$ [kcal/m² h deg]	292	391	483	580	642	680	686	702	685	650

to this system where $\sigma[S]$ is given by Eq. (312) takes into account the presence of all the above processes.

In this connection the attempt of Cimmermanis and coworkers [646] to consider the structure formation by introducing an additional thermodynamic function, the amplitude of the structure formation, seems to be unsuitable. Indeed, if a system is closed, the amplitude of the structure formation (determined by porosity) is a function of the other thermodynamic parameters which per se characterize the state. Therefore the amplitude of the structure formation does not enter phase Eq. (308) as a new independent variable. Introducing it implies, essentially, openness of a system, i.e., it is a measure of the external effect on the system (moisture exchange).

In conclusion it may be said that a broad investigation of silicate systems by methods of the non-equilibrium thermodynamics is a matter of future. Our task in this chapter was to stimulate such investigations.

References

1. Akopyan AA (1963) The chemical thermodynamics. Vys Shk Moscow
2. Alekseyev AG (1960) Investigation of glass crystallization in the system Na_2O-SiO_2. In: Proceedings of the IIIrd Conference on the Vitreous State, Leningrad, p 194
3. Althaus E (1966) Phase border andalusit/sillimanit, Naturwiss 53:129
4. Appen AA (1970) The chemistry of glass. Khimiya, Leningrad
5. Appen AA, Glushkova VB, Kayalova SS (1965) Calculation of the bond energy in oxides. Izv Akad Nauk SSSR Neorg Mater 1(4):576–583
6. Aramaki S, Roy R (1959) Revised equilibrium diagram for the system $Al_2O_3-SiO_2$. Nature 184(4686):631–632
7. Arlyuk BI (1966) Calculation of the fusion heats of silicates. Zh Prikl Khim Leningrad 39(5):1196–1198
8. Arlyuk BI (1966) On the thermodynamic analysis of the solid state reactions of binary compounds formation. Zh Prikl Khim Leningrad 39(2):454–455
9. Assarsson GO (1957, 1958) Hydrothermal reactions between calcium hydroxide and silica: the reactions between 120° and 160° and 180°–220°. J Phys Chem 61(4): 473–479, 62(2):223–228
10. Assarsson GO, Rydberg E (1956) Hydrothermal reactions between calcium hydroxide and amorphous silica. J Phys Chem 60(4):397–404
11. Aukrust E, Muan A (1963) The stabilities of $CoO \cdot Al_2O_3$, $CoO \cdot Cr_2O_3$ and $2CoO \cdot SiO_2$. J Am Ceram Soc 46(7):358
12. Avgustinik AI (1947) Physical chemistry of silicates. Goskhimizdat, Leningrad
13. Avgustinik AI, Mchedlov-Petrossyan OP (1947) The heat of formation and reaction kinetics of barium oxide monoaluminate after the reaction between barium sulphate and aluminium oxide. Zh Prikl Khim Leningrad 20:584, 1125
14. Avgustinik AI, Mchedlov-Petrossyan OP (1952) On the thermodynamics of mullite formation. Zh Prikl Khim Leningrad 25:216
15. Avgustinik AI (1975) Ceramics, 2nd edn. Stroyizdat, Leningrad
16. Babushkin VI (1968) Physico-chemical processes of concrete and reinforced concrete corrosion. Stroyizdat, Moscow
17. Babushkin VI (1960) Application of thermodynamics to the technology of silicate materials. In: Proceedings of KhIIT 39, Kharkov, p 97
18. Babushkina MI, Babushkin VI (1975) Thermodynamic investigation of acidproof compositions at high temperature and pressure. Zh Prikl Khim Leningrad 48(5): 1039–1044
19. Babushkin VI, Kolomatsky AS (1979) Approximate method of evaluation of thermodynamic data for calcium hydroferrite. In: Proceedings of MISI and BTISM, Moscow, pp 87–92
20. Babushkin VI, Matveyev GM, Mchedlov-Petrossyan OP (1972) Thermodynamics of silicates. Gosstroyizdat, Moscow
21. Babushkin VI, Mchedlov-Petrossyan OP (1959) The thermodynamic study of solid phase reactions in the system calcium oxide – aluminium oxide. Zh Prikl Khim Leningrad 32(1):46–50

22. Babushkin VI, Mchedlov-Petrossyan OP (1962) Silicate water-resistant products. Gosstroyizdat Ukr SSR, Kiev
23. Babushkin VI, Mchedlov-Petrossyan OP (1966) About the connection of thermodynamics and kinetics of some mineral-building reaction. In: Proceedings of the VIIth Conference on the Experimental and Technical Mineralogy and Petrography. Nauka, Moscow, pp 17–23
24. Babushkin VI, Mokritskaya LP, Novikova SP, Zinov VG (1974) The investigation of the physico-chemical processes during the hydration and hardening of expansive cements. In: Proceedings of the VIth International Congress on the Chemistry of Cement, vol 3. Stroyizdat, Moscow, pp 187–189
25. Babushkin VI, Mokritskaya LP, Mchedlov-Petrossyan OP (1962) The study of clinker minerals hydrolysis by pH-metry methods. In: Proceedings of KhIIT 54, Kharkov, pp 13–18
26. Babushkin VI, Mchedlov-Petrossyan OP (1959) On the thermodynamics of the reactions in the systems $Ca(OH)_2-SiO_2-H_2O$, β-C_2S-H_2O and C_3S-H_2O in hydrothermal conditions. Silikattechnik 10(12):605–608
27. Babushkin VI, Novikova SP, Zinov WG (1973) Electroosmothic effect of the volume changing in the cement stone and concrete. In: Internationale Baustoff- und Silikattagung, 5 Ibausil, sect 2, part 2. Hochschule für Bauwesen und Architektur, Weimar, pp 323–334
28. Babushkin VI, Novikova SP (1973) Colloid-chemical phenomena in cement stone and concrete volume changing. In: Proceedings of VNII VODGEO 'Technology, structure and properties of hydraulic concretes for agricultural building', Kharkov, pp 133–144
29. Babushkin VI, Vandalovskaya LA (1973) The thermodynamics of gypsum and ettringite in hardened concrete. Zh Prikl Khim Leningrad 46(2):246–251
30. Babushkin VI, Zenchenko JN (1967) Thermodynamic reactions of calcium hydrosulphoaluminate in hardened concrete. Zh Prikl Khim Leningrad 40(10):2236–2242
31. Babushkin VI, Zinov VG (1978) A study of the effect of the salt physical corrosion on the moisture state of cement stone. In: 'Gidratatsiya i tverdenive tsementov', vol 3. Ural NIIS Stromproyekt, Chelyabinsk, pp 27–32
32. Barany R, King EG, Todd SS (1957) Heats of formation of crystalline silicates of strontium and barium. J Am Chem Soc 79(14):3639–3641
33. Barenboym AM, Taliyeva TM, Ginzburg DB et al. (1964) Heat calculations of kilns and dryers of the silicate industry. Gosstroyizdat, Moscow
34. Bart T (1956) Theoretical petrology. Inostrannaya Liter, Moscow
35. Barta R (1961) Chemistry and technology of cement. Naklad Česk Akad Ved Praha
36. Barta R, Bartuska M (1957) Technology of silicates. Sb Vys Pr Praha, IV:146–183
37. Barta R, Satava V (1954) Theoretical foundations of glasses and ceramics, vol II. SPN, Praha
38. Batler J (1973) Ionic equilibria. Khimiya, Leningrad
39. Bell PM (1963) Aluminium silicate system: experimental determination of the triple point. Science 139(3559):1055–1056
40. Belyankin DS, Lapin VV, Toropov NA (1954) Physico-chemical systems of silicate technology. Nauka, Moscow
41. Benz, R, Schmalzried NZ (1961) On measuring the free enthalpy formation of the solid phase in the system $PbO-SiO_2$ by means of galvanic chains. Phys Chem Neue Folge 29(1–2):77–82
42. Berezky A (1962) About the durability of concrete. Baustoffindustrie 5:300–301
43. Berezhnoy AS (1970) Multicomponent oxide systems. Naukova Dumka, Kiev
44. Berezhnoy AS (1958, 1962, 1963) Proceedings of VNIIO. Metallurgizdat 2:5, 6:5, 7:115–161, 173–203

45. Berkovich TM, Grachova OI, Kupreyeva NI (1960) On thermodynamic stability of calcium hydrosulphoaluminate. Dokl Akad Nauk SSSR 130(5):1085 – 1088
46. Berman HA, Newman ES (1960) Heat of calcium trisulphoaluminate formation at 25 °C. In: IVth International Symposium on the Chemistry of Cement. National Bureau of Standards, Washington, pp 247 – 257
47. Berman HA, Newman ES (1963) Heat of formation of calcium aluminatesulphate. J Res NBS 57A(1):1 – 13
48. Bernal J (1952) The structure of cement hydration compounds. In: The IIIrd International Symposium on the Chemistry of Cement. Cement and Concrete Association, London, pp 216 – 236
49. Bernal J (1960) The structure of liquids. Sci Am 8:125 – 134
50. Bliudzius GS, Kičas PV (1971) Solid phase reactions in the system $CaO - SiO_2$. Sb Trudow NII Teploisolatia, Wilnjus 5:228 – 249
51. Boekris JO, Mellors GM (1956) Electric conductance in liquid lead silicates and borates. J Phys Chem 60:9
52. Bonikson KR (1955, 1954, 1955) High temperature heat contents of aluminates of calcium and magnesium; of calcium and magnesium ferrites; of some titanates of aluminium, iron and zink. J Phys Chem 59(3):220 – 221; J Am Chem Soc 76(6):1480 – 82, 77(8):2152 – 2154
53. Botvinkin OK (1955) Physical chemistry of silicates. Promstroyizdat, Moscow
54. Botvinkin OK, Avramenko YI, Golba TY, Okhotin MV (1941) Kinetics of the glass solidification. Gizlegprom, Moscow
55. Botvinkin OK, Cherevkova YV (1961) The technique for determining the e.m.f. and structure of the melts in the system $PbO - SiO_2$. Bull Gos NII Stekla 4
56. Bowen NL (1943) Petrology and silicate technology. J Am Ceram Soc 26(9):285 – 301
57. Bowen NL, Tuttle OF (1949) The system $MgO - SiO_2 - H_2O$. Bull Geol Soc Am 60(3):439 – 460
58. Braga Reis MO (1981) Formation of expansive calcium sulphoaluminate. Cem Concr Res 11(4):541 – 547
59. Brandenberger E (1936) Intercrystalline structure and the chemistry of cement. Schweiz Arch Wiss Techn 2:45 – 58
60. Bratchikov SG, Yesin OA, Sryvalin IT (1959) Thermochemistry of melting lead silicates. Izv Vuzow (Metallurgy) (2):32 – 37
61. Bredig M (1954) Discussion. In: Proceedings of the IIIrd Symposium on the Chemistry of Cement. Cement and Concrete Association, London, p 82
62. Brindley GW, Nakahira M (1959) The kaolinite-mullite reaction series: I. A survey of outstanding problems. II. Metakaolin. III. High temperature phases. J Am Ceram Soc 42(7):1 – 14
63. Brintzinger H, Brintzinger W (1931) Investigation of molecular dispersed silicon and titan acids. Z Anorg Allg Chem 96:44
64. Britske EV, Kapustinsky AF et al. (1949) Thermic constants of inorganic substances. Akad Nauk SSSR, Moscow
65. Brodsky AI (1947) The calculations of thermodynamic functions. Metallurgizdat, Moscow
66. Brophy JH, Rose RM, Wulff J (1964) Thermodynamics of structure. Wiley, New York London Sydney
67. Brunauer S, Greenberg S (1960) The hydration of C_3S and β-C_2S at room temperature. In: IVth International Symposium on the Chemistry of Cement. National Bureau of Standards, Washington
68. Brunauer S, Hayes JC, Hoss WE (1954) The heat of hydration of C_3S and β-C_2S. J Phys Chem 58:279 – 285

69. Brunauer S, Kantro DL, Weise CM (1956) The Heat of Decomposition of C_3S into β-C_2S and calcium oxide. J Phys Chem 60:771 – 774
70. Brunauer S, Kantro DL, Weise CM (1959) The surface energy of tobermorite. Can J Chem 37:714 – 724
71. Buckle ER, Gard JA, Taylor HKW (1958) Tricalcium silicate hydrate. J Chem Soc 4:1351 – 1355
72. Buckle ER, Taylor HKW (1959) The hydration of tricalcium and β-dicalcium silicates in pastes under normal and steam curing conditions. J Appl Chem 9:163 – 172
73. Budnikov PP, Berezhnoy AS, Bulavin IA, Kalliga GP, Kukolev GV, Poluboyarinov DN (1962) Production of ceramics and refractories. Gosstroyizdat, Moscow, pp 338 – 341
74. Budnikov PP, Ved' EI, Bludov BF (1968) Thermodynamic analysis of decomposition of some magnesium hydrosilicates. Izv. Akad Nauk SSSR Sibirsk Otd (5):3 – 7
75. Budnikov PP, Ginstling AM (1961) Reactions in solid mixtures. Gosstroyizdat, Moscow
76. Budnikov PP, Ginstling AM (1965) Reactions in solid mixtures. Stroyizdat, Moscow
77. Budnikov PP, Kerbe FG, Matveyev GM (1967) A thermodynamic analysis of solid phase reactions of the formation of the noble spinel $MgAl_2O_4$. In: Vitreous systems and materials. Zinatne, Riga, p 67
78. Budnikov PP, Kravchenko IV (1959) Study of hydration processes of calcium monoaluminate. Kolloidny Zh 21(1):9 – 17
79. Budnikov PP, Mchedlov-Petrossyan OP (1960) Thermodynamics of kaolinite change at heating. Dokl Akad Nauk SSSR 134(2):349 – 350
80. Budnikov PP, Petrovykh NV (1957) The influence of the grain size, and temperature of hydrothermal treatment on silicate materials. In: Proceedings of MKhTI 24. Promstroyizdat, Moscow, pp 96 – 110
81. Budnikov PP, Cherepanov AM (1950) On the synthesis of phenakite. Dokl Akad Nauk SSSR 74(5):1110
82. Bulakh AG (1968) Thermodynamic methods in mineralogy. Nedra, Leningrad
83. Bulakh AG, Bulakh KG (1978) Physico-chemical properties of minerals and components of hydrothermal solutions. Nedra, Leningrad
84. Butt YM (1954) Interaction between calcium hydroxide and crystalline silica at normal temperatures. In: Proceedings of RosNIIMS 6. Stroyizdat, Moscow, pp 115 – 122
85. Butt YM, Grachyova OM, Rashkovich LN, Mayer AA, Kheyker DM (1959) Synthesis of individual phases in the system $CaO - SiO_2 - H_2O$. In: Proceedings of a High-school Conference on the Study of Autoclaved Materials. LISI, Leningrad, pp 274 – 280
86. Butt YM, Mayer AA, Varshal BG (1961) Stability of calcium hydrosulphoaluminate. Dokl Akad Nauk SSSR 136(2):398 – 400
87. Butt YM, Rashkovich LN (1965) Binder hardening at elevated temperatures. Gosstroyizdat, Moscow
88. Butt YM, Timashev VV (1967) Portland cement clinker. Stroyizdat, Moscow
89. Carlson ET (1964) Action of water on calcium aluminoferrite. J Res NBS-A-Phys Chem 68(5):453 – 463
90. Carlson RW, Forbrich LR (1938) Correlation of methods for measuring heat of hydration of cement. Ind Eng Chem Anal Ed 10:382 – 386
91. Carter PT, MacFarlane TG (1957) Thermodynamics of slag systems. J Iron Steel Inst 185(1):54 – 66
92. Cerchez MV, Marinescu (1969) Aspects of formation and crystallization of multicomponent compound in ceramics. Rev Constr Mater Constr 11:600 – 607
93. Chang L, Derge G (1948) Met Technol NY 10:194

94. Chao-Hsia C (1975) A method of estimation of standard free energies of formation of silicate minerals at 298, 15K. Am J Sci 275(7):801 – 817
95. Chernyavsky VL (1981) Late period of cement hydration. Izv Vuzov (Str i Arkh) 12:70 – 73
96. Chester J (1966) The theory of irreversible processes. Nauka, Moscow
97. Chipman I (1961) The free energy of silica. J Am Chem Soc 83(7):1762 – 1763
98. Christian JW (1965) The theory of transformations in metal and alloys. Pergamon, New York
99. Clark SP (1961) A redetermination of equilibrium relations between kyanite and silimanite. Am J Sci 259:641 – 650
100. Clark SP, Robertson EC, Birch F (1957) Experimental determination of kyanite-silimanite equilibrium relations at high temperatures and pressures. Am J Sci 255:628 – 640
101. Cochran CN, Foster LM (1962) The stability of silica. J Phys Chem 66(2):380
102. Coing-Boyat J (1962) Structure of the high-temperature form of anhydrous magnesium sulfate, β-$MgSO_4$. CR Acad Sci 255:1962 – 1964
103. Cooper AC, Kay DAR, Taylor J (1961) The free energy of formation of mullite. Trans Br Ceram Soc 60(2):124 – 134
104. Coughlin JP (1955) High temperature heat contents, heats of transition and heats of fusion of anhydrous sodium sulphate. J Am Chem Soc 77(4):868 – 870
105. Coughlin JP (1956) Heats of formation of crystalline $CaO \cdot Al_2O_3$, $2CaO \cdot Al_2O_3$ and $3CaO \cdot Al_2O_3$. J Am Chem Soc 78:5479 – 5482
106. Coulson CA (1951) Critical survey of the method of ionic homopolar resonance. Proc R Soc London 207(1088):63 – 73
107. Curtis CE, Sowman HG (1952) Investigation of the thermal dissociation, reassociation and synthesis of zircon. J Am Ceram Soc 36:190 – 198
108. Danielson A (1950) The equilibrium of calcite-wollastonite. Geochim Cosmochim Acta 1:55 – 69
109. Darken LS, Gurri RV (1953) Physical chemistry of metals. McGraw-Hill, New York London
110. D'Ans J, Eick H (1953) The system $CaO \cdot Al_2O_3 - CaSO_4 - H_2O$ at 20°C. Zement-Kalk-Gips 6(9):302 – 311
111. Davis RE, Carlson RW et al. (1934) Cement investigation of Boulder Dam with results up to the age of one year. J ACI 5:485 – 497
112. Dayal RR, Johnson RE, Muan A (1967) Stability of mullite as divided from equilibria in the system $CaO - Al_2O_3 - SiO_2$. J Am Ceram Soe 51(10):527
113. Dear SP (1957) X-ray diffraction data for silicates, aluminates and aluminosilicates of strontium. Bull Va Polytech Inst 50, N6; Eng Expt Sta Ser No 117:16
114. Denbig K (1954) Thermodynamics of stationary irreversible processes. Inostran Literat, Moscow
115. De Groot SR (1956) Thermodynamics of irreversible processes. Gostekhizdat, Moscow
116. De Groot SR, Mazur P (1964) Non-equilibrium thermodynamics. Mir, Moscow
117. Dietzel A (1935) Ellucidating the problem of enamaling iron tin. Sprechsaal 68:3 – 6, 20 – 23, 34 – 36, 53 – 56, 67 – 69, 84 – 85
118. Dodge BF (1950) Chemical thermodynamics applied to chemical processes and chemical technology. Inostran Literat, Moscow
119. Dreving VP (1954) The phase rule. Izd Mosovskogo Univers
120. Drozin NP (1952) On the calculation of the enthropy of inorganic compounds. Z Prikl Khim Moscow 25:1109
121. D'yakonov GK (1956) The problems of the theory of similarity in the physicochemical processes. Izd Akad Nauk SSSR, Moscow

122. D'yachenko SS, Materova YA (1977) Determination of the activity of Ca^{2+} in the cement suspension by means of ionselective electrode. Z Prikl Khim Leningrad 6:1269–1270
123. Dyckerhoff W (1950) On the mineral formation process in mixtures of CaO, SiO_2, Al_2O_3, 2nd edn. Bauverlag, Wiesbaden
124. Eitel W (1952) Thermochemical methods in silicate investigation. Rutgers, New Brunswick, New Jersey
125. Eitel W (1952) The physical chemistry of the silicates. The University Press, Chicago, Illinois
126. Eitel W (1957) Recent investigations of the system Lime-alumina-calcium sulphate-water and its importance in building research problems. J Am Concr Inst 28:679–698
127. Eitel W (1966) Silicate science, vol V. Academic Press, London New York, p 263
128. Eitel W, Schwiete HE (1934) Equilibria on colloidal boundary phases in aqueous silicate systems. Naturwissenschaften 22:403–406
129. Elliott IF (1955) Activities of the iron oxide-silica-lime system. J Metals 7, AIME Trans 203:485–488
130. Evans MG, Polanyi M (1938) Enertia and driving force of chemical reactions. Trans Faraday Soc 34:11–24, 203:485–488
131. Fayf WS, Terner F, Verhoogen JP (1962) Metamorphic reactions and metamorphic facies. Inostran Literat, Moscow
132. Fersman AY (1958) Lattice energy and the energy of geochemical processes. Selected works, vol IV. Akad Nauk SSSR, Moscow, p 105
133. Fierens P, Verhoogen JP (1976) Hydration of tricalcium silicate by water vapour. Cem Concr Res 6(1):103–111
134. Filipovich VN (1965) Structural transformations in glasses at elevated temperatures. Nauka, Moscow, pp 15–29
135. Findley A (1932) Phase rule and its application. Gostekhizdat, Moscow
136. Flint EP, McMurdie HF, Wells LS (1938) Formation of hydrated calcium silicates at elevated temperatures and pressures. J Res NBS 21(5):617
137. Flint EP, Wells LS (1939) Analogy of hydrated calcium aluminate to hydrated calcium aluminate sulphate. J Res NBS 21(6):751
138. Flörke O (1967) The modification of SiO_2. Fortschr Mineralogie 44:181–230
139. Flood H, Knapp WI (1957) Stability of the aluminium silicates. J Am Ceram Soc 40(6):206–208
140. Frenkel' YI (1945) The kinetic theory of liquids. Izd Akad Nauk SSSR, Moscow
141. Freund F (1960) The interpretation of the exothermal reaction of kaolinate. Ber Dtsch Keram Ges 37:209–218
142. Fridrikhsberg YA (1974) A course in colloid chemistry. Khimiya, Leningrad
143. Funk H, Schreppel B, Thilo E (1960) Chemical investigation of silicates. XXIV. About the primary hydration products of β-C_2S... Z Anorg Allg Chem 304(1–2):12
144. Gard JA, Howison JW, Taylor HFW (1959) Synthetic compounds related to tobermorite: an electron-microscope, X-ray, and dehydration study. Mag Concr Res 11(33):151–158
145. Garrels RM (1960) Some values of free energy. In: 'Thermodynamics of geochemical processes'. Inostran Literat, Moscow, pp 122–136
146. Garrels RM (1962) Mineral equilibria. Inostran Literat, Moscow
147. Geller RF, Creamer AS, Bunting EN (1934) System PbO – SiO_2. J Res Nat Bur Stand B2
148. Gerasimenko P, Speit G (1950) Ionic theory of slag-metal equilibrium. J Iron Steel Inst 166:169

149. Gerasimov YI, Krestovnikov AN, Shakhov AS (1960) Chemical thermodynamics in non-ferrous metallurgy, part I. Metallurgizdat, Moscow
150. Gerasimov YI, Dreving VP, Yeryomin YN, Kiselyov AV, Lebedev VP, Penchenkov GM, Shlygin AI (1963) A course in physical chemistry, vol 1. Goskhimisdat, Moscow
151. Gibbs JW (1950) Thermodynamic papers, Gosizd tekhn-teor literat, Moscow Leningrad
152. Ginnings DC (1947) Heat contents of Al_2O_3. J Res NBS 38:593
153. Gittermann AK, Bobrov BS, Malinin JuS (1969) Hydration kinetics of anhydrous calcium sulfoaluminate. In: Uralniistromprojekt Chelyabinsk, p 22
154. Ginzburg II (1963) Questions of reaction energetics ... In: 'Erosion crust'. 5. Izd Akad Nauk SSSR, pp 87 – 119
155. Glensdorf P, Prigozhin I (1973) A thermodynamic theory of structure, stability and fluctuations. Mir, Moscow
156. Glushkova VB (1957) Formation conditions of barium silicates IV Values evaluation of thermodynamic functions. Zh Neorg Khim 2(10):2436 – 2449
157. Glushkova VB (1963) Thermodynamic calculations of solid phase reactions. In: 'Silicates and oxides in the high temperature chemistry'. Nauka, Moscow, p 308 – 321
158. Golutvin YM (1956) Formation heats of titanium silicate. Zh Phys Khim Moscow 30:2251
159. Gonikberg MG (1953) Chemical equilibrium and rate of reactions at high pressures. Izd Akad Nauk SSSR, Moscow
160. Good WD (1962) The heat of formation of silica. J Phys Chem 66:381
161. Gorshkov VS, Timashev VV (1963) Methods of the physico-chemical analysis of binders. Vyssh Shkola, Moscow
162. Govorov IN (1977) Thermodynamics of ionomineral equilibria and mineralogeny of hydrothermal deposites. Nauka, Moscow
163. Grebenshchikov RG (1963) Thermal investigation of silicates and aluminates. In: 'Silicates and oxides in the high temperature chemistry'. Nauka, Moscow, pp 290 – 302
164. Grebenshchikov RG (1964) The energetics of the crystalline lattice. Zh Neorg Khim 9:1038 – 1048
165. Grebenshchikov RG (1968) On the interconnection between some thermodynamic properties of silicates and fluorborates. Zh Neorg Khim 13(10):2627 – 2632
166. Green KT (1960) IVth International Symposium on the Chemistry of Cement, vol 1. National Bureau Standards, Washington, pp 359 – 385
167. Greenberg SA (1957) Thermodynamic functions for the solution of silica in water. J Phys Chem 61(2):197
168. Greenberg SA, Copeland LE (1960) The thermodynamic functions for the solution of calcium hydroxide. J Phys Chem 64(8):1057 – 1059
169. Greig JW (1926) Formation of mullite from kyanite, andalusite and silimanite. Am J Sci 11(61):22
170. Gruner IW (1930) Hydrothermal oxidation and leaching experiments; their bearing on the origin of Lake Superior hematit-limonite ores. J Econ Geol 25(7):697 – 719
171. Guinier A, Regourd M (1968) Structure of portland cement minerals. In: Principal paper on the Vth International Symposium on the Chemistry of Cement, vol 1. Cement Association of Japan, Tokyo, pp 1 – 32
172. Gulinova LG, Kornilovich YY, Skatynsky VI (1958) The production of autoclaved building materials. Gosstroyizdat Ukr, Kiev
173. Gurmai M (1964) A thermodynamic study of reactions taking plase in glass-making during silicate formation. Építoanyag 6:222 – 228
174. Gurov AP (1978) Phenomenological thermodynamics of irreversible processes. Nauka, Moscow
175. Gutzov I (1965) Collection 'Vitreous state'. Nauka, Moscow, pp 158 – 176

176. Gutzov I, Konstantinov I, Kaishev R (1969) Izvest Akad Nauk Bulg Otd Chim Nauk 2:249 – 260
177. Gutzov I (1962) On the zeropoint entropy of the glassy state. Z Phys Chem 221(3/4):158 – 176
178. Gutzov I (1971) The temperature dependence of the vapour pressure and solubility of glass-forming melts. Amorphous Mat, pp 159 – 170
179. Gutzov I (1972) On the steam pressure and solubility of overcooled melts and glasses. Z Phys Chem Neue Folge 81:195 – 212
180. Gutzov I, Konstantinov I, Kaishev R (1972) Thermodynamics of crystallization from undercooled melts and glasses: model statistical treatment. Izvest Akad Nauk Bulg Otd Chim Nauk 5(3):433 – 448
181. Haggerty JS, Cooper AK, Heasley JH (1968) Heat capacity of three inorganic glasses and supercold liquids. Phys Chem Glasses 9(12):47 – 51
182. Hala E, Reisner A (1960) Physical chemistry I. NCSAV, Praha
183. Handbook of chemistry and physics (1956) 37th ed. Cleveland, Ohio, Chem Rubber Publ, p 3206
184. Handbook of chemistry, vol 1, 3rd edn. (1962) Khimiya, Moscow
185. Handbook of chemistry, vol 3 (1964) Khimiya, Moscow Leningrad
186. Harker RI (1964) Dehydration series in the system $CaSiO_3 - SiO_2 - H_2O$. J Am Ceram Soc 47(10):521 – 529
187. Harker RI, Tuttle OF (1955) The thermal dissociation of calcite, dolomite and magnesite. Am J Sci 253:209 – 224
188. Harman RW (1927) Aqueous solutions of sodium silicates. VII Silicate ions. Electrometric titrations. Diffusion. Colorimetric estimation. J Phys Chem 31:616 – 625
189. Harrington RV, Hatchins JR, Sherman JD (1962) Advances in glass technology. Plenum, New York, p 65
190. Hartmann H, Brand H (1953) On the study of the mean heat capacities of some technically important types of glass I. Glastech Ber 26:29 – 33
191. Hartmann H, Brand H (1954) On the study of the mean heat capacities of some technically important types of glass II. Glastech Ber 27(N3):12 – 15
192. Hartmann H, Kiessling K (1957) On the study of the mean heat capacities of some technically important types of glass III. Glastech Ber 30:186 – 188
193. Haul RAW (1954) In: Proceedings of the International Symposium on the Reactivity of Solids. Elsevier, Göteborg, p 431
194. Haveman R (1957) Introduction to the chemical thermodynamics. Verlag der Wissenschaften, Berlin
195. Hedvall IA (1952) Introduction to the chemistry of solids. Vieweg, Braunschweig
196. Heller L (1954) In: Proceedings of the IIIrd Symposium on the Chemistry of Cement. Cement and Concrete Association, London, p 237
197. Heller L, Taylor HFW (1952) Hydrated calcium silicates. III. Hydrothermal reactions of lime: silica molar ration 3:2. J Chem Soc 2535:1018 – 1020
198. Hench LL, Frieman SW, Kinser DL (1971) The early stages of crystallisation in a $Li_2O - 2SiO_2$ glass. Phys Chem glasses 12(2):58 – 63
199. Henderson D, Taylor J (1966) Thermodynamic properties in the $CaO - MgO - SiO_2$ and $MgO - Al_2O_3 - SiO_2$ system. J Iron Steel Inst 1:39 – 43
200. Henning F (1955) Temperature measuring, 2nd edn. Barth, Leipzig
201. Hildebrand JH, Foster AH, Beebe CW (1919) Vapor pressures of Cd, Pb, and Sn amalgams. J Am Chem Soc 39:2293
202. Hinz W (1970) Silikate, vol 2. Verlag für Bauwesen, Berlin
203. Hoff van't JH (1887, 1890) About solid solutions. Z Phys Chem 1:165; 5:325
204. Hoffmann H (1936) Calcium ferrite hydrates. Zement 25:130 – 132, 675 – 680, 693 – 698, 711 – 716

205. Hoffman JD (1958) Thermodynamic driving force in nucleation and growth processes. J Chem Phys 29(5):1192–1193
206. Holm JL, Kleppa OJ (1966) The thermodynamic properties of the aluminium silicates. J Phys Chem 70(5):1690
207. Holster WI (1954) Fugasity of water at high temperature and pressure. J Phys Chem 58(4):316–317
208. Houtepen SDM, Steyn GN (1976) In: Proceedings of the VI International Congress on the Chemistry of Cement, vol 2, book 1. Stroiizdat, Moscow, pp 192–194
209. Hoppe R (1958) Connections between formation heats in inorganic compounds. Z Anorg Allg Chem 296(1–6):104–116
210. Humphrey GL, King EG (1952) Heats of quarz and crystobalite formation. J Am Chem Soc 74(8):2041–2042
211. Iler RK (1979) The chemistry of silica. Wiley, New York
212. Ioshii T, Sudo G (1958) Review of the 12th General Meeting. Jpn Cement Eng Assoc, Tokyo
213. Ivanov NK, Arbuzov AM, Maksimova IP(1978) Polimerization of silicate anions in aqueous solutions. Zh Prikl Khim Leningrad 3:572–577
214. Ivanova LI (1961) Connection between heat capacity of solids and the temperature of the first phase transition. Zh Fiz Khim 35(9):1035, 2120–2121
215. Izmaylov NA (1959) Electrochemistry of solutions. Izd KhGU, Kharkov
216. Jander G, Heudeshoven W (1931) Amphoteric oxyhydrates, their aqueous solutions and crystalline compounds. XIII. Silicic acid and silicon dioxyhydrate in alkaline and in acid solutions. Z Anorg Allg Chem 201:361–382
217. Jander G, Jahr KF (1934) Recent observations concerning the hydrolysis of inorganic salts and the chemistry of the hydrolytic products of high molecular weight (the isopoly and heteropoly compounds inclusive). Kolloid Beihefte 41:1–57
218. Jander W, Wuhrer J (1934) Parallel determination of barium oxide and barium silicate. Z Anorg Allg Chem 218:211–217
219. Jander W, Wuhrer J (1936) Reactions in the solid state at higher temperatures (XIII) Reaction course in the formation of Mg, Sr and Ba silicates. Z Anorg Allg Chem 226:225–247
220. Janz GJ, Neuenschwander E, Kelly FJ (1963) High temperature heat contents and similar properties for Li_2CO_3, Na_2CO_3, K_2CO_3 and the ternary eutectic mixture. Trans Faraday Soc 59(4):841–845
221. Jeffery JW (1954) The tricalcium silicate phase. In: Proceedings of the IIIrd International Symposium on the Chemistry of Cement. Cement and Concrete Association, London, pp 30–48
222. Johanson OK, Thorvaldson T (1934) Thermochemistry of the compounds occurring in the system $CaO-Al_2O_3-SiO_2$. V. Heat of formation of C_3S and C_2S. J Am Chem Soc 56:2327–2330
223. Jones FE (1944) The quaternary system $CaO-Al_2O_3-CaSO_4-H_2O$ at 25°. J Phys Chem 48:311–356
224. Jones FE (1960) In: IVth International Symposium on the Chemistry of Cement. National Bureau of Standards, Washington, pp 205–242
225. Joseph AF, Oakley HB (1925) Action of silica on electrolytes (II); anomalous flocculation of clay. J Chem Soc 127:2813
226. Joung JF, Tong HS, Berger RL (1977) Compositions of solutions in contact with hydrating tricalcium silicate pastes. J Am Ceram Soc 60(5–6):193–198
227. Jouquet E (1921) The laws of chemical dynamics. J Ecole Polytech Paris 21(61): 181–194
228. Kasanskaya AS, Skoblo WA (1974) Calculation of chemical equilibria. High School, Moscow

229. Kakabadze VM, Chachanidze GD (1953, 1954) To the problem of obtaining barium hydroxide. Tr Gruzinskogo Politekh Inst 28(3):110; Zh Fiz Khim 25:1013
230. Kalousek GL (1955) Tobermorite and related phases in the system $CaO-SiO_2-H_2O$. ACI J 26(10):989–1011
231. Kalousek GL, Logindice JS, Dodson VH (1954) Studies in the lime-rich crystalline solid phases in the system lime-silica-water. J Am Ceram Soc 37(1):7–13
232. Kalousek GL, Prebus AF (1958) Crystal chemistry of hydrous calcium silicates. III. Morphology and other properties of tobermorite and related phases. J Am Ceram Soc 41(4):124–132
233. Kapranov VV (1976) Hardening of binders and of products based on them. Yuzhno-Ural'skoye knizh izdat, Chelyabinsk
234. Kapustinsky AF (1935) Thermodynamics of chemical reactions and its application to inorganic technology. ONTI, Moscow
235. Kapustinsky AF (1943) Universal equation of the lattice energy of ionic crystals. Zh Obshch Khim 13(7–8):497–502
236. Kapustinsky AF (1948) Thermochemistry and atomic structure. Izv Akad Nauk SSSR Otd Khim Nauk (6):568–589
237. Kapustinsky AF (1956) On the theory of earth. In: Collection 'Problems of geochemistry and mineralogy'. Izd Akad Nauk SSSR, Moscow, pp 37–71
238. Kapustinsky AF, Yatsimirsky KB (1948) Entropy of ions in the crystall and the solubility of salts. Zh Fiz Khim 22:1271–1279
239. Karakozov ES, Kopylov YN, Matveyev GM, Shorshorov MK (1969) On bonding quartz with metal. Fiz i Khim Obr Mat 2:72–77
240. Karakozov ES, Kotyolkin AS, Matveyev GM, Merkulov VA, Shorshorov MK (1968) On the mechanism of the formation of bonding while welding in solid-state ceramics of aluminium oxide with niobium. Fiz i Khim Obr Mat 3:123–129
241. Karakozov ES, Matveyev GM, Musin RA, Shorsporov MK (1973) Thermodynamic calculation of the diffusion welding temperature of ceramics with copper. Fiz i Khim Obr Mat 5:132–137
242. Karapet'yants MKh (1953) Chemical thermodynamics. Goskhimizdat, Moscow
243. Karapet'yants MKh (1954) On the approximate methods of entropy calculation. In: Proceedings of MKhTI 18, Izd 'Sov Nauka', pp 27–44
244. Karapet'yants MKh (1954) An approximate method of calculating the Gibbs free energies and heats of formation of various substances. Zh Fiz Khim 28:186–187
245. Karapet'yants MKh (1955) On methods of comparative calculations of various properties. Zh Fiz Khim 29(7):1328–1333
246. Karapet'yants MKh (1958) On the approximate calculation of heat capacity temperature dependance. Zh Fiz Khim 32(8):1763–1773
247. Karapet'yants MKh, Karapet'yants ML (1961) Tables of some thermodynamic properties of different substances. In: Proceedings of MKhTI 34, Izd 'Sov Nauka', pp 1–165
248. Karapet'yants MKh, Karapet'yants ML (1968) The main thermodynamic constants of inorganic and organic substances. 'Khimiya', Moscow
249. Karpov IK, Kashik SA, Pampura VD (1968) The constants of substances for thermodynamic calculations in geochemistry and petrology. Nauka, Novosibirsk
250. Karsh KH, Schwiete HE (1963) An adiabatic calorimeter for determining the heat of hydration of cement. Zement-Kalk-Gips 16(5):165–169
251. Kay DAR, Taylor J (1960) Activities of silica in the lime + alumina + silica system. Trans Faraday Soc 56:1372
252. Keil F (1964) Zement-Herstellung und Eigenschaften. Springer, Berlin Heidelberg New York

253. Keler EK, Glushkova VB (1956) Conditions of barium silicate formation. Zh Neorg Khim 1(10):2283–2293
254. Keler EK, Glushkova VB (1957) Study of formation of strontium silicates. Zh Prikl Khim Leningrad 30(4):517–523
255. Kelli A (1966) The strengthening of metals with disperse particles. Collection 'Mechanical properties of new materials'. Mir, Moscow
256. Kelley KK (1949, 1950) Data on theoretical metallurgy: X High-temperature, heat-content, heat-capacity and entropy data for inorganic compounds. US Bur Mines Bull 476:241; 477:147
257. Kelley KK (1960) Contributions to the data on theoretical metallurgy: XIII High-temperature, heat-content, heat-capacity and entropy data for the elements and inorganic compounds. US Bur Mines Bull 584:160, 161
258. Kelley KK (1960) High-temperature, heat-content, heat-capacity and entropy data for elements and inorganic compounds. US Bur Mines Bull, Washington
259. Kelley KK, King EG (1961) Contributions to the data on theoretical metallurgy: XIV Entropies of the elements and inorganic compounds. US Bur Mines Bull 592:149
260. Kelley KK, Southard JC, Anderson CT (1944) Thermodynamic properties of gypsum and its dehydration products. US Bur Mines Tech Paper 625:73
261. Kennedy GC (1944) Hydrothermal solubility of SiO_2. Econ Geol 39(1):25
262. Kennedy GC (1950) A portion of the system silica-water. Econ Geol 45(7):629–653
263. Kern R, Vaysbrod A (1966) The fundamentals of thermodynamics for mineralogists, petrographers and geologists. Mir, Moscow
264. Kheynman AS (1939,1946) Molecular states of Martin slags and distribution of oxygen and sulphur in the system slag-metal. Izv Akad Nauk SSSR OTN 10:1439–1458
265. Khitarov NI, Pughin VA, Chzhao-Bin, Slutski AB (1963) Correlations between andaluzite, cianite and sillimanite under moderate pressures and temperatures. Geokhimiya 3:219–228
266. Kingery W (1951) Kinetics of high-temperature processes. Massachusetts Institute of Technology
267. Kingery WD (1964) Introduction to ceramics. Moscow, pp 192–194
268. Kind VA, Okorokov SD, Vol'fson SL (1937) Hardening heat of portland cement of various chemical composition. Cement 7:12–17
269. Kind VV (1955) Cement and concrete corrosion in hydraulic structures. Gosenergoizdat, Leningrad
270. King EG (1951) Heats of formation of crystalline calcium orthosilicate, tricalcium silicate and zinc orthosilicate. J Am Chem Soc 73(2):656–658
271. King EG (1955) Heat capacities at low temperatures and entropies at 298.16 °K of crystalline calcium and magnesium aluminates. Phys Chem 59(3):218
272. King EG, Coughlin IP (1950) High temperature heat contents of some zirconium-containing substances. J Am Chem Soc 72(5):2262–2265
273. Kingery WD (1960) Introduction to ceramics. Wiley, New York
274. Kireyev VA (1955) A course in physical chemistry. Goskhimizdat, Moscow
275. Kireyev VA (1960) Thermodynamics of chemical equilibria in similar reactions. Zh Fiz Khim 34(5):945–956
276. Kireyev VA (1963) Double comparison method for similar reactions and compounds. Zh Fiz Khim 37(2):452–456
277. Kireyev VA (1975) Methods of practical calculations in the thermodynamics of chemical reactions. 'Khimiya', Moscow
278. Kirilin VA, Sheyndlin AE (1963) Investigation of thermodynamic properties of substances. Gosenergoizdat, Moscow-Leningrad

279. Kittel C (1961) Introduction to solid state physics. Wiley, New York London
280. Kleber W (1965) Adsorption und Stufenregel. Akademie, Berlin
281. Kozheurov VA (1949) On thermodynamics of basic metallurgical slags. Zh Fiz Khim 23(4):484–496
282. Kozheurov VA (1950) Distribution of manganum and phosphor among liquid iron and basic slags. Zh Prikl Khim 23:233
283. Kozheurov VA (1952) Methods of estimation of the nearest order in solutions. Zh Fiz Khim 26(4):479–483
284. Kozheurov VA (1954) The phenomena of limited solubility in binary silicate melts. Zh Fiz Khim 28(4):729–734
285. Kozheurov VA (1956) The thermodynamics of metallurgic slags. Metallurgizdat, Sverdlovsk
286. Kozheurov VA, Laptev DM (1954) On certain metallurgical equilibria with participation of acid slags. Zh Fiz Khim 28:814–819
287. Kokovin GA, Testova NA, Kuznetsov FA (1978) A thermodynamic analysis of the systems $M-SiO_2$ (M = Al, Ge, Ti, V, Ta, Cr, Mo) in connexion with the problem of the stability of MOP-structures. Izv Akad Nauk SSSR, Sibirskoye otdeleniye, Inst Neorg Khim, dep 527
288. Kolesov VP, Skuratov SM, Zaykin ID (1959) The enthalpy of the formation of lithium oxide. Zh Neorg Khim 4(6):1237–1241
289. Komar' NP (1955) The principles of the qualitative chemical analysis. Izd KhGU, Khar'kov
290. Kondo R, Daymon M (1976) Phase composition of hardened cement stone. In: The VIth International Congress on the Chemistry of Cement, vol 2, book 1. Stroyizdat, Moscow, pp 244–257
291. Korzhinsky DS (1935) Thermodynamics and geology of some metamorphic reactions with gaseous phase released. Zap Vses Miner Obshch 64/1(1):1–20
292. Korzhinsky DS (1956) Thermodynamics and geology of some metamorphic reactions with gaseous phase released. Dokl Akad Nauk SSSR 106(2):295
293. Korzhinsky DS (1957) Physico-chemical principles of the analysis of the paragenesis of minerals. Izd Akad Nauk SSSR, Moscow
294. Korzhinsky DS (1959) Basic-acid interaction of components in silicate melts. Dokl Akad Nauk SSSR 128(2):383–386
295. Kottrell T (1956) The strength of chemical bonds. Inostrannaya literatura, Moscow
296. Kracek FC (1951) Phase transformations in one-component silicate systems. In: Phase transformations in solids. Wiley, New York; Chapman, London, pp 257–277
297. Krakau KA, Mukhin YY, Genrikh MS (1937) Diagram of the ternary system $Na_2SiO_3-PbSiO_3-SiO_2$. Dokl Akad Nauk SSSR 14(5):281–285
298. Krakau KA, Mukhin YY, Genrikh MS (1949) Equilibrium diagram of the system $Na_2O-PbO-SiO_2$. Physico-chemical properties of the ternary system $Na_2O-PbO-SiO_2$. Akad Nauk SSSR
299. Krasil'nikov KG (1956) Study of the system $CaO-SiO_2-H_2O$. In: Proceedings of the Conference on the Chemistry of Cement. Gosstroyizdat, Moscow, p 351
300. Kreidl NI (1942) Zirconium oxide and thorium oxide in ceramics. J Am Ceram Soc 25(5):129
301. Krestov GA (1973) Thermodyamics of ionic processes in solutions. 'Khimiya', Leningrad
302. Krestovnikov AN, Vigdorovich VN (1973) Chemical thermodynamics. Metallurgiya, Moscow
303. Krestovnikov AN, Vladimirov LP, Gulyanitskiy GS, Fisher AY (1965) A handbook of the calculations of equilibria of metallurgic reactions. Metallurgizdat, Moscow
304. Krichevskiy IR (1962) The ideas and principles of thermodynamics. Goskhimizdat, Moscow

305. Krisement O, Schädler H, Trömel G, Obst KH (1957) Calorimetric investigations of the $\alpha \rightleftarrows \beta$ transformation of crystobalite. Arch Eisenhüttenw 28:523–530
306. Kröger C (1953) Theoretical heat input for the fusion of glass, melting process. Glastech Ber 26(7):202–214
307. Kröger C, Fingas E (1933) Action of quartz and alkali silicates upon alkali carbonates. Z Anorg Allg Chem 213:12–57
308. Kröger C, Fingas E (1935) The systems alkali-oxide $-CaO-Al_2O_3-SiO_2-CO_2$. V. Stable equilibrium in the system $Na_2O-SiO_2-CO_2$. Z Anorg Allg Chem 225:1–25
309. Kröger C, Kreitlow G (1956) Die Lösung- und Bildungswärmen der Natronkalksilikate. Glastech Ber 29:393–400
310. Kuan-Han S (1947) Fundamental conditions of glass formation. J Am Ceram Soc 30:277
311. Kubaschewski O, Evans E, Alcock CB (1958) Metallurgical thermochemistry. Pergamon, Oxford
312. Kubaschewski O (1966) Thermochemistry of metal-refractory interactions. Rev Hautes Temper Refract 3:229–234
313. Kubo R (1970) Thermodynamics. Mir, Moscow
314. Kühl H (1961) Zement-Chemie, Bd III. Verlag Technik, Berlin
315. Kukolev GC (1966) The chemistry of silicon and physical chemistry of silicates. Izd 'Vysshaya Shkola', Moscow
316. Kulikov IS (1955) Some questions of the theory of metallurgical slags. Izv Akad Nauk SSSR OTN (2):113–119
317. Kulikov IS (1969) The thermic dissociation of compounds. 'Metallurgiya', Moscow
318. Kulikov IS, Matveyev GM, Rudoy BL, Shorshorov MK (1968) Oxide phases for dispersed hardening. Fiz i Khim Obr Mat 4:120–124
319. Kurczyk HG, Schwiete HE (1960) Electron microscope and thermochemical studies of the hydration of calcium silicates C_3S and β-C_2S and the effect of calcium chloride and gypsum on the hydration process. Tonind Z 84:585–598
320. Landiya NA (1962) A calculation of high temperature capacities of inorganic substances by standard entropy. Akad Nauk Gruz SSR, Tbilissi
321. Landiya NA, Mchedlov-Petrossyan OP (1952) Thermodynamic study of the solid phase reactions in the system $CaO-SiO_2$. Zh Fiz Khim 26:1785
322. Larionova ZM, Nikitina LV, Gorshin VR (1977) The phase composition of the microstructure and the strength of cement stone and concrete. Stroyizdat, Moscow
323. Latimer VM (1954) Oxidation states of elements and their potentials in water solutions. Inostrannaya Literatura, Moscow
324. Latiu E (1958) Crystal energy in solid state silicate reactions. Acad RPR Stiinte Chim 5(3–4):55–63
325. Lea FM (1970) The chemistry of cement and concrete. Arnold, Glasgow
326. Lebedev VI (1957) Fundamentals of the energy analysis of geochemical processes. Izd LGU, Leningrad
327. Lenyov LM, Novokhatskiy IA (1966) A study of the thermodynamic properties of cobalt and nickel chromite. Zh Fiz Khim 40(9):2030
328. Lerch W (1946) The influence of gypsum on the hydration and properties of portland cement pastes. Proceedings of ASTM 46:1252–1292
329. Lerch W, Bogue RH (1934) Hydrolysis of compounds which may occur in portland cement. J Res NBS 12:645–664
330. Levin EM, Robbins CR, McMurdie HF (1964) Phase diagrams for ceramist. American Ceramic Society, Columbus, Ohio
331. Levitskiy VA, Popov SG, Ratiani DD, Lebedev BG (1972) Thermodynamic proper-

ties of ferric orthosilicate, metatitanate, aluminate and chromite at elevated temperatures. Term i Kin Protz Vosstan Mat 3(6):35 – 40
332. Levitskiy VA, Rezukhina TN (1966) Thermodynamic properties of cobalt and nickel aluminates starting from the electromotive forces at elevated temperatures. Izv Akad Nauk SSSR Neorg Mat 2(1):145
333. Lifshits IM (1939) To the theory of solid solutions. Zh Eks i Teor Fiz 9(4):481 – 511
334. Liszi J, Kocsis G, Kiraly J (1968) Calculation of thermodynamic properties of the $Na_2O - CaO - SiO_2$ (soda-lime-silica) ternary silicate glass. Építoanyag 20(10): 362 – 367
335. Locher FW (1964) The chemical reactions of the hardening of cement. Zement-Kalk-Gips 17:175 – 187
336. Lozhkin LN, Gerasimenko LN (1957) A study of the system plumbum monoxide – plumbum silicate by means of measuring e.m.f. In: Proceedings of the Leningrad Polytechnical Inst. Electrometallurgy of non-ferrous metals. Èlektrometall Tsvet Metall, vol 188
337. Ludwig U (1976) Study of clincer minerals hydration mechanism. In: VIth International Congress on the Chemistry of Cement, vol 2(1). Stroyizdat, Moscow, pp 104 – 121
338. Lur'ye YY (1971) A handbook of analtical chemistry. Khimiya, Moscow
339. Lykov AV, Mikhailov YA (1959) The theory of the energy and substance transfer. Izd Akad Nauk SSSR, Moscow
340. Machin IS, Hamma DL (1945) Viscosity studies of system $CaO - MgO - Al_2O_3 - SiO_2$. J Am Cer Soc 28(11):310
341. Makarov YS (1973) The isomorphism of atoms in crystals. Atomizdat, Moscow
342. Malkin VI, Khokhlov SF, Shvartsman LA (1956) Determination of cation transfer numbers in melts $CaO \cdot Na_2O \cdot 4SiO_2$. Dokl Akad Nauk SSSR 106(3):491 – 493
343. Malkin VI, Khokhlov SF (1958) Determination of calcium ion transfer number in melts of the system $CaO - MgO - SiO_2$ and $CaO - MgO - Al_2O_3 - SiO_2$. Izv Akad Nauk SSSR OTN 3:108
344. Mamedov KS, Belov NV (1956) Crystal structure of minerals of the wollastonite group. Zap Vses Miner Obshch 85(1):13 – 38
345. Mamedov KS, Belov NV (1958) Crystal structure of calcium hydrosilicates (hillebrandite, foshagite, okenite et al.). Dokl Akad Nauk SSSR 104(4):615; 121(4):720; 121(5):901; 123(1):163; 123(4):741
346. Mannchen W (1965) Einführung in die Thermodynamik der Mischphasen. Verlag für Grundstoffindustrie, Leipzig
347. Marakushev AA (1968) The thermodynamics of the metamorphic hydration of minerals. Nedra, Moscow
348. Markholiya TP, Yudin BF, Voronin NI (1967) The interaction of high-melting metals with alumina and silica. The chemistry of high-temperature materials. Nauka, Leningrad
349. Maslov PG, Maslov YP (1959) Thermodynamic characteristics of crystalline compounds containing litium. Zh Fiz Khim 33(8):1687 – 1691
350. Massazza F (1966) Thermochemical investigations in the system $3CaO \cdot MgO \cdot 2SiO_2 - 3SrO \cdot MgO \cdot 2SiO_2$. Ann Chim 56(11):1298 – 1305
351. Massazza F, Riva B (1966) Thermochemical investigations in the system $2CaO \cdot MgO \cdot 2SiO_2 - 2SrO \cdot MgO \cdot 2SiO_2$. Ann Chim 56(11):1278 – 1286
352. Massazza F (1962) Thermochemical investigations in the system $2CaO - Al_2O_3 \cdot SiO_2 - 2SrO \cdot Al_2O_3 \cdot SiO_2$. Ann Chim 52(7):589 – 619
353. Matingnon C (1923) The influence of elevated temperatures on some refractory substances. Compt Rend 117(24):1290 – 1295

354. Matousek J (1975) Stability of oxides at elevated temperatures. I Evaporation and dissociation. II Oxidation and reduction. Sklař a Keramik 10:289 – 291; 11:333 – 336
355. Matveyev GM (1957) A thermodynamic analysis of solid phase reactions in the system $CaO - Al_2O_3$. Collection of works on the chemistry and technology of silicates. Promstroyizdat, Moscow, p 94
356. Matveyev GM (1957) On the interconnection of structure and thermodynamic properties of silicates. Proceedings of MKhTI 24:233 – 236
357. Matveyev GM (1958) On the calculation of thermodynamic properties of binar compounds in silicate technology. Izv Vuzov Ser Khim i Tekhn 1(2):135 – 141
358. Matveyev GM (1963) On the thermodynamic analysis of solid phase reactions in the system $BaCO_3 - BaO - SiO_2$. In: Silicates and oxides in the high-temperature chemistry. Akad Nauk SSSR i VKhO im DI Mendeleyeva, Moscow, pp 322 – 324
359. Matveyev GM (1964) A thermodynamic evaluation of the effect of some oxides on the chemical stability of glass. Collection 'Vitreous state', vol III/4. Minsk, p 140
360. Matveyev GM (1967) On the problem of chemical stability of vitreous and ceramic materials. Zh VKhO im DI Mendeleyeva 12(6):711
361. Matveyev GM (1958) Dissertation, Prague
362. Matveev MA (1962) A thermographic study of sodium and lithium silicates. Proceedings of MKhTI 37:49
363. Matveyev GM, Agarkov AS (1972) Thermodynamic analysis of ternary compounds formation in the system $Na_2O - CaO - SiO_2$ through solid state reactions. Silikattechnik 23(4):122 – 124
364. Matveyev GM, Agarkov AS, Kerbe F (1967) Thermodynamic analysis of mullite formation from oxides. Silikattechnik 18(10):324 – 326
365. Matveyev GM, Agarkov AS, Kerbe F (1969) Thermodynamic analysis of zircon formation from oxides. Silikattechnik 20(3):86 – 88
366. Matveyev GM, Agarkov AS (1968) Thermodynamic study of the solid state reactions in the system $BaSO_4 - SiO_2$. Izv Akad Nauk SSSR Neogr Mat 4(6):957 – 959
367. Matveyev GM, Elkin GB, Kerbe F (1969) Thermodynamic analysis of reactions of formation of plumbum silicates in the system $PbO - SiO_2$ and determination of heat required for plumbum silicate melts. Silikattechnik 20(7):225 – 227
368. Matveyev GM, Elkin GB (1968, 1970) Estimation of heat required for glass fusion. In: Technical information of TsNIITEstrom. Ser Glass Ind 11:27 – 31; 10:10 – 13
369. Matveyev GM, Ganikirzh VA (1959, 1962) Thermodynamic analysis of solid state reactions in the system $Na_2O - SiO_2$. Sb Tr Khim-Tekhnol Inst Prague 3(1):397 – 403; Tr MKhTI 37:44 – 48
370. Matveyev GM, Matveyev MA (1964) A thermodynamic analysis of solid phase reactions in the system $BeO - SiO_2$. Zh VKhO im DI Mendeleyeva 9(2):344 – 355
371. Matveyev GM, Matveyev MA (1965) Thermodynamics of silicate-formation reactions from the oxides of bivalent metals and silica. Zh VKhO im DI Mendeleyeva 10(5):590 – 591
372. Matveyev GM, Novosadov VS, Nuzhdin AA, Rudoy BL, Tykachinskiy ID, Khill GP, Shorshorov MK (1970) An analysis of the contact zone of the aluminium melt with silicates. Fiz Khim Obr Mat 1:140
373. Matveyev GM, Rudoy BL, Shorshorov MK (1969) Thermodynamic analysis of the possibility of direct interaction of Al, Ni, Co, Fe, Nb with silicates of different composition. Fiz i Khim Obr Mat 6:94 – 98
374. Matveyev GM, Rudoy BL, Shorshorov MK (1972) A study of the process of chemical interaction of a disperse vitreous phase with the nickel matrix and of refractory properties of the composite material. In: Proceedings of the Scientific and Technical Conference of VIAM 'Metallic composite materials', pp 145 – 152

375. Matveyev GM, Udovenko NG (1967) Thermodynamic analysis of glass crystallization in the system $CaO-Al_2O_3-SiO_2$. In: Bor-free, alkali-free and low-alkali vitreous systems and new glasses based on them. TsNIITEstrom, Moscow, pp 64–69
376. Matveyev MA, Frenkel' BN (1967) The kinetics of reactions in mixtures of caesium carbonate with quartz. In: Vitreous systems and materials, 4. Izd Zinatne, Riga, pp 17–24
377. Matveyev MA, Frenkel' BN, Matveyev GM (1964) Reactions in mixtures of quartz with Na_2CO_3 and sulphate and their thermodynamics. Proceedings of MKhTI 45:125–138
378. Matveyev MA, Frenkel' BN, Matveyev GM (1965) Thermodynamics of alkali-silicates formation. Izv Akad Nauk SSSR Neorg Mat 1(8):1426–1436
379. Matveyev MA, Frenkel' BN, Matveyev GM (1965) Reactions in mixtures of quartz with Na_2CO_3 and K_2SO_4 and their thermodynamic study. In: Studies in the chemistry of silicates and oxides. Nauka, Moscow Leningrad, pp 49–57
380. Matveyev MA, Frenkel' BN, Matveyev GM (1966) Thermodynamics of reactions in mixture Li_2O-SiO_2. Proceedings of MKhTI 50:84–91
381. Matveyev MA, Frenkel' BN, Matveyev GM (1965) Thermodynamics of reactions of formation of alkali silicates. In: Actes du VIIième Congrès International du Verre. Gordon-Breach, New York, pp 5966–5981
382. Matveyev MA, Kaminskaya VS (1964) On glass-formation in strontium-sulphate batch. Izv Akad Nauk SSSR Ser Fiz-Tekhn Nauk 3:80–85
383. Matveyev MA, Matveyev GM (1963) On thermodynamic analysis of solid state reactions. In: Silicates and oxides in the high-temperature chemistry. Izd VKhO im DI Mendeleyeva, Moscow pp 303–307
384. Matveyev MA, Matveyev GM (1963) On thermodynamic analysis of solid state reactions in the system $SrCO_3-SrO-SiO_2$. Izv Akad Nauk Belorus SSR Ser Fiz-Tekhn Nauk 3:59–62
385. Matveyev MA, Matveyev GM (1964) Thermodynamic analysis of glass crystallization processes in the system Na_2O-SiO_2. Zh Prikl Khim Leningrad 37:2357–2361
386. Matveyev MA, Matveyev GM, Elkin GB (1970) A thermodynamic analysis of reactions of silicate formation in the system $PbO-SiO_2$. Zhurn Prikl Khim 43(2):267–273
387. Matveyev MA, Matveyev GM, Kharitonov FY (1966) The strength of ceramics prepared from pure oxides. Izv Akad Nauk SSSR Neorg Mat 2(2)
388. Matveyev MA, Novikov VN, Matveyev GM (1966) Thermodynamic analysis of solid state reactions in the system $SrSO_4-SrO-SiO_2$. Izv Akad Nauk SSSR Neorg Mat 2(9):1720–1723
389. Matveyev MA, Novikov VN, Matveyev GM (1969) Thermodynamic analysis of solid state reactions of ternary compounds formation in the system $MgO-SrO-SiO_2$. Proceedings of MKhTI 59:41–46
390. Mattson S (1934) Soil colloids. Sel'khozizdat, Moscow
391. Matusita K, Tashiro M (1973) Rate of homogeneous nucleation in alkali disilicate glasses. J Non-Cryst Solids 11:471–484
392. Mazurin OV, Filipovich VN, Shul'ts MM (1977) On some peculiarities of temperature dependance of isobar-isothermal potentials of glass-building substances. Fiz i Khim Stekla 3(1):39
393. McConnell JDC (1955) A chemical optical and X-ray study of scawtite from Ballycraigy, Larne, N. Ireland. Am Mineralogist 40:510–514
McConnell JDC, Long JVP (1959) A mineralogical application of x-ray absorption microspectroscopy: the hydration of larnite. Mineral Mag 32:117–127
394. Mchedlov-Petrossyan OP (1950) On the thermoanalysis of serpentine and caolin. Dokl Akad Nauk SSSR 74(4):799–802

395. Mchedlov-Petrossyan OP (1950) Thermodynamic study of the solid state reactions in magnesium silicate systems. Zh Fiz Khim 24(11):1299 – 1301
396. Mchedlov-Petrossyan OP (1953) On the origin of the binding properties. Dokl Akad Nauk SSSR 89(1):137 – 139
397. Mchedlov-Petrossyan OP (1956) Thermodynamics of solid state reactions in silicate systems. In: Physico-chemical fundamentals of ceramics. Promstroyizdat, Moscow, pp 499 – 503
398. Mchedlov-Petrossyan OP (1956) Crystallochemistry of bindung properties. In: Proceedings of the Conference on the Chemistry of Cement. Promstroyizdat, Moscow, pp 63 – 77
399. Mchedlov-Petrossyan OP (1957) Crystallochemical origin of the thermal effects of clay minerals. In: Proceedings of the Conference on the Study and Usage of Clays. L'vov University Press, Lvov, pp 745 – 750
400. Mchedlov-Petrossyan OP et al. (1958) Collection 'Thermography of cements'. Gosstroyizdat USSR, Kiev, pp 1 – 28
401. Mchedlov-Petrossyan OP (1958) On the usage of thermal analysis in the study of mineral binders. Silikattechnik 9(12):556 – 560
402. Mchedlov-Petrossyan OP, Babushkin VI (1958) On the thermodynamic study of solid state reactions in silicate systems. Silikattechnik 9(5):209 – 212
403. Mchedlov-Petrossyan OP, Babushkin VI (1959) Crystallochemical data for thermodynamic analysis of hydrothermal synthesis of calcium hydrosilicates. Dokl Akad Nauk SSSR 128(2):348 – 351
404. Mchedlov-Petrossyan OP, Babushkin VI (1959) Thermodynamic study of reactions in the system $Ca(OH)_2 - SiO_2 - H_2O$ at hydrothermal conditions. In: Proceedings of High-School Conference on the Study of Autoclaved Materials. LISI, Leningrad, pp 253 – 259
405. Mchedlov-Petrossyan OP, Babushkin VI (1960) Thermodynamic methods of investigation of solid state reactions in silicate systems. In: Up-to-date methods of the investigation of silicates and building materials. Gosstroyizdat, Moscow, pp 28 – 40
406. Mchedlov-Petrossyan OP, Babushkin VI (1960) On the thermodynamics of silicates and hydrosilicates of calcium. Silikattechnik 11(8):384 – 386
407. Mchedlov-Petrossyan OP, Babushkin VI (1960) Thermodynamics of the hardening processes of cement. In: Proceedings of the 4th International Symposium on the Chemistry of Cement, vol 1. US Covernment Printing Office, Washington, pp 533 – 544
408. Mchedlov-Petrossyan OP, Babushkin VI (1960) Energy aspects of physico-chemical analysis of silicates. In: Summary of the reports of the IVth All-union Conference on the Physico-chemical Analysis. Izd Akad Nauk SSSR, Moscow, pp 99 – 100
409. Mchedlov-Petrossyan OP, Babushkin VI (1960) Discussion. In: The IIIrd All-union Conference on the Vitreous State. Izd Akad Nauk SSSR, Moscow Leningrad, p 131
410. Mchedlov-Petrossyan OP, Babushkin VI (1960) Thermodynamic foundation of the lime-sand products hardening process. In: Materials for the Scientific-technical Conference on Silicalcite. MPSM Èston SSR, Tallin, pp 344 – 384
411. Mchedlov-Petrossyan OP, Usherov-Marshak AW, Urshenko AM (1984) Heat evolution of binders and concrete. Stroiizdat, Moscow
412. Mchedlov-Petrossyan OP, Babushkin VI (1960) Chemical thermodynamics for the technology of building materials. Zh Vses Khim Obshch DI Mendeleyeva 5(2):126 – 133
413. Mchedlov-Petrossyan OP, Babushkin VI (1961) Thermochemistry and thermodynamics of unhydrated and hydrated cement materials. Zh Vses Khim Obshch DI Mendeleyeva 6(6):677 – 680

414. Mchedlov-Petrossyan OP, Babushkin VI (1961) On the role of structural analogy and stoichiometry in thermodynamic study of silicates. Kristallografiya 6(6):933–936
415. Mchedlov-Petrossyan OP, Matveyev GM, Sofronov VS (1965) Study of the energy of the crystallization process as a method of investigation of glass structure. In: Proceedings of the IVth All-union Conference on the Vitreous State. Nauka, Moscow Leningrad, pp 339–341
416. Mchedlov-Petrossyan OP, Chernyavskiy VL (1967) Influence of low temperatures on the hydration of portland cement. Silikattechnik 18(3):72–76
417. Mchedlov-Petrossyan OP, Chernyavskiy VL (1968) Destructive processes and hardening of cement-concrete at low temperatures. Baustoffind 8:260–264
418. Mchedlov-Petrossyan OP, Chernyavskiy VL (1968) Physico-chemical investigation of cement hydration at low temperatures. Zh Prikl Khim Leningrad 41(12):2625–2630
419. Mchedlov-Petrossyan OP, Salop GP, Sidorovich YY (1969) Cements and concretes hardening observation. Budivel'nik, Kiev
420. Mchedlov-Petrossyan OP, Babushkin VI (1969) Thermodynamic study concerning the autoclaving and corrosion reactions of lime-silica materials. In: International Symposium on Autoclaved Lime-silica Materials. Preliminary summary of organizing committee Hannover, 1, paper 44:1–10
421. Mchedlov-Petrossyan OP, Chernyavskiy VL (1969) A method of differential calorimetric analysis (DCA) in the cement chemistry. In: Proceedings of Ist International Conference on Calorimetry and Thermodynamics. Arkady, Warsaw, pp 403–408
422. Mchedlov-Petrossyan OP (1970) Chemistry of inorganic building materials. Stroyizdat, Moscow
423. Mchedlov-Petrossyan OP, Polak AF (1970) On the law of phasebuilding in chemical reactions. Silikattechnik 22(2):59–60
424. Mchedlov-Petrossyan OP, Babushkin VI (1974) Thermodynamics and thermochemistry of cement. In: Proceedings of VIth International Congress on the Chemistry of Cement, vol 2(1). Stroyizdat, Moscow, pp 6–16
425. Mchedlov-Petrossyan OP, Sapoznikova NI, Shchyotkina TY (1975) Thermodynamical methods of investigation of the reactivity of raw material mixtures and cement. Tsement 10:17–19
426. Mchedlov-Petrossyan OP, Karyakina EL (1979) Evolution of opinions of the clays change at heating in the recent 50 years. In: Petrologomineralogical peculiarities of rocks and technical stones. Nauka, Moscow, pp 51–57
427. Megow HD (1949, 1952) Hydroxyl groups in afwillite. Structure of afwillite. Acta Cryst 2:419–420; 5:477–491
428. Mel'nik YP (1972) A handbook of the thermodynamic constants for the analysis of the conditions of the iron ore formation. Naukova Dumka, Kiev
429. Mel'nik YP (1972) Physico-chemical conditions of the formation of precambrian ferruginous quartzites. Naukova Dumka, Kiev
430. Meneret MJ (1957) Effect of alkalies and earth-alkalies on the physico-chemical properties of kaolines. Bull Soc Franc Ceram 35:35–40
431. Metyolkin II, Makarkin AY, Pavlova MA (1967) The welding of ceramic materials with metals. Svarochnoye Proizvodstvo 6:10
432. Metyolkin II, Shmelyov AY (1972) The brazing of ceramics by active metals. Fiz Khim Obr Mat 4:123
433. Meyer K (1968) Physico-chemical crystallography. Grundstoffindustrie, Leipzig
434. Minenko VI, Ivanova NS (1963) Thermodynamic properties of molten lead silicate. Izv Vysshikh Uchebn Zavedeniy, Tsvetn Met 6(3):64–69
435. Minenko VI, Petrov SM, Ivanova NS (1960, 1961) Electromotive forces in the fused

lead oxide-silicon dioxide system. Zhurn Vsesoyuz Khim Obshchestva im DI Mendeleyeva 5:230–231. Izv. Vuzov Khim i Khim Technolog 4(1):3–6
436. Mishchenko KP (1959) Thermodynamic properties of water in electrolyte solutions. In: Thermodynamics and structure of solutions. Izd Akad Nauk SSSR, Moscow, pp 97–105
437. Mishchenko KP, Poltoratskiy GM (1968) Problems of the thermodynamics and structure of aqueous and non-aqueous solution of electrolytes. Khimiya, Leningrad, p 56
438. Miyashiro A (1958) Regional metamorphism of the Gosaisyo-Takanuki District in the Central Abukuma Plateau. J Fac Sci Univ Tokyo, Sect II, 11, part 2:219–272
439. Moore J, Sharp DE (1958) Calculation of effect of temperature and composition on specific heat of glass. J Am Ceram Soc 41:461–463
440. Morey GW (1954) Properties of glass. Reinold, New York
441. Mosesman MA, Pitzer KS (1941) Thermodynamic properties of the crystalline forms of silica. J Am Chem Soc 63:2348–2356
442. Moshchanskiy (1962) The increase of the stability of building materials and structures effected by aggressive media. Gosstroyizdat, Moscow
443. Moskvin VM (1952) Corrosion of concrete. Gosstroyizdat, Moscow
444. Mozzhukhin YI (1968) High-temperature materials. Metallurgiya, Moscow
445. Mulert O (1912) About the thermodynamic of SiO_2 and silicates. Z Anorg Chem 15:198
446. Murakami K, Tanaka H, Ikehata T, Suga F (1959) Reactions of $CaSO_4 \cdot 0.5\,H_2O$ with $3CaO \cdot Al_2O_3$, $4CaO \cdot Al_2O_3 \cdot Fe_2O_3$ and $2CaO \cdot Fe_2O_3$. J Ceram Assoc Japan Jogyo Kyokai Shi 67(766):326–333
447. Müller RL (1960) Vitreous state. Izd Akad Nauk SSSR, Leningrad
448. Naumov GB, Ryzhenko BN, Khodakovskiy PL (1971) A handbook of thermodynamic values (for geologists). Atomizdat, Moscow
449. Nemilov SV (1976) Entropy and valent structure of glass. Fiz Khim Stekla 2(2): 97–109
450. Newman ES (1956) Heats of formation of xonotlite, hillebrandite, and foshagite. J Res NBS 57(1):27–30
451. Niggli P (1946) Magma and its products. Gosgeolizdat, Moscow Leningrad
452. Nikolayev VA (1952) On the calculation of the thermodynamic potential increase... In: Collection of mineralogical papers of the Lvov geological society 6:49–60
Nikolayev VA (1952) Limited solubility of binary systems with volatile components. Izv Akad Nauk SSSR Ser Geol 1952:122–131
453. Nikolayev VA (1957) On the application of thermodynamics to some petrological processes. Zap Vses Miner Obshch 86(2):223–238
454. Nikol'skiy SS (1965) To the thermodynamics of solid solutions. In: 'High-temperature inorganic compounds'. Naukova Dumka, Kiev
455. Novokhatstkiy IA, Lenyov LM (1966) Thermodynamic properties of Cr_2O_3 and $FeCr_2O_4$ at high temperatures. Zhur Neorg Khim 11(9):2014
456. Nurse RW (1960) Phase equilibria and constitution of portland cement clinker. In: 4th International Symposium on the Chemistry of Cement, vol 1. National Bureau of Standards, Washington, pp 9–21
457. Numerical data for the calculation of the theoretical heat input for the fusion of glass (1954) Glass Industry 10(11)
458. Oka L, Kawagaki K, Kadoya R (1943) J Chem Soc Japan 64:718
459. Okorokov SD, Golynko-Vol'fson SL (1961) Improvement of technical properties of binders. In: Conference on the chemistry and technology of cement. Summaries of reports. Stroyizdat, Moscow, pp 18–20
460. Okura T (1951) Colorimetric determination of silica. J Chem Soc Japan 72:927–930

461. Ol'shanskiy YI (1951) Equilibria of two immiscible liquids in systems of alkaline earth silicates. Doklady Akad Nauk SSSR 76:93–96
462. Onsager L (1931) Reciprocal relations in irreversible processes, part I and II. Phys Rev 37:405; 38:2265–2278
463. Orlow VA (1981) Method for the calculation of thermodynamic parameters of high modulus liquid glasses. Zh Prikl Khim Leningrad 54(7):1577–1599
464. Ormont BF (1950) Structure of inorganic substances. Gostekh-TeorIzdat, Moscow Leningrad
Van't Hoff HI (1887, 1890) Z Phys Chem 1(5):165, 5:325
465. Osin BV (1954) Quicklime as a new binder. Promstroyizdat, Moscow
466. Oshcherin BN (1965) Calculation of the energy of dissociation of carbide and nitride interstitial phases. 'High-temperature inorganic compounds'. Akad Nauk Ukr SSR Kiev, pp 196–198
Oshcherin BN (1965) Some problems of the thermodynamics of carbidic interstitial phases in the homogeneity regions. 'High-temperature inorganic compounds'. Akad Nauk Ukr SSR:157–165
467. Ostwald W (1897) Studies about the formation and transformation of solids. Z Phys Chem 22(3):289–330
468. Palatnik LS, Landau AI (1961) Phase equilibria in multi-component systems. Izd KhGU, Kharkov
469. Parmelee CW, Badger AE (1934) Determination of mean specific heats at high temperatures of some commercial glasses. Univ Illinois Bull 271:24
470. Parimbetov BP (1978) Building materials from industrial mineral waste-products. Stroyizdat, Moscow
471. Patzak I, Konopicky K (1969) Transformation behavior of different minerals with the composition $Al_2O_3 \cdot SiO_2$. Tonind Z 93(2):43–49
472. Pauwels L (1971) Theories on the formation of inorganic glasses. II. Kinetic theory. Ind Chim Belge 36(5):375–379
Pauwels L (1971) Theories of the formation of inorganic glasses. I. Structural theories. Ind Chim Belge 36(4):284–290
473. Pavlushkin NM, Zhuravlyov AK (1970) Low-melting glasses. Energiya, Moscow
474. Pechuro SS (1949) Thermodynamic properties of gypsum. Promstroyizdat, Moscow
475. Peppler RB (1955) System of $CaO-SiO_2-H_2O$ at 180°C. J Res NBS 54(4):205–211
476. Petit G, Bourlange C (1957, 1958) The relation between fusion entropy and crystalline structure. Cr 245:1788; 246:2865
477. Petit G, Cremien MC (1956) Fusion heats of fluor earth alkalines. Cr 243:360
478. Petit G, Jaeger M (1957) Determination of fusion heats of some alkaline borates. Cr 245:1734
479. Petit G, Jaeger M (1957) The determination of the fusion heats of some alkaline molybdates. Cr 244:1900
480. Petrov YS (1957) Thesis, University of Novosibirsk
481. Petzold A (1955) Enamel. Technik, Berlin
482. Pham Huu Thanh (1967) Fusion entropy of selenium and selenium oxide. Chim Mod 12(77):53–66
483. Plyushchyov VY, Markovskaya NF (1954) Fusion diagram of the system caesium sulphate-magnesium sulphate. Dokl Akad Nauk SSSR 95(3):555–558
484. Polak AF (1963) Elementary processes of hardening of monomineral binders. In: Proceedings of BashNIIStroy, III. Gosgortekhizdat, Ufa, pp 179–307
485. Polak AF (1966) The hardening of monomineral binders. Stroyizdat, Moscow
486. Poling L (1964) General chemistry. Mir, Moscow
487. Pople JA (1951) A theory of the structure of water. Proc R Soc London 205:163–178

488. Popov MM (1954) Thermometry and calorimetry. Izd MGU, Moscow
489. Popov MM, Ginzburg DM (1956) Specific heats of Na_2CO_3, Na_2SO_4, NaOH at high temperatures. Zh Obshch Khim 26(4):971 – 980
490. Powers TC, Brownyard TL (1946) The physical properties of hardened portland cement paste. J Am Concr Inst 18:101 – 132, 249 – 336, 496 – 504, 549 – 602, 669 – 712, 845 – 880, 933 – 992
491. Prigogine I, Defay R (1962) Chemische Thermodynamik. Grundstoffindustrie, Leipzig
492. Prigogine I (1960) Introduction to the thermodynamics of irreversible processes. Inostrannaya Lit, Moscow
493. Prins JA (1965) Structure of non-crystalline solids. In: Physics of non-crystalline solids. North-Holland, Amsterdam
494. Prod'home M (1959) Specific heat of glasses. Verres Réfract 13:3 – 16, 21 – 31
495. Pryadilov VE (1980) Protection of reinforcement from corrosion in heavy ash-concrete. Beton Zhelezobeton 9:30 – 31
496. Pshenitsyn PA (1960) Automation of thermostatic method of the cement hardening heat determination. In: Methods of the study of silicates and building materials. Gosstroyizdat, Moscow, p 154
497. Puzyryov EI, Metyolkin II, Shapiro YM, Doronina AA (1972) Welding of metals with aluminooxide ceramics containing no vitreous phase. Elektronika SVCh 10:89
498. Ralko AV (1958, 1960) Experimental investigation of heat-mass-transfer during burning; Kinetics of the burning process. Inzh Fiz Zh 1:88 – 92; 3:12 – 16
499. Ramberg G (1957) Thermodynamics and kinetics of petrogenesis. In: Earth crust. Inostrannaya Lit, Moscow, p 447
500. Rassonskaya IS, Semendyayeva NK (1963) Phase transformations of some lithium salts. Zh Neorg Khim 8(6):1419 – 1425
501. Ratinov VB, Zabezhinskiy YL, Rosenberg TI (1957) To the theory of hardening of mineral binders. Papers of VNIIZhelezobeton. Promstroyizdat, Moscow, pp 3 – 35
502. Ratinov VB, Kucheryaeva GD, Melent'yeva GG (1961) Thermodynamical and diffusion characteristics of silicate cement components at solution in water. Dokl Akad Nauk SSSR 136(4):875 – 878
503. Ratinov VB, Rosenberg TI (1950) Basic regularities of film building at hardening. Papers of NIIZhelezobeton. Promstroyizdat, Moscow, pp 7 – 36
504. Ratinov VB, Todes OM (1960) A spontaneous crystallization of dissociated salts. Dokl Akad Nauk SSSR 132(N2):402 – 405
Ratinov VB, Todes OM (1961) Spontaneous crystallization of substances dissociating in a solution hardening of binders. Proceedings of the Institute of Reinforced Concrete, Building Materials and Rocks N4:154 – 170
505. Ratinov VB, Ivanov FM (1970) Chemistry in building. Stroyizdat, Moscow
506. Rehbinder PA (1958) Urgent problems of the colloid chemistry. Koll Zh 20(5):527
507. Rehbinder PA (1958) Physico-chemical mechanics. Izd Znan, Moscow
508. Rey M (1949) Thermodynamic activity of silica and of oxides in silicate melts. Discussions Faraday Soc 4:257 – 265
509. Reichelt W. Gross B (1966) Methoden zur Berechnung von Schmelzdiagrammen binärer Oxid-Systeme. Ber Forschungsarb Inst Kerntechn, Techn Univ Berlin (West-Berlin) 2:945
510. Reisman A (1958) Reactions of the group VB pentoxides with alkali oxides and carbonates. IX. DTA study of alkali metal carbonates. J Am Chem Soc 80(14):3558 – 3561
511. Repa AG (1952) Thermodynamics of reactions of silicate systems. Nauchno-tekhn Inform Bull VNIIStekla 2:4 – 15

512. Reznitskiy AA (1961) A method of approximate calculation of heat of formation of inorganic compounds. Zh Fiz Khim 35:1853 – 1859
513. Richardson FD (1953) The physical chemistry of metals. Inst Min Met, London, pp 75 – 95
514. Richardson FD, Jeffers JN, Withers G (1950) Thermodynamics of substances of interest in iron and steel making. II. Compounds between oxides. J Iron Steel Inst 166(5):213 – 245
515. Richardson FD, Webb LE (1955) Oxygen in molten lead and the thermodynamics of lead oxide silica melts. Bull Inst Min Met 584:529 – 564
516. Richter H (1942) The thermochemistry of portland cement formation. Diss Techn Hochschule, Berlin, Zementverlag
517. Rita RA, Bergeron CG, Lukacs JM (1973) Enthalpy of fusion of lead silicate. J Am Ceram Soc 56(1):47 – 48
518. Roberts MH (1957) New calcium aluminate hydrates. J Appl Chem (London) 7:543 – 546
519. Robson TD (1968) Chemistry of calcium aluminates and their relating compounds. In: 5th International Symposium on the Chemistry of Cement, Tokyo. Cem Assoc, Japan
520. Rode TV (1953) Thermoanalytical study of lithium carbonate. Dokl Akad Nauk SSSR 91:313 – 314
521. Rodigina EI, Gomel'skiy KZ (1961) Heat-contents of berillium and lithium oxides at high temperatures. NSFKh 35(8):1828 – 1831
522. Rodnikova VV, Sakhno VN (1966) The dependance of heat capacities of glass on the chemical constitution and temperature. Vestnik KPI Ser Khim Mashin 3:10 – 14
523. Rodnikova VV, Sakhno VN (1967) Investigation of the heat capacity of barium glasses as a function of their chemical composition and temperature. Optiko-mekh Prom 4:32 – 34
524. Rodnikova VV, Sakhno VN, Gorodov VS (1969) Study of the influence of glass ingredients on its heat capacity at high temperatures. Vestnik KPI Ser Khim Mash 6:23
525. Roller PS, Ervin G (1940) The system calcium oxide-silica-water at 30°. The association of silicate ion in diluted alkali solution. J Am Chem Soc 62(3):461 – 471
526. Rosen E, Muan M (1965) Stability of $ZrSiO_4$ at high temperature as derived from equilibrium measurements in $CoO-ZrO_2-SiO_2$ and $CoO-SiO_2$ systems. J Am Ceram Soc 48(11):603 – 604
527. Rossini FD (1950) Chemical thermodynamics. Wiley, New York
528. Rossini FD, Wagman DD (1952) Selected values of chemical thermodynamic properties. NBS, Washington
529. Rossini FD (ed) (1955) Thermodynamics and physics of matter. Princeton Univ Press, Princeton
530. Rossoti F, Rossoti K (1965) The determination of the stability constants and other equilibrium constants in solutions. Mir, Moscow
531. Roth W (1942) Heats of formation of calcium aluminates. Z Elektrochem 48: 264 – 267
532. Roy DM, Johnson AM (1967) Investigations of stabilities of calcium silicate hydrates at elevated temperatures and pressures. Autoclaved Calcium Silicate Bldg Prod Symp London, pp 114 – 120
533. Roy R, Osborn E (1954) Studies in the system $Al_2O_3-SiO_2-H_2O$. In: Experimental studies in the field of petrography and ore formation. Inostrannaya Lit, Moscow, pp 71 – 76
534. Royak SM, Royak GS (1969) Special cements. Stroyizdat, Moscow

535. Rumyantsev PF, Khotimchenko VS, Nikushchenko VM (1974) The hydration of calcium aluminates. Nauka, Leningrad
536. Rusanov AI (1967) Phase equilibria and surface phenomena. Khimiya, Leningrad, p 316
537. Ryabin VA, Ostroumov MA, Svit TF (1974) A handbook of the thermodynamic properties of substances. Khimiya, Leningrad, p 392
538. Rykalin NN, Shorshorov MK, Krasulin YI (1965) Physical and chemical problems of combining different materials. Izv Akad Nauk SSSR Neorg Mat 1(1):29
539. Rykov VI, Yakovleva GS, Slonovskiy LF, Sheynfel'd VL (1968) Collection 'Surface phenomena in melts'. Naukova Dumka, Kiev, pp 47–63
540. Ryschkevitsch E (1944) Linear shrinkage and E-modul depending on the burning temperature. Ber Dtsch Keram Ges 25:105–112
541. Saduakasov AS, Bondarenko AI, Akhmetov SF (1969) Formation of calcium silicates during the heating of mixtures of wollastonite rock with calcium carbonate. Tr Khim-Met Inst Akad Nauk Kaz SSR 11:89–112
542. Sakharov BP, Solomatin MS, Dolzhanskaya YB (1968) Technical and economical information of the industry of chemical agents and hyperpure substances. Khimiya, Moscow, pp 377–383
543. Salmang H, Scholze H (1968) Die physikalischen und chemischen Grundlagen der Keramik, 5. Aufl. Springer, Berlin Heidelberg New York
544. Samoylov OY (1957) The structure of aqueous solutions of electrolytes and the hydration of ions. Izd Akad Nauk SSSR, Moscow
545. Samsonov GV, Shulishova OI (1965) Energy of crystalline lattice. In: High-temperature inorganic compounds. Naukova Dumka, Kiev, pp 116–140
546. Šatava V (1965) Introduction to the physical chemistry of silicates. SNTL, Praha
547. Saull VA (1955) Chemical energy and metamorphism. Geochim Cosmochim Acta 8:86–106
548. Savitskiy YM (Gen Ed) (1966) High-melting metallic materials for cosmic technique. Mir, Moscow
549. Schairer JE (1951) Phase transformation in polycomponent silicate systems. Phase transformations in solids. Chapman, London, pp 278–295
550. Schmalzried H, Navrotsky A (1978) Festkörperthermodynamik. Chemie des festen Zustandes. Akad Verlag, Berlin
551. Scholze H (1955) To the silimanite–mullit problem. Der DKG 32:381–385
552. Shomate CH, Cohen AJ (1955) High-temperature heat content and entropy of lithium oxide and lithium hydroxide. J Am Chem Soc 77:285–286
553. Schumann R Jr, Ensio PJ (1951) Thermodynamics of iron silicate slags: slags saturated with γ iron. J Metals 3:401–411
554. Schwiete HE, Hummel C (1958) Thermodynamics investigation in the system Na_2O-SiO_2. Glastechn Ber 32(10):413–420
Forschungsberichte des Wirtschafts- und Verkehrsministeriums Nordrhein-Westfalen, Heft 515. Westdeutscher Verlag, Köln Opladen
555. Schwiete HE, Müller-Hesse H, Orhun OT (1962) Forschungsberichte des Landes Nordrhein-Westfalen, Heft 1077. Westdeutscher Verlag, Köln Opladen
556. Schwiete HE, Müller-Hesse H, Planz IE (1961) Investigation of the solid phase reaction in the system $BaO-Al_2O_3-SiO_2$ by means of infrared spectroscopy. Forschungsberichte des Landes Nordrhein-Westfalen, Heft 998. Westdeutscher Verlag, Köln Opladen
557. Schwiete HE, Tan Tik-Ien A (1960) Contribution to the determination of the heat of hydration of clinker minerals. In: 4th International Symposium on the Chemistry of Cement. NBS, Washington

558. Schwiete HE, Wagner H (1932) The specific heat of bottle glasses. Glastech Ber 10:26 – 30
559. Schwiete HE, Ziegler G (1958) Foundations and application fields of the differential microcalorimetry. Ber Dtsch Keram Ges 35(6):193 – 204
560. Segalova EE, Izmaylova VN (1958) Structurebuilding in the process of hydration hardening of calcium sulphati semihydrate. Koll Zh 20(5):601 – 610
561. Segalova EE, Rehbinder PA (1960) Contemporary physico-chemical conceptions of the process of binding materials hardening. Stroitel'nyye materialy (1):21
562. Seij BK (1969) Thermodynamics of multicomponent systems. Nedra, Moscow
563. Semyonov NN (1959) The main problems of the chemical kinetics. Izd Akad Nauk SSSR, Moscow
564. Shakhparonov MI (1956) An introduction to the molecular theory of solutions. Gostekhteorizdat, Moscow
565. Sharp DE, Ginther LB (1951) Effect of composition and temperature on the specific heat of glass. J Am Ceram Soc 34:260 – 270
566. Shchukaryov SA (1945) On the thermodynamic stability of manganum and iron oxides. Uch Zap LGU Ser Khim Nauki 7(79):197 – 254
567. Sheller T, Kuzel' KY (1976) Study of the hydrates of dicalcium aluminate hydrate. In: 6th International Congress on the Chemistry of Cement, vol 2(1). Stroyizdat, Moscow, pp 217 – 220
568. Sherman J (1932) Crystal energies of ionic compounds and thermochemical application. Chem Rev 11:93 – 170
569. Sheybany HA (1948, 1949) Structure of mixed alkali-silicate glasses. Verres Réfract 2:127 – 145, 229 – 242, 363 – 375 (1948); 3:27 – 39 (1949)
570. Shikiryanskiy AM (1976) Discussion. In: VIth International Congress on the Chemistry of Cement, vol 2(1). Stroyizdat, Moscow, p 186
571. Shteynberg YG (1967) Strontium glazes. Gosstroyizdat, Moscow
572. Shul'ts MM (1969) Some thermodynamic aspects in liquation phenomena. In: 1st All-union Symposium on Luquation Phenomena in Glasses. Nauka, Leningrad, pp 22 – 26
573. Shul'ts MM (1976) Thermodynamic functions of the formation of binary and more complicated compounds in the light of equilibrium stability criteria. In: Proceedings of the Seminar 'Thermodynamics and properties of condensed silicate and oxide systems'. Veda Izd Slov Akad Nauk, Bratislava, pp 9 – 17
574. Shul'ts MM (1981) Evolution of fundamental studies of silicates and refractory oxides. In: XIIth Conference on General and Applied Chemistry of the Mendeleyev Society. Synopses of papers N5. Silicate chemistry and technology. Nauka, Moscow
575. Shvartsman LA, Tomilin IA (1960) Acid-basic properties of metallurgical slags. In: Problems of slag reprocessings. Chelyabinsk, p 94
576. Silvermann SM (1956) Phase transformation of quartz. J Chem Phys 25(5):1081
577. Simon F, Lange F (1926) Zur Frage der Entropie amorpher Substanzen. Z Phys 38:227
578. Sivertsev GN (1956) Some experimental premises for creation a universal theory of the hardening of binders on the colloid-chemical base. In: Proceedings of the Conference on the Chemistry of Cement. Promstroyizdat, Moscow, pp 201 – 220
579. Siyenko M, Pleyn R, Khester R (1968) Structural inorganic chemistry. Mir, Moscow
580. Skuratov SM, Kolesov VP, Vorob'yov AF (1964) Thermochemistry, vol 1. Izd MGU, Moscow
581. Smit FG (1968) Physical geochemistry. Izd Nedra, Moscow
582. Sobolev VS (1949) Introduction to the mineralogy of silicates. Izd L'vovskogo Universiteta, L'vov

583. Sobolev VS, Tsvetkov AI (1956) On thermal effects during thermal analysis of dysten and andalusite. Geokhimiya 4:45
584. Sokol YY, Babushkin VI (1970) Thermodynamic studies of the hardening and corrosion reactions of acid-resistant cements and concretes effected by acid media. Zh Prikl Khim Leningrad 43(2):262–267
585. Solomin NV (1940) Viscosity and structure of molten quartz glass. Zh Fiz Khim 14:235–243
586. Sosman RB (1965) The phases of silica. Rutgers University Press, New Brunswick, New York
587. Sryvalin IT, Yesin OA (1959) Activities of the components of melts of the system $CaO-Al_2O_3-SiO_2$. Izv Vyssh Ucheb Zav 8:9–16
588. Sryvalin IT, Yesin OA (1960) Thermodynamic properties of silicate melts in systems $CaO-FeO-SiO_2$ and $CaO-Al_2O_3-SiO_2$. In: Papers of 3rd All-union Conference on Vitreous State. Akad Nauk SSSR, Moscow Leningrad, p 130–131
589. Sryvalin IT, Yesin OA, Khlynov VV (1959) On the deviation of molten silicates from ideal solutions. Zh Neorg Khim 4:877–883
590. Starosel'skiy AA, Ol'ginskiy AG, Spirin YA (1976) Electrokinetic properties of cement stone. In: VIth International Congress on the Chemistry of Cement. Stroyizdat, Moscow, pp 192–195
591. Stein HN (1976) Discussion. In: VIth International Congress on the Chemistry of Cement, vol 2(2). Stroyizdat, Moscow, p 207
592. Stein HN (1980) The colloid chemistry of calcium aluminate hydrates. In: Proceedings of the VIIth Congress on the Chemistry of Cement, Paris. Ed Septima 4:449–454
593. Steinherz AR, Welcman N (1958) Reaction of portland cement with water. Rev Mat Constr Trav Publ 517:265
594. Steinour HH (1952) The reactions and thermochemistry of cement hydration at ordinary temperature. In: Proceedings of the 3rd International Symposium on the Chemistry of Cement.Cement and Concrete Association, London, pp 261–289
595. Stevens CG, Turkgodan ES (1955) Heat of formation of sodium di-silicate. Trans Faraday Soc 51:356
596. Stott VH, Hilliard A (1949) Zircon refractories. Trans Br Ceram Soc 48(4):133–139
597. Stranskiy IN, Totomanov D (1933) Speed of embryo-formation and the Ostwald step rule. Z Phys Chem A 163:399–408
598. Strassen zur H (1957) Theoretical needed heat for cement burning. Zement-Kalk-Gips 10:1–12
599. Strelkov MI (1956) Change of the real composition of the liquid phase at cement hardening. In: Proceedings of the Conference on the Chemistry of Cement. Promstroyizdat, Moscow, pp 183–200
600. Stroganov YF, Matveyev GM, Khokonov KB (1960) Application of low-temperature calorimetry in the study of silicates. In: Up-to-date methods of investigation of silicates and building materials. Promstroyizdat, Moscow, pp 52–66
601. Sychyov MM (1974) The hardening of binders. Stroyizdat, Leningrad
602. Syrkin YK, Dyatkina MY (1946) The chemical bond and the structure of molecules. Goskhimizdat, Moscow
603. Syrota NN (1969) Thermodynamics and statistical physics. Vysheishaya Shk, Minsk
604. Takahashi K, Yoshio T (1970) Thermochemical investigations on glass. J Ceram Soc Jpn 78(1):29–32; 10:329–337
605. Takahashi K, Yoshio T (1973) Thermodynamic quantities of alkali silicates in the temperature range from 25 °C to melting point. Yogyo-Kyokai-Shi 81(12):524–533
606. Talaber Y (1976) Alumina cements. In: 6th International Congress on the Chemistry of Cement. Stroyizdat, Moscow, pp 124–133

607. Tamás F, Pál J (1970) Phase equilibria spatial diagram. Akad Kiado, Budapest
608. Tammann G (1924) To the thermodynamic of the reaction in the mixtures of two crystal types. Ges Wiss Nachr Math-Phys Klasse 2:159
609. Tammann G (1925) Reactivity of solids. Z Anorg Allg Chem 149:21 – 29
610. Tanaka M (1960) Thermodynamics of the formation of silicate glass. Yogyo Kyokai Shi 68(9):204 – 210
611. Tarasov VV (1959) New problems of the physics of glass. Gosstroyizdat, Moscow
612. Taylor HFW (1950) Hydrated calcium silicates. I. Compounds formation at ordinary temperatures. J Chem Soc 1950:3682 – 3690
613. Taylor HFW (1953) Hydrated calcium silicates. V. The water content of calcium silicate hydrate. J Chem Soc 1953:163 – 171
614. Taylor HFW (1955) The hydration of portland cement. Ind Chim Belge 20:63 – 66
615. Taylor HFW (1964) The chemistry of cements. Academic Press, London New York
616. Taylor HFW (1968) The calcium silicate hydrates. In: 5th International Symposium on the Chemistry of Cement, Tokyo, vol 1, pp 21 – 26
617. Taylor HFW (1976) The crystallochemistry of the products of the cement hydration. In: 6th International Congress on the Chemistry of Cement, vol II, book 1. Stroyizdat, Moscow, pp 192 – 206
618. Taylor JE (1968) The heat of formation of xonotlite, tobermorite, hillebrandite and afwillite. In: Proceedings of the 9th Conference on the Silicate Industry. Akad Kiado, Budapest, pp 179 – 184
619. Taylor RW, Schmalzried H (1964) The free energy of formation of some titanates, silicates, and magnesium aluminates measurements made with galvanic cells involving solid electrolytes. J Phys Chem 68(9):2444 – 2449
620. Thermal analysis. Proc of Intern Conferences 1 – 7 (1965 – 1981)
621. Thermic constants of substances. A handbook (1972) Izd Akad Nauk SSSR, part VI, Moscow, p 192
622. Thermodynamics. The office of the scientific and technical terminology of the academy of sciences of the USSR. A book of recommended terms (1973), vol 85. Nauka, Moscow, p 55
623. Thermodynamic properties of individual substances (1962) A handbook edited by Acad VP Glushko. Izd Akad Nauk SSSR, Moscow
624. Thermodynamic properties of inorganic substances (1965) A handbook. Atomizdat, Moscow
625. Thilo E, Funk H, Wichman EM (1950) Abh Dtsch Akad Wiss Berl Kl Mat Allg Naturwiss 4(1)
626. Thomas SB, Parks GS (1931) Studies on glass. VI. Some specific heat data on boron trioxide. J Phys Chem 35:2091 – 2102
627. Thompson JB (1955) The thermodynamic basis for the mineral facies concept. Am J Sci 253:65
628. Thorvaldson T, Brown WG (1930) Thermochemistry of the compounds occurring in the system: $CaO - Al_2O_3 - SiO_2$(II) Heat of solution of $Ca(OH)_2$ in $HCl \cdot 200H_2O$. J Am Chem Soc 32:910 – 915
Thorvaldson T, Brown WG, Peaker CR (1930) Thermochemistry of the compounds occurring in the system: $CaO - Al_2O_3 - SiO_2$(III) Heat of hydration of calcium oxide. J Am Chem Soc 32:3927 – 3936
629. Todd SS (1951) Low-temperature heat capacities and entropies at 298 K of crystalline calcium orthosilicate, zinc, orthosilicate and tricalcium silicate. J Am Chem Soc 73(7):3277 – 3278
630. Tool AQ, Eichlin CG (1920) Absorption of heat in glass. J Optic Soc Am 4:1920

631. Tool AQ, Eichlin CG (1925) Variations in glass caused by heat treatment. J Am Ceram Soc 8:3
632. Tool AQ, Eichlin CG (1931) Variations caused in the heating curves of glass by heat treatment. J Am Ceram Soc 14:265 – 275
633. Torgeson DR, Sahama TG (1948) A hydrofluoric acid solution calorimeter and the determination of the heats of formation of Mg_2SiO_4, $MgSiO_3$, $CaSiO_3$. J Am Chem Soc 70:2156 – 2160
634. Torkar K (1961) On the thermodynamics of solids. In: The Proceedings of the 4th International Symposium on the Reactivity of Solids. Elsevier, Amsterdam, pp 400 – 408
635. Toropov NA (1956) Chemistry of cements. Promstroyizdat, Moscow
636. Toropov NA, Barzakovskiy VP, Lapin VV, Kurtseva NN (1969) Phase diagrams of silicate systems. Nauka, Leningrad
637. Toropov NA, Barzakovskiy VP, Bondar' JA, Udalov JP (1970) Phase diagrams of silicate systems. A handbook 11. Nauka, Leningrad
638. Toropov HA, Barzakovskiy VP, Lapin VV, Kurtseva NN, Boukova AJ (1972) Phase diagrams of silicate systems. A handbook III. Nauka, Leningrad
639. Toropov NA, Barzakovskiy VP (1963) The high-temperature chemistry of silicate and other oxide systems. Izd Akad Nauk SSSR, Moscow Leningrad
640. Toropov NA, Nikogosyan KS, Boykova AI (1960) Formation of α-Ca_2SiO_4 hydrate. Dokl Akad Nauk SSSR 135(1):98 – 100
641. Tret'yakov YD (1967) Thermodynamics of ferrites. Khimiya, Leningrad
642. Tripp HP, King BW (1955) Thermodynamic data on oxides at elevated temperatures. J Am Ceram Soc 38(12):432 – 437
643. Troitzsch H (1936) Dissertation. Braunschweig
644. Trzebiatowski W, Damm J, Romotowski T (1956) Reactions of anhydrite with quartz (I) thermodynamics, (II) reactions in the system $CaSO_4 + SiO_2$. Roczn Chem 30(2):431 – 443
645. Tsagareyshvili DS, Gvelesiani GC (1968) On plotting equations of temperature dependence of isobaric potentials of reactions of formation of inorganic compounds. Soobshcheniya Akad Nauk Gruzinskoy SSR, vol 2, pp 429 – 434
646. Tsimermanis LB, Genkin AR, Shtakel'berg DI (1976) Thermodynamic analysis of the hardening of mineral binders in a closed system. In: Proceedings of the VIth International Congress on the Chemistry of Cement (1974) Th vol of the hydration and hardening of cement. Stroyizdat, Moscow, pp 25 – 28
647. Turkdogan ET, Pearson I (1953) Thermodynamic function of elements and compounds: entropies of inorganic substances. J Appl Chem 3:495 – 501
648. Turnbull D (1956) Solid state physics, vol 3. Academic Press, London New York
649. Turritziani R (1969) Calcium hydroaluminates and cognate compounds. In: Chemistry of cement. Stroyizdat, Moscow, pp 167 – 213
650. Tydlint V, Blazek A, Endrys J, Stanek J (1972) Measurement of enthalpy and calculation of specific heat of glasses. Glastech Ber 45(8):352 – 359
651. Tyomkin MI (1946) Mixtures of melted salts as ion mixtures. Zh Fiz Khim 20:105
652. Tyomkin MI, Shvartsman LA (1948) Subsidiary tables for calculations in chemical thermodynamics. Usp Khim 17(2):259 – 262
653. Ubbelohde AR (1950) Q Rev Chem Soc 4:356
654. Uhlmann DR Chalmers B (1966) Nucleation phenomena. American Chemical Society, Washington
655. Uhlmann DR (1969) Kinet React Ion Syst, New York, pp 172 – 195
656. Uhlmann DR, Hays JE, Turnlull D (1966) Effect of high pressure on B_2O_3 crystallization, densification and the crystallization anomaly. Phys Chem Glasses 7:159

657. Uiks KY, Blok FY (1965) The thermodynamic properties of 65 elements, of their oxides, halides, carbides and nitrides. Metallurgiya, Moscow
658. Ulikh G (1933, 1940) The chemical thermodynamics. An introduction to the theory of the chemical affinity and equilibria. Goskhimizdat Usp Khim 9(2 – 3):214
659. Urusov VS (1977) The theory of isomorphic miscibility. Nauka, Moscow, p 252
660. Urosovskaya LN, Yakhkind AK (1969) Thermodynamic analysis of chemical reactions taking place during glass-formation in the system $Al(PO_3)_3-NaF$. Zh Prikl Khim Moscow 42(8):1732 – 1739
661. Vail JG (1952) Soluble silicates in industry. American Chemical Society Monograph Series, The Chemical Catalog Company, New York
662. Valenkov NN, Poray-Koshits YA (1935) On chemical compounds in the system $PbO-SiO_2$ in crystalline state. Zh Fiz Khim 6(6):757 – 761
663. Van Byuren (1962) Defects in crystals. Inostrannaya Liter, Moscow
664. Vaysburd SY, Novikova NN (1969) In: Proceedings of 4th All-union Conference on the Physical Chemistry and Electrochemistry of Melted Salts and Slags, part I. Naukova Dumka, Kiev
665. Ved' YI, Baklanov GM, Zharov YF (1966) Physico-chemical principles of the production of autoclaved building materials. Budivel'nik, Kiev
666. Ved' YI, Zharov YE, Baklanov GM, Breshchenko AA, Gordeyev YV, Golodovnikova AP (1972) Early stages hardening of portland cement. Budivel'ni Mat i Konstr 3:30 – 31
667. Venner R (1950) Thermodynamical calculations. Inostrannaya Liter, Moscow
668. Verbeck GJ, Helmuth RH (1968) Structures and physical properties of cement paste. In: 5th International Symposium on the Chemistry of Cement, vol 3. Nat Bur Stand, Washington, pp 1 – 32
669. Verbek GJ, Foster CW (1950) Proc Am Soc Testing Mat 50:1235 – 1262
670. Vernadskiy VI, Kurbatov SM (1937) Earth silicates, alumino-silicates and their analogues. ONTI, Moscow
671. Veryatin UD et al. (1965) Thermodynamic properties of inorganic substances. Khimiya, Moscow
672. Veynik AI (1956) Technical thermodynamics and the fundamentals of heat transfer. Metallurgizdat, Moscow
673. Vladimirov LP (1956) Accelerated methods of the thermodynamic calculations of equilibria of chemical reactions. Izd L'vovsk Gosuniversiteta, L'vov
674. Vol'kenshteyn MV (1959) Configurational statistics of polymer chains. Izd Akad Nauk SSSR, Moscow
675. Volmer M (1939) Kinetik der Phasenbildung. Steinkopff, Dresden Leipzig
676. Volzhenskiy AV, Burov YS, Kolokol'nikov VS (1966) Mineral binders. Stroyizdat, Moscow
677. Von Wartenberg CH, Gurr W (1931) Melting-point diagrams of highly inert oxides. Z Anorg Allg Chem 196:279 – 280
678. Voskresenskaya NK, Banashek YI (1954) Thermodynamic properties of unhydrous Li^- sulphate at high temperatures. Izv Sekt Fiz Khim Anal IONKh NS Kurnakova 25:150 – 158
679. Vyrodov IP (1963) Theoretical and experimental studies of the processes of the mineral binder hardening. Tr Kubanskogo Sel'skokhoz Inst 11(39), Krasnodar
680. Vysotskiy ZZ (1971) An outline of the history of the chemistry of disperse silicas. Naukova Dumka, Kiev
681. Wagman DD et al. (1968) Selected values of chemical thermodynamic properties. Techn Notes N 270-3, 264 pp; N 270-4, 141 pp

682. Wagman DD, Evans WH, Parker VB et al. (1971) Selected values of chemical thermodynamic properties. NBS, Techn Note 270-5, 37 pp
683. Wagner H (1932) The thermochemistry of the silicates of calcium and magnesium and of diopside. Z Anorg Allg Chem 208(1):1 – 22
684. Waldbaum DR (1965) Thermodynamic properties of mullite, andalusite, kyonite and sillimanite. Am Min 50:186 – 195
685. Washburn EW, Libman EE (1920) Approximate determination of the system ZrO_2-SiO_2. J Am Ceram Soc 3:634 – 640
686. Weitz E, Frank H, Schuchard M (1950) Chem Ztg 74:256 – 257
687. Weyl WA (1951) Transitions in glass. In: Phase transformations in solids. Wiley, London, pp 296 – 331
688. Weyl WA (1962) The constitution of glasses. Internal Publ, New York London
689. White J (1974) Refractories research IV. The thermodynamic stability of oxide refractories. J Aust Ceram Soc 10(1):1 – 4
690. Winkelmann A (1893) On the specific heat of glasses of different composition. J Am Phys Chem 49:401
691. Wilson RN (1962) In: 5th International Congress on Electron Microscopy NI. Academic Press, London New York
692. Wise SS, Margave IR, Teder HM, Hubbard WN (1962) The heat of formation of silica and silicon tetrafluoride. J Phys Chem 66:380
693. Wietzel R (1921) The relative stability of the glass and crystal phases of silicon dioxide. Z Anorg Allg Chem 116:71 – 95
694. Woytowitch RF (1971) Thermodynamic characteristics of high temperature melting compounds. Naukowa Dumka, Kiew
695. Woods H, Steinour HH, Starke HR (1932) The heat evolved by cement during the hardening. Eng News-Record 109:404 – 407
696. Woods H, Steinour HH, Starke HR (1932) The heat evolved by cement in relation to strength. Ind Eng Chem 24:1207 – 1214
697. Wygant IF, Kingery WD (1952) Applications of thermodynamics in ceramics. II Summary, bibliography and source of data. Am Ceram Soc Bull 31:5 – 10
698. Wyrodow IP, Mchedlov-Petrossyan OP (1967) Determination of the direction of reactions and interconnection between kinetic and thermodynamic parameters of the system. Silikattechnik 18(8):252 – 253
699. Yageman VD, Pchelyakov KA, Matveyev GM (1976) Some problems of the investigation of the heat capacity of glass forming melts. In: Production and investigation of glass and silicate materials, vol 5. Yaroslavl, pp 133 – 141
700. Yatsimirskiy KB (1951) Thermochemistry of complex compounds. Akad Nauk SSSR, Moscow
701. Yatsimirskiy KB, Krestov GA (1960) Lattice entropy of compounds with polyatomic ions. Zh Fiz Khim 34:2248 – 2453
702. Yesin OA (1948) Effect of silicic acid on the equilibrium between liquid iron and the simplest basic slags. Zh Fiz Khim 22:617 – 623; Izv Akad Nauk SSSR OTN 6:561
703. Yesin OA (1954) Ionic theory of liquid slags. Papers of Ural Politechn Inst Sverdlovsk 49:4
704. Yesin OA (1960) On the effective ion charge in slags. Izv Vuz Chyornaya Met 8:5 – 14
705. Yesin OA (1960) On the nature of liquid slags. In: Problems of slag reprocessing. Metallurgizdat, Chelyabinsk, p 81
706. Yesin OA, Vatolin NA (1957) In: Proceedings of 3rd Conference on the Physicochemical Principles of the Steel Production. Izv Akad Nauk SSSR p 263
707. Yesin OA, Kozheurov VA (1948) Distribution of silicon between metal and basic slag from the point of view of the ionic theory. Zh Prikl Khim Moscow 21:765

708. Yesin OA, Lepinskikh BM (1954) Influence of cations on the strength of anions in melted slags. Dokl Akad Nauk SSSR 95:135
709. Yesin OA, Lepinskikh BM (1954) Investigation of the properties of the melted slags components by the method of electromotive forces. Izv Akad Nauk SSSR OTN 2:60
710. Yesin OA, Sryvalin IT, Khlynov VV (1957) Study of properties of melts in the system $PbO-Na_2O-SiO_2$ by the method of electromotive forces. Zh Neorg Khim 2(10):2429
711. Yesin OA, Gel'd PV (1966) Physical chemistry of pyrometallurgical processes, 2nd edn, part II. Metallurgiya, Moscow
712. Yevstrop'yev KS, Skornyakov MM (1949) Physico-chemical properties of the ternary system $Na_2O-PbO-SiO_2$. Sborn Statey Pod Red Grebenshchikova
713. Yudin BF (1960) 'The Mendeleyev Rule' and its application in calculation of physico-chemical values. Sborn Tr Gos Inst Prikl Khim 46:102–115
714. Yudin BF (1967) Chemistry of high-temperature materials. In: Proceedings of the 2nd All-union Conference on the Chemistry of Oxides at High Temperatures. Nauka, Leningrad, p 32
715. Zagar L (1962) Application of thermodynamics in the field of glass. Silicates Industriels 27(6):279–283
716. Zharkova LA (1962) Determination of the functions of titanates and silicates by the method of comparative calculation. Zh Fiz Khim 36(8):1819–1821
717. Zhirnova N (1934) A fusibility diagram of the system SrO_2-SiO_2. Zh Prikl Khim Leningrad 4:1464–1470
718. Zhukhovitskiy AA, Finkel'shtein BN, Kulikov IS (1951) On the theory of real solutions. Dokl Akad Nauk SSSR 81(2):227–230
719. Zhukhovitskiy AA, Shvartsman LA (1968) Physical chemistry. Metallurgizdat, Moscow

Appendix 1. Standard Enthalpies, Gibbs Free Energies and Heat Capacity Equations of Some Elements, Simple Substances and Compounds

The tables given in Appendix 1 comprise the mutually correlated values of three thermodynamic properties of substances at 25 °C: standard enthalpy of the formation from elements ΔH^0_{298} [kcal/mol], standard Gibbs free energy of the formation from elements ΔG^0_{298} [kcal/mol] and standard entropy S^0_{298} [cal/mol deg]. In addition equations of the temperature dependence of heat capacity of the type

$$C_p = a + bT + CT^{-2}$$

are given.

Elements and their compounds are alphabetically ranged. Indices accompanying the formulae of substances reveal their physical state: gas: gaseous, liq: liquid, cr: crystalline, am: amorphous, gl: vitreous.

Each value is followed by a reference to the literature from which it was borrowed, the references occupying a column of their own. Index c points to the values calculated by the authors at the time the book was written.

Standard Enthalpies, Gibbs Free Energies and Heat Capacity Equations of Some Elements, Simple Substances and Compounds

Formulae of compounds and ions	State	Heat of formation from elements $-\Delta H^0_{298}$ [kcal/mol]	References	Gibbs free energy of formation from elements $-\Delta G^0_{298}$ [kcal/mol]	References	Entropy S^0_{298} [cal/deg mol]	References	$C_p = f(T)$ a	$b \cdot 10^3$	$c \cdot 10^{-5}$	Temp. range in K	References
Al	gas	−78.00	[120]	−68.30	[120]	39.30	[120]	–	–	–	–	–
Al	cr	0.00	[90]	0.00	[90]	6.77	[90]	4.94	2.96	–	298–932	[90]
Al^{3+}	gas	−1310.70	[120]	–	–	–	–	–	–	–	–	–
Al_4C_3	cr	49.9	[120]	46.9	[120]	21.26	[120]	–	–	–	–	–
$AlCl_3$	cr	168.3	[120]	150.3	[120]	26.45	[120]	13.25	28.0	–	298–Tm	[90]
AlF_3	cr	359.5	[120]	340.6	[120]	15.88	[120]	20.93	3.0	–	727–1400	[90]
AlN	cr	76.0	[120]	68.6	[120]	4.82	[120]	–	–	–	–	–
Al_2O_3-α	cr	400.48	[90]	378.2	[90]	12.17	[90]	27.43	3.06	−8.47	298–1800	[32]
Al_2O_3-γ	cr	395.0	[120]	–	–	–	–	16.37	11.1	–	–	[88]
γ-AlO(OH) (boehmite)	cr	236.09	[90]	218.26	[90]	11.57	[90]	14.43	4.20	–	298–500	[35]
γ-$Al(OH)_3$ (gibbsite)	cr	308.8	[29]	275.8	[29]	16.34	[29]	8.65	45.6	–	298–425	[35]
$Al(OH)_3$	am	305.0	[90]	273.0	[90]	19.8	[90]	–	–	–	–	–
$Al_2O_3 \cdot 2SiO_2 \cdot 2H_2O$ (kaolinite)	cr	984.0	[29]	907.4	[29]	48.5	[90]	57.47	35.30	−7.87	298–1000	[85]
$Al_2O_3 \cdot 2SiO_2 \cdot 2H_2O$ (dickite)	cr	983.5	[29]	906.6	[29]	47.1	[90]	–	–	–	–	–
$Al_2O_3 \cdot 2SiO_2 \cdot 2H_2O$ (halloysite)	cr	979.5	[29]	902.9	[29]	48.6	[90]	–	–	–	–	–
$Al_2O_3 \cdot SiO_2$ (andalusite)		619.57	[30]	584.7	[90]	22.28	[30]	41.22	6.24	−12.22	298–1600	[98]
$Al_2O_3 \cdot SiO_2$ (kyanite)	cr	620.11	[30]	585.91	[90]	22.03	[30]	41.05	6.98	−12.46	298–1500	[98]
$Al_2O_3 \cdot SiO_2$ (sillimanite)	cr	618.83	[30]	583.75	[90]	22.99	[30]	39.30	8.04	−11.02	298–1500	[98]
$3Al_2O_3 \cdot 2SiO_2$ (mullite)	cr	1631.62	[67]	1539.98	[67]	60.8	[90]	115.9	11.2	−37.0	298–2000	[67]

$Al_2O_3 \cdot 2SiO_2$ (metakaolinite)	cr	807.46	[85]	758.46	[85]	29.68	[85]	7.06	5.1	−4.87	298−1700	[c]
$Al_2O_3 \cdot TiO_2$	cr	629.9	[5]	–	–	26.2	[33]	43.63	5.30	−11.21	298−1800	[7]
$Al_2(SO_4)_3$	cr	820.98	[104]	738.99	[104]	57.2	[104]	88.09	14.80	−27.12	–	[32]
$Al_2(SO_4)_3 \cdot 6H_2O$	cr	1268.14	[104]	1105.14	[104]	112.1	[104]	–	–	–	–	–
$AlVO_4$	cr	371.37	[95]	–	–	36.68	[95]	29.49	9.76	−2.80	298−968	[95]
As_4O_6 (octahedron)	cr	314.04	[120]	275.46	[120]	51.2	[120]	–	–	–	–	–
As_2O_5	cr	221.05	[120]	187.0	[120]	25.2	[120]	–	–	–	–	–
B	gas	−97.2	[104]	−86.7	[104]	36.649	[104]	–	–	–	–	–
B	cr	0.0	[104]	0.0	[104]	1.56	[104]	–	–	–	–	–
B_2O_3	cr	304.2	[120]	285.3	[120]	12.91	[120]	15.88	14.81	−0.49	298−723	[90]
B_2O_3	gl	299.84	[120]	282.6	[120]	18.6	[38]	2.28	42.10	–	298−723	[37]
B_4C	cr	17.0	[120]	17.0	[120]	6.48	[120]	22.99	5.40	−10.72	298−1700	[32]
BCl_3	liq	102.1	[120]	92.6	[120]	49.3	[120]	16.96	2.86	−2.44	298−1100	[32]
BF_3	cr	271.75	[120]	267.77	[120]	60.71	[120]	–	–	–	–	–
BN	cr	60.8	[120]	54.6	[120]	3.54	[120]	–	–	–	–	–
Ba	gas	−41.96	[104]	−34.60	[104]	40.699	[104]	–	–	–	–	–
Ba	cr	0.00	[104]	0.00	[104]	16.00	[104]	–	–	–	–	–
Ba^{2+}	gas	−395.705	[104]	–	–	–	–	–	–	–	–	–
$BaCl_2$	cr	205.56	[104]	193.80	[104]	30.0	[104]	17.0	3.34	–	278−1198	[32]
$BaCO_3$	cr	291.30	[104]	272.20	[104]	26.80	[104]	20.77	11.70	−2.86	298−1079	[52a]
BaF_2	cr	286.0	[52a]	–	–	23.0	[37]	–	–	–	–	–
$BaMoO_3$	cr	289.0	[100]	270.3	[100]	33.7	[100]	–	–	–	–	–
$BaMoO_4$	cr	368.5	[100]	343.0	[100]	35.4	[100]	25.37	13.38	–	–	[100]
BaO	cr	133.40	[104]	126.30	[104]	16.80	[104]	12.74	1.04	−1.984	298−1270	[52a]
$BaO \cdot Al_2O_3$	cr	557.88	[99]	–	–	–	–	–	–	–	–	–
$3BaO \cdot Al_2O_3$	cr	845.58	[99]	–	–	–	–	–	–	–	–	–
$BaO \cdot Fe_2O_3$	cr	306.6	[58]	–	–	–	–	–	–	–	–	–
$2BaO \cdot Fe_2O_3$	cr	421.6	[58]	–	–	–	–	–	–	–	–	–
$7BaO \cdot 2Fe_2O_3$	cr	1324.5	[58]	–	–	–	–	–	–	–	–	–
$Ba(OH)_2$	cr	226.2	[104]	204.7	[29]	–	–	–	–	–	–	–
$BaO \cdot 2SiO_2$	cr	610.18	[4]	577.08	[66]	36.8	[66]	41.28	3.125	−6.918	–	[66]
$2BaO \cdot 3SiO_2$	cr	1002.47	[4]	949.27	[66]	63.6	[66]	50.53	9.18	−0.168	–	[66]

Table (continued)

Formulae of compounds and ions	State	Heat of formation from elements $-\Delta H^0_{298}$ [kcal/mol]	Refer-ences	Gibbs free energy of forma-tion from elements $-\Delta G^0_{298}$ [kcal/mol]	Refer-ences	Entro-py S^0_{298} [cal/deg mol]	Refer-ences	$C_p = f(T)$ a	$b \cdot 10^3$	$c \cdot 10^{-5}$	Temp. range in K	Refer-ences
$BaO \cdot SiO_2$	cr	389.18	[4]	369.08	[66]	26.8	[66]	24.5	5.9	–	–	[66]
$2BaO \cdot SiO_2$	cr	549.03	[4]	521.83	[66]	43.5	[66]	34.5	8.0	–	–	[66]
$3BaO \cdot SiO_2$	cr	708.88	[66]	674.58	[66]	60.4	[66]	44.46	10.9	–	–	[66]
Ba_2Si_2	cr	367	[52a]	–	–	–	–	–	–	–	–	–
Ba_2Si_3	cr	107	[104]	–	–	–	–	–	–	–	–	–
$BaSO_4$	cr	350.2	[104]	323.38	[c]	31.6	[37]	9.89	1.37	–	298 – 1551	[c]
$BaSO_4$	cr	–	–	–	–	–	–	34.93	6.7	–	1551 – 1853	[c]
$2BaO \cdot TiO_2$	cr	538.0	[33]	511.1	[33]	–	–					
$BaO \cdot TiO_2$	cr	397.6	[33]	376.4	[33]	25.32	[33]	–	–	–	–	–
$2BaO \cdot ZrO_2$	cr	506	[25]	–	–	–	–	–	–	–	–	–
$BaO \cdot ZrO_2$	cr	383.4	[25]	–	–	–	–	–	–	–	–	–
$BeCl_2$	cr	122.3	[104]	118.8	[29]	205.6	[52a]	–	–	–	–	–
$BeCO_3$	cr	234.6	[75]	215.3	[75]	12.5	[75]	–	–	–	–	–
BeF_2	cr	227.0	[10]	–	–	13.7	[54]	–	–	–	–	–
BeO	cr	143.1	[11]	136.15	[75]	3.4	[37]	8.45	4.0	– 3.45	298 – 1200	[37]
$BeO \cdot SiO_2$	cr	367.98	[75]	347.15	[75]	12.9	[75]	22.8	2.08	– 10.95	–	[75]
$2BeO \cdot SiO_2$	cr	515.95	[52a]	488.4	[75]	15.4	[52a]	24.63	5.65	– 14.05	–	[75]
$BeSO_4$	cr	286.41	[34]	260.20	[34]	21.5	[34]	–	–	–	–	–
Ca(α)	cr	0.0	[104]	0.0	[104]	9.97	[104]	–	–	–	–	–
Ca	gas	42.6	[120]	– 34.5	[120]	36.99	[120]	–	–	–	–	–
Ca^+	gas	– 185.05	[120]	–	–	–	–	–	–	–	–	–
Ca^{2+}	gas	– 460.29	[120]	–	–	–	–	–	–	–	–	–

$CaCl_2$ (hydrophilite)	cr	189.8	[90]	179.05	[90]	27.2	[90]	17.18	3.04	−0.6	298−1055	[35]
$CaCl_2 \cdot 2CaO$	cr	505.0	[104]	–	–	–	–	–	–	–	–	–
$CaCl_2 \cdot 3CaO$	cr	654.0	[104]	–	–	–	–	–	–	–	–	–
$CaCl_2 \cdot 3CaO \cdot 3H_2O$	cr	910.6	[104]	–	–	–	–	–	–	–	–	–
$CaCl_2 \cdot 3CaO \cdot 16H_2O$	cr	1833.0	[104]	–	–	–	–	–	–	–	–	–
$CaOCl_2$	cr	178.6	[104]	–	–	–	–	–	–	–	–	–
$CaBr_2$	cr	161.8	[104]	156.8	[104]	31	[104]	–	–	–	–	–
CaI_2	cr	127.8	[104]	126.6	[104]	34	[104]	–	–	–	–	–
CaF_2	cr	290.30	[104]	277.7	[104]	15.46	[104]	14.3	7.28	−0.47	–	[32]
$CaCO_3$ (calcite)	cr	288.45	[104]	269.78	[104]	22.2	[104]	24.98	5.24	−6.2	298−1200	[29]
$CaCO_3$ (aragonite)	cr	288.49	[104]	269.58	[104]	21.2	[104]	20.13	10.24	−3.34	298−600	[35]
$CaCO_3 \cdot MgCO_3$	cr	557.29	[104]	518.48	[104]	–	–	37.33	19.23	−5.16	–	[88]
$CaMoO_3$	cr	299.4	[20]	279.9	[20]	24.7	[20]	–	–	–	–	–
$CaMoO_4$	cr	372.3	[100]	346.6	[100]	28.6	[100]	–	–	–	–	–
CaO		151.9	[104]	144.4	[104]	9.50	[104]	11.67	1.08	−1.56	298−1800	[36]
$CaO \cdot Al_2O_3$	cr	556.18	[17]	527.7	[17]	27.3	[43]	36.01	9.98	−7.96	298−1800	[8]
$3CaO \cdot Al_2O_3$	cr	851.0	[17]	808.4	[17]	49.1	[43]	62.28	4.58	−12.09	298−1800	[8]
$12CaO \cdot 7Al_2O_3$	cr	4640.16	[17]	4410.0	[17]	249.7	[43]	301.96	65.5	−55.3	298−1310	[8]
$CaO \cdot 2Al_2O_3$	cr	957.06	[c]	907.1	[c]	42.5	[43]	66.09	5.48	−17.8	298−1800	[8]
$3CaO \cdot Al_2O_3 \cdot 6H_2O$	cr	1326.0	[42]	1198.4	[2]	96.7	[2]	68.91	127.17	–	–	[2]
$2CaO \cdot Al_2O_3 \cdot 8H_2O$	cr	1299.2	[2]	1151.54	[2]	106.4	[2]	68.38	153.45	–	–	[2]
$CaO \cdot Al_2O_3 \cdot 10H_2O$	cr	1271.4	[2]	1103.7	[2]	116.1	[2]	67.85	182.46	–	–	[2]
$4CaO \cdot Al_2O_3 \cdot 13H_2O$	cr	1988.0	[2]	1756.17	[2]	171.6	[2]	109.6	242.19	–	–	[2]
$4CaO \cdot Al_2O_3 \cdot 19H_2O$	cr	2411.0	[2]	2096.33	[2]	228	[2]	122.36	394.0	–	–	[2]
$CaO \cdot Al_2O_3 \cdot 6SiO_2$	cr	1907.0	[120]	–	–	–	–	–	–	–	–	–
$3CaO \cdot Al_2O_3 \cdot 2SiO_2$	cr	1329.0	[120]	–	–	–	–	–	–	–	–	–
$CaO \cdot Al_2O_3 \cdot 2SiO_2$ (anorthite)	cr	1013.33	[29]	959.4	[29]	47.6	[24]	64.42	13.7	−16.89	298−1700	[35]
$2CaO \cdot Al_2O_3 \cdot SiO_2$ (gehlenite)	cr	957.1	[29]	909.6	[29]	50.1	[29]	63.74	8.0	−15.12	298−1700	[90]
$CaO \cdot Al_2O_3 \cdot 2SiO_2$	gl	996.2	[29]	945.0	[29]	56.6	[29]	–	–	–	–	–
$CaO \cdot Al_2O_3 \cdot 2SiO_2 \cdot 2H_2O$ (lawsonite)	cr	1165.2	[29]	1080.8	[29]	41.87	[24]	–	–	–	–	–

Table (continued)

Formulae of compounds and ions	State	Heat of formation from elements $-\Delta H^0_{298}$ [kcal/mol]	References	Gibbs free energy of formation from elements $-\Delta G^0_{298}$ [kcal/mol]	References	Entropy S^0_{298} [cal/deg mol]	References	$C_p=f(T)$ a	$b \cdot 10^3$	$c \cdot 10^{-5}$	Temp. range in K	References
$2CaO \cdot Al_2O_3 \cdot 8SiO_2 \cdot 7H_2O$ (leonhardite)	cr	3402.0	[29]	3151.0	[29]	220.4	[120]	–	–	–	–	–
$4CaO \cdot Al_2O_3 \cdot Fe_2O_3$	cr	1214.2	[87]	1144	[87]	78	[87]	89.49	17.4	–	–	–
$6CaO \cdot 2Al_2O_3 \cdot Fe_2O_3$	cr	1908.86	[91]	–	–	–	–	–	–	–	–	–
$CaO \cdot B_2O_3$	cr	485.41	[120]	459.87	[120]	25.06	[120]	31.02	9.76	–8.07	298–Tm	[29]
$CaO \cdot 2B_2O_3$	cr	803.12	[120]	756.96	[120]	32.2	[120]	51.34	19.16	–17.16	298–Tm	[29]
$2CaO \cdot B_2O_3$	cr	653.54	[120]	620.62	[120]	34.68	[120]	43.75	11.5	–10.69	298–804	[29]
$2CaO \cdot 3B_2O_3$	cr	1290	[120]	–	–	–	–	–	–	–	–	–
$3CaO \cdot B_2O_3$	cr	819.57	[120]	779.14	[120]	43.9	[120]	56.55	10.42	–13.02	298–Tm	[29]
$2CaO \cdot B_2O_3 \cdot 2SiO_2 \cdot H_2O$	cr	1170.0	[120]	–	–	–	–	–	–	–	–	–
$CaO \cdot Fe_2O_3$	cr	363.47	[90]	337.65	[90]	34.7	[45]	39.42	4.79	–3.66	298–1510	[35]
$2CaO \cdot Fe_2O_3$	cr	507.7	[91]	474.4	[2]	45.1	[45]	59.24	11.68	–	298–1750	[35]
$Ca(OH)_2$	cr	235.33	[90]	214.39	[90]	19.93	[120]	19.07	10.80	–	298–700	[35]
$CaO \cdot MgO \cdot 2SiO_2$ (diopside)	cr	766.29	[90]	724.67	[90]	34.2	[33]	52.87	7.84	–15.74	298–1600	[36]
$CaO \cdot MgO \cdot SiO_2$ (monticellite)	cr	540.91	[90]	–	–	–	–	34.54	9.21	6.81	–	[90]
$2CaO \cdot MgO \cdot 2SiO_2$ (akermanite)	cr	926.69	[90]	879.5	[90]	–	–	–	–	–	–	–
$3CaO \cdot MgO \cdot 2SiO_2$ (merwinite)	cr	1091.67	[90]	1037.34	[90]	60.5	[90]	–	–	–	–	–

$Ca(NO_3)_2$	cr	224.28	[120]	177.63	[120]	46.2	[104]	29.37	36.8	−4.13	298 − 800	[36]
$CaO \cdot SiO_2$ β-wollastonite	cr	390.95	[114]	370.55	[114]	19.6	[104]	26.64	3.6	−6.52	298 − 1450	[36]
$CaO \cdot SiO_2$ α-wollastonite	cr	389.75	[114]	369.75	[114]	20.9	[104]	25.85	3.94	−5.65	298 − 1700	[36]
$2CaO \cdot SiO_2$-β	cr	551.74	[114]	524.19	[114]	30.5	[114]	36.25	8.83	−7.24	298 − 1200	[52a]
$2CaO \cdot SiO_2$-γ	cr	552.72	[114]	525.19	[114]	28.8	[44]	27.17	19.6	–	298 − 948	[36]
$3CaO \cdot SiO_2$	cr	700.43	[46]	665.47	[88]	40.3	[114]	49.85	8.62	−10.15	298 − 1800	[36]
$3CaO \cdot 2SiO_2$	cr	938.72	[88]	890.80	[88]	50.4	[44]	64.0	9.05	−16.6	–	[88]
$2CaO \cdot SiO_2 \cdot 1.17H_2O$	cr	637.15	[86]	592.9	[86]	38.4	[1]	41.4	22.4	−7.4	–	[86]
$3CaO \cdot 2SiO_2 \cdot 3H_2O$	cr	1143.2	[86]	1052.95	[86]	74.6	[1]	81.54	45.1	−14.67	–	[86]
$4CaO \cdot 3SiO_2 \cdot 1,5H_2O$	cr	1439.9	[86]	1347.9	[86]	78.95	[1]	87.95	3.95	−13.48	–	[86]
$6CaO \cdot 6SiO_2 \cdot H_2O$	cr	2396.7	[86]	2259.4	[86]	121.3	[1]	132.25	65.2	−18.35	–	[86]
$5CaO \cdot 6SiO_2 \cdot 3H_2O$	cr	2375.0	[86]	2215.0	[86]	122.65	[1]	143.55	74.7	−20.82	–	[86]
$5CaO \cdot 6SiO_2 \cdot 5.5H_2O$	cr	2556.3	[86]	2361.45	[86]	146.15	[1]	110.6	189.0	–	–	[86]
$5CaO \cdot 6SiO_2 \cdot 10.5H_2O$	cr	2911.25	[86]	2647.3	[86]	193.15	[1]	132.2	270.0	–	–	[86]
$CaO \cdot 2SiO_2 \cdot 2H_2O$	cr	750.3	[86]	686.4	[86]	40.9	[1]	44.81	18.7	−10.35	–	[86]
$2CaO \cdot 3SiO_2 \cdot 2.5H_2O$	cr	1175.85	[86]	1085.65	[86]	64.0	[1]	79.47	36.3	−17.55	–	[86]
$CaO \cdot Al_2O_3 \cdot CaCO_3 \cdot 11H_2O$	cr	1961.6	[c]	1743.6	[c]	156.6	[c]	82.1	260.3	–	–	[c]
$3CaO \cdot Al_2O_3 \cdot 3CaCO_3 \cdot 30H_2O$	cr	3876.0	[120]	–	–	–	–	–	–	–	–	–
$3CaO \cdot Al_2O_3 \cdot CaSO_4 \cdot 12H_2O$	cr	2096.44	[2]	1859.14	[2]	178.6	[2]	113.71	246.84	–	–	[2]
$3CaO \cdot Al_2O_3 \cdot 3CaSO_1 \cdot 31H_2O$	cr	4130.82	[2]	3577.52	[2]	408.2	[2]	204.31	721.8	–	–	[2]
$3CaO \cdot Al_2O_3 \cdot 3CaSO_4 \cdot 32H_2O$	cr	4201.32	[2]	3634.26	[2]	417.6	[2]	208.0	740.93	–	–	[2]
CaS	cr	115.3	[104]	114.1	[104]	13.5	[104]	–	–	–	–	–
$3CaO \cdot Fe_2O_3 \cdot 6H_2O$	cr	1110.8	[c]	984.79	[c]	101.5	[c]	78.09	116.26	–	–	[c]
$4CaO \cdot Fe_2O_3 \cdot 13H_2O$	cr	1772.8	[c]	1542.57	[c]	176.4	[c]	118.15	234.67	–	–	[c]
$3CaO \cdot Fe_2O_3 \cdot CaSO_4 \cdot 12H_2O$	cr	1881.24	[c]	1645.53	[c]	183.4	[c]	122.83	239.14	–	–	[c]

Table (continued)

Formulae of compounds and ions	State	Heat of formation from elements $-\Delta H^0_{298}$ [kcal/mol]	Refer-ences	Gibbs free energy of formation from elements $-\Delta G^0_{298}$ [kcal/mol]	Refer-ences	Entropy S^0_{298} [cal/deg mol]	Refer-ences	$C_p = f(T)$ a	$b \cdot 10^3$	$c \cdot 10^{-5}$	Temp. range in K	Refer-ences
$3CaO \cdot Fe_2O_3 \cdot 3CaSO_4 \cdot 32H_2O$	cr	3986.12	[c]	3420.55	[c]	422.4	[c]	217.93	730.92	–	–	[c]
$CaSO_4$ (anhydrite)	cr	342.42	[104]	315.56	[104]	25.5	[104]	18.52	21.97	−1.57	298−1570	[32]
$CaSO_4$ (α-soluble form)	cr	340.64	[120]	313.93	[120]	25.9	[120]	–	–	–	–	–
$CaSO_4$ (β-soluble form)	cr	339.58	[120]	312.87	[120]	25.9	[120]	–	–	–	–	–
$CaSO_4 \cdot 0.5H_2O$-β	cr	376.35	[120]	343.18	[120]	32.1	[120]	11.48	61.00	–	298−450	[90]
$CaSO_4 \cdot 0.5H_2O$-α	cr	376.85	[120]	343.41	[120]	31.2	[120]	16.95	3.00	–	298−450	[90]
$CaSO_4 \cdot 2H_2O$	cr	483.42	[120]	429.60	[120]	46.36	[120]	21.84	76.0	–	298−400	[90]
$CaSO_3 \cdot 2H_2O$	cr	418.9	[120]	371.7	[120]	44.00	[120]	–	–	–	–	–
$CaO \cdot TiO_2$-α	cr	–	–	–	–	22.41	[37]	30.47	1.36	−6.69	298−1530	[37]
$CaO \cdot TiO_2$-β	cr	396.9	[33]	376.5	[33]	–	–	32.03	–	–	1530−1800	[37]
$3CaO \cdot 2TiO_2$	cr	944.1	[33]	896.4	[33]	–	–	–	–	–	–	–
$2CaO \cdot TiO_2$	cr	549.6	[25]	–	–	–	–	–	–	–	–	–
$CaO \cdot TiO_2 \cdot SiO_2$	cr	613.95	[33]	580.15	[33]	30.9	[33]	–	–	–	–	–
$CaWO_4$	cr	381.5	[118]	357.0	[118]	33.5	[118]	–	–	–	–	–
$CaO \cdot ZrO_2$	cr	400.2	[52]	–	–	–	–	–	–	–	–	–
$2CaO \cdot ZrO_2 \cdot 4SiO_2$	cr	–	–	–	–	–	–	96.40	13.42	−28.5	298−1473	[110]
$3CaO \cdot ZrO_2 \cdot 2SiO_2$	cr	–	–	–	–	–	–	77.04	10.04	−15.24	298−1573	[110]
CH_3COOH	liq	116.4	[104]	93.8	[104]	38.2	[104]	–	–	–	–	–
C_2H_5OH	liq	66.356	[104]	41.77	[104]	38.40	[104]	–	–	–	–	–

$CoCr_2O_4$	cr	342.1	[55]	317.2	[55]	29.1	[55]	–	–	–	–	–
CO	gas	26.42	[90]	32.78	[90]	47.2	[90]	6.79	0.98	−0.11	298−2500	[90]
CO_2	gas	94.05	[104]	94.26	[104]	51.06	[104]	10.55	2.16	−2.04	298−2500	[36]
Cd	gas	−26.77	[120]	−18.51	[120]	40.066	[120]	–	–	–	–	–
Cd-α	cr	0.0	[120]	0.0	[120]	12.37	[120]	–	–	–	–	–
Cd^{2+}	gas	−627.04	[120]	–	–	–	–	–	–	–	–	–
$CdCl_2$	cr	93.57	[120]	82.21	[120]	27.55	[120]	18.30	4.30	−1.75	273−871	[32]
$CdCO_3$	cr	179.40	[120]	160.0	[120]	22.1	[120]	–	–	–	–	–
CdF_2	cr	167.4	[120]	154.8	[120]	18.5	[120]	–	–	–	–	–
CdO	cr	61.7	[120]	54.6	[120]	13.10	[120]	9.78	2.02	–	298−2086	[36]
$Cd(OH)_2$	cr	134.0	[120]	113.2	[120]	23.0	[120]	–	–	–	–	–
$CdO \cdot SiO_2$	cr	284.20	[120]	264.20	[120]	23.3	[33]	–	–	–	–	–
$CdSO_4$	cr	223.06	[120]	196.65	[120]	29.41	[120]	–	–	–	–	–
$CdSO_4 \cdot H_2O$	cr	296.26	[120]	255.46	[120]	36.81	[120]	–	–	–	–	–
$CdSO_4 \cdot 8/3H_2O$	cr	413.33	[120]	350.22	[120]	54.88	[120]	–	–	–	–	–
Cr	gas	−80.5	[104]	−69.8	[104]	41.637	[104]	–	–	–	–	–
Cr	cr	0.0	[104]	0.0	[104]	5.68	[104]	–	–	–	–	–
Cr^{3+}	gas	−1310.0	[104]	–	–	–	–	–	–	–	–	–
$CrCl_3$	cr	123.4	[90]	106.6	[90]	29.39	[90]	19.44	7.03	–	298−1220	[90]
$CrCl_2$	cr	94.52	[90]	85.16	[90]	27.56	[90]	17.2	3.90	1.20	298−1000	[90]
Cr_2O_3	cr	272.7	[90]	253.2	[90]	19.4	[90]	28.53	2.20	−3.74	298−1800	[32]
$Cr(OH)_3$	cr	242.35	[105]	207.43	[105]	–	–	–	–	–	–	–
CsCl	cr	103.5	[104]	–	–	23.9	[120]	–	–	–	–	–
Cs_2CO_3	cr	267.4	[104]	246.72	[70]	45.1	[70]	31.68	8.91	−7.77	298−1065	[70]
Cs_2CO_3	liq	–	–	–	–	–	–	35.5	5.0	–	1065−1800	[70]
CsF	cr	126.9	[104]	–	–	19.8	[120]	–	–	–	–	–
Cs_2O	cr	75.9	[104]	–	–	–	–	–	–	–	–	–
$Cs_2O \cdot SiO_2$	cr	365.64	[70]	343.70	[70]	44.0	[70]	27.18	17.84	−2.12	298−1100	[70]
$Cs_2O \cdot SiO_2$	liq	–	–	–	–	–	–	46.5	–	–	1100−1800	[70]
$Cs_2O \cdot 2SiO_2$	cr	586.70	[70]	551.80	[70]	54.0	[70]	44.30	15.65	−5.98	298−1400	[70]
$Cs_2O \cdot 2SiO_2$	liq	–	–	–	–	–	–	66.0	–	–	1400−1800	[70]
$Cs_2O \cdot 4SiO_2$	cr	1030.80	[70]	969.98	[70]	74.0	[70]	62.64	29.05	−5.41	298−1200	[70]
$Cs_2O \cdot 4SiO_2$	liq	–	–	–	–	–	–	93.0	–	–	1200−1800	[70]

Table (continued)

Formulae of compounds and ions	State	Heat of formation from elements $-\Delta H^0_{298}$ [kcal/mol]	References	Gibbs free energy of formation from elements $-\Delta G^0_{298}$ [kcal/mol]	References	Entropy S^0_{298} [cal/deg mol]	References	$C_p = f(T)$			Temp. range in K	References
								a	$b \cdot 10^3$	$c \cdot 10^{-5}$		
$Cs_2O \cdot ZrO_2 \cdot 2SiO_2$	cr	857.2	[111]	–	–	–	–	–	–	–	–	–
Cs_2SO_4-β	cr	339.38	[70]	310.71	[70]	51.1	[70]	22.77	23.86	–	298 – 913	[70]
Cs_2SO_4-α	cr	–	–	–	–	–	–	45.71	1.36	–	913 – 1292	[70]
Cs_2SO_4	liq	–	–	–	–	–	–	47.5	–	–	1292 – 1800	[70]
Fe-α	gas	−99.50	[120]	−88.60	[120]	43.11	[120]	–	–	–	–	–
Fe-α	cr	0.0	[90]	0.0	[90]	6.49	[90]	3.04	7.58	−0.60	298 – 1033	[35]
Fe-β	cr	–	–	–	–	–	–	11.13	–	–	1033 – 1183	[35]
Fe-γ	cr	–	–	–	–	–	–	5.80	1.98	–	1183 – 1673	[35]
Fe-δ	cr	–	–	–	–	–	–	6.74	1.60	–	1673 – Tm	[35]
Fe^{2+}	gas	−657.8	[120]	–	–	–	–	–	–	–	–	–
Fe^{3+}	gas	−1365.9	[120]	–	–	–	–	–	–	–	–	–
$FeCl_2$	cr	81.7	[90]	72.43	[90]	28.19	[90]	18.94	2.08	−1.17	298 – 950	[90]
$FeAl_2O_4$	cr	10.8[a]	[57]	5.6[a]	[57]	22.4	[57]	–	–	–	–	–
$FeCl_3$	cr	95.48	[120]	79.84	[120]	34.00	[120]	–	–	–	–	–
FeS (troilite)	cr	24.13	[89]	24.22	[89]	14.42	[89]	–	–	–	–	–
FeS_2 (pyrite)	cr	41.00	[89]	38.30	[89]	12.65	[89]	–	–	–	–	–
FeOOH-α (goethite)	cr	133.85	[89]	117.04	[89]	14.70	[89]	–	–	–	–	–
FeOOH-α (lepidocrocite)	cr	–	–	112.67	[89]	–	–	–	–	–	–	–
$Fe(OH)_3$	am	–	–	169.19	[89]	–	–	–	–	–	–	–
$Fe(OH)_2$	am	–	–	115.84	[89]	–	–	–	–	–	–	–
$FeCO_3$	am	–	–	158.9	[89]	–	–	–	–	–	–	–

$Fe(CO)_5$	cr	185.4	[89]	166.3	[89]	105.0	[89]	–	–	–	–	–
$7FeO \cdot 8SiO_2 \cdot H_2O$	cr	2294.0	[89]	2141.0	[89]	187.0	[89]	–	–	–	–	–
$3FeO \cdot 4SiO_2 \cdot H_2O$	cr	1152.6	[89]	1069.9	[89]	85.2	[89]	–	–	–	–	–
$FePO_4 \cdot 2H_2O$	cr	451.3		396.2	[120]	40.93	[120]	–	–	–	–	–
$FeCO_3$	cr	179.78	[89]	162.40	[89]	22.97	[89]	11.63	26.8	–	298 – 885	[32]
$FeCr_2O_4$	cr	349.3	[93]	325.2	[93]	35.0	[93]	–	–	–	–	–
FeF_2	cr	168.0	[10]	–	–	20.0	[104]	–	–	–	–	–
FeF_3	cr	235.0	[10]	–	–	–	–	–	–	–	–	–
$Fe_{0.95}O$ (wüstite)	cr	63.44	[89]	58.40	[89]	13.76	[36]	11.66	2.0	– 0.67	298 – 1651	[36]
$Fe_{0.95}O$	liq	–	–	–	–	–	–	14.91	–	–	1651 – 1800	[36]
FeO	cr	65.0	[120]	–	–	–	–	–	–	–	–	–
Fe_2O_3-α (hematite)	cr	196.31	[89]	176.77	[89]	21.5	[104]	23.49	18.6	– 3.55	298 – 950	[36]
Fe_2O_3-β	cr	–	–	–	–	–	–	36.0	–	–	950 – 1050	[36]
Fe_2O_3-γ	cr	–	–	–	–	–	–	31.70	1.76	–	1050 – 1750	[36]
Fe_3O_4-α (magnetite)	cr	267.9	[104]	242.32	[104]	35.0	[104]	21.88	48.2	–	298 – 900	[90]
Fe_3O_4-β	cr	–	–	–	–	–	–	48.0	–	–	900 – 1800	[36]
$Fe(OH)_2$	cr	137.0	[90]	117.84	[89]	22.1	[90]	–	–	–	–	–
$Fe(OH)_3$	cr	201.8	[90]	170.8	[90]	23.0	[90]	–	–	–	–	–
$FeO \cdot SiO_2$	cr	285.89	[89]	267.32	[89]	22.19	[89]	25.88	5.12	– 4.23	298 – 1413	[c]
$FeO \cdot SiO_2$	liq	–	–	–	–	–	–	33.44	–	–	1413 – 1800	[c]
$2FeO \cdot SiO_2$	cr	354.83	[89]	330.75	[c]	34.7	[89]	36.51	9.36	– 6.70	298 – 1493	[36]
$2FeO \cdot SiO_2$	liq	–	–	–	–	–	–	57.50	–	–	1493 – 1800	[36]
$FeSO_4$	cr	221.9	[90]	196.1	[90]	25.7	[90]	–	–	–	–	–
$FeSO_4 \cdot 7H_2O$	cr	720.48	[90]	599.95	[90]	97.8	[90]	96.2	–	–	291 – 319	[90]
$Fe_2(SO_4)_3$	cr	617.10	[90]	537.0	[90]	61.9	[90]	66.2	–	–	273 – 373	[90]
$FeO \cdot TiO_2$	cr	295.55	[33]	277.05	[33]	25.3	[37]	27.87	4.36	– 4.79	298 – 1640	[36]
$2FeO \cdot TiO_2$	cr	–	–	–	–	39.0	[116]	33.34	15.08	– 8.40	298 – 1600	[7]
$Fe_2O_3 \cdot TiO_2$	cr	–	–	–	–	37.4	[116]	46.03	5.26	– 7.41	298 – 1700	[7]
$FeVO_4$	cr	277.72	[95]	–	–	40.5	[95]	27.09	14.10	– 0.99	298 – 1113	[95]
Ge	cr	0	[120]	0	[120]	7.43	[120]	–	–	–	–	–
GeO (brown)	cr	50.7	[120]	56.7	[120]	12	[120]	–	–	–	–	–

[a] From oxides at $T = 1273$ K

Table (continued)

Formulae of compounds and ions	State	Heat of formation from elements $-\Delta H^0_{298}$ [kcal/mol]	References	Gibbs free energy of formation from elements $-\Delta G^0_{298}$ [kcal/mol]	References	Entropy S^0_{298} [cal/deg mol]	References	$C_p = f(T)$ a	$b \cdot 10^3$	$c \cdot 10^{-5}$	Temp. range in K	References
GeO (yellow)	cr	–	–	49.5	[120]	–	–	–	–	–	–	–
GeO_2	cr	131.7	[120]	118.8	[120]	13.21	[120]	–	–	–	–	–
H	gas	–52.095	[120]	–48.581	[120]	27.391	[120]	–	–	–	–	–
H_2	gas	0.0	[120]	0.0	[120]	31.208	[120]	–	–	–	–	–
H^{1+}	gas	–367.161	[120]	–	–	–	–	–	–	–	–	–
H_3BO_3	cr	260.20	[104]	230.20	[104]	21.41	[104]	–	–	–	–	–
HCOOH	liq	97.80	[104]	82.40	[104]	30.82	[104]	–	–	–	–	–
HF	gas	64.2	[104]	64.7	[104]	41.47	[104]	6.43	0.82	–	273–2000	[32]
HNO_3	liq	41.404	[104]	19.10	[104]	37.19	[104]	–	–	–	–	–
H_2O	gas	57.796	[90]	54.634	[90]	45.104	[90]	7.30	2.46	–	298–2750	[90]
H_2O	liq	68.315	[90]	56.687	[90]	16.71	[90]	12.65	11.38	1.73	273–373	[90]
HCl	gas	22.06	[90]	22.774	[90]	44.646	[90]	6.27	–1.24	–0.30	298–2000	[90]
H_2SiO_3	cr	284.10	[120]	261.10	[120]	32	[120]	–	–	–	–	–
H_4SiO_4	cr	354.00	[120]	318.60	[120]	46	[120]	–	–	–	–	–
$H_2Si_2O_5$	cr	499.20	[120]	464.50	[120]	46.0	[120]	–	–	–	–	–
$H_6Si_2O_7$	cr	638.0	[120]	579.8	[120]	79.0	[120]	–	–	–	–	–
K	gas	–21.51	[104]	–14.62	[104]	38.296	[104]	–	–	–	–	–
K	cr	0.0	[104]	0.0	[104]	15.20	[104]	–	–	–	–	–
K^+	gas	123.07	[104]	–	–	–	–	–	–	–	–	–
$KAl_2[AlSi_3O_{10}] \cdot (OH)_2$ (muscovite)	cr	1427.4	[29]	1337.6	[29]	73.1	[29]	–	–	–	–	–
$KMg_3 \cdot AlSi_3O_{10}F_2$	cr	1526.8	[29]	1445.6	[29]	80.3	[29]	–	–	–	–	–

KCl	cr	104.2	[104]	97.59	[104]	19.76	[104]	9.89	5.20	0.77	–	[37]
K_2CO_3	cr	273.93	[71]	253.5	[71]	35.7	[71]	19.19	26.06	–	298 – 1174	[71]
K_2CO_3	liq	–	–	–	–	–	–	36.95	10.64	–	1174 – 1800	[71]
KF	cr	134.5	[104]	–	–	15.9	[121]	11.02	3.17	–	298 – Tm	[36]
K_2O	cr	86.40	[104]	76.97	[29]	23.5	[52a]	–	–	–	–	–
$K_2O \cdot Al_2O_3 \cdot 2SiO_2$	cr	1013.4	[29]	958.0	[29]	63.6	[29]	–	–	–	–	–
$K_2O \cdot Al_2O_3 \cdot 4SiO_2$ (leucite)	cr	1450.49	[29]	1373.6	[29]	95.6	[29]	–	–	–	–	–
$K_2O \cdot Al_2O_3 \cdot 4SiO_2$	gl	1442.39	[104]	–	–	–	–	–	–	–	–	–
$K_2O \cdot Al_2O_3 \cdot 6SiO_2$ (microclin)	cr	1895.2	[29]	1787.4	[29]	102.2	[29]	–	–	–	–	–
$K_2O \cdot Al_2O_3 \cdot 6SiO_2$ (high-temp. sanidine)	cr	1891.2	[29]	1786.2	[29]	111.2	[29]	–	–	–	–	–
$K_2O \cdot Al_2O_3 \cdot 6SiO_2$	gl	1869.8	[29]	1769.0	[29]	124.8	[29]	–	–	–	–	–
$K_2O \cdot Al_2O_3 \cdot 6SiO_2$ (adular)	cr	1917.48	[104]	–	–	–	–	–	–	–	–	–
$K_2O \cdot 23CaO \cdot 12SiO_2$	cr	6494.1	[92]	–	–	–	–	–	–	–	–	–
KOH	cr	101.78	[104]	89.5	[104]	14.2	[29]	–	–	–	–	–
$K_2O \cdot SiO_2$	cr	372.5	[71]	350.15	[71]	33.0	[71]	32.62	5.85	–5.15	298 – 1249	[71]
$K_2O \cdot SiO_2$	liq	–	–	–	–	–	–	40.0	–	–	1249 – 1800	[71]
$K_2O \cdot 2SiO_2$-γ	cr	593.5	[71]	558.23	[71]	43.5	[71]	32.69	35.17	–	298 – 523	[71]
$K_2O \cdot 2SiO_2$-β	cr	–	–	–	–	–	–	39	25.05	–	523 – 1318	[71]
$K_2O \cdot 2SiO_2$	liq	–	–	–	–	–	–	72	–	–	1318 – 1800	[71]
$K_2O \cdot 4SiO_2$-β	cr	1035	[71]	978.82	[71]	63.5	[71]	60.52	38.09	–	298 – 865	[71]
$K_2O \cdot 4SiO_2$-α	cr	–	–	–	–	–	–	93.54	3.87	–	865 – 1038	[71]
$K_2O \cdot 4SiO_2$	liq	–	–	–	–	–	–	98	–	–	1038 – 1800	[71]
$2K_2O \cdot SiO_2$	gl	512.0	[82]	475.2	[82]	64.0	[82]	–	–	–	–	–
$K_2O \cdot SiO_2$	gl	370.5	[82]	348.79	[82]	35.55	[82]	26.93	14.08	0.64	298 – 1249	[83]
$K_2O \cdot 2SiO_2$	gl	590.5	[82]	556.92	[83]	49.22	[82]	21.31	41.63	–9.02	298 – 1318	[83]
$K_2O \cdot 3SiO_2$	gl	798.0	[82]	749.7	[82]	58.2	[82]	–	–	–	–	–
$K_2O \cdot 4SiO_2$	gl	1010.0	[82]	950.3	[83]	68.58	[82]	10.07	126.73	–66.98	298 – 1038	[83]
K_2SiF_6	cr	671.0	[104]	651.0	[29]	54.70	[29]	–	–	–	–	–
K_2SO_4-β	cr	342.66	[104]	314.62	[104]	42.0	[104]	28.77	23.80	4.26	298 – 856	[36]

Table (continued)

Formulae of compounds and ions	State	Heat of formation from elements $-\Delta H^0_{298}$ [kcal/mol]	References	Gibbs free energy of formation from elements $-\Delta G^0_{298}$ [kcal/mol]	References	Entropy S^0_{298} [cal/deg mol]	References	$C_p = f(T)$ a	$b \cdot 10^3$	$c \cdot 10^{-5}$	Temp. range in K	References
K_2SO_4-α	cr	–	–	–	–	–	–	33.60	13.40	–	856 – Tm	[36]
K_2SO_4	liq	–	–	–	–	–	–	47.8	–	–	1342 – 1800	[71]
$LiAlO_2$	cr	284.33	[18]	–	–	12.7	[47]	–	–	–	–	–
$LiAlSiO_4$ (evcryptite)	cr	507.1	[29]	479.8	[104]	24.7	[29]	–	–	–	–	–
$LiAlSi_2O_6$ (α-spodumene)	cr	729.2	[29]	687.8	[104]	30.8	[29]	–	–	–	–	–
$LiAlSi_2O_6$ (β-spodumene)	cr	722.5	[29]	682.9	[29]	36.8	[29]	–	–	–	–	–
LiCl	cr	96.9	[104]	–	–	13.9	[99]	11.0	3.40	–	298 – Tm	[36]
Li_2CO_3	cr	290.54	[52a]	270.66	[104]	21.6	[104]	3.82	42.7	–	298 – 996	[72]
Li_2CO_3	liq	–	–	–	–	–	–	30.85	13.5	–	996 – 1800	[72]
LiF	cr	146.3	[104]	–	–	8.5	[15]	9.14	5.19	–	298 – Tm	[21]
$LiFeO_2$	cr	–	–	–	–	18.0	[47]	–	–	–	–	–
Li_2O	cr	142.8	[72]	134.2	[72]	9.1	[104]	14.54	6.08	– 8.38	298 – 1800	[72]
LiOH	cr	116.5	[104]	105.9	[29]	10.2	[29]	–	–	–	–	–
$Li_2O \cdot SiO_2$	cr	390.00	[72]	368.7	[72]	20.0	[52a]	26.9	8.38	– 3.65	298 – 1474	[72]
$Li_2O \cdot SiO_2$	liq	–	–	–	–	–	–	39.0	–	–	1474 – 1800	[72]
$Li_2O \cdot 2SiO_2$	cr	602.5	[72]	568.4	[72]	30.0	[72]	36.35	18.1	–	298 – 1306	[72]
$Li_2O \cdot 2SiO_2$	liq	–	–	–	–	–	–	60	–	–	1306 – 1800	[72]
$2Li_2O \cdot SiO_2$	cr	552.5	[72]	522.8	[72]	30.0	[72]	32.82	16.33	2.35	298 – 1528	[72]
$2Li_2O \cdot SiO_2$	liq	–	–	–	–	–	–	57.87	–	–	1528 – 1800	[72]

$2Li_2O \cdot SiO_2$	gl	530.0	[83]	501.09	[83]	32.3	[81]	38.69	15.23	6.41	298 – 1528	[83]
$Li_2O \cdot SiO_2$	gl	371.0	[83]	350.23	[83]	21.72	[81]	20.82	16.60	−6.89	298 – 1474	[83]
$Li_2O \cdot 3SiO_2$	gl	793.0	[81]	745.8	[81]	45.0	[81]	–	–	–	–	–
Li_2SiF_6	cr	688.1	[104]	–	–	–	–	–	–	–	–	–
Li_2SO_4-β	cr	342.83	[104]	315.4	[72]	27.3	[72]	33.34	15.64	0.09	298 – 859	[72]
Li_2SO_4-α	cr	–	–	–	–	–	–	52.3	–	–	859 – 1133	[72]
Li_2SO_4	liq	–	–	–	–	–	–	48.4	–	–	1133 – 1800	[72]
Mg	gas	−35.30	[120]	−27.04	[120]	35.502	[120]	–	–	–	–	–
Mg	cr	0.0	[120]	0.0	[120]	7.78	[89]	–	–	–	–	–
Mg^{2+}	gas	−561.788	[104]	–	–	–	–	–	–	–	–	–
$MgAl_2O_4$	cr	548.76	[102, 12]	518.94	[12]	19.26	[43]	36.80	6.40	−9.78	298 – 1800	[8]
$MgCl_2$	cr	153.28	[90]	141.44	[90]	21.4	[90]	18.9	1.42	−2.06	298 – 987	[90]
$MgCO_3$	cr	266.04	[89]	246.11	[89]	15.7	[89]	18.62	13.80	−4.16	298 – 750	[32]
$MgCr_2O_4$	cr	418.54	[52a]	–	–	25.3	[37]	–	–	–	–	–
MgF_2	cr	263.50	[104]	250.80	[104]	13.68	[104]	16.93	2.52	−2.20	298 – Tm	[32]
$MgFe_2O_4$	cr	349.9	[33]	322.9	[33]	28.3	[33]	–	–	–	–	–
$MgMoO_3$	cr	282.7	[100]	262.9	[100]	22.2	[100]	–	–	–	–	–
$MgMoO_4$	cr	335.5	[100]	310.5	[100]	29.0	[100]	25.19	12.60	–	–	[100]
MgO	cr	143.84	[104]	136.13	[104]	6.4	[104]	10.18	1.74	−1.48	298 – 2100	[32]
$Mg(OH)_2$	cr	221.77	[89]	200.05	[89]	15.09	[89]	13.04	15.80	–	298 – 600	[90]
$MgO \cdot SiO_2$	cr	370.25	[104]	349.48	[104]	16.2	[104]	24.55	4.74	−6.28	298 – 1600	[36]
$2MgO \cdot SiO_2$	cr	520.53	[46]	492.15	[c]	22.70	[104]	35.81	6.54	−8.52	298 – 1800	[96]
$3MgO \cdot Al_2O_3 \cdot 3SiO_2$	cr	1484.87	[14]	–	–	–	–	–	–	–	–	–
$2MgO \cdot 2Al_2O_3 \cdot 5SiO_2$ (cordierite)	cr	2177.0	[120]	2055.0	[120]	97.3	[120]	–	–	–	–	–
$3MgO \cdot 2SiO_2 \cdot 2H_2O$	cr	1043.34	[89]	965.77	[89]	54.96	[89]	68.64	33.49	−13.89	–	[6]
$3MgO \cdot 4SiO_2 \cdot H_2O$	cr	1415	[89]	1323.28	[89]	62.33	[89]	82.27	34.02	−11.9	–	[6]
$7MgO \cdot 8SiO_2 \cdot H_2O$	cr	2889.56	[89]	2717.87	[89]	133.65	[89]	154.2	74.0	−35.5	–	[6]
$Mg_3(PO_4)_2$	cr	961.5	[104]	904.0	[88]	56.8	[88]	–	–	–	–	–
Mg_2Si	cr	13.60	[104]	–	–	–	–	–	–	–	–	–
$Mg_3Si_2O_5(OH)_4$ (chrysotile)	cr	1041.7	[29]	963.4	[29]	52.8	[29]	–	–	–	–	–
$Mg_3Si_4O_{10}(OH)_2$ (talc)	cr	1412.9	[29]	1322.2	[29]	62.2	[29]	–	–	–	–	–

Table (continued)

Formulae of compounds and ions	State	Heat of formation from elements $-\Delta H^0_{298}$ [kcal/mol]	References	Gibbs free energy of formation from elements $-\Delta G^0_{298}$ [kcal/mol]	References	Entropy S^0_{298} [cal/deg mol]	References	$C_p = f(T)$ a	$b \cdot 10^3$	$c \cdot 10^{-5}$	Temp. range in K	References
$MgSO_4$	cr	307.10	[120]	279.80	[120]	21.90	[104]	26.18	8.56	−5.08	298 − Tm	[12]
$MgTiO_3$	cr	375.9	[33]	354.8	[33]	17.80	[67]	28.29	3.28	−6.53	298 − 1800	[36]
Mg_2TiO_4	cr	517.45	[33]	488.5	[33]	24.8	[115]	35.96	8.54	−6.89	298 − 1800	[97]
$MgTi_2O_5$	cr	599.8	[25]	566.45	[33]	30.4	[115]	40.68	9.20	−7.35	298 − 1800	[97]
Mn	gas	68.34	[104]	−58.23	[104]	41.493	[104]	–	–	–	–	–
Mn	cr	0.0	[104]	0.0	[104]	7.59	[104]	–	–	–	–	–
Mn^{2+}	gas	601.45	[104]	–	–	–	–	–	–	–	–	–
MnC_2	cr	115.3	[104]	105.5	[104]	28.0	[104]	18.04	3.16	−1.37	298 − 923	[32]
$MnCl_2$	cr	113.0	[10]	–	–	28.0	[37]	–	–	–	–	–
$MnCO_3$	cr	213.9	[104]	195.40	[104]	20.5	[104]	21.99	9.3	−4.69	298 − 700	[32]
MnF_2	cr	190.0	[10]	–	–	22.25	[104]	–	–	–	–	–
MnF_3	cr	238.0	[10]	–	–	28.0	[25]	–	–	–	–	–
$MnFe_2O_4$	cr	292.8	[103]	–	–	37.0	[103]	–	–	–	–	–
MnO	cr	92.0	[104]	86.8	[104]	14.4	[104]	11.11	1.94	−0.88	298 − 1800	[32]
MnO_2	cr	124.2	[104]	111.1	[29]	12.7	[29]	16.60	2.44	−3.88	298 − 780	[32]
Mn_2O_3	cr	229.4	[11]	212.7	[52a]	26.4	[48]	24.73	8.38	−3.23	298 − 1350	[36]
Mn_2O_7	cr	174.1	[15]	–	–	–	–	–	–	–	–	–
Mn_3O_4	cr	331.4	[104]	306.0	[104]	35.5	[104]	34.64	10.82	−2.20	298 − 1445	[36]
$Mn(OH)_2$	cr	166.8	[29]	146.9	[29]	21.1	[104]	–	–	–	–	–
$Mn(OH)_3$	cr	212.0	[104]	181.0	[29]	23.8	[29]	–	–	–	–	–
$MnO \cdot SiO_2$	cr	315.65	[49]	296.55	[68]	21.3	[37]	26.42	3.88	−6.16	298 − 1500	[36]
$2MnO \cdot SiO_2$	cr	413.55	[31]	389.58	[68]	36.9	[68]	34.38	11.25	−3.92	298 − Tm	[68]

$MnSO_4$	cr	254.24	[104]	228.48	[104]	26.8	[104]	29.26	8.92	−7.04	298 − 1100	[32]
$Mn_2(SO_4)_3$	cr	666.9	[104]	580.9	[29]	61.7	[29]	–	–	–	–	–
Mo	cr	0	–	0	–	6.83	[100]	5.69	1.88	−0.503	–	[100]
MoO_2	cr	140.4	[100]	–	–	11.06	[100]	17.56	1.47	−3.6	–	[100]
MoO_3	cr	178.1	[100]	–	–	18.58	[100]	20.07	5.90	−3.68	–	[100]
Na	gas	−25.98	[104]	−18.67	[104]	36.715	[104]	–	–	–	–	–
Na	cr	0.0	[104]	0.0	[104]	12.20	[104]	–	–	–	–	–
Na^+	gas	−146.015	[104]	–	–	–	–	–	–	–	–	–
$NaAlO_2$	cr	271.3	[29]	–	–	16.9	[47]	–	–	–	–	–
$NaAlF_6$	cr	759.60	[104]	–	–	–	–	–	–	–	–	–
$NaAlSi_3O_8$ (albite)	cr	939.8	[29]	886.4	[29]	49.5	[29]	65.49	10.48	−18.94	298 − 1400	[32]
$NaAlSi_3O_8$ (analbite)	cr	937.2	[29]	883.8	[29]	54.0	[29]	–	–	–	–	–
$NaAlSi_3O_8$	gl	925.5	[29]	875.4	[29]	60.1	[29]	–	–	–	–	–
$NaAlSi_2O_6$ (jadeite)	cr	–	–	–	–	31.8	[37]	–	–	–	–	–
$NaAlSi_2O_6 \cdot H_2O$ (analcime)	cr	790.5	[29]	738.4	[29]	55.9	[29]	–	–	–	–	–
$NaAlSiO_4$ (nepheline)	cr	504.0	[29]	476.6	[29]	29.6	[29]	–	–	–	–	–
NaCl	cr	98.232	[104]	91.785	[104]	17.30	[104]	10.98	3.90	–	298 − 1073	[32]
Na_2CO_3-β	cr	270.30	[104]	250.40	[104]	32.5	[104]	14.48	39.02	–	298 − 723	[73]
Na_2CO_3-α	cr	–	–	–	–	–	–	32.197	8.43	–	723 − 1123	[73]
Na_2CO_3	liq	–	–	–	–	–	–	−312.63	0.56	−21.94	1123 − 1800	[73]
NaF	cr	136.4	[104]	–	–	13.1	[37]	9.66	4.50	–	298 − Tm	[36]
$NaFeO_2$	cr	–	–	–	–	21.1	[47]	–	–	–	–	–
$NaHCO_3$	cr	226.50	[104]	203.60	[104]	24.40	[104]	–	–	–	–	–
$NaHPO_3$	cr	388.0	–	–	–	–	–	–	–	–	–	–
Na_2HPO_4	cr	417.40	[104]	–	–	–	–	–	–	–	–	–
$NaNO_2$	cr	85.9	[104]	67.80	[29]	25.3	[29]	–	–	–	–	–
Na_2O	cr	99.40	[104]	90.00	[104]	17.4	[104]	15.40	5.40	–	298 − 1100	[32]
$Na_2O \cdot Al_2O_3 \cdot 3SiO_2$	cr	1242.0	[104]	–	–	–	–	–	–	–	–	–
$Na_2O \cdot Al_2O_3 \cdot 4SiO_2$	cr	1442.4	[104]	–	–	–	–	–	–	–	–	–
$Na_2O \cdot B_2O_3$	cr	471.8	[27]	–	–	–	–	–	–	–	–	–
$Na_2O \cdot 2B_2O_3$	cr	786.4	[27]	–	–	–	–	–	–	–	–	–
$Na_2O \cdot 3B_2O_3$	cr	1101.4	[27]	–	–	–	–	–	–	–	–	–

Table (continued)

Formulae of compounds and ions	State	Heat of formation from elements $-\Delta H^0_{298}$ [kcal/mol]	References	Gibbs free energy of formation from elements $-\Delta G^0_{298}$ [kcal/mol]	References	Entropy S^0_{298} [cal/deg mol]	References	$C_p = f(T)$			Temp. range in K	References
								a	$b \cdot 10^3$	$c \cdot 10^{-5}$		
$Na_2O \cdot 4B_2O_3$	cr	1413.2	[27]	–	–	–	–	–	–	–	–	–
$Na_2O \cdot 2CaO \cdot 3SiO_2$	cr	1167.2	[69]	1103.80	[69]	66.4	[69]	83.87	14.08	−16.31	298−1557	[69]
$2Na_2O \cdot CaO \cdot 3SiO_2$	cr	1138.39	[69]	1073.0	[69]	74.0	[69]	88.46	16.24	−15.05	298−1414	[69]
$Na_2O \cdot 3CaO \cdot 6SiO_2$	cr	1998.99	[69]	1890.43	[69]	110.4	[69]	110.29	42.13	−11.57	298−1320	[69]
$Na_2O \cdot 2CaO \cdot 3SiO_2$	gl	1153.40	[83]	1090.95	[83]	68.75	[83]	66.81	38.76	2.79	298−1557	[83]
$2Na_2O \cdot CaO \cdot 3SiO_2$	gl	1117.60	[83]	1053.81	[83]	78.70	[83]	71.40	40.91	1.54	298−1414	[83]
$Na_2O \cdot 3CaO \cdot 6SiO_2$	gl	1980.52	[83]	1873.43	[83]	113.88	[83]	76.17	91.48	−15.46	298−1320	[83]
$Na_2O \cdot 8CaO \cdot 3Al_2O_3$	cr	2577.38	[93]	–	–	–	–	–	–	–	–	–
$NaOH$	cr	101.99	[104]	90.10	[29]	12.50	[29]	19.20	–	–	298−593	[32]
$2Na_2O \cdot SiO_2$	cr	507.25	[73]	476.05	[73]	46.8	[84]	38.86	17.74	–	298−1000	[74]
$Na_2O \cdot SiO_2$	cr	372.06	[73]	351.26	[73]	27.2	[104]	31.14	9.60	−6.47	298−1361	[36]
$Na_2O \cdot 2SiO_2$	cr	591.35	[73]	559.35	[73]	39.4	[104]	44.38	16.86	−10.67	298−1147	[36]
$Na_2O \cdot 3SiO_2$	cr	809.65	[c]	766.05	[c]	51.6	[76]	48.25	26.69	−4.98	298−1000	[93]
$2Na_2O \cdot SiO_2$	gl	500.25	[83]	470.21	[83]	50.5	[83]	43.35	20.17	4.23	298−1393	[83]
$Na_2O \cdot SiO_2$	gl	363.06	[83]	341.35	[83]	29.55	[83]	25.45	17.83	1.96	298−1361	[83]
$Na_2O \cdot 2SiO_2$	gl	584.35	[83]	550.57	[83]	42.55	[83]	32.91	33.32	1.64	298−1147	[83]
$Na_2O \cdot 3SiO_2$	gl	801.75	[83]	775.76	[83]	55.08	[83]	40.55	44.20	1.34	298−1023	[83]
$Na_2O \cdot SiO_2 \cdot 5H_2O$	cr	732.35	[104]	–	–	–	–	–	–	–	–	–
$Na_2O \cdot SiO_2 \cdot 9H_2O$	cr	1014.35	[104]	–	–	–	–	–	–	–	–	–
$Na_2O \cdot ZrO_2 \cdot SiO_2$	cr	–	–	–	–	43.5	[108]	–	–	–	–	–
$Na_2O \cdot ZrO_2 \cdot 2SiO_2$	cr	865.93	[109]	–	–	–	–	–	–	–	–	–
$2Na_2O \cdot 2ZrO_2 \cdot 3SiO_2$	cr	1519.3	[109]	–	–	–	–	–	–	–	–	–

Na_2SiF_6	cr	677.0	[104]	610.4	[29]	51.3	[29]	–	–	–	–	–
Na_2S	cr	89.2	[104]	–	–	–	–	–	–	–	–	–
Na_2SO_4-β	cr	330.90	[104]	302.78	[104]	35.73	[104]	13.94	5.4	–	298 – 518	[73]
Na_2SO_4-α	cr	–	–	–	–	–	–	31.75	15.48	–	518 – 1157	[73]
Na_2SO_4	liq	–	–	–	–	–	–	136.09	72.0	–	1157 – 1800	[73]
$Na_2SO_4 \cdot 10H_2O$	cr	1033.5	[104]	870.9	[104]	141.7	[104]	–	–	–	–	–
Na_2TiO_3	cr	380.0	[101]	–	–	29.1	[37]	25.18	20.72	–	298 – 560	[36]
Na_2TiO_3	cr	–	–	–	–	–	–	25.95	17.00	–	560 – 1303	[36]
$Na_2Ti_2O_5$	cr	–	–	–	–	41.5	[99]	49.32	7.06	– 4.60	298 – 1258	[36]
$Na_2Ti_3O_7$	cr	850.0	[101]	–	–	55.9	[99]	63.46	10.64	– 5.64	298 – 1403	[36]
$Na_2Ti_3O_7$	liq	–	–	–	–	–	–	94.15	–	–	1403 – 1700	[101]
$Na_8Ti_5O_{14}$	cr	1750.0	[101]	–	–	130.0	[101]	131.16	33.0	1.22	298 – 1293	[101]
$(NH_4)_2SiF_6$	cr	629.7	[104]	–	–	–	–	–	–	–	–	–
NaV_6O_{15}	cr	1200	[23]	–	–	103	[23]	–	–	–	–	–
$Na_2V_5O_{13.3}$	cr	1100	[23]	–	–	102	[23]	–	–	–	–	–
$NaVO_3$	cr	274.04	[23]	–	–	27.2	[23]	–	–	–	–	–
$Na_4V_2O_7$	cr	697.7	[23]	–	–	76.1	[23]	–	–	–	–	–
Na_3VO_4	cr	420.2	[23]	–	–	45.3	[23]	–	–	–	–	–
$NiCr_2O_4$	cr	342.0	[55]	315.8	[55]	29.1	[55]	–	–	–	–	–
$NiAl_{2.27}C_{4.41}$	cr	5.5[b]	[57]	6.1	[57]	22.9	[57]	–	–	–	–	–
NiO	cr	57.3	[112]	50.57	[112]	9.08	[112]	–	–	–	–	–
$2NiO \cdot SiO_2$	cr	334.35	[56]	307.49	[56]	26.7	[56]	–	–	–	–	–
NO	gas	– 21.6	[32]	20.72	[32]	50.34	[32]	7.03	0.92	– 0.14	298 – 2500	[32]
NO_2	gas	– 8.09	[32]	12.39	[32]	57.47	[32]	10.26	2.04	– 1.61	298 – 2000	[32a]
O	gas	– 59.553	[120]	– 55.389	[120]	38.467	[120]	–	–	–	–	–
O_2	gas	0.0	[120]	0.0	[120]	49.003	[120]	8.2	0.26	– 1.88	–	[106]
OH	gas	– 9.31	[120]	– 8.18	[120]	43.890	[120]	–	–	–	–	–
OH^{1-}	gas	33.67	[120]	–	–	–	–	–	–	–	–	–
P_4O_6	cr	392.0	[120]	–	–	–	–	–	–	–	–	–
P_4O_{10}	cr	713.2	[120]	644.8	[120]	54.70	[120]	8.37	–	–	298 – 631	[36]
Pb	gas	46.34	[104]	– 38.47	[104]	41.89	[104]	–	–	–	–	–
Pb	cr	0.0	[104]	0.0	[104]	15.51	[104]	–	–	–	–	–

[b] From oxides $T = 1273$ K

Table (continued)

Formulae of compounds and ions	State	Heat of formation from elements $-\Delta H^0_{298}$ [kcal/mol]	References	Gibbs free energy of formation from elements $-\Delta G^0_{298}$ [kcal/mol]	References	Entropy S^0_{298} [cal/deg mol]	References	$C_p = f(T)$			Temp. range in K	References
								a	$b \cdot 10^3$	$c \cdot 10^{-5}$		
Pb^{2+}	gas	−566.89	[104]	–	–	–	–	–	–	–	–	–
$PbCO_3$	cr	167.3	[104]	149.7	[104]	31.30	[104]	12.39	28.6	–	298−800	[32]
PbO (red)	cr	52.4	[104]	45.25	[104]	15.6	[50]	10.60	4.00	–	298−762	[36]
PbO (yellow)	cr	52.07	[104]	45.05	[104]	16.1	[50]	9.05	6.40	–	298−1000	[36]
PbO_2	cr	66.1	[104]	52.34	[104]	18.3	[67]	12.7	7.80	–	298−1000	[36]
Pb_3O_4	cr	175.6	[104]	147.6	[104]	50.5	[67]	44.13	7.75	−11.13	–	–
$Pb(OH)_2$	cr	123.0	[104]	100.6	[104]	2.0	[104]	–	–	–	–	–
$PbO \cdot PbCO_3$	cr	220.0	[104]	195.6	[104]	48.50	[104]	–	–	–	–	–
$2PbO \cdot PbCO_3$	cr	273.0	[104]	242.0	[104]	65.0	[104]	–	–	–	–	–
$PbO \cdot SiO_2$	cr	274.97	[117]	254.91	[117]	26.2	[39]	18.5	14.4	–	298−858	[c]
$PbO \cdot SiO_2$	cr	–	–	–	–	–	–	5.87	27.2	–	858−1037	[c]
$PbO \cdot SiO_2$	liq	–	–	–	–	–	–	31.10	–	–	1037−1800	[c]
$PbO \cdot SiO_2$	gl	273.04	[80]	253.3	[80]	27.26	[80]	35.07	1.98	−10.43	298−1037	[80]
$2PbO \cdot SiO_2$	cr	327.39	[117]	300.90	[117]	44.6	[39]	26.63	20.7	–	298−893	[c]
$2PbO \cdot SiO_2$	cr	–	–	–	–	–	–	34.12	10.9	–	893−1016	[c]
$2PbO \cdot SiO_2$	liq	–	–	–	–	–	–	45.19	–	–	1016−1800	[c]
$2PbO \cdot SiO_2$	gl	323.79	[80]	297.83	[80]	46.4	[80]	37.81	15.32	−8.06	298−1016	[80]
$4PbO \cdot SiO_2$	cr	431.42	[117]	390.92	[117]	77.6	[c]	53.22	20.9	−4.04	298−998	[c]
$4PbO \cdot SiO_2$	liq	–	–	–	–	–	–	73.7	–	–	998−1800	[c]
$4PbO \cdot SiO_2$	gl	426.13	[83]	386.55	[83]	80.73	[83]	43.29	42.0	3.32	298−998	[83]
$PbSO_4$	cr	219.5	[104]	193.9	[104]	35.20	[104]	10.91	31.0	−4.20	298−1100	[32]
RbCl	cr	102.9	[104]	–	–	21.9	[120]	–	–	–	–	–

Rb_2CO_3-β	cr	269.6	[120]	249.6	[70]	40.9	[70]	21.62	22.75	–	298 – 676	[70]
Rb_2CO_3-α	cr	–	–	–	–	–	–	26.21	17.46	–	676 – 1140	[70]
Rb_2CO_3	liq	–	–	–	–	–	–	40.0	5.0	–	1140 – 1800	[70]
RbF	cr	131.3	[104]	–	–	18.0	[52a]	–	–	–	–	–
Rb_2O	cr	78.9	[104]	–	–	–	–	–	–	–	–	–
$Rb_2O \cdot SiO_2$	cr	366.38	[70]	345.25	[70]	36.2	[70]	29.56	11.76	– 4.44	298 – 1143	[70]
$Rb_2O \cdot SiO_2$	liq	–	–	–	–	–	–	42.5	–	–	1143 – 1800	[70]
$Rb_2O \cdot 2SiO_2$	cr	587.50	[70]	553.60	[70]	51.0	[70]	46.19	12.04	– 7.98	298 – 1363	[70]
$Rb_2O \cdot 2SiO_2$	liq	–	–	–	–	–	–	62.0	–	–	1363 – 1800	[70]
$Rb_2O \cdot 4SiO_2$	cr	1030.00	[70]	972.87	[70]	71.0	[70]	61.03	29.77	– 4.59	298 – 1173	[70]
$Rb_2O \cdot 4SiO_2$	liq	–	–	–	–	–	–	95.0	–	–	1173 – 1800	[70]
Rb_2SO_4-β	cr	340.50	[70]	313.71	[70]	46.9	[70]	21.9	25.4	–	298 – 923	[70]
Rb_2SO_4-α	cr	–	–	–	–	–	–	47.0	0.9	–	923 – 1347	[70]
Rb_2SO_4	liq	–	–	–	–	–	–	42.5	–	–	1347 – 1800	[70]
S	gas	– 66.636	[120]	– 56.951	[120]	40.084	[120]	–	–	–	–	–
S (rhombic)	cr	0	[120]	0	[120]	7.60	[120]	–	–	–	–	–
S^{2+}	gas	– 848.20	[120]	–	–	–	–	–	–	–	–	–
S^{4+}	gas	– 2750.0	[120]	–	–	–	–	–	–	–	–	–
S^{1-}	gas	– 17.42	[120]	–	–	–	–	–	–	–	–	–
SO_2	gas	70.96	[104]	71.79	[104]	59.4	[104]	10.38	2.54	– 1.42	298 – 1800	[32]
SO_3	gas	94.58	[120]	88.69	[120]	61.34	[120]	13.70	6.42	– 3.12	298 – 1200	[32]
Sb_4O_6	cr	338.7	[120]	299.5	[120]	58.8	[120]	–	–	–	–	–
Sb_2O_5	cr	232.3	[120]	198.2	[120]	29.9	[37]	–	–	–	–	–
Si	gas	– 108.9	[120]	– 98.3	[120]	40.12	[120]	–	–	–	–	–
Si	cr	0.0	[104]	0.0	[104]	4.50	[120]	5.664	0.79	– 1.04	298 – 1690	[113]
Si^+	gas	298.35	[120]	–	–	–	–	7.1	–	–	1690 – 5800	[113]
Si (amorphous)	am	– 1.0	[120]	–	–	–	–	–	–	–	–	–
$SiBr_4$	gas	99.3	[120]	103.2	[120]	90.29	[120]	–	–	–	–	–
$SiBr_4$	liq	109.3	[120]	106.1	[120]	66.4	[120]	–	–	–	–	–
SiC-β	cr	15.6	[120]	15.0	[120]	3.97	[120]	8.93	3.0	– 3.07	298 – 1700	[32]
SiC-α	cr	15.0	[120]	14.4	[120]	3.94	[120]	–	–	–	–	–
$Si(C_2H_5O)_4$	liq	334.0	[120]	–	–	–	–	–	–	–	–	–
$SiCl_4$	gas	157.03	[120]	147.47	[120]	79.02	[120]	24.25	1.64	– 2.75	298 – 1000	[32]

Table (continued)

Formulae of compounds and ions	State	Heat of formation from elements $-\Delta H^0_{298}$ [kcal/mol]	References	Gibbs free energy of formation from elements $-\Delta G^0_{298}$ [kcal/mol]	References	Entropy S^0_{298} [cal/deg mol]	References	$C_p = f(T)$			Temp. range in K	References
								a	$b \cdot 10^3$	$c \cdot 10^{-5}$		
$SiCl_4$	liq	164.2	[120]	148.16	[120]	57.3	[120]	–	–	–	–	–
SiF_4	gas	385.98	[120]	375.88	[120]	67.49	[120]	22.95	2.65	−4.72	298–2000	[32]
SiH_4	gas	8.2	[120]	13.6	[120]	48.88	[120]	11.95	8.78	−3.05	298–1800	[32]
Si_3N_4-α	cr	177.7	[120]	153.6	[120]	24.2	[120]	16.83	23.60	–	298–900	[32]
SiO	gas	23.8	[120]	30.2	[120]	50.55	[120]	–	–	–	–	–
SiO_2 α-quartz	cr	217.60	[104, 119]	–	–	–	–	14.41	1.94	–	848–2000	[36]
SiO_2 β-quartz	cr	217.75	[119, 107]	204.75	[119]	10.0	[104]	11.22	8.20	−2.70	298–848	[36]
SiO_2 γ-tridymite	cr	216.43	[119, 107]	203.58	[119, 107]	10.4	[104]	3.27	24.80	–	298–390	[36]
SiO_2 α-tridymite	cr	216.50	[104, 119]	–	–	–	–	13.64	2.64	–	390–2000	[36]
SiO_2 β-cryst.	cr	215.95	[119, 107]	203.0	[119, 107]	10.19	[104]	4.28	21.06	–	298–523	[36]
SiO_2 α-cryst.	cr	215.64	[104, 119]	–	–	–	–	14.40	2.04	–	298–2000	[36]
SiO_2 glass	gl	215.48	[119, 107]	202.83	[119, 107]	11.2	[104]	13.38	3.68	−3.45	298–2000	[36]
SiO_3^{2-}	gas	181.5	[c]	–	–	–	–	–	–	–	–	–
SiO_4^{4-}	gas	−350.0	[122]	–	–	–	–	–	–	–	–	–
$Si_6O_{17}^{10-}$	gas	1478.8	[c]	–	–	–	–	–	–	–	–	–

$Si_{12}O_{31}^{14-}$	gas	3949.0	[c]	–	–	–	–	–	–	–	–	–
$Si_6O_{15}^{6-}$	gas	2115	[c]	–	–	–	–	–	–	–	–	–
SiS_2	cr	34.70	[104]	–	–	–	–	–	–	–	–	–
Sn	gas	–72.0	[104]	–64.0	[104]	40.245	[104]	–	–	–	–	–
Sn (white)	cr	0.0	[104]	0.0	[104]	12.30	[104]	–	–	–	–	–
Sn^+	gas	–243.04	[120]	–	–	–	–	–	–	–	–	–
Sn^{2+}	gas	–581.95	[120]	–	–	–	–	–	–	–	–	–
Sn^{4+}	gas	–2228.3	[120]	–	–	–	–	–	–	–	–	–
$SnCl_2$	cr	77.7	[120]	72.2	[29]	29.3	[29]	15.1	9.26	–	273–520	[32]
$SnCl_4$	liq	122.2	[120]	105.2	[120]	61.8	[120]	25.57	0.2	–1.87	298–1000	[32]
SnO	cr	68.3	[120]	61.4	[120]	13.50	[120]	9.40	3.62	–	273–1273	[32]
SnO_2	cr	138.3	[120]	124.2	[120]	12.50	[120]	17.66	2.40	–5.16	298–1500	[32]
$Sn(OH)_2$	cr	134.1	[120]	117.5	[120]	37.0	[120]	–	–	–	–	–
$Sn(OH)_4$	cr	265.3	[120]	227.5	[120]	29.0	[29]	–	–	–	–	–
$Sn(SO_4)_2$	cr	393.4	[104]	346.8	[104]	37.1	[104]	–	–	–	–	–
Sr	gas	–39.20	[104]	26.30	[104]	39.325	[104]	–	–	–	–	–
Sr	cr	0.0	[104]	0.0	[104]	13.0	[104]	–	–	–	–	–
Sr^{2+}	gas	–427.75	[104]	–	–	–	–	–	–	–	–	–
$SrCO_3$	cr	291.20	[104]	271.90	[104]	23.2	[104]	21.42	8.56	–3.396	298–1197	[53]
SrF_2	cr	289.0	–	–	–	–	–	–	–	–	–	–
$SrMoO_3$	cr	295.6	[100]	277.0	[100]	30.7	[100]	–	–	–	–	–
$SrMoO_4$	cr	371.3	[100]	345.6	[100]	29.2	[100]	25.21	14.52	–	–	[100]
SrO	cr	141.10	[104]	138.80	[104]	13.0	[104]	12.34	1.12	1.806	298–1270	[53]
$SrO \cdot Al_2O_3$	cr	554.98	[61]	–	–	–	–	–	–	–	–	–
$3SrO \cdot Al_2O_3$	cr	844.79	[62]	–	–	–	–	–	–	–	–	–
$4SrO \cdot Al_2O_3$-α	cr	976.08	[63]	–	–	–	–	–	–	–	–	–
$4SrO \cdot Al_2O_3$-β	cr	975.08	[63]	–	–	–	–	–	–	–	–	–
$2SrO \cdot Al_2O_3 \cdot SiO_2$	cr	941.83	[64]	–	–	–	–	–	–	–	–	–
$7SrO \cdot 5Fe_2O_3$	cr	2032.7	[65]	–	–	–	–	–	–	–	–	–
$2SrO \cdot Fe_2O_3$	cr	509.6	[65]	–	–	–	–	–	–	–	–	–
$3SrO \cdot Fe_2O_3$	cr	684.6	[65]	–	–	–	–	–	–	–	–	–
$2SrO \cdot 6Fe_2O_3$	cr	1338.1	[65]	–	–	–	–	–	–	–	–	–
$2SrO \cdot MgO \cdot 2SiO_2$	cr	921.74	[64]	885.11	[78]	56.62	[78]	61.4	13.5	–10.59	298–1900	[78]

Table (continued)

Formulae of compounds and ions	State	Heat of formation from elements $-\Delta H^0_{298}$ [kcal/mol]	References	Gibbs free energy of formation from elements $-\Delta G^0_{298}$ [kcal/mol]	References	Entropy S^0_{298} [cal/deg mol]	References	$C_p = f(T)$ a	$b \cdot 10^3$	$c \cdot 10^{-5}$	Temp. range in K	References
$3SrO \cdot MgO \cdot 2SiO_2$	cr	1058.94	[60]	1021.43	[78]	75.84	[78]	76.82	11.48	12.9	298–1900	[78]
$SrO \cdot MgO \cdot 2SiO_2$	cr	760.34	[78]	724.12	[78]	38.7	[78]	58.65	2.37	−15.32	298–1900	[78]
$3SrO \cdot MgO \cdot 4SiO_2$	cr	1540.54	[78]	1473.36	[78]	83.68	[78]	103.51	21.41	−22.58	298–1900	[78]
$Sr(OH)_2$	cr	229.3	[104]	207.8	[29]	21.0	[29]	–	–	–	–	–
$SrO \cdot SiO_2$	cr	390.09	[4]	374.64	[77]	22.5	[29]	26.76	4.59	−7.24	–	[77]
$2SrO \cdot SiO_2$	cr	549.99	[4]	532.39	[77]	36.0	[77]	36.8	6.7	−7.52	–	[77]
$3SrO \cdot SiO_2$	cr	709.89	[77]	689.99	[77]	49.0	[77]	47.02	8.43	−8.14	–	[77]
$SrO \cdot TiO_2$	cr	400.9	[25]	–	–	–	–	28.23	2.04	−4.58	298–1800	[20]
$SrO \cdot Ti_2O_3$	cr	554.0	[25]	–	–	–	–	38.45	3.84	−4.67	298–1800	[20]
$SrSO_4$	cr	345.30	[104]	318.50	[104]	29.1	[104]	26.25	11.3	−2.53	298–1800	[79]
$SrZrO_3$	cr	422.6	[41]	402.4	[41]	–	–	–	–	–	–	–
ThO_2	cr	293.2	[52a]	–	–	15.6	[52a]	15.84	2.88	1.60	298–1800	[52a]
Ti	gas	−95.0	[104]	−84.0	[104]	43.069	[104]	–	–	–	–	–
Ti	cr	0.0	[104]	0.0	[104]	7.24	[104]	–	–	–	–	–
$TiCl_2$	cr	114.0	[104]	96.0	[29]	26.0	[29]	–	–	–	–	–
$TiCl_3$	cr	165.0	[104]	148.0	[29]	30.5	[29]	–	–	–	–	–
$TiCl_4$	liq	179.3	[104]	161.2	[104]	60.4	[104]	–	–	–	–	–
Ti_2O_3-α	cr	362.9	[104]	342.23	[34]	18.8	[36]	7.31	53.52	–	298–473	[36]
Ti_2O_3-β	cr	–	–	–	–	–	–	34.68	1.30	−10.20	473–1800	[36]
Ti_3O_5-α	cr	587.65	[34]	553.78	[34]	30.9	[36]	35.47	29.5	–	298–450	[36]
Ti_3O_5-β	cr	–	–	–	–	–	–	41.60	8.0	–	450–1400	[36]
TiO_2-rutile	cr	225.5	[104]	212.28	[34]	12.0	[36]	17.97	0.28	−4.35	298–1800	[36]

TiO_2-anatase	cr	218.1	[34]	204.86	[34]	11.93	[34]	17.83	0.50	−4.23	298−1300	[36]
TiO-α	cr	123.9	[34]	116.89	[34]	8.3	[36]	10.57	3.60	−1.86	298−1264	[36]
TiO-β	cr	–	–	–	–	–	–	11.85	3.00	–	1264−1800	[36]
$TiSiO_3$	cr	320.0	[34]	–	–	–	–	–	–	–	–	–
Tl_2O_3	cr	97.5	[34]	76.0	[34]	35.4	[34]	–	–	–	–	–
Tl_2SiO_3	cr	265.0	[34]	–	–	–	–	–	–	–	–	–
V_2O_3	cr	299.4	[52a]	–	–	23.5	[36]	29.35	4.76	−5.42	298−1800	[36]
V_2O_5	cr	380.6	[52a]	–	–	31.3	[36]	46.54	−3.90	−13.22	298−Tm	[36]
Zn	gas	−31.245	[120]	−2.748	[120]	38.45	[120]	–	–	–	–	–
Zn	cr	0.0	[104]	0.0	[104]	9.95	[104]	–	–	–	–	–
Zn^{2+}	gas	−665.1	[120]	–	–	–	–	–	–	–	–	–
$ZnCl_2$	cr	99.20	[120]	88.30	[120]	26.64	[120]	–	–	–	–	–
$ZnCO_3$	cr	194.26	[120]	174.85	[120]	19.7	[120]	9.3	33.0	–	–	[40]
$ZnFe_2O_4$	cr	279.8	[22]	–	–	–	–	–	–	–	–	–
ZnO	cr	83.24	[104]	76.08	[104]	10.43	[120]	11.71	1.22	−2.18	298−1600	[36]
$Zn(OH)_2$-β	cr	153.42	[120]	132.31	[120]	19.4	[120]	–	–	–	–	–
$ZnO \cdot SiO_2$	cr	306.95	[104]	287.15	[29]	21.4	[34]	–	–	–	–	–
$2ZnO \cdot SiO_2$	cr	391.09	[46]	363.98	[c]	31.4	[114]	–	–	–	–	–
$ZnO \cdot TiO_2$	cr	309.1	[34]	–	–	32.8	[48]	39.82	5.54	−7.69	298−1800	[7]
$ZnSO_4 \cdot H_2O$	cr	311.78	[120]	270.58	[120]	33.1	[120]	–	–	–	–	–
$ZnSO_4 \cdot 6H_2O$	cr	663.83	[120]	555.64	[120]	86.9	[120]	–	–	–	–	–
$ZnSO_4 \cdot 7H_2O$	cr	735.6	[120]	612.59	[120]	92.9	[120]	–	–	–	–	–
$ZnSO_4$	cr	234.9	[120]	209.0	[120]	28.6	[120]	17.07	20.80	–	298−1000	[32]
Zr	gas	−125.0	[104]	−115.0	[104]	43.313	[104]	–	–	–	–	–
Zr-α	cr	0.0	[104]	0.0	[104]	9.18	[104]	6.83	1.12	−0.91	298−1135	[52a]
Zr-β	cr	–	–	–	–	–	–	7.27	–	–	1135−1400	[52a]
Zr^{4+}	gas	–	–	−141.0	[29]	–	–	–	–	–	–	–
ZrC	cr	45.0	–	36.5	[33]	–	–	–	–	–	–	–
ZrC_4	cr	230.0	[104]	209.0	[104]	44.5	[104]	32.85	−3.82	–	298−550	[32]
ZrO_2-β	cr	261.5	[113]	247.7	[c]	12.12	[113]	16.64	1.80	−3.36	298−1478	[51]
ZrO_2-α	cr	–	–	–	–	–	–	17.8	–	–	1478−1850	[52a]
$Zr(OH)_4$	cr	411.2	[104]	370.0	[29]	31.0	[29]	–	–	–	–	–
$ZrSiO_4$	cr	483.3	[c]	456.95	[c]	20.2	[36]	31.48	3.92	−8.08	298−1800	[51]
$ZrO \cdot TiO_2$	cr	429.0	[25]	–	–	–	–	–	–	–	–	–

Appendix 2. Thermodynamic Properties of Ions and Molecules in Aqueous Solutions

Formulae of ions and molecules	$-\Delta H^0_{298}$ [kcal/mol]	$-\Delta G^0_{298}$ [kcal/mol]	S^0_{298} [cal/degmol]	C_{p298} [cal/degmol]	References
Ag-Silver					
Ag^+	25.23	−18.43	+17.42	+5.9	[13]
Ag^{2+}	−64.20	−64.30	− 2.1	−	[13]
$Ag(OH)^0$	−	21.9	−	−	[13]
$Ag(OH)_2^0$	−	62.20	−	−	[13]
$AgCl^0$	17.42	17.46	+37.0	−	[13]
$AgCl_2$	58.57	51.48	+55.3	−	[13]
AgF^0	57.40	49.00	+ 6.2	−	[13]
Al, Aluminium					
Al^{3+}	127.0	117.59	−71.6	+9.1	[13]
$Al(OH)^{2+}$	183.32	167.46	−37.5	−	[13]
$Al(OH)_2^+$	−	216.87	−	−	[13]
$Al(OH)_3^0$	−	265.55	−	−	[13]
$Al(OH)_4^-$	356.20	312.0	+34.5	−	[13]
AlF^{2+}	205.73	194.03	−39.5	−	[13]
AlF_2^+	284.60	268.62	−14.0	−	[13]
AlF_3^0	364.21	341.07	+ 1.8	−	[13]
AlF_4^-	443.96	411.24	+ 9.5	−	[13]
AlF_5^{2-}	521.10	480.00	+11.2	−	[13]
AlF_6^{3+}	605.40	547.60	+ 5.0	−	[13]
$AlSO_4^+$	342.03	299.48	−45.9	−	[13]
$Al(SO_4)_2^-$	558.57	479.84	−30.5	−	[13]
$Al_2(SO_4)_3^0$	897.10	735.31	−	−	[104]
AlO_2^-	219.6	196.8	− 5.0	−	[120]
$Al(NO_3)_2^0$	276.0	196.0	28.1	−	[120]
As, Arsenic					
AsO^+	−	37.41	−	−	[13]
$H_3AsO_3^0$	178.90	154.40	+46.65	+48.9	[13]
$H_2AsO_3^-$	164.70	125.30	− 3.6	−66.4	[13]
AsO_3^{3-}	158.65	107.00	−44.6	−113.6	[13]
$H_3AsO_4^0$	216.70	184.04	+43.7	+50.6	[13]
$H_2AsO_4^-$	218.41	181.04	+28.0	−13.1	[13]
$HAsO_4^{2-}$	217.64	171.53	− 1.3	−67.9	[13]
AsO_4^{3-}	213.29	155.83	−39.4	−116.9	[13]

Table (continued)

Formulae of ions and molecules	$-\Delta H^0_{298}$ [kcal/mol]	$-\Delta G^0_{298}$ [kcal/mol]	S^0_{298} [cal/degmol]	C_{p298} [cal/degmol]	References
Au, Gold					
Au^+	+53.10	+42.70	+30.5	+ 0.7	[13]
$AuCl_2^-$	41.70	36.13	+61.5	–	[13]
$AuCl_4^-$	77.60	56.17	+61.6	–	[13]
$AuCl_3OH^-$	–	73.10	–	–	[13]
$AuCl_2(OH)_2^-$	–	88.70	–	–	[13]
$AuCl(OH)_3^-$	–	103.10	–	–	[13]
$Au(OH)_4^-$	–	116.50	–	–	[13]
$Au(OH)_5^{2-}$	–	155.0	–	–	[13]
$Au(OH)_3^0$	–	75.90	–	–	[13]
B, Boron					
$B(OH)_3^0$	256.05	231.51	+39.4	–	[13]
$B(OH)_4^-$	321.00	275.99	+25.1	–	[13]
$B(OH)_5^{2-}$	–	315.50	–	–	[13]
$B(OH)_6^{3-}$	–	353.90	–	–	[13]
BF_4^-	376.13	356.45	+47.9	–	[13]
$BF_3(OH)^-$	364.16	338.32	+26.0	–	[13]
$BF_2(OH)_2^-$	–	318.90	–	–	[13]
Ba, Barium					
Ba^{2+}	125.25	130.86	+2.1	–8.3	[13]
$Ba(OH)^+$	179.12	171.48	+13.4	–	[13]
$Ba(OH)_2^0$	238.58	209.20	–2.0	–	[13]
$BaSO_4^0$	–	311.86	–	–	[13]
$BaCl_2$	208.72	196.70	29.0	–	[1]
Be, Beryllium					
Be^{2+}	–96.50	–91.10	–47.00	+23.6	[13]
$Be(OH)_3^-$	–	229.30	–	–	[13]
$Be(OH)_4^{2-}$	–	266.60	–	–	[13]
$Be_3(OH)_3^{3+}$	–	431.20	–	–	[13]
BeF^+	176.50	166.20	–23.7	–	[13]
BeF_2^0	254.80	238.50	–3.8	–	[13]
BeF_3^-	–	309.20	–	–	[13]
BeF_4^{2-}	–	379.20	–	–	[13]
Bi, Bismuth					
Bi^{3+}	–19.26	–21.95	–42.2	–10.5	[13]
BiO	–	34.90	–	–	[13]
$Bi(OH)^{2+}$	–	32.60	–	–	[13]
$Bi(OH)_4^0$	–	174.40	–	–	[13]
$BiCl^{2+}$	16.70	14.09	+0.3	–	[13]
$BiCl_2^+$	–	47.60	–	–	[13]

Table (continued)

Formulae of ions and molecules	$-\Delta H^0_{298}$ [kcal/mol]	$-\Delta G^0_{298}$ [kcal/mol]	S^0_{298} [cal/degmol]	C_{p298} [cal/degmol]	References
$BiCl_3^0$	–	80.50	–	–	[13]
$BiCl_4^-$	–	114.60	–	–	[13]
Br, Bromine					
Br^-	29.05	24.91	+19.9	−35.7	[13]
BrO_3^-	15.45	−4.84	+39.2	−20.0	[13]
Br_2^0	0.2	−0.96	+32.5	–	[13]
Br_3	31.6	–	–	–	[13]
C, Carbon					
CO^0	28.91	28.66	−25.0	+34.1	[13]
$H_2CO_3^0$	167.22	148.94	+44.8	+50.1	[13]
CO_2^0	98.90	92.26	+28.9	+31.5	[13]
HCO_3^-	165.39	140.26	+21.8	−8.9	[13]
CO_3^{2-}	161.84	126.17	−13.6	−59.7	[13]
CH_4^0	21.28	8.22	+20.0	+36.7	[13]
$C_2O_4^{2-}$	193.2	159.55	+19.1	–	[13]
CH_3COOH^0	116.74	95.51	–	–	[54]
CH_3COO^-	115.84	89.02	–	–	[54]
$C_2H_2OH^0$	68.85	42.40	–	–	[54]
Ca, Calcium					
Ca^{2+}	129.70	132.10	−13.2	+0.2	[13]
$Ca(OH)^+$	182.67	171.36	−3.5	–	[13]
$CaSO_4^0$	346.21	313.20	+4.9	–	[13]
$CaCO_3^0$	–	262.64	–	–	[13]
$CaCl_2^0$	209.64	195.04	+14.30	–	[120]
CaF_2^0	287.1	264.34	−17.8	–	[104]
$Ca(OH)_2^0$	239.68	207.49	+17.8	–	[120]
$Ca(NO_3)_2^0$	228.86	185.52	–	–	[120]
Cd, Cadmium					
Cd^{2+}	18.06	18.61	−17.0	+1.0	[13]
$Cd(OH)^+$	–	64.90	–	–	[13]
$Cd(OH)_2^0$	–	106.70	–	–	[13]
$Cd(OH)_3^-$	–	144.60	–	–	[13]
$Cd(OH)_4^{2-}$	–	181.60	–	–	[13]
$CdCl_2^0$	96.75	85.00	26.2	–	[13]
$Cd(HS)_2^0$	–	35.40	–	–	[13]
$Cd(SO_4)^0$	234.40	199.50	+0.9	–	[13]
Ce, Cerium					
Ce^{3+}	166.80	161.50	−48.0	−6.6	[13]

Table (continued)

Formulae of ions and molecules	$-\Delta H^0_{298}$ [kcal/mol]	$-\Delta G^0_{298}$ [kcal/mol]	S^0_{298} [cal/degmol]	C_{p298} [cal/degmol]	References
Ce^{4+}	137.90	121.30	−101.7	−0.5	[13]
$Ce(OH)^{3+}$	194.10	179.10	−40.7	–	[13]
$CePO_4^0$	–	430.30	–	–	[13]
Cl, Chlorine					
Cl^-	39.95	31.37	+13.5	−32.6	[13]
Cl_2^0	5.60	−1.65	+29.0	–	[13]
ClO^-	25.60	8.80	+10.00	−1.0	[13]
ClO_2^-	15.90	−4.10	+24.2	−10.6	[13]
ClO_3^-	24.88	2.03	+39.1	−19.7	[13]
ClO_4^-	30.91	2.12	+43.8	−24.4	[13]
$HClO^0$	28.90	19.10	+35.0	+56.6	[13]
$HClO_2^0$	12.40	−6.80	+45.0	+50.0	[13]
Co, Cobalt					
Co^{2+}	14.00	13.41	−26.0	+8.9	[13]
$Co(OH)_2^0$	–	101.13	–	–	[13]
$Co(OH)_3^-$	–	140.50	–	–	[13]
$CoSO_4^0$	230.82	194.41	−9.3	–	[13]
$CoS_2O_9^0$	–	138.00	–	–	[13]
Co^{3+}	−6.00	−18.69	−75.9	+11.8	[13]
Cr, Chromium					
Cr^{3+}	56.43	53.35	−51.53	–	[105]
$Cr_2O_7^{2-}$	356.57	312.00	64.67	–	[105]
CrO_4^{2-}	210.85	174.46	12.91	–	[105]
Cu, Copper					
Cu^+	−17.24	−11.94	+10.1	+14.3	[13]
Cu^{2+}	−15.70	−15.60	−22.9	+5.6	[13]
$Cu(OH)^+$	–	31.07	–	–	[13]
$Cu(OH)_3^-$	–	117.65	–	–	[13]
$Cu(OH)_4^{2-}$	–	156.46	–	–	[13]
$CuCl_2^-$	–	58.36	–	–	[13]
$CuCl^+$	–	17.07	–	–	[13]
$CuCO_3^0$	–	120.00	–	–	[13]
$Cu(CO_3)_2^{2-}$	–	250.60	–	–	[13]
$CuSO_4^0$	200.4	165.26	−4.3	–	[13]
F, Fluorine					
F^-	79.79	66.92	−3.35	−25.1	[13]
HF^0	76.61	71.25	+21.85	+9.4	[13]
HF_2^-	157.90	138.98	+16.2	–	[13]

Table (continued)

Formulae of ions and molecules	$-\Delta H^0_{298}$ [kcal/mol]	$-\Delta G^0_{298}$ [kcal/mol]	S^0_{298} [cal/degmol]	C_{p298} [cal/degmol]	References
Fe, Iron					
Fe^{2+}	22.13	22.05	−25.0	+7.9	[13]
Fe^{3+}	12.13	4.27	−66.7	+5.9	[13]
$Fe(OH)^+$	78.45	66.91	−7.7	–	[89]
$Fe(OH)^0_2$	–	109.25	–	–	[89]
$Fe(OH)^-_3$	–	148.00	–	–	[89]
$Fe(OH)^{2-}_4$	–	185.00	–	–	[89]
$Fe(OH)^{2+}$	69.75	57.46	−25.8	–	[89]
$Fe(OH)^+_2$	131.77	109.28	−4.3	–	[89]
$Fe(OH)^0_3$	–	160.30	–	–	[89]
$Fe(OH)^-_4$	–	201.20	–	–	[89]
$Fe(HS)^0_2$	–	28.50	–	–	[13]
$Fe(HS)^-_3$	–	28.40	–	–	[13]
$FeCl^+_2$	–	69.44	–	–	[13]
$FeCl^0_2$	101.2	81.59	−5.9	–	[120]
$FeCl^{2+}$	46.48	37.66	−27.7	–	[13]
$FeCl^0_3$	–	99.45	–	–	[13]
$FeSO^0_4$	238.89	202.83	−8.8	–	[13]
$FeSO^+_4$	223.25	187.71	−22.7	–	[13]
$Fe(NO_3)^0_3$	160.3	80.9	29.5	–	[120]
$Fe(HPO_4)^+$	324.9	279.57	−37.7	–	[13]
FeO^{2-}_4	114.6	–	–	–	[13]
H, Hydrogen					
H^+	0	0	0	0	[13]
H_2	1.00	−4.2	+13.8	+41.6	[13]
H_2O^0	68.32	56.69	+16.71	+17.995	[13]
Hg, Mercury					
Hg^{2+}_2	−41.20	−36.70	+20.2	–	[13]
Hg^{2+}	−40.90	−39.30	−7.7	–	[13]
$Hg(OH)^+$	20.2	12.50	+17.0	–	[13]
$Hg(OH)^0_2$	84.90	65.70	+34.0	–	[13]
$Hg(OH)^-_3$	–	121.20	–	–	[13]
$HgCl^+$	4.5	1.3	+18.0	–	[13]
$HgCl^0_2$	51.70	41.40	+37.0	–	[13]
$HgCl^-_3$	92.90	73.90	+50.0	–	[13]
$Hg(HS)^0_2$	–	6.40	–	–	[13]
HgS^{2-}_2	–	−10.0	–	–	[13]
$Hg(S_2O_3)^{2-}_2$	–	248.30	–	–	[13]
$HgSrO^0_3$	–	123.40	–	–	[13]

Table (continued)

Formulae of ions and molecules	$-\Delta H^0_{298}$ [kcal/mol]	$-\Delta G^0_{298}$ [kcal/mol]	S^0_{298} [cal/degmol]	C_{p298} [cal/degmol]	References
I, Iodine					
I^-	13.79	12.33	+24.6	−40.1	[13]
IO_3^-	52.51	30.10	+28.0	−19.0	[13]
I_2	−5.20	−3.93	+32.2	–	[13]
I_3^-	12.55	12.34	+56.5	–	[13]
K, Potassium					
K^+	60.30	67.56	+24.18	+5.2	[13]
KCl^0	100.06	98.82	37.70	–	[104]
KSO_4^-	276.61	246.36	+35.2	–	[13]
La, Lanthanum					
La^{3+}	169.00	164.00	−50.0	−13.6	[13]
$LaCO_3^+$	–	300.70	–	–	[13]
LaF^{2+}	–	235.80	–	–	[13]
$LaSO_4^+$	382.82	346.56	−18.1	–	[13]
Li, Lithium					
Li^+	66.56	69.94	+2.7	+16.8	[13]
Mg, Magnesium					
Mg^{2+}	110.36	108.81	−28.6	+9.2	[89]
$Mg(OH)^+$	–	152.50	–	–	[c]
$MgCl_2^0$	190.46	171.69	−1.9	–	[104]
$MgCO_3^0$	–	239.60	–	–	[13]
$MgHCO_3^+$	–	251.1	–	–	[89]
$MgSO_4^0$	327.17	298.63	−12.5	–	[13]
Mn, Manganese					
Mn^{2+}	57.70	54.96	−16.0	+3.0	[13]
$Mn(OH)^+$	106.60	97.20	0.6	–	[13]
$Mn(OH)_3^-$	–	178.40	–	–	[13]
Mn^{3+}	24.00	20.40	−51.0	−4.6	[13]
MnO_4^{2-}	–	118.10	–	–	[13]
MnO_4^-	127.4	105.23	+46.9	−25.9	[13]
Mo, Molybdenum					
MoO_4^{2-}	238.50	200.30	+7.9	−74.0	[13]
$HMoO_4^-$	–	207.13	–	–	[13]
$H_2MoO_4^0$	–	209.64	–	–	[13]
MoO_2^{2+}	110.40	98.30	−16.0	–	[13]
$MoO_2(OH)^+$	–	154.37	–	–	[13]

Table (continued)

Formulae of ions and molecules	$-\Delta H^0_{298}$ [kcal/mol]	$-\Delta G^0_{298}$ [kcal/mol]	S^0_{298} [cal/degmol]	C_{p298} [cal/degmol]	References
N, Nitrogen					
N_2^0	2.52	4.34	+22.8	+35.6	[13]
NO_3^-	49.56	26.63	+35.1	−20.7	[13]
NH_4^+	31.67	18.82	+26.6	+19.1	[13]
NH_4OH^0	87.51	62.90	+42.9	–	[13]
NO_2^-	25.00	0.89	+33.5	−23.3	[13]
Na, Sodium					
Na^+	57.47	62.67	+14.1	+11.1	[13]
$NaCl^0$	97.30	93.94	27.6	–	[104]
$NaHCO_3^0$	222.50	202.89	37.1	–	[104]
$Na_2SO_4^0$	331.46	302.52	32.9	–	[104]
$NaOH^0$	112.24	100.18	11.9	–	[104]
Nb, Niobium					
NbO_3^-	243.80	225.40	+36.0	−18.4	[13]
$HNbO_3^0$	248.30	235.50	+55.0	+43.8	[13]
NbO_2^+	–	178.00	–	–	[13]
Ni, Nickel					
Ni^{2+}	12.80	10.78	−30.83	+1.4	[13]
$Ni(OH)^+$	–	52.57	–	–	[13]
$Ni(OH)_2^0$	–	99.90	–	–	[13]
$Ni(OH)_3^-$	–	141.24	–	–	[13]
$NiSO_4^0$	229.71	191.73	−14.6	–	[13]
$NiS_2O_3^0$	–	136.70	–	–	[13]
O, Oxygen					
O_2	2.8	−3.9	+26.5	+33.2	[13]
OH^-	54.97	37.59	−2.57	−35.5	[13]
OH	–	−8.53	–	–	[88]
P, Phosphorus					
$H_3PO_4^0$	307.92	273.10	+37.8	+54.8	[13]
$H_2PO_4^-$	309.82	270.17	+21.6	−10.5	[13]
HPO_4^{2-}	308.83	260.34	−8.0	−63.5	[13]
PO_4^{3-}	305.30	243.5	−53.0	−80.3	[13]
$H_3PO_3^0$	230.60	–	–	–	[13]
$H_2PO_3^-$	231.70	–	–	–	[13]
HPO_3^{2-}	231.60	–	–	–	[13]

Table (continued)

Formulae of ions and molecules	$-\Delta H^0_{298}$ [kcal/mol]	$-\Delta G^0_{298}$ [kcal/mol]	S^0_{298} [cal/degmol]	C_{p298} [cal/degmol]	References
Pb, Lead					
Pb^{2+}	0.33	5.83	+2.7	−12.6	[13]
$Pb(OH)^+$	–	54.10	–	–	[13]
$Pb(OH)^0_2$	–	95.80	–	–	[13]
$Pb(OH)^-_3$	–	137.60	–	–	[13]
PbF^+	–	74.5	–	–	[13]
PbF^0_2	–	142.80	–	–	[13]
$PbSO^0_4$	–	187.20	–	–	[13]
$Pb(HS)^0_2$	–	20.90	–	–	[13]
$Pb(HS)^-_3$	–	19.80	–	–	[13]
Pd, Palladium					
Pd^{2+}	–	−42.20	–	–	[13]
$PdCl^+$	–	−2.64	–	–	[13]
$PdCl^0_2$	–	35.00	–	–	[13]
$Pd(OH)^+$	–	12.70	–	–	[13]
$Pd(OH)^0_2$	–	68.60	–	–	[13]
Pt, Platinum					
Pt^{2+}	–	−44.40	–	–	[13]
$PtCl^{2-}_4$	12.03	88.10	+40.0	–	[13]
$PtCl^-_3$	–	54.60	–	–	[13]
$PtCl^0_2$	–	19.10	–	–	[13]
$PtCl^{2-}_6$	161.0	117.0	+52.6	–	[13]
S, Sulphur					
H_2S^0	9.50	6.66	+29.0	+31.5	[13]
HS^-	4.20	−2.80	+15.0	−33.6	[13]
S^{2-}	−7.90	−20.50	−3.5	−95.7	[13]
HSO^-_4	212.42	180.48	+29.7	−17.0	[13]
SO^{2-}_4	217.32	177.78	+4.2	−71.6	[13]
$H_2SO^0_3$	145.51	128.56	+55.5	+43.0	[13]
HSO^-_3	149.67	126.15	+33.4	−16.7	[13]
SO^{2-}_3	151.90	116.30	−7.0	−64.1	[13]
SO^0_2	77.19	71.87	+38.7	+25.0	[13]
$H_2S_2O^0_3$	144.45	125.97	+57.9	+41.3	[13]
$HS_2O^-_3$	149.30	125.15	+38.9	−20.3	[13]
$S_2O^{2-}_3$	155.90	122.80	+8.9	−71.8	[13]
Sb, Antimony					
$Sb(OH)^+_2$	–	99.60	–	–	[13]
$Sb(OH)^0_3$	184.90	154.70	+30.0	–	[13]
$Sb(OH)^-_4$	–	195.30	–	–	[13]

Table (continued)

Formulae of ions and molecules	$-\Delta H^0_{298}$ [kcal/mol]	$-\Delta G^0_{298}$ [kcal/mol]	S^0_{298} [cal/degmol]	C_{p298} [cal/degmol]	References
$H_3SbO_4^0$	216.80	–	–	–	[13]
Se, Selenium					
Se^{2-}	–15.30	–30.90	–11.0	–90.7	[13]
HSe^-	–3.80	–10.50	+19.0	–36.3	[13]
H_2Se^0	–4.60	–5.30	+39.1	+24.7	[13]
$HSeO_4^-$	137.70	107.67	+38.9	–20.3	[13]
SeO_4^{2-}	143.20	105.50	+12.9	–77.4	[13]
SeO_3^{2-}	121.70	86.95	–1.7	–67.6	[13]
$HSeO_3^-$	122.98	98.30	+32.1	–16.5	[13]
$H_2SeO_3^0$	121.29	101.87	+49.7	+46.4	[13]
Si, Silicon					
$H_4SiO_4^0$	349.11	312.72	+42.9	+51.4	[89]
$H_3SiO_4^-$	340.55	299.39	+26.9	–17.3	[89]
$H_2SiO_4^{2-}$	333.53	283.49	–3.1	–67.0	[89]
SiO_4^{4-}	318.10	250.69	–96.1	–	[89]
SiO_2^0	214.4	–	–	–	[120]
SiO_3^{2-}	–	224.6	–	–	[89]
$HSiO_3^-$	–	240.7	–	–	[89]
$Si_6O_{17}^{10-}$	–	1343.53	–	–	[c]
$Si_{12}O_{31}^{14-}$	–	2631.81	–	–	[c]
$Si_6O_{15}^{6-}$	–	1307.52	–	–	[c]
SiF_6^{2-}	572.56	527.50	+29.9	–	[13]
Sn, Tin					
Sn^{2+}	5.83	6.48	–16.7	+2.3	[13]
$Sn(OH)^+$	–	60.3	–	–	[13]
$Sn(OH)_2^0$	–	110.80	–	–	[13]
$Sn(OH)_3^-$	–	154.50	–	–	[13]
$SnCl^+$	–	39.90	–	–	[13]
$SnCl_2^0$	–	72.30	–	–	[13]
$SnCl_3^-$	–	103.40	–	–	[13]
Sn^{4+}	–	0.65	–	–	[29]
Sr, Strontium					
Sr^{2+}	133.00	136.58	–6.5	–5.0	[13]
$Sr(OH)^+$	186.80	175.30	–1.4	–	[13]
Ta, Tantalum					
TaO_3^-	275.20	256.40	–36.0	–18.4	[13]
$HTaO_3^0$	282.6	269.50	+55.0	–43.8	[13]
TaO_2^+	–	214.20	–	–	[13]

Table (continued)

Formulae of ions and molecules	$-\Delta H^0_{298}$ [kcal/mol]	$-\Delta G^0_{298}$ [kcal/mol]	S^0_{298} [cal/degmol]	C_{p298} [cal/degmol]	References
Te, Tellurium					
Te^{2-}	−24.20	−41.6	−15.3	−87.8	[13]
HTe^-	−16.60	−25.0	+14.7	−33.4	[13]
H_2Te^0	−18.6	−21.40	+33.7	+28.4	[13]
$H_2TeO_3^0$	132.6	113.40	52.20	+45.2	[13]
$HTeO_3^-$	133.0	108.10	33.2	−15.5	[13]
TeO_3^{2-}	127.30	93.50	+3.2	−70.9	[13]
$Te(OH)^+$	–	61.40	–	–	[13]
U, Uranium					
UO_2^{2+}	244.80	229.78	−20.6	+9.4	[13]
$UO_2(OH)^+$	300.62	279.45	+14.4	–	[13]
$UO_2(OH)_2^0$	362.10	326.19	+20.8	–	[13]
$UO_2(OH)_3^-$	–	367.65	–	–	[13]
$UO_2(SO_4)^0$	457.10	411.27	+12.9	–	[13]
UO_2^+	–	231.20	–	–	[13]
U^{4+}	141.10	139.50	−82.6	−13.0	[13]
U^{3+}	117.40	116.9	−36.5	−14.0	[13]
V, Vanadium					
V^{2+}	54.3	51.89	−32.1	+12.7	[13]
V^{3+}	62.90	57.77	−57.0	−0.6	[13]
$H_3VO_4^0$	–	249.33	–	–	[13]
W, Tungsten					
$H_2WO_4^0$	–	230.60	–	–	[13]
HWO_4^-	–	227.60	–	–	[13]
WO_4^{2-}	256.54	222.60	23.3	84.0	[13]
Zn, Zinc					
Zn^{2+}	36.73	35.18	−26.45	+8.8	[13]
$Zn(OH)^+$	–	79.65	–	–	[13]
$Zn(OH)_2^0$	–	127.96	–	–	[13]
$Zn(OH)_3^-$	–	168.43	–	–	[13]
$Zn(OH)_4^{2-}$	–	208.24	–	–	[13]
$Zn(HS)_2^0$	–	49.80	–	–	[13]
$ZnSO_4^0$	253.42	216.20	−9.3	–	[13]

References to Appendices 1 and 2

1. Babushkin VI (1960) Trudy KhIITa 39
2. Babushkin VI, Kolomatskiy NS (1979) Calcium Aluminate Hydration. Trudy Kubanskogo politekhnicheskogo instituta. Krasnodar
3. Babushkin VI, Kolomatskiy AS (1980) Tentative Methods of Evaluating the Thermodynamic Data of Calcium Hydroferrites. Proceedings of BTISM. A Complex Utilization of the Rocks of the Kursk Magnetic Anomaly in Building. Moscow
4. Barany R, King EG, Todd SS (1957) J Am Chem Soc 79:363
5. Berezhnoy AS (1962) Doklady Akad Nauk USSR 5/I
6. Bludov BF (1969) Termodinamĭchnĭ rozrakhunki v sistemĭ MgO-SiO_2-H_2O. Izd KhGU
7. Bonnikson KR (1955) J Am Chem Soc 77:2152
8. Bonnikson KR (1955) J Phys Chem 59:220
9. Bonnikson KR (1954) J Am Chem Soc 76:1480
10. Brewer L et al. Natl Nuclear Sci Ser, Div IV, 19 B:76
11. Brewer L (1953) Chem Reb 1:52
12. Budnikov PP, Kerbe FT, Matveyev GM (1967) Vitreous Systems and Materials. Zinatne 67 – 76, Riga
13. Bulakh AG, Bulakh KG (1978) Physico-Chemical Properties of Minerals and Components of Hydrothermal Solutions. Nedra, Leningrad, pp 134 – 151
14. Charlu TV, Newton RC, Kleppa OJ (1975) Geochim Cosmochim Acta 39/II:1487 – 1497
15. Clemser O, Schröder H (1953) Z Anorg Allg Chemie 271:293
16. Clusius K, Coldmann J, Perlik A (1949) Z Naturforsch 4a:424
17. Coughlin JP (1956) J Am Chem Soc 78:5479
18. Coughlin JP (1957) J Am Chem Soc 79:2397
19. Coughlin JP (1955) J Am Chem Soc 77:868
20. Coughlin IP, Orr RL (1953) J Am Chem Soc 75:530
21. Douglas TB, Dever JL (1954) J Am Chem Soc 76:4826
22. Fischer M (1966) Z Anorg Allg Chemie B 345:134 – 136
23. Fotiyev AA (1977) Zhurn Neorg Khim XXII/9:2531 – 2533
24. Garrels R (1962) Mineral'nyye ravnovesiya. Inostrannaya Literatura
25. Glushkova VB (1963) Silicates and Oxides in the High-Temperature Chemistry, pp 308 – 321. Moscow
26. Govorov IN (1977) Thermodynamics of Ion-Mineral Equilibria and Mineralogeny of Hydrothermal Deposits. Nauka, Moscow, pp 194 – 222
27. Grenier G, White D (1957) J Phys Chem 61:1681
28. Halla F, Van Tassel R (1968) Radex-Rundschau vol I, pp 27 – 30
29. Hemingway BS, Robie RA (1977) J Res US Geol Surv 5/4:413 – 429
30. Holm JL, Kleppa OJ (1966) J Phys Chem 70/5:1690
31. Jeffes JHE, Richardson FD, Pearson J (1954) Trans Farad Soc 50:364
32. Karapet'yants MK (1953) Chemical Thermodynamics. Goskhimizdat
33. Karapet'yants MK, Karapet'yants ML (1961) Trudy MKhTI im. DI Mendeleyeva 34
34. Karapet'yants MK, Karapet'yants ML (1968) The Main Thermodynamic Constants of Inorganic and Organic Substances. Khimiya, Moscow
35. Kelley KK (1960) US Bur Mines Bull 584
36. Kelley KK (1949) US Bur Mines Bull 476
37. Kelley KK (1950) US Bur Mines Bull 477
38. Kelley KK, King EG (1961) US Bur Mines Bull 500

39. Kelley KK, King EG (1961) US Bur Mines Bull 592
40. Kelley KK, Anderson CT (1935) Bull US Bur Mines 384
41. Khekimov Y, Levitskiy VA, Shevchenko NN, Gerasimov Y (1973) Doklady Akad Nauk SSSR 211/2:393 – 398
42. Khoutepen SD, Steyn GN (1976) The Enthalpy of Formation and Dehydration of Calcium Hydroaluminates with Univalent Anions. The VIth International Congress on the Chemistry of Cement, vol 2, book I. Moscow, Stroyizdat, pp 17 – 25
43. King EG (1955) J Phys Chem 59:218
44. King EG (1957) J Am Chem Soc 79:5437
45. King EG (1954) J Am Chem Soc 76:5849
46. King EG (1951) J Am Chem Soc 73:656
47. King EG (1955) J Am Chem Soc 77:3189
48. King EG (1954) J Am Chem Soc 76:3289
49. King EG (1952) J Am Chem Soc 74:4446
50. King EG (1958) J Am Chem Soc 80:2400
51. King EC, Coughlin IP (1950) J Am Chem Soc 72:2262
52. Korneyev VR, Glushkova VB, Keler EK (1971) Neorganicheskiye Materialy 7/5:886 – 887
52a. Kubaschewski O, Evans E (1956) Metallurgical Thermochemistry, London
53. Lander II (1951) J Am Chem Soc 73:5794
54. Latimer WM (1951) J Am Chem Soc 73:180
55. Lenyov LM, Novokhatskiy IA (1966) Zhurn Fiz Khim XL/9:2030 – 2034
56. Levitskiy VA, Golovanova YG, Popov SG, Chentsov VN (1975) Zhurn Fiz Khim 49/7:1645 – 1649
57. Levitskiy VY, Rezukhina TN (1966) Izv Akad Nauk SSSR, Neorganicheskiye Materialy 2/1:145 – 150
58. Massazza F (1962) Ann Chim 52:51 – 58
59. Massazza F, Riva B (1966) Ann Chim 56/II:1278 – 1286
60. Massazza F (1966) Ann Chim 56/II:1298 – 1305
61. Massazza F (1961) Ann Chim 51:1375 – 1381
62. Massazza F, Cannes M (1961) Ann Chim 51:904 – 911
63. Massazza F (1961) Ann Chim 51:898 – 903
64. Massazza F, Virdis P (1965) Ann Chim 55:1176 – 1183
65. Massazza F, Fadda R (1964) Ann Chim 54:95 – 102
66. Matveyev GM (1963) Silicates and Oxides in the High-Temperature Chemistry. Moscow, pp 322 – 334
67. Matwejew GM, Agarkow AS, Kerbe F (1967) Silikattechnik 10:324 – 326
68. Matveyev GM, Agarkov AS, Zhuravlyov AK (1972) Inorganic Materials, VIII/2:399 – 400
69. Matwejew GM, Agarkow AS (1972) Thermodynamische Analyse der Bildung ternärer Verbindungen im System Na_2O-CaO-SiO_2 durch Festkörperreaktion. Silikattechnik, vol 23/4, pp 122 – 124
70. Matveyev MA, Frenkel' BN, Matveyev GM (1965) Izv Akad Nauk SSSR, Neorganicheskiye Materialy I/8:1426 – 1436
71. Matveyev MA, Frenkel' BN, Matveyev GM (1965) Studies in the Chemistry of Silicates and Oxides. Nauka, Moscow-Leningrad, pp 49 – 57
72. Matveyev MA, Frenkel' BN, Matveyev GM (1968) Zhurn Prikl Khim XLI/8:1652 – 1662
73. Matveyev MA, Frenkel' BN, Matveyev GM (1964) Trudy MKhTI Im. DI Mendeleyeva XLV/125. Moscow
74. Matveyev GM, Ganikirzh VY (1962) Trudy MKhTI Im. DI Mendeleyeva 37:44

75. Matveyev GM, Matveyev MA (1964) Zhurn VKhO Im. DI Mendeleyeva 9/2:354
76. Matveyev MA, Matveyev GM (1956) Physico-chemical Foundations of Ceramics, vol I, pp 504–506. Promstroyizdat
77. Matveyev GM, Matveyev MA (1963) Izv Akad Nauk SSSR 3, Seriya Fiziko-Tekhnicheskikh Nauk 59–62
78. Matveyev MA, Novikov VN, Matveyev GM (1969) Silicates. Trudy MKhTI Im. DI Mendeleyeva LIX:41–46. Moscow
79. Matveyev MA, Novikov VN, Matveyev GM (1966) Izv Akad Nauk SSSR, Neorganicheskiye Materialy 2/9:1720–1723
80. Matveyev GM, Zhuravlyov AK (1977) Thermodynamic Properties of Some Vitreous Silicates at Elevated Temperatures. Deposited in the VNIIESM of the Ministry of Building Materials of the USSR, vol 272
81. Matveyev GM, Zhuravlyov AK, Kuznetsova OM (1977) A Thermodynamic Analysis of the Reactions of Glass Formation in the System Li_2O-SiO_2. Deposited in the VNIIESM of the Ministry of Building Materials of the USSR, vol 273
82. Matveyev GM, Zhuravlyov AK, Kuznetsova OM (1978) A Thermodynamic Analysis of the Reactions of Glass Formation in the System K_2O-SiO_2. Deposited in the VNIIESM of the Ministry of Building Materials of the USSR, vol 303
83. Matveyev GM, Zhuravlyov AK (1975) Thermodynamic Properties of Some Vitreous Silicates at Elevated Temperatures. Zhurn VKhO Im. DI Mendeleyeva XX/4:474–475
84. McCready NW (1948) J Phys Colloid Chem 52:1277
85. Mchedlov-Petrosyan OP (1971) The Chemistry of Inorganic Building Materials. Moscow, Stroyizdat, p 274
86. Mchedlow-Petrosyan OP, Babushkin VI (1959) Doklady Akad Nauk SSSR 128/2
87. Mchedlov-Petrosyan OP, Babushkin VI (1976) Thermodynamics and Thermochemistry of Cement. The 6th International Congress on the Chemistry of Cement, vol 2, book I. Moscow, Stroyizdat, pp 7–16
88. Mtschedlow-Petrossjan OP, Babuschkin WJ (1958) Silikattechnik, vol 9/5
89. Mel'nik YP (1972) Thermodynamic Constants for Analyzing the Conditions of Formation of Iron Ores. A Handbook. Kiev, Naukova Dumka
90. Naumov GB, Ryzhenko BN, Khodakovskiy IL (1971) A Handbook of Thermodynamic Values for Geologists. Moscow, Atomizdat
91. Newman KS, Hoffman R (1956) J Res Nat Bur Stand 56/6:313
92. Newman ES (1959) J Res Nat Bur Stand 62/5:207
93. Newman ES (1958) J Res Nat Bur Stand 61/2:75
94. Novokhatskiy IA, Lenyov LM (1966) Zhurn Neorg Khim XI/9:2014–2020
95. Oralov TA, Yegorova IV, Malyshev VP, Buketov YA (1974) Manuscript deposited in VINITI, vol 928–974. Dep on Apr. 11
96. Orr RL (1953) J Am Chem Soc 75:528
97. Orr RL, Coughlin JP (1952) J Am Chem Soc 74:3186
98. Pankratz LB, Kelley KK (1964) US Bur Mines Rep, p 6370
99. Peppler RB, Newman ES (1951) J Res Bur Stand 47:439
100. Rezukhina TN, Levitskiy VA (1967) Izv Akad Nauk SSSR, Neorganicheskiye Materialy 3/I:138–145
101. Robrova AM, Fotiyev AA, Karapet'yants MK (1966) Zhurn Fiz Khim XL/8:1835–1841
102. Rosen E, Muan A (1966) J Am Ceram Soc 49/2:107–108
103. Roshchupkin VI, Lavrent'yev (1967) Izv Akad Nauk SSSR, Neorganicheskiye Materialy 3/3:551–556
104. Rossini FD et al. (1952) Selected values of chemical thermodynamic properties. Circular of the National Bureau of Standards 500, Washington

105. Ryabov VA, Ostroumov MA, Svit TF (1974) A Handbook of the Thermodynamic Properties of Substances. Leningrad, Khimiya
106. Šatava V (1959) Teoretické základy technologie silikátů, II. Termodynamika, SNTL. Praha
107. Schwiete HE, Hummel C (1958) Forschungsber Wirtsch- und Verkehrsmin Nordrhein-Westfalen 515/110
108. Shibanov YV, Chukhlantsev VG (1973) Manuscript deposited in VINITI on Oct. 9, vol 6991 – 73
109. Shibanov YV, Chukhlantsev VG (1973) The editorial board of the 'Zhurn Fiz Khim' of Akad Nauk SSSR, Moscow. Manuscript deposited in VINITI on Sept. 25, vol 6776 – 73
110. Shibanov YV, Chukhlantsev VG (1977) Zhurn Fiz Khim Akad Nauk SSSR, Moscow. The manuscript deposited in the VINITI on Dec. 14, vol 4470 – 77
111. Shibanov YV, Chukhlantsev VG, Alyamovskaya KV (1977) Zhurn Fiz Khim Akad Nauk SSSR, Moscow. The manuscript deposited in the VINITI on Dec. 14, vol 4469 – 77
112. Termicheskiye konstanty veshchestv. General Editor Acad CP Glushko (1972) VI Moscow
113. Thermodynamic Properties of Individual Substances. A Handbook edited by acad VP Glushko. Izv Akad Nauk SSSR I
114. Todd SS (1951) J Am Chem Soc 73:3277
115. Todd SS (1952) J Am Chem Soc 74:4669
116. Todd SS, King EC (1953) J Am Chem Soc 75:4547
117. Toropov PA, Barzakovskiy VP (1963) The High-Temperature Chemistry of Silicate and Other Oxide Systems. Izv Akad Nauk SSSR. Moscow-Leningrad, p 92
118. Vasil'yeva IA, Mudretsova SN, Nikitina SD (1967) Zhurn Fiz Khim XIII/I:49 – 53
119. Vise S, Margave JR, Feder HM, Hubbard I (1962) J Phys Chem 66:380
120. Wagman DD et al. (1968; 1969; 1971; 1971; 1973) Selected values of chemical thermodynamic properties, NBS, Technical note 270 – 273; 270 – 274; 270 – 275; 270 – 277; 270 – 277
121. Westrum EF, Pitzer KS (1949) J Am Chem Soc 71:1940
122. Yatsimirskiy KB (1951) Thermochemistry of Complex Compounds. Izd Akad Nauk SSSR

Appendix 3. The Heats of Polymorphic Transitions and Melting

Substance	T_{tr} [K]	ΔH_{tr} [cal/mol]	References	T_m [K]	ΔH_m^0 [cal/mol]	References
Ag	–	–	–	1234	2855	[18]
AgBr	–	–	–	691	2100	[2]
AgCl	–	–	–	725	2900	[2]
AgI	423	1470	[18]	831	2250	[19]
$AgNO_3$	433	610	[18]	484	2960	[14]
Ag_2S	452	1010	[18]	1115	2700	[19]
Ag_2SO_4	406	1700	[6]	–	–	–
Al	–	–	–	932	2570	[18]
$AlBr_3$	–	–	–	370.6	2710	[18]
$AlCl_3$	–	–	–	465.6	8500	[18]
AlF_3	723	150	[18]	–	–	–
$AlF_3 \cdot 3NaF$	845	2160	[18]	1300	27640	[18]
AlI_3	–	–	–	464	3980	[18]
Al_2O_3	1273	20600	[19]	2303	26000	[19]
$3Al_2O_3 \cdot 2SiO_2$ (mullite)	–	–	–	2123	27000	[31]
$AsBr_3$	–	–	–	304	2800	[19]
$AsCl_3$	–	–	–	257	2400	[19]
AsF_3	–	–	–	266	2490	[19]
AsF_5	–	–	–	193	2700	[19]
AsI_3	–	–	–	415	2200	[19]
As_4O_6	240	1350	[19]	582	8800	[19]
Au	–	–	–	1336	3050	[19]
B	–	–	–	2300	5300	[18]
B_2O_3	–	–	–	723	5500	[18]
Ba	643	150	[18]	983	1830	[18]
$BaBr_2$	–	–	–	1130	7630	[10]
$BaCO_3$	1079	4490	[18]	–	–	–
	1241	730	[18]	–	–	–
BaF_2	–	–	–	1617	5625	[26]
BaI_2	–	–	–	984	6340	[10]
$Ba(NO_3)_2$	–	–	–	868	6000	[6]
BaO	–	–	–	2198	13800	[19]
$Ba(OH)_2$	–	–	–	681	3400	[19]
$Ba_3(PO_4)_2$	–	–	–	2000	18600	[6]
$BaSO_4$	–	–	–	1623	9700	[19]
$BaTiO_3$	278	16	[19]	–	–	–
	393	47	[19]	–	–	–

Table (continued)

Substance	T_{tr} [K]	ΔH_{tr} [cal/mol]	References	T_m [K]	ΔH^0_m [cal/mol]	References
Be	–	–	–	1556	2800	[18]
$BeBr_2$	–	–	–	761	4500	[19]
$BeCl_2$	–	–	–	683	3000	[19]
BeI_2	–	–	–	753	4500	[19]
BeO	–	–	–	2803	17000	[19]
Bi	–	–	–	544.5	2600	[18]
$BiBr_3$	–	–	–	–	5190	[33]
$BiCl_3$	–	–	–	506.7	5680	[34]
Bi_2O_3	–	–	–	1090	6800	[19]
Bi_2S_3	–	–	–	1023	8900	[19]
Bi_3Ti_2	–	–	–	490	8700	[19]
C	–	–	–	4700	33000	[19]
Ca	713	270	[18]	1123	2070	[18]
$CaBr_2$	–	–	–	1015	6900	[10]
CaC_2	720	1330	[18]	–	–	–
$CaCO_3$	323	45	[19]	–	–	–
$CaCl_2$	–	–	–	1045	6700	[10]
CaF_2	1424	1140	[18]	1683	7250	[8]
CaI_2	–	–	–	1052	10000	[10]
$CaFe_2O_5$	–	–	–	1753	36100	[19]
$CaFe_2O_4$	–	–	–	1513	25900	[19]
$Ca(NO_3)_2$	–	–	–	834	5100	[6]
CaO	–	–	–	2873	19000	[19]
$3CaO \cdot B_2O_3$	–	–	–	1760	34490	[18]
$2CaO \cdot B_2O_3$	804	1100	[18]	1585	24090	[18]
$CaO \cdot B_2O_3$	–	–	–	1435	17670	[18]
$CaO \cdot 2B_2O_3$	–	–	–	1260	27060	[18]
$Ca_3(PO_4)_2$	1373	3700	[18]	–	–	–
$Ca_2P_2O_7$	1413	1600	[18]	1626	24100	[18]
$CaAlSi_2O_8$ (anorthite)	–	–	–	1823	29400	[17]
$CaMgSi_2O_6$ (diopside)	–	–	–	1665	23000	[12]
$Ca_2MgSi_2O_7$	–	–	–	1727	39600	[29]
$CaSiO_3$	1463	1300	[19]	1813	14300	[19]
Ca_2SiO_4	970	440	[18]	–	–	–
	1710	3390	[18]	–	–	–
$CaSO_4$ (anhydrite)	–	–	–	1673	6700	[6]
$CaTiO_3$	1533	550	[19]	–	–	–
$CaTi[SiO_4]O$ (sphene)	–	–	–	1670	29590	[24]
Cd	–	–	–	594	1450	[18]
$CdBr_2$	–	–	–	840	7970	[33]
$CdCl_2$	–	–	–	841	7220	[33]
CdF_2	–	–	–	1383	5400	[19]

Table (continued)

Substance	T_{tr} [K]	ΔH_{tr} [cal/mol]	References	T_m [K]	ΔH^0_m [cal/mol]	References
CdI_2	–	–	–	663	4950	[33]
Ce	1003	700	[32]	1077	1238	[32]
CeF_3	–	–	–	1732	13200	[18]
CeI_3	–	–	–	1033	12400	[10]
Co	700	105	[18]	1768	4100	[18]
$CoCl_2$	–	–	–	1113	14100	[19]
Co_2Si	–	–	–	1603	16500	[19]
CoSi	–	–	–	1673	16000	[19]
$CoSi_3$	–	–	–	1579	34000	[19]
Cr	–	–	–	2176	5000	[18]
$CrCl_2$	–	–	–	1088	7700	[19]
Cs	–	–	–	302	510	[18]
CsBr	–	–	–	909	5640	[9]
CsCl	–	–	–	918	4840	[9]
CsF	–	–	–	976	5190	[9]
Cs_2CO_3	–	–	–	1065	3000	[23]
CsI	–	–	–	890	5640	[9]
$CsNO_3$	–	–	–	690	3250	[6]
CsOH	496	1760	[6]	545	1610	[6]
Cs_2SiO_3	–	–	–	1100	9500	[23]
$Cs_2Si_2O_5$	–	–	–	1400	12500	[23]
$Cs_2Si_4O_9$	–	–	–	1200	10500	[23]
Cs_2SO_4	913	1000	[23]	1277	9577	[7]
Cu	–	–	–	1356	3100	[19]
CuBr	653	1400	[19]	–	–	–
	738	700	[19]	761	2300	[19]
CuCl	–	–	–	703	2620	[18]
CuI	–	–	–	861	2600	[14]
CuO	–	–	–	1720	13300	[24]
Cu_2O	–	–	–	1515	15350	[24]
Cu_2S	376	920	[19]	–	–	–
	623	200	[19]	1403	2600	[19]
Fe	1033	0	[18]	–	–	–
	1183	215	[18]	–	–	–
	1673	165	[18]	1812	3670	[18]
Fe_3C	463	180	[18]	1500	12300	[19]
$FeCl_2$	–	–	–	950	10280	[18]
$FeCl_3$	–	–	–	577	10300	[18]
$Fe_{0.947}O$	–	–	–	1650	7490	[18]
Fe_2O_3	950	160	[18]	–	–	–
Fe_3O_4	–	–	–	1870	33000	[19]
FeS	411	570	[18]	–	–	–
FeS	598	120	[18]	1468	7730	[18]
Fe_2SiO_4	–	–	–	1490	22030	[18]
$FeSiO_3$	–	–	–	1413	14600	[c]
$FeTiO_3$	–	–	–	1640	21670	[18]

Table (continued)

Substance	T_{tr} [K]	ΔH_{tr} [cal/mol]	References	T_m [K]	ΔH_m^0 [cal/mol]	References
Ge	–	–	–	1213	7700	[19]
$GeBr_4$	–	–	–	299	2900	[19]
GeO_2	–	–	–	1389	10500	[19]
K	–	–	–	336.4	558	[18]
KBO_2	–	–	–	1220	5660	[17]
KBr	–	–	–	1007	6340	[17]
K_2CO_3	695	700	[23]	1169	6540	[15]
KCl	–	–	–	1043	6340	[9]
$KClO_4$	573	3290	[6]	–	–	–
K_2CrO_4	938	2450	[6]	1245	8800	[27]
$K_2Cr_2O_7$	–	–	–	671	8510	[18]
KF	–	–	–	1130	6750	[18]
KI	–	–	–	954	5770	[9]
KHF_2	469	2660	[18]	512	1570	[18]
$KHSO_4$	437	490	[6]	–	–	–
$KMg_3AlSi_3O_{10}F_2$ (fluoroflogopite)	–	–	–	1670	73800	[18]
K_2MoO_4	–	–	[18]	1164	10800	[27]
KNO_3	401	1400	[18]	611	2800	[18]
KOH	522	1500	[19]	673	1800	[19]
K_2S	419	85	[6]	–	–	–
K_2SO_4	856	2140	[18]	1342	9060	[18]
K_2SiO_3	–	–	–	1249	1150	[23]
$K_2Si_2O_5$	523	1000	[23]	1309	7600	[21]
$K_2Si_4O_9$	865	770	[34]	1038	11700	[34]
Li	–	–	–	453.7	715	[18]
$LiAlSiO_4$ (eucryptite)	1300	300	[24]	–	–	–
LiBr	–	–	–	823	4220	[9]
Li_2CO_3	683	1000	[23]	996	10700	[15]
LiCl	–	–	–	883	4760	[9]
LiF	–	–	–	1121	6470	[18]
LiI	–	–	–	742	3500	[9]
$LiNO_3$	–	–	–	627	6100	[6]
Li_2SO_4	848	6800	[19]	1130	2373	[7]
Li_2SiO_3	–	–	–	1450	7210	[17]
Li_4SiO_4	–	–	–	1522	7430	[17]
$Li_2Si_2O_5$	1212	150	[23]	1306	14600	[21]
Li_2TiO_3	1485	2750	[18]	1820	26330	[18]
LiOH	–	–	–	744	5000	[18]
Mg	–	–	–	923	2140	[18]
$MgBr_2$	–	–	–	983	8300	[19]
$MgCl_2$	–	–	–	987	10300	[18]
MgF_2	–	–	–	1525	13150	[26]
$MgFe_2O_4$	665	0	[18]	–	–	–
	1230	350	[18]	–	–	–

Table (continued)

Substance	T_{tr} [K]	ΔH_{tr} [cal/mol]	References	T_m [K]	ΔH_m^0 [cal/mol]	References
Mg_3N_2	823	110	[18]	–	–	–
	1061	220	[18]	–	–	–
MgO	–	–	–	3073	18500	[19]
$6MgO \cdot MgCl_2 \cdot 8B_2O_3$	538	2290	[18]	–	–	–
Mg_2Si	–	–	–	1373	20500	[19]
$MgSiO_3$	–	–	–	1797	14700	[17]
Mg_2SiO_4	–	–	–	2163	25200	[3]
$MgSO_4$	773	–	[5]	1400	3500	[5]
Mn	1000	535	[18]	–	–	–
	1374	545	[18]	–	–	–
	1410	430	[18]	1517	3500	[18]
Mn_3C	1313	3570	[18]	–	–	–
$MnCl_2$	–	–	–	923	8970	[18]
MnO	–	–	–	2058	13000	[19]
Mn_3O_4	1445	4970	[18]	–	–	–
MnS	–	–	–	1803	6240	[18]
$MnSiO_3$	–	–	–	1545	8000	[6]
Mn_2SiO_4	–	–	–	1618	26150	[3]
$MnTiO_3$	–	–	–	1677	7960	[17]
Mo	–	–	–	2800	6650	[18]
$MoCl_5$	–	–	–	467	1800	[6]
MoF_6	263.3	1960	[4]	290.7	1060	[4]
MoO_3	–	–	–	1068	12500	[18]
NH_4Br	410.2	882	[1]	–	–	–
NH_4Cl	456.1	1073	[1]	–	–	–
NH_4NO_3	305.3	380	[18]	–	–	–
	357.1	320	[18]	–	–	–
	398.4	1010	[18]	442.8	1300	[18]
NH_4HSO_4	–	–	–	417	3410	[18]
Na	–	–	–	371	622	[18]
Na_3AlF_6	–	–	–	1282	25640	[20]
$NaAlSiO_4$	980	204	[24]	–	–	–
$NaBO_2$	–	–	–	1239	8660	[17]
NaBr	–	–	–	1020	6240	[9]
Na_2CO_3	723	700	[28]	1123	7303	[28]
NaCl	–	–	–	1073	6690	[9]
NaCN	172	150	[6]	–	–	–
	288	700	[6]	835	400	[6]
NaF	–	–	–	1268	8030	[25]
$NaFeO_2$	870	0	[18]	–	–	–
	1270	520	[18]	1620	11760	[18]
$NaHSO_4$	–	–	–	456	2480	[30]
NaI	–	–	–	933	5640	[9]
Na_2MoO_4	713	14600	[19]	960	3600	[19]
$NaNO_3$	549.2	810	[18]	579	3490	[18]

Table (continued)

Substance	T_{tr} [K]	ΔH_{tr} [cal/mol]	References	T_m [K]	ΔH_m^0 [cal/mol]	References
NaOH	506	1520	[18]	593	1520	[18]
Na_2S	–	–	–	1223	1600	[19]
Na_2SO_4	514	1790	[18]	1165	5670	[18]
Na_4SiO_4	–	–	–	1393	13800	[23]
Na_2SiO_3	–	–	–	1361	12470	[18]
$Na_2Si_2O_5$	953	1700	[19]	1147	8500	[19]
$Na_2Si_3O_7$	–	–	–	1023	6600	[23]
Na_2TiO_3	560	400	[18]	1155	5670	[18]
$Na_2Ti_2O_5$	–	–	–	1258	26230	[18]
$Na_2Ti_3O_7$	–	–	–	1401	37100	[18]
Na_2WO_4	861	8400	[19]	968	5700	[19]
Nb	–	–	–	2770	6400	[26]
$NbCl_5$	–	–	–	478	6900	[19]
Nb_2O_5	–	–	–	1785	24590	[18]
Nd	1135	713	[32]	1297	1705	[32]
$NdCl_3$	–	–	–	1032	12000	[10]
NdI_3	847	3400	[10]	1060	9700	[10]
Ni	633	0	[18]	1725	4210	[18]
$NiBr_2$	–	–	–	1236	13000	[19]
$NiCl_2$	–	–	–	1303	18470	[18]
Ni_3S_2	–	–	–	1063	5800	[19]
Ni_2Si	–	–	–	1591	12000	[19]
NiSi	–	–	–	1265	10600	[19]
OsO_4 (white)	–	–	–	315	2300	[19]
OsO_4 (yellow)	–	–	–	329	4100	[19]
P	–	–	–	317.4	600	[18]
PCl_3	–	–	–	264	1080	[19]
P_4O_6	–	–	–	297	3400	[19]
P_4O_{10}	–	–	–	843	11500	[19]
Pb	–	–	–	600	11400	[18]
$PbBr_2$	–	–	–	640	5000	[2]
$PbCl_2$	–	–	–	768	5700	[2]
PbI_2	–	–	–	685	6010	[18]
PbO (yellow)	–	–	–	1159	7000	[18]
PbS	–	–	–	1385	4200	[19]
PbF_2	–	–	–	1093	4160	[27]
$PbSO_4$	1139	4100	[19]	1363	9600	[19]
$PbTiO_3$	763	1150	[19]	–	–	–
Rb	–	–	–	313	524	[24]
RbBr	–	–	–	965	5270	[9]
RbCl	–	–	–	995	5670	[9]
RbF	–	–	–	1068	6150	[9]
RbI	–	–	–	920	5270	[9]
$RbNO_3$	–	–	–	589	4400	[6]
Rb_2CO_3	576	500	[23]	1146	5000	[23]
Rb_2SiO_3	–	–	–	1143	10000	[23]

Table (continued)

Substance	T_{tr} [K]	ΔH_{tr} [cal/mol]	References	T_m [K]	ΔH_m^0 [cal/mol]	References
$Rb_2Si_2O_5$	–	–	–	1363	13000	[23]
$Rb_2Si_4O_9$	–	–	–	1173	11000	[23]
Rb_2SO_4	923	1500	[23]	1347	10000	[23]
S (rhombic)	368	885	[18]	–	–	–
S (monoclinic)	–	–	–	392	335	[18]
$SO_3(\alpha)$	–	–	–	290	500	[19]
$SO_3(\beta)$	–	–	–	305.5	2500	[19]
$SO_3(\gamma)$	–	–	–	335	6100	[19]
Sb	–	–	–	903	4740	[18]
$SbBr_3$	–	–	–	470	3510	[6]
$SbCl_3$	–	–	–	346	3100	[19]
SbI_3	–	–	–	443	4200	[19]
Sb_4O_6	843	3400	[19]	929	26000	[19]
Sb_4S_6	–	–	–	819	30000	[19]
Se	–	–	–	490	1300	[18]
Si	–	–	–	1690	11950	[16]
SiO_2 (quartz)	848	290	[18]	1883	3400	[22]
SiO_2 (cristobalite)	523	200	[2]	1713	2200	[35]
SiO_2 (tridymite)	390	40	[18]	–	–	–
Sn	286	500	[19]	505	1720	[18]
$SnBr_2$	–	–	–	505	1720	[19]
$SnBr_4$	–	–	–	303	2850	[19]
$SnCl_2$	–	–	–	520	3050	[19]
SnI_2	–	–	–	593	3000	[19]
SnI_4	–	–	–	418	4600	[18]
SnO_2	683	450	[19]	–	–	–
	813	300	[19]	–	–	–
SnS	875	160	[18]	1153	7550	[18]
Sr	862	200	[18]	1043	2400	[18]
$SrBr_2$	918	2900	[10]	930	2500	[10]
$SrCO_3$	1197	4700	[18]	–	–	–
$SrCl_2$	1003	1650	[10]	1146	3880	[10]
SrF_2	–	–	–	1735	7135	[26]
$Sr(OH)_2$	–	–	–	808	5490	[18]
SrO	–	–	–	2733	16700	[27]
Te	621	130	[6]	723	4500	[18]
$TeCl_4$	–	–	–	497	4500	[18]
TeF_4	–	–	–	383	6350	[19]
TeO_2	–	–	–	1006	7050	[24]
Th	1673	670	[18]	1698	4500	[18]
$ThBr_4$	–	–	–	953	9500	[19]
$ThCl_4$	–	–	–	1038	22500	[19]
ThI_4	–	–	–	839	8000	[19]

Table (continued)

Substance	T_{tr} [K]	ΔH_{tr} [cal/mol]	References	T_m [K]	ΔH_m^0 [cal/mol]	References
Ti	1155	950	[18]	1940	4460	[18]
$TiBr_4$	–	–	–	311	3080	[18]
$TiCl_3$	–	–	–	1003	5000	[19]
$TiCl_4$	–	–	–	248	2240	[19]
TiF_4	–	–	–	839	8000	[19]
TiI_4	–	–	–	423	4180	[18]
TiO	1264	820	[18]	2293	14000	[19]
Ti_2O_3	473	215	[18]	–	–	–
Ti_3O_5	450	2240	[18]	–	–	–
TiO_2	1264	820	[6]	–	–	–
Tl	507	90	[18]	577	1020	[18]
$TlBr_2$	–	–	–	733	4120	[18]
$TlBi_2$	–	–	–	5380	5380	[18]
Tl_2Bi_3	–	–	–	–	8130	[18]
TlCl	–	–	–	700	3950	[18]
TlI	–	–	–	713	3100	[19]
$TlNO_3$	334	70	[18]	417	990	[18]
Tl_2SO_4	703	0	[19]	905	5500	[19]
U	935	700	[18]	–	–	–
	1045	1145	[18]	1406	3250	[18]
UH_3	–	3000	[11]	–	–	–
UF_6	–	–	–	337	4590	[18]
V	–	–	–	2190	5050	[18]
V_2O_4	345	2050	[18]	1818	27210	[18]
V_2O_5	–	–	–	943	15560	[18]
W	–	–	–	3653	8400	[18]
WCl_5	–	–	–	513	4200	[19]
WCl_6	500	3400	[6]	557	2300	[6]
WF_6	265	1600	[19]	273	500	[19]
WO_3	1050	410	[24]	1745	17550	[24]
Zn	–	–	–	692.7	1765	[18]
$ZnBr_2$	–	–	–	667	4000	[19]
$ZnCl_2$	–	–	–	581	5500	[19]
ZnS (sphalerite)	1293	3200	[19]	–	–	–
Zr	1135	915	[18]	2130	4900	[18]
ZrO_2	1478	1420	[18]	2988	20800	[24]

References to Appendix 3

1. Arell A (1960) Suomalais Tiedeakat Toimituks, Sar A-VI, vol 57
2. Blanc M (1914) Ann Phys 44:615 – 633
3. Bradley RS (1962) Am J Sci 260:550 – 554
4. Brady AP, Myers OE, Claus JK (1960) J Phys Chem 64:588 – 594
5. Budnikov PP, Kerbe FG, Matveyev GM (1967) Sbornik 'Stekloobraznyye sistemy i materialy'. Zinatne, Riga, pp 67 – 76
6. Chemist's Handbook (1962) 2nd edn, vol I. Goskhimizdat, Moscow
7. Darmois G, Petit G (1958) Bull Soc Chim France 4:511 – 517
8. Delbove F (1961) C R Acad Sci 252:2192 – 2194
9. Dworkin AS, Bredig MA (1960) J Phys Chem 64:269 – 272
10. Dworkin AS, Bredig MA (1963) J Phys Chem 67:697 – 698
11. Flotow HE, Lohr HR, Abrahams BM, Osborne DW (1959) J Am Chem Soc. 81:3529 – 3533
12. Goranson RW (1942) Section 16 of Special Paper N 36. Geological Society of America: Handbook of Physical Constants
13. Janz GJ, Goodkin J (1959) J Phys Chem 63:1975 – 1976
14. Janz GJ, James DW, Goodkin J (1960) J Phys Chem 64:937 – 938
15. Janz EL, Neuenschwander E, Kelly FL (1963) Trans Far Soc 59:841 – 845
16. Kantor PB, Kisil' OM, Fomichow SM (1960) Ukr Fiz Zhurn 5:358 – 360
17. Kelley KK (1936) US Bur Mines Bull 393
18. Kelley KK (1960) US Bur Mines Bull 584
19. Kubaschewski O, Evans EL (1958) Metallurgical Thermochemistry. London
20. Matiasovsky K, Fellner P (1974) Vesmir vol 53/9, pp 265 – 267
21. Matusita K, Tashiro M (1973) J of Non-Crystalline Solids II/5:471 – 484
22. Matveyev MA, Frenkel' BN, Matveyev GM (1965) Studies in the Chemistry of Silicates and Oxides. Moscow-Leningrad, Nauka, pp 49 – 57
23. Matveev MA, Frenkel BN, Matveev GM (1965) Thermodynamique des réactions de formation des silicates alcalins. Actes de VIIième Congress International du Verre. Gordon-Breach, NY, pp 5966 – 5981
24. Naumov GB, Ryzhenko BN, Khodakovskiy IL (1971) A Handbook of Thermodynamic Values. Atomizdat, Moscow
25. O'Brien CJ, Kelley KK (1957) J Am Chem Soc 79:5616
26. Petit G, Delbove F (1962) CR Acad Sci 254:1388 – 1390
27. Pham Huu Thanh (1967) Chim Mod 12/77:53 – 66
28. Popov MM, Ginzburg DM (1956) Zhurn Obshch Khim 36/88/4
29. Próks I, Eliášová M, Kosa L (1977) The heat of fusion of akermanite. Silikáty, vol XXI/1, pp 3 – 9
30. Reznitskiy LA, Khomyakov KG (1959) Vestnik MGU, Ser Matem, Mekh, Fiz, Astron, Khimiya, vol 2, pp 196 – 202
31. Risbud Subhash H, Pask JA (1977) Calculated thermodynamic data and metastable immiscibility in the system $SiO_2 - Al_2O_3$. Am Ceram Soc 60/9 – 10:418 – 424
32. Spedding FH, McKeown JJ, Daane AH (1960) J Phys Chem 64:289 – 294
33. Topol LE, Ransom LD (1960) J Phys Chem 64:1339 – 1340
34. Topol LE, Ransom LD (1960) J Phys Chem 64:862 – 865
35. Yesin OA, Bratchikov SG (1959) Izv VUZov, Khimiya i Khim Tekhnologiya 2:247 – 253

Author Index

Subject Index

MIX
Papier aus verantwortungsvollen Quellen
Paper from responsible sources
FSC® C105338

If you have any concerns about our products,
you can contact us on
ProductSafety@springernature.com

In case Publisher is established outside the EU,
the EU authorized representative is:
Springer Nature Customer Service Center GmbH
Europaplatz 3, 69115 Heidelberg, Germany

Printed by Libri Plureos GmbH
in Hamburg, Germany